TRANSLATIONS OF
MATHEMATICAL
MONOGRAPHS

VOLUME 43

Ju. L. Daleckiĭ
M. G. Kreĭn

Stability of Solutions of Differential Equations in Banach Space

American Mathematical Society

УСТОЙЧИВОСТЬ РЕШЕНИЙ ДИФФЕРЕНЦИАЛЬНЫХ УРАВНЕНИЙ В БАНАХОВОМ ПРОСТРАНСТВЕ

Ю. Л. ДАЛЕЦКИЙ и М. Г. КРЕЙН

Издательство "Наука"
Главная Редакция
Физико-Математической Литературы
Москва 1970
Translated from the Russian by S. Smith

2000 *Mathematics Subject Classification*. Primary 34-XX.

Library of Congress Cataloging-in-Publication Data

Daleckiĭ, Ĭurii L'vovich.
Stability of solutions of differential equations in banach space.
(Translations of mathematical monographs, v. 43)
Translation of Ustoĭchivost' resheniĭ differentsial'nykh uravneniĭ v banakhovom prostranstve.
Bibliography: p.
1. Differential equations. 2. Banach spaces. 3. Linear operators. I. Kreĭn, Mark Grigor'evich, 1907– joint author. II. Title. III. Series.
QA371.D25513 515'.73 74-8403
ISBN 0-8218-1593-0
AMS softcover ISBN 978-0-8218-3238-7

10 9 8 7 6 5 4 3 07 06 05 04 03 02

Table of Contents

AUTHORS' PREFACE TO ENGLISH EDITION

We were delighted to learn that immediately following the appearance of our book in Russian the American Mathematical Society decided to put out an English edition of it.

We are very grateful to the Editor of Translations of the American Mathematical Society, Dr. S. H. Gould, for the initiative, energy and interest which he displayed in this regard.

We are well aware of the fact that we have not fully succeeded in observing the precept of that wise poet, Samuel Maršak (see the epigraph): the "warm exuberance of the rough draft" is far from being everywhere in evidence, while the slips characteristic of a rough draft have turned out to be more numerous than we would have liked. However, thanks to the efforts of the translator, Dr. S. F. Smith, the number of slips has been substantially reduced. Unfortunately, it was not possible for us to note everywhere where we should have the considerable amount of work performed by Dr. Smith; the statements and proofs of theorems were sometimes changed as a result of it. We take pleasure in expressing our sincere gratitude to Dr. Smith for his responsible, careful and conscientious attitude toward the work of the translation.

Odessa, Accadia
July 17, 1972

The authors

«Но лучше, если строгая строка хранит
веселый жар черновика»
"It's good if printed line, austere and strict,
retains the lively spirit of the script."
*S. Maršak**

PREFACE

The development of this book into its present form has been largely accidental. One might say that it was created by the method of successive approximations applied over intervals spanning many years.

We begin by noting that neither of the authors has considered himself to be a true specialist in the field of stability theory.

At any rate this field is now a sort of "hobby" for each of us. First the senior and then the junior author developed the desire to interpret the various methods and situations in stability theory from the point of view of the infinite-dimensional reaches of function spaces, the operators acting in them, their norms and their definite and indefinite metrics.

This "hobby" has turned out to be quite fascinating. The attempts directly to generalize the traditional methods which make essential use of the phenomena of linear algebra (e.g. the discreteness of the spectrum, Jordan cells, etc.) run up against insurmountable difficulties because of the lack of a spectral theory for general nonselfadjoint operators. One is therefore compelled to seek other approaches, which on the one hand frequently lead to new problems in functional analysis and on the other hand shed new light on known facts. As so often happens, a view from a different perspective permits one to make new discoveries in an old field.

It all began as follows.

In the Spring of 1947 M. G. Kreĭn commenced a short lecture course for undergraduates, graduate students and faculty members of the University of Odessa on the theory of linear operators in Banach spaces and the theory of normed rings. This course was to have been completed by the Fall of 1947. And so it would have been if M. G. Kreĭn had not attempted to show in his lectures that a number of the methods and situations in the Ljapunov theory of stability for solutions of differential equations can be carried over and generalized to the case of differential equations in Banach spaces. This project met with such a lively response from the students that the course was completely given over to its further development and was consequently extended to April of 1948.

*The foregoing is a free translation of Maršak into English by O. G. Dorohin, who is a teacher of English in the U.S.S.R.

During this same period M. G. Kreĭn gave a paper (M. G. Kreĭn [2]) at a meeting of the Moscow Mathematical Society in which he told of his first attempts to develop a stability theory for systems with an infinite number of degrees of freedom and of the problems encountered in this connection.

M. G. Kreĭn later supplemented the lecture course with talks at his seminars. A number of results along the lines laid down by these talks were subsequently obtained in the master's thesis of D. L. Kučer [1, 2], as well as in the work of M. A. Rutman [1].

In 1954 Ju. L. Daleckiĭ, who at the time was occupied with investigations in the field of asymptotic methods for operator equations with a small parameter, undertook to give a systematic account of the material that had accumulated from the course of M. G. Kreĭn. In the process he improved and enlarged the discussion of a number of points and in addition wrote a chapter on his own results.

This account, on which the "Lectures" (M. G. Kreĭn [9]) were subsequently based, was finished by the Spring of 1955. It was originally planned to submit it for publication in the journal, "Uspehi Matematičeskih Nauk," as soon as some apparently minor tasks were performed: the writing of an introduction, the compilation of a bibliography and a comparison of the results with those then known for equations in a finite-dimensional space. But the need for this tedious addition to the "hobby" brought matters to a standstill for ten years.

At the same time individual copies of the future "Lectures" were retyped and circulated among scientists in Odessa, Voronezh, Kiev, Kishinev, Leningrad, Moscow and possibly other cities.

The second stage in the writing of the book dates from 1964, when in response to a growing interest in the "Lectures" it was decided to have them mimeographed.

The previous ten years, however, had seen a very intensive development of the theory of differential equations. A comparison with the literature was now less advantageous to us and much more difficult (see the 69 pages of the frighteningly large bibliography in the well-known survey of L. Cesari [1] originally published back in 1959). We decided on the basis of this work to give up and publish the "Lectures" after only partially revising them.

The most important changes had the following history. In the previously referred to paper given at a 1948 meeting of the Moscow Mathematical Society M. G. Kreĭn called attention to the fact that the remarkable investigations of A. M. Ljapunov on second order linear differential equations with periodic coefficients had not been extended to equations in Hilbert space or even to the n-dimensional case. After two years M. G. Kreĭn [3–8] succeeded in finding a new approach to the establishment of criteria for the boundedness of the solutions of a canonical linear system of differential equations with a periodic Hamiltonian (in particular, of a certain class of systems of second order equations). This approach received further development in the papers of various Soviet writers (see the survey of M. G. Kreĭn and V. A. Jakubovič [1]). Several results in this direction were presented in the mimeographed edition of the "Lectures" for the case of Hilbert spaces directly.

An unpolished form of the "Lectures" appeared at the end of 1964 and instantly became a literary rarity (an edition of 400 copies). Beginning at this point the authors became more receptive to the idea of getting out a much larger edition. Nowadays no one is astonished by a functional analytic approach to differential equations nor by the equations themselves being in Banach spaces; these phenomena have penetrated even university texts (see, for example, the book of J. Dieudonné [1])!

We have ceased to be monopolists even in the matter of discussing the stability of solutions of differential equations in a Banach space. During this time there appeared the fundamental book of J. Massera and J. Schäffer [1], which summarized their investigations on the stability of solutions of linear differential equations in Banach spaces. There also appeared the short but excellent book of the Australian mathematician W. Coppel [1] and the fundamental book of P. Hartman [1], in both of which the presentation is conducted by applying the methods of functional analysis in a finite-dimensional space but on a level often permitting an almost direct extension to a Banach space. A number of interesting questions are discussed in the books of E. A. Barbašin [1] and N. N. Krasovskiĭ [1]. We note further the well written intensive university text of B. P. Demidovič [3].

In recent years members of the Kiev school of N. N. Bogoljubov on nonlinear mechanics (Ju. O. Mitropol'skiĭ, O. B. Likova and others) have become occupied with extending their results to the case of equations in Hilbert space. Closely related investigations have been carried out by the Czechoslovakian mathematician J. Kurzweil [1].

All of these sources have in one way or another had an influence on our selection of material for the enlarged edition and occasionally on the presentation.

During this time a substantial contribution to stability theory was discovered in the works of the remarkable Riga mathematician P. G. Bohl (1865–1921).[1] who has dwelt so long in obscurity. We found it necessary as a result to substantially revise the brief bibliographical notes appearing in the "Lectures".

Thus there arose the present book, which is more than three times as long as the "Lectures".

Unfortunately, our book suffers from various "limitations". To keep its size within bounds we omitted many important topics. In particular, our treatment is limited to bounded operators only.

The authors clearly recognize that the arena of infinite-dimensional spaces requires the presence of unbounded operators, without which the present stability theory for systems with an infinite number of degrees of freedom would be impossible.

Such a theory has not yet been created. Not so long ago there did not even exist

---

[1] For more details see the Notes at the end of Chapter III as well as at the ends of Chapters IV and VII.

a basis for it, viz. a theory of linear differential equations with unbounded operator coefficients in Banach spaces. During the past decade and a half the situation has radically changed and it can now be said that such a basis exists (see the book of S. G. Kreĭn [1], where the many investigations in this field are summarized).

Individual papers have already appeared in the field of stability theory for unbounded operator coefficients (S. Agmon and L. Nirenberg [1], Z. I. Halilov [1, 2], Ju. I. Domslak [1, 2], V. I. Derguzov [1, 2], V. N. Fomin [1, 5] and others). We would like to hope that our book might serve as a jumping off point for some of the investigations in this more advanced field (in some of the exercises a generalization to the case of unbounded operators can be trivially obtained without particular difficulty).

When raw lectures are processed into a book there can arise a danger which in another connection was wonderfully depicted by H. Weyl: "The stringent precision attainable for mathematical thought has led many authors to a mode of writing which must give the reader the impression of being shut up in a brightly illuminated cell where every detail sticks out with the same dazzling clarity, but without relief. I prefer the open landscape under a clear sky with its depth of perspective, where the wealth of sharply defined nearby details gradually fades away towards the horizon."*

Few possess a gift of picturing "landscapes under a clear sky ..." similar to the gift of H. Weyl, that indefatigable seeker and discoverer of harmony in mathematics. The authors had to settle for a system of shadows and halfshadows by means of the standard tools known as remarks, exercises and notes.

To this we nevertheless add that on the horizon of our landscapes, besides the above mentioned mountainous massif of differential equations with unbounded operators, there can be seen in the distance smooth manifolds with dynamic streams and cascades on them.

We consider it our pleasant duty to note that during the last stage of work on the manuscript, for a period of about a year, essential help was rendered us by the editor of the book, M. G. Sonis, to whom we express our very sincere gratitude.

Odessa, Arcadia
August 6, 1969

The authors

---

*Preface to the first edition of *The classical groups*.

Introduction

In the book, as the title suggests, we are concerned with differential equations for vector functions with values in infinite-dimensional spaces. Therefore, to completely understand its contents, the reader should be familiar with what is known about the geometry of Hilbert and Banach spaces and the theory of linear operators acting in them.

In the first chapter we recall the basic definitions and classical theorems of the theory of Banach spaces. It is assumed, however, that the fundamentals of the geometry of Hilbert spaces are known to the reader. We have attempted to avoid any use of the theory of spectral resolutions of selfadjoint and unitary operators, although we do not completely succeed in doing this (see § 8).

At the same time we note that a considerable part of the apparatus used in the book, which, beginning with § 3, is presented with more or less complete proofs, has not yet entered even the fundamental courses of functional analysis. This applies especially to questions in the theory of spaces with an indefinite metric, but also, for example, to the methods of solving linear operator equations. And what is more, even the interpretation in finite-dimensional spaces of a number of propositions cited by us has not managed to find a place for itself in books on linear algebra (i.e. the theory of matrices).

We wish to emphasize here that a reader unfamiliar with functional analysis can read the book by assuming everywhere that it is dealing with finite-dimensional spaces. We hope that even in this case a reader qualified in the theory of differential equations will find something new in the book.

The easiest chapter to understand is the second chapter, in which stationary linear equations are considered. It contains the simplest version of many of the ideas and results presented in the other chapters. A reader becoming acquainted with the subject for the first time should first thoroughly study this chapter. For this purpose he will not need the results concerning an indefinite metric presented in §§ 7 and 8 of Chapter I (except for the simplest definitions these results are used only in the exercises) nor the contents of § 9 of Chapter I.

In the third and fourth chapters we present the general theory of nonstationary linear equations; in particular (Chapter IV), we carry out a detailed investigation of the important notion of an exponential splitting of the solutions of an equation.

More restricted classes of linear equations are studied in Chapter V (equations with periodic coefficients) and in Chapter VI (equations with a regular singularity in the complex plane). In particular, a large part of Chapter V is given over to the study of conditions for the strong stability of canonical equations. In this connection we make essential use of the methods of the theory of operators in spaces with an indefinite metric. The seventh chapter is devoted to the consideration of various problems connected with the behavior of the solutions of nonlinear equations.

Finally, in the last, the eighth, chapter we return to linear equations and consider special asymptotic developments of the solutions with respect to a large parameter of the equation.

The reader can obtain a more precise picture of the contents of the book from the Table of Contents and the brief introductions which preface the chapters of the book. In addition to these introductions each chapter is provided with a set of exercises and some bibliographical notes.

The exercises are of a diverse character and pursue different ends. Some of them are routine drill problems for the main text. Others together with hints to their solution contain brief developments of important additional questions which were not included in the main text in order to avoid unduly increasing the size of the book. Finally, many exercises contain additional historical and bibliographical commentary and formulations of the results of various authors. Sometimes the notions introduced in the exercises are subsequently used (with an appropriate reference) in the main text of the book.

A serious analysis of the material contained in the exercises will enable the reader to go more deeply into various aspects of the theory being studied and sometimes to understand its connection with other branches of functional analysis. Furthermore, it is only in the exercises that we occasionally give a mechanical interpretation of the results being presented and explain the role of individual methods in the theory of oscillations.

At the present time, in addition to this book, only the book of J. Massera and J. Schäffer [1] is especially devoted to the study of differential equations in Banach spaces. Its contents overlap those of the third, (more particularly) the fourth and to some extent the fifth chapters of this book. For the reader who wishes to become deeply acquainted with the notion of a dichotomy of the solutions of a linear differential equation the book of Massera and Schäffer will serve as an important supplement to ours (although we do have something new to say in this field). On the other hand, we present a more complete account of the theory of periodic linear equations and a number of other questions. Moreover, the book of Massera and Schäffer considers only linear equations.

To better partition the material and to more clearly distinguish the places of greater importance we enumerate not only the lemmas and theorems but also a large number of remarks and corollaries. They are always numbered by two nonnegative integers (for example, Lemma 3.5) the first of which indicates the section and the second, the lemma, theorem, etc. in the given section. We also number formulas in the same way, a formula number of the form (0.3), for example, referring to a formula in the exercises. In referring to a formula, theorem or section of another chapter we add a Roman numeral to indicate the chapter (for example, Theorem I. 3.5).

CHAPTER I

SOME INFORMATION FROM THE THEORY OF BOUNDED OPERATORS IN BANACH SPACES

In this chapter we present various bits of information from the theory of bounded operators in Banach spaces, which will be systematically used in the sequel. Some of this information has long since found a place for itself in the textbooks and much of the rest is well known to the specialist. But some of the facts presented here are quite new.

In § 1 we present a number of basic definitions and theorems concerning Banach spaces and the operators acting in them.

We then use Cauchy's formula to define a function of an operator (§ 2) and to consider the most important operator function for our purposes, viz. the exponential operator function e^{At} (§ 4). This function is a one parameter commutative group of bounded operators. Its order of growth at infinity has a simple connection with the spectrum of the operator A. In § 6 we consider bounded commutative groups of operators in Hilbert space that are similar to groups composed of unitary operators.

Many questions connected with differential equations reduce to the consideration of "algebraic" linear operator equations. We consider such equations in § 3, where we find conditions for their solvability and integral formulas for their solutions. Other expressions for the solutions of these equations can sometimes be obtained with the use of the exponential operator function (§ 4).

In a number of cases the behavior of the solutions of differential equations becomes more visible if the norm of the space is replaced by one of a special form. Such norms are introduced in § 4 with the use of the exponential operator function e^{At}. When the space is a Hilbert space there correspond to such norms positive quadratic forms that can be calculated by means of linear operator equations if the spectrum of the operator A lies in the interior of the left halfplane (§ 5). We assume that the reader is familiar with the fundamentals of the geometry of Hilbert spaces.

An essential role in the investigations in the sequel is played by the indefinite metric. In §§ 7 and 8 we study in a Hilbert space possessing an indefinite metric introduced with the use of a certain Hermitian operator W the important classes of W-dissipative, W-contraction and W-unitary operators.

The results in §§ 7 and 8 are of a less elementary nature. But except for the simplest definitions connected with an indefinite metric they are used only in the fifth chapter. The reader can therefore postpone the study of these sections until he reaches Chapter V.

In § 9 we present some well-known facts from nonlinear analysis that play an important role in the study of nonlinear equations.

At the end of the chapter we give a rather large number of exercises of various types and pursuing different ends. Some of these are routine drill problems for a better understanding of the text, while others present important notions and results that will frequently be used in subsequent chapters in the main text (with an appropriate reference).

§ 1. General propositions on the geometry of Banach spaces and their linear mappings

1. *Normed and Banach spaces.* A set $\mathfrak{L}$ is called a *real* (*complex*) *normed space if*

1) It is a linear vector space over the field of real (complex) numbers, and

2) for each vector $x \in \mathfrak{L}$ there is defined a nonnegative number $\|x\|$, viz. the *norm* of x, having the following properties:

a) $\| \alpha x \| = |\alpha| \, \| x \|$ for any $x \in \mathfrak{L}$ and any real (complex) number α;

b) $\| x + y \| \leqq \| x \| + \| y \|$ for any $x, y \in \mathfrak{L}$ (the triangle inequality);

c) $\|x\| = 0$ if and only if $x = 0$.

The function $\rho(x, y) = \| x - y \|$ in a normed space defines a metric, so that $\mathfrak{L}$ is a metric space.

A sequence $\{ x_n \} \subset \mathfrak{L}$ is called a *Cauchy sequence* if $\lim_{n,m\to\infty} \| x_n - x_m \| = 0$.

A normed space is called a *Banach space* if every Cauchy sequence in it has a limit point x for which $\lim_{n\to\infty} \| x_n - x \| = 0$ (in other words, a Banach space $\mathfrak{L}$ ($= \mathfrak{B}$) is complete in the metric $\rho(x, y) = \| x - y \|$).

Every incomplete normed space $\mathfrak{L}$ can be completed, i.e. can be "identified" with a dense linear manifold in a Banach space $\mathfrak{B}$. This is done in a well-known way by considering the set of all Cauchy sequences in $\mathfrak{L}$.

If another norm $\| x \|_2$ is given in a normed space $\mathfrak{L}$ with norm $\| x \|_1$, every convergent (in the norm $\| x \|_1$) sequence converges in the norm $\| x \|_2$ if and only if there exists a positive constant c_1 such that $\| x \|_2 \leqq c_1 \| x \|_1$.

Two norms $\| x \|_1$ and $\| x \|_2$ given in a linear space $\mathfrak{L}$ are said to be *topologically equivalent* if convergence in one of them implies convergence in the other. For this to be so it is necessary and sufficient that there exist positive constants c_1 and c_2 such that

$$c_2 \leqq \| x \|_2 / \| x \|_1 \leqq c_1 \qquad (x \neq 0). \tag{1.1}$$

Two Banach spaces are said to be *isomorphic* if there exists a one-to-one continuous linear map of one space onto the other. This map is called an *isomorphism*. If it preserves the norm, it is called an *isometric isomorphism*.

Clearly, a Banach space $\mathfrak{B}$ goes over into an isomorphic space if its norm is replaced by a topologically equivalent one.

2. *Linear operators.* Let $\mathfrak{B}_1$ and $\mathfrak{B}_2$ be Banach spaces. A map $A: \mathfrak{B}_1 \to \mathfrak{B}_2$ is called a *linear operator* if $A(\alpha x + \beta y) = \alpha A x + \beta A y$ for any numbers α, β and any $x, y \in \mathfrak{B}_1$.

A linear operator is continuous if it is continuous at the point $x = 0$.

Continuity is equivalent to the property of boundedness of an operator A, i.e. of finiteness of the quantity

$$\begin{aligned} \| A \| &\stackrel{\text{def}}{=} \sup \{ \| Ax \|_2 / \| x \|_1 \mid x \in \mathfrak{B}_1,\ x \neq 0 \} \\ &= \sup \{ \| Ax \|_2 \mid x \in \mathfrak{B}_1,\ \| x \|_1 = 1 \}. \end{aligned} \tag{1.2}$$

The set of bounded linear operators $A: \mathfrak{B}_1 \to \mathfrak{B}_2$ will be denoted by $[\mathfrak{B}_1, \mathfrak{B}_2]$. This set is a Banach space with norm (1.2) under the natural definitions of the operations of adding operators and multiplying an operator by a number.

$$(A + B)x = Ax + Bx; \quad (\alpha A)x = \alpha(Ax).$$

If $A: \mathfrak{B}_2 \to \mathfrak{B}_3$ and $B: \mathfrak{B}_1 \to \mathfrak{B}_2$, the formula $Cx = A(Bx)$ $(x \in \mathfrak{B}_1)$ defines an operator $C: \mathfrak{B}_1 \to \mathfrak{B}_3$, which is called the *product* $C = AB$ of the operators A and B.

If the operators A and B in this connection are bounded ($A \in [\mathfrak{B}_2, \mathfrak{B}_3]$, $B \in [\mathfrak{B}_1, \mathfrak{B}_2]$). Then the operator C is bounded ($C \in [\mathfrak{B}_1, \mathfrak{B}_3]$) and $\| C \| = \| AB \| \leqq \|A\| \, \|B\|$.

An operator $B: \mathfrak{B}_2 \to \mathfrak{B}_1$ is called the *inverse* of an operator $A: \mathfrak{B}_1 \to \mathfrak{B}_2$, and one writes $B = A^{-1}$, if $AB = I_2$ and $BA = I_1$, where I_k is the identity operator in $\mathfrak{B}_k$: $I_k x = x$ for any $x \in \mathfrak{B}_k$ $(k = 1, 2)$.

The set of bounded linear operators mapping a space $\mathfrak{B}$ into itself is briefly denoted by $[\mathfrak{B}]$. Thus $[\mathfrak{B}] \stackrel{\text{def}}{=} [\mathfrak{B}, \mathfrak{B}]$.

Under the operation of multiplication defined above $[\mathfrak{B}]$ is not only a Banach space but also a Banach algebra.

We formulate without proof the following fundamental proposition.

THEOREM 1.1 (BANACH). *Suppose an operator $A \in [\mathfrak{B}_1, \mathfrak{B}_2]$ is a one-to-one mapping of a Banach space $\mathfrak{B}_1$ onto a Banach space $\mathfrak{B}_2$. Then its inverse A^{-1} is bounded and linear: $A^{-1} \in [\mathfrak{B}_2, \mathfrak{B}_1]$.*

This justifies the above definition of isomorphic Banach spaces.

COROLLARY 1.1. *Suppose given in a linear space $\mathfrak{L}$ two norms $\| x \|_1$ and $\| x \|_2$ that convert it into Banach spaces $\mathfrak{B}_1$ and $\mathfrak{B}_2$ respectively, and suppose $\| x \|_2 \leqq c_1 \| x \|_1$ $(x \in \mathfrak{L})$, where c_1 is a positive constant.*

Then there exists a positive constant c_2 such that $\| x \|_1 \leqq c_2 \| x \|_2$, and consequently the norms $\| x \|_1$ and $\| x \|_2$ are topologically equivalent.

For by Banach's theorem the identity mapping $x \to x$ of $\mathfrak{B}_1$ onto $\mathfrak{B}_2$, being bounded, has a bounded inverse.

The method by which Theorem 1.1 is normally proved permits one to prove the following important proposition.

THEOREM 1.2 (uniform boundedness principle). *Suppose $\mathfrak{A}$ is a set of operators from $[\mathfrak{B}_1, \mathfrak{B}_2]$ such that for each $x \in \mathfrak{B}_1$*

$$\sup \{ \| Ax \| \mid A \in \mathfrak{A} \} < \infty.$$

Then $\mathfrak{A}$ is bounded, i.e. $\sup \{ \| A \| \mid A \in \mathfrak{A} \} < \infty$.

From this theorem it follows in particular that if a sequence of operators $A_n \in$

$[\mathfrak{B}_1, \mathfrak{B}_2]$ converges at each element $x \in \mathfrak{B}_1$, the equality $Ax = \lim_{n\to\infty} A_n x$ defines a continuous linear operator $A \in [\mathfrak{B}_1, \mathfrak{B}_2]$.

3. *Direct sums of subspaces and projections.* A closed linear manifold of a Banach space $\mathfrak{B}$ is called a *subspace* of $\mathfrak{B}$.

One says that a Banach space $\mathfrak{B}$ decomposes into a *direct sum of subspaces* $\mathfrak{B}_1$ and $\mathfrak{B}_2$:

$$\mathfrak{B} = \mathfrak{B}_1 \dotplus \mathfrak{B}_2, \tag{1.3}$$

if each element $x \in \mathfrak{B}$ admits a unique representation in the form

$$x = x_1 + x_2, \tag{1.4}$$

where $x_1 \in \mathfrak{B}_1$ and $x_2 \in \mathfrak{B}_2$:

Each of the subspaces $\mathfrak{B}_1$ and $\mathfrak{B}_2$ is said to be a *direct complement* of the other.

The decomposition (1.3) defines two linear operators $P_k: \mathfrak{B} \to \mathfrak{B}_k$ $(k = 1, 2)$ by means of the equalities $P_k x = x_k$, where x_1 and x_2 are the *components* of x in the decomposition (1.4). The operators P_1 and P_2 obviously have the properties

$$P_k^2 = P_k; \quad P_1 + P_2 = I; \quad P_1P_2 = P_2P_1 = 0. \tag{1.5}$$

They are also continuous: $P_k \in [\mathfrak{B}]$. To prove this let

$$\| x \|_1 = \| x_1 \| + \| x_2 \| \, (= \| P_1x \| + \| P_2x \|)$$

for any $x \in \mathfrak{B}$. Since $x = x_1 + x_2$, we have $\| x \| \leqq \| x_1 \| + \| x_2 \| = \| x \|_1$.

It is not difficult to see that the completeness of the spaces $\mathfrak{B}_1$ and $\mathfrak{B}_2$ implies the completeness of $\mathfrak{B}$ in the norm $\| x \|_1$. (Therefore Corollary 1.1 implies the existence of a constant $c > 0$ such that $\| x \|_1 \leqq c\| x \|$, whence $\| P_kx \| \leqq c\| x \|$, i.e. $P_k \in [\mathfrak{B}]$ $(k = 1, 2)$.

An operator $P \in [\mathfrak{B}]$ is called a *projection* if $P^2 = P$.

If an operator P_1 is a projection then the operator $P_2 = I - P_1$ is also a projection, all of the relations (1.5) being satisfied. The operators P_1 and P_2 are called *mutually complementary projections.*

As noted above, every direct decomposition (decomposition into a direct sum) defines a pair of mutually complementary projections. Conversely, every pair of mutually complementary projections defines a direct decomposition (1.3), where $\mathfrak{B}_1 = P_1\mathfrak{B}$ and $\mathfrak{B}_2 = P_2\mathfrak{B}$.

A subspace $\mathfrak{L}_1 \subset \mathfrak{B}$ is said to be *complemented* if there exists *at least* one subspace $\mathfrak{L}_2 \subset \mathfrak{B}$ such that $\mathfrak{B} = \mathfrak{L}_1 \dotplus \mathfrak{L}_2$.

From what has been said it follows that a subspace $\mathfrak{L}_1 \subset \mathfrak{B}$ is complemented if and only if there exists at least one projection $P \in [\mathfrak{B}]$ that projects $\mathfrak{B}$ onto $\mathfrak{L}_1$: $P\mathfrak{B} = \mathfrak{L}_1$.

It is easily seen that every finite-dimensional subspace as well as every subspace $\mathfrak{L}$ of finite codimension (i.e. such that $\dim \mathfrak{B}/\mathfrak{L} < \infty$) is complemented.

Clearly, in a Hilbert space each subspace $\mathfrak{L}$ is complemented and, what is more, has a unique orthogonal complement $\mathfrak{L}^{\perp}$.

An analogous definition exists for the decomposition of a Banach space $\mathfrak{B}$ into a direct sum $\mathfrak{B} = \mathfrak{B}_1 \dotplus \cdots \dotplus \mathfrak{B}_n$ of several subspaces, and it can be shown that every such decomposition defines and is defined by a *decomposition of the identity into a sum of projections*: $I = P_1 + \cdots + P_n$, *that are pairwise disjoint*:

$$P_jP_k = P_kP_j = \delta_{jk}P_k$$

$$\left(\delta_{jk} = \begin{cases} 1, j = k \\ 0, j \neq k \end{cases} \text{ is the Kronecker symbol}\right). \tag{1.6}$$

We also note that $\| P \| \leqq \| P \|^2$, i.e. $\| P \| \geqq 1$, for any projection P by virtue of the equality $P^2 = P$.

4. *Direct sums of Banach spaces.* Let $\mathfrak{B}_1, \cdots, \mathfrak{B}_n$ be Banach spaces with norms $\| \cdot \|_1, \cdots, \| \cdot \|_n$. These spaces can be regarded as subspaces of a Banach space $\mathfrak{B}$ such that

$$\mathfrak{B} = \mathfrak{B}_1 \dotplus \mathfrak{B}_2 \dotplus \cdots \dotplus \mathfrak{B}_n. \tag{1.7}$$

The space $\mathfrak{B}$ in this connection will be defined to within an isomorphism.

In fact, let $\mathfrak{L}$ denote the linear space of ordered collections[1)]

$$x = (x_1, x_2, \cdots, x_n) \qquad (x_k \in \mathfrak{B}_k;\ k = 1,2,\cdots,n)$$

with the natural definition of a sum of two collections and multiplication by a scalar.

We convert $\mathfrak{L}$ into a Banach space $\mathfrak{B}$ by defining a norm $\| \cdot \|$ in $\mathfrak{L}$ so that

$$\| x \| = \| x_k \|_k \tag{1.8}$$

for every collection x all of whose components except the kth are equal to zero. This can always be done by putting $\| x \| = \| x \|_M$, for example, where

$$\| x \|_M = \sum_{k=1}^{n} \| x_k \|_k. \tag{1.9}$$

After this we can isometrically imbed the spaces $\mathfrak{B}_k$ in $\mathfrak{L}$ by identifying any element $x_k \in \mathfrak{B}_k$ with the collection $(0, \cdots, 0, x_k, 0, \cdots, 0)$. The decomposition (1.7) is then clearly obtained.

We note that of all of the norms $\| x \|$ satisfying condition (1.8) the norm (1.9) is the *maximal norm*, i.e. one always has $\| x \| \leqq \| x \|_M$ for each $x \in \mathfrak{B}$. For if the decomposition (1.7) holds then $x = x_1 + \cdots + x_n$, and hence

$$\| x \| \leqq \| x_1 \| + \| x_2 \| + \cdots + \| x_n \| = \| x_1 \|_1 + \| x_2 \|_2 + \cdots + \| x_n \|_n = \| x \|_M.$$

It is not difficult to give a description of all of the norms satisfying condition

1) We will find it more convenient at times to write this collection in the form of a column or in the form $x = x_1 \dotplus \cdots \dotplus x_n$ (we recall that this sum depends on the order of the summands).

(1.8) by making use of the norms in an n dimensional space (the Minkowski norms).

We confine ourselves to mentioning the family of norms

$$\| x \|_{(p)} = \left(\sum_{k=1}^{n} \| x_k \|_k^p \right)^{1/p} \qquad (1 \leqq p \leqq \infty).^{2)}$$

5. *Direct (orthogonal) sum of Hilbert spaces.* If the $\mathfrak{B}_k$ $(k = 1, \cdots, n)$ are Hilbert spaces $\mathfrak{H}_k$, a scalar product can be defined in their direct sum $\mathfrak{B}$ so that $\mathfrak{B}$ is also a Hilbert space $\mathfrak{H}$ and the $\mathfrak{H}_k$ are mutually orthogonal subspaces of $\mathfrak{H}$. To this end we put

$$(x, y) = \sum_{k=1}^{n} (x_k, y_k)_k \qquad (x, y \in \mathfrak{H}).$$

Clearly, this definition of a scalar product is the only one possible.

One then calls $\mathfrak{H}$ an *orthogonal sum of Hilbert spaces* and writes

$$\mathfrak{H} = \mathfrak{H}_1 \oplus \mathfrak{H}_2 \oplus \cdots \oplus \mathfrak{H}_n.$$

From what has been said we can make the following useful

REMARK 1.1. Suppose $I = P_1 + \cdots + P_n$ is a decomposition of the identity into a sum of pairwise disjoint projections acting in a Hilbert space $\mathfrak{H}$. Then the scalar product in $\mathfrak{H}$ can be replaced by a topologically equivalent one with respect to which all of the projections P_k are orthogonal[3)] projections.

To this end it is necessary to construct the orthogonal sum

$$\mathfrak{H}' = P_1 \mathfrak{H} \oplus P_2 \mathfrak{H} \oplus \cdots \oplus P_n \mathfrak{H}$$

and identify it in the appropriate manner with $\mathfrak{H}$. This leads to the following definition of the desired scalar product: $(x,y)' = \sum_1^n (P_k x, P_k y)$.

6. *Complex hull of a real space.* It is often useful to consider in conjunction with a real Banach space $\mathfrak{B}$ a complex Banach space $\tilde{\mathfrak{B}}$, which is called the *complexification* of $\mathfrak{B}$. The space $\tilde{\mathfrak{B}}$ consists by definition of all possible pairs (x_1, x_2) of elements of $\mathfrak{B}$, with algebraic behavior defined by the symbolic form $(x_1, x_2) = x_1 + ix_2$ and with norm

$$\left|\left|\left| x_1 + ix_2 \right|\right|\right| = \max_{\theta} \| x_1 \cos \theta + x_2 \sin \theta \|.$$

It is easy to verify that under this definition $\tilde{\mathfrak{B}}$ is actually a complex Banach space.

The complex hull $\tilde{\mathfrak{H}}$ of a real Hilbert space $\mathfrak{H}$ is more naturally defined by specifying, instead of a norm, a scalar product in such a way that $\tilde{\mathfrak{H}}$ will be a complex Hilbert space. This can be done by setting

$$(x_1 + ix_2, y_1 + iy_2) = (x_1, y_1) + (x_2, y_2) + i[(x_2, y_1) - (x_1, y_2)].$$

---

2) $\| x \|_\infty \overset{\text{def}}{=} \lim_{p\to\infty} \| x \|_{(p)} = \max \{ \| x_1 \|_1, \| x_2 \|_2, \cdots, \| x_n \|_n \}.$

3) A projection $P \in [\mathfrak{H}]$ is called an *orthogonal projection* if $x - Px \perp Px$ or, equivalently, $P = P^*$.

§ 2. Functions of bounded linear operators

1. *Spectrum and resolvent.* Let $\mathfrak{B}$ be a complex Banach space.[4]

A point λ of the complex plane is called a *regular point* of an operator $A \in [\mathfrak{B}]$ if $[\mathfrak{B}]$ contains the operator (the value of the *resolvent* of A at λ)

$$R_\lambda = (A - \lambda I)^{-1}.$$

The set $\rho(A)$ of all regular points of an operator A is open. Its complement $\rho(A)$ is called the *spectrum* of A. The spectrum $\sigma(A)$ is always nonempty, closed and lies in the disk $|\lambda| \leqq \|A\|$. More precisely, *the spectrum $\sigma(A)$ lies in the disk of radius*

$$r_A = \lim_{n\to\infty} \sqrt[n]{\|A^n\|}$$

(the existence of the limit readily follows from the relation $\|A^{m+n}\| \leqq \|A^m\| \cdot \|A^n\|$).

In fact, when $|\lambda| > r_A$ the series $\sum_0^\infty \lambda^{-(k+1)} A^k$ converges absolutely in the metric of $[\mathfrak{B}]$ inasmuch as the corresponding series of norms is ultimately (for k sufficiently large) majorized by the geometric progression $\{(r_A + \varepsilon)^k / |\lambda|^{k+1}\}$ for any $\varepsilon > 0$. Upon multiplying this series by $\lambda I - A$, we obtain I.

Thus, when $|\lambda| > r_A$, the resolvent always exists and, moreover,

$$R_\lambda = -\sum_{k=0}^{\infty} \lambda^{-(k+1)} A^k. \tag{2.1}$$

It can be shown that the circle $|\lambda| = r_A$ always contains a point of the spectrum $\sigma(A)$. Therefore the limit $\lim_{n\to\infty} \sqrt[n]{\|A^n\|}$ is called the *spectral radius* of A.

As can be directly verified, the *resolvent equation*

$$R_\lambda - R_\mu = (\lambda - \mu) R_\lambda R_\mu \qquad (\lambda, \mu \in \rho(A)) \tag{2.2}$$

holds for R_λ.

The resolvent R_λ is an analytic function of λ in a neighborhood of each regular point μ, the absolutely convergent expansion

$$R_\lambda = R_\mu + \sum_{k=1}^{\infty} (\lambda - \mu)^k R_\mu^{k+1}$$

being valid for $|\lambda - \mu| < 1/\|R_\mu\|$.

An expansion of the resolvent in a neighborhood of the point at infinity has just been indicated in (2.1).

We end this subsection with the following special rule, notable for its simplicity: *if $\|A\| < 1$ (more generally, $r_A < 1$), then the operator $I - A$ is invertible and*

$$(I - A)^{-1} = \sum_{k=0}^{\infty} A^k.$$

2. *Behavior of the spectrum and resolvent of an operator under small perturbations.*

---

[4] Analogous notions can be introduced for a real space by going over to its complex hull.

Let G be an open set covering the spectrum $\sigma(A)$ of an operator $A \in [\mathfrak{B}]$. Then in every case the resolvent $R_\lambda(A) = (A - \lambda I)^{-1}$ is continuous on the complement F of G in the closed complex plane. This implies the existence of

$$M = \max \{ \| R_\lambda(A) \| \mid \lambda \notin G \} (< \infty).$$

We consider a δ-neighborhood $U_\delta(A) = \{ X \mid \| A - X \| < \delta \}$ of A in $[\mathfrak{B}]$. Since for any $X \in [\mathfrak{B}]$

$$\begin{aligned} X - \lambda I &= (A - \lambda I) + (X - A) \\ &= (A - \lambda I)[I - R_\lambda(A)(A - X)] \qquad (\lambda \notin G), \end{aligned}$$

$X - \lambda I$ will be invertible whenever $\| R_\lambda(A)(A - X) \| < 1$. Consequently, if $\delta < 1/M$, an operator $X \in U_\delta(A)$ will have a resolvent $R_\lambda(X)$ for all $\lambda \notin G$. Thus, under the indicated choice of δ, the spectrum $\sigma(X) \subset G$ for any $X \in U_\delta(A)$.

Let us estimate the difference $R_\lambda(A) - R_\lambda(X)$ for $\lambda \notin G$. From the above representation for $X - \lambda I$ it follows that

$$R_\lambda(X) = [I - R_\lambda(A)(A - X)]^{-1} R_\lambda(A).$$

For $X \in U_\delta(A)$ and $\lambda \notin G$ we have

$$\| R_\lambda(A)(A - X) \| < M\delta \, (< 1),$$

so that

$$\| R_\lambda(X) - R_\lambda(A) \| = \left\| \sum_{k=1}^{\infty} R_\lambda^{k+1}(A)(A - \lambda I)^k \right\| \leqq \frac{M^2\delta}{1 - M\delta}.$$

We have arrived at the following proposition.

THEOREM 2.1. *Suppose $A \in [\mathfrak{B}]$ and G is an open set covering the spectrum $\sigma(A)$. Then there exists a δ-neighborhood $U_\delta(A)$ of A such that $\sigma(X) \subset G$ for all $X \in U_\delta(A)$. Moreover, for any $\varepsilon > 0$ there exists a δ such that $\| R_\lambda(X) - R_\lambda(A) \| < \varepsilon$ for $X \in U_\delta(A)$ and $\lambda \notin G$.*

3. *Analytic functions of operators.* Let $x(t)$ $(a < t < b)$ be a vector function with values in $\mathfrak{B}$. If this function is continuous, the integral

$$\int_a^b x(t)\, dt$$

exists in the usual (Cauchy) sense as a limit of integral sums.

Similarly, if $A(t)$ $(a \leqq t \leqq b)$ is a continuous operator function with values in $[\mathfrak{B}_1, \mathfrak{B}_2]$, we have the existence of $\int_a^b A(t)dt$, which is an element of $[\mathfrak{B}_1, \mathfrak{B}_2]$.

The following estimates hold in this connection:

$$\left\| \int_a^b x(t)\, dt \right\| \leqq \int_a^b \|x(t)\| dt, \quad \left\| \int_a^b A(t)\, dt \right\| \leqq \int_a^b \|A(t)\|\, dt.$$

Analogous assertions hold for integrals of vector and operator functions along curves in the complex plane.

The classical theory of Cauchy on the vanishing of an integral around a closed contour can be extended to analytic vector and operator functions.

Suppose $A \in [\mathfrak{B}]$. Let K_A denote the class of all functions $\phi(\lambda)$ of a complex variable that are *piecewise analytic on the spectrum* $\sigma(A)$. Theis means that a function $\phi \in K_A$ if it has the following properties.

1) The domain of definition D_ϕ of $\phi(\lambda)$ consists of a finite number of open connected components whose union contains the spectrum $\sigma(A)$ of A, each component containing at least one point of the spectrum.

2) The function $\phi(\lambda)$ is piecewise holomorphic, i.e. holomorphic in each component of its domain of definition D_ϕ.

Clearly, if two functions $\phi_1(\lambda), \phi_2(\lambda) \in K_A$ coincide on an open neighborhood of the spectrum $\sigma(A)$, they will be analytic continuations of each other. Let us agree not to distinguish between such functions.

The class K_A can be converted into an algebra with unit by introducing in a natural way the operations of addition, multiplication and multiplication by a scalar.

According to a well-known principle of F. Riesz there exists an algebraic isomorphism between the algebra K_A and a certain commutative algebra of operators under which the function $\phi(\lambda) \equiv \lambda$ corresponds to the operator A (and hence $\phi(\lambda) = 1/(\lambda - z)$ $(z \notin \sigma(A))$ corresponds to the resolvent $R_z = (A - zI)^{-1}$).

This isomorphism is established in the following way. For $\phi(\lambda) \in K_A$ there can always be found a smooth and, in general, composite contour Γ_A that surrounds the spectrum $\sigma(A)$. The latter statement means that Γ_A decomposes into a finite number of boundaries of certain open sets whose union belongs to D_ϕ and covers $\sigma(A)$. Each of the Jordan contours making up Γ_A can be oriented so that the corresponding open set remains on the left as the contour is traversed in the positive direction.

After this, we put for $\phi(\lambda) \in K_A$

$$\phi(A) = -\frac{1}{2\pi i} \oint_{\Gamma_A} \phi(\lambda)\, R_\lambda d(\lambda). \tag{2.3}$$

It follows from Cauchy's theorem that the integral in (2.3) is independent of the choice of Γ_A.

The correspondence $\phi(\lambda) \leftrightarrow \phi(A)$ is clearly linear. Its multiplicativeness follows from the more general formula (2.4), which we prove below.

LEMMA 2.1. *Let $F_1(\lambda)$ and $F_2(\lambda)$ be operator functions that are piecewise analytic on the spectrum $\sigma(A)$ and whose values are operators in* $[\mathfrak{B}]$.[5] *Then*

[5] The definition of an operator function that is piecewise analytic on the spectrum $\sigma(A)$ is analogous to the definition in the scalar case.

$$\left\{-\frac{1}{2\pi i}\oint_{\Gamma_A} F_1(\lambda)\, R_\lambda\, d\lambda\right\}\cdot\left\{-\frac{1}{2\pi i}\oint_{\Gamma_A} R_\lambda F_2(\lambda)\, d\lambda\right\}$$
$$= -\frac{1}{2\pi i}\oint_{\Gamma_A} F_1(\lambda)\, R_\lambda F_2(\lambda)\, d\lambda. \tag{2.4}$$

PROOF. Let Γ_1 and Γ_2 be two contours of the type of Γ_A, the contour Γ_1 surrounding Γ_2. From (2.2) it follows that

$$\frac{1}{2\pi i}\oint_{\Gamma_1} F_1(\lambda)\, R_\lambda\, d\lambda\, \frac{1}{2\pi i}\oint_{\Gamma_2} R_\mu\, F_2(\mu)\, d\mu$$
$$= -\frac{1}{4\pi^2}\oint_{\Gamma_1}\oint_{\Gamma_2} F_1(\lambda)\, R_\lambda\, R_\mu\, F_2(\mu)\, d\lambda\, d\mu$$
$$= -\frac{1}{4\pi^2}\oint_{\Gamma_1}\oint_{\Gamma_2} F_1(\lambda)\, \frac{R_\lambda}{\lambda-\mu}\, F_2(\mu)\, d\lambda\, d\mu$$
$$+\frac{1}{4\pi^2}\oint_{\Gamma_1}\oint_{\Gamma_2} F_1(\lambda)\, \frac{R_\mu}{\lambda-\mu}\, F_2(\mu)\, d\lambda\, d\mu$$
$$= -\frac{1}{4\pi^2}\oint_{\Gamma_1} F_1(\lambda)\, R_\lambda \oint_{\Gamma_2} \frac{F_2(\mu)}{\lambda-\mu}\, d\mu\, d\lambda$$
$$+\frac{1}{4\pi^2}\oint_{\Gamma_1}\left\{\oint_{\Gamma_2} \frac{F_1(\lambda)}{\lambda-\mu}\, d\lambda\right\} R_\mu\, F_2(\mu)\, d\mu.$$

From here we get formula (2.4), since

$$\oint_{\Gamma_2} \frac{F_2(\mu)}{\lambda-\mu}\, d\mu = 0 \quad \text{for} \quad \lambda\in\Gamma_1 \quad \text{while} \quad \frac{1}{2\pi i}\oint_{\Gamma_1} \frac{F_1(\lambda)}{\lambda-\mu}\, d\lambda = F_1(\mu) \quad \text{for} \quad \mu\in\Gamma_2.$$

The lemma is proved.

By choosing $\phi_1(\lambda)I$, $\phi_2(\lambda)I$ as $F_1(\lambda)$, $F_2(\lambda)$, we get that $\phi(\lambda) = \phi_1(\lambda)\,\phi_2(\lambda)$ implies $\phi(A) = \phi_1(A)\phi_2(A)$.

The correspondence $\lambda \leftrightarrow A$ follows from the relation

$$-\frac{1}{2\pi i}\oint_{\Gamma_A} \lambda\,(A-\lambda I)^{-1}\, d\lambda = A,$$

which is obtained from (2.1) by taking as Γ_A any Jordan contour surrounding the disk $|\lambda| \leqq r_A$.

From the linearity and multiplicativeness of the correspondence $\phi(\lambda) \leftrightarrow \phi(A)$ it follows that $\phi(\lambda) = \sum_0^n c_k\lambda^k$ implies the formula

$$\phi(A) = \sum_{k=0}^{n} c_k\, A^k.$$

By passing to the limit, this formula can be extended to power series that converge in a disk containing the spectrum $\sigma(A)$. In particular, the operator

$$e^{At} = \sum_{k=0}^{\infty} \frac{A^k t^k}{k!} = -\frac{1}{2\pi i} \oint_{\Gamma_A} e^{\lambda t} R_\lambda \, d\lambda$$

corresponds to the function $e^{\lambda t}$.

It is known that there exists a simple connection between the spectrum $\sigma(A)$ of an operator A and a spectrum $\sigma(\phi(A))$, where $\phi(\lambda)$ is a function of class K_A. This connection is described by the relation (the *spectral mapping theorem*)

$$\sigma(\phi(A)) = \phi(\sigma(A)). \tag{2.5}$$

We note a particular consequence of the latter formula.

LEMMA 2.2. *If* $\| e^A \| = q(\| e^A \| < q)$, *the spectrum* $\sigma(A)$ *of* A *lies in (in the interior of) the closed halfplane* $\operatorname{Re} \lambda \leqq \ln q$.

For according to (2.5)

$$\sigma(e^A) = \{ \mu \mid \mu = e^\lambda, \lambda \in \sigma(A) \}.$$

On the other hand, by hypothesis, $\sigma(e^A)$ lies in (in the interior of) the disk $|\mu| \leqq q$.

4. *The F. Riesz integral.* Let us consider an operator $A \in [\mathfrak{B}]$ whose spectrum forms a disconnected set. A closed subset of $\sigma(A)$ whose complement in $\sigma(A)$ is also closed is called a *spectral* set. Suppose

$$\sigma(A) = \bigcup_{k=1}^{m} \sigma_k(A),$$

where the $\sigma_k(A)$ $(k = 1, \cdots, m)$ are disjoint spectral sets. In considering such a situation below we will assume that the contour Γ_A consists of disjoint parts Γ_k each of which is the boundary of a domain G_k containing the corresponding spectral set $\sigma_k(A)$.

Let

$$\phi_k(\lambda) = \begin{cases} 1 \text{ for } \lambda \in G_k, \\ 0 \text{ for } \lambda \in G_j, \ j \neq k. \end{cases}$$

The functions $\phi_k(\lambda)$ $(k = 1, \cdots, m)$ belong to the class K_A, and we can therefore let

$$P_k = \phi_k(A) = -\frac{1}{2\pi i} \oint_{\Gamma_A} \phi_k(\lambda) R_\lambda \, d\lambda = -\frac{1}{2\pi i} \oint_{\Gamma_k} R_\lambda \, d\lambda. \tag{2.6}$$

Inasmuch as the relations

$$\phi_k(\lambda)\phi_j(\lambda) = \delta_{kj} \phi_k(\lambda); \quad \sum_{k=1}^{m} \phi_k(\lambda) \equiv 1$$

(δ_{kj} is the Kronecker symbol) are clearly satisfied in the domain $G = \bigcup_{k=1}^m G_k$, we have the equalities

$$P_k P_j = 0 \ (k \neq j); \quad P_k^2 = P_k; \ \sum_{k=1}^{m} P_k = I. \tag{2.7}$$

Thus the operators P_k form a decomposition of the identity into a sum of pairwise disjoint projections.

From formula (2.6) it follows that the P_k commute with A;

$$P_k A = A P_k \qquad (k = 1,2,\cdots, m). \tag{2.8}$$

We recall that a subspace $\mathfrak{L} \subset \mathfrak{B}$ is said to be *invariant* under an operator A if $A\mathfrak{L} \subset \mathfrak{L}$. It is easily seen that if $\mathfrak{L}$ is a complemented subspace: $\mathfrak{B} = \mathfrak{L} \dotplus \mathfrak{M}$, i.e. if there exists a projection $P_{\mathfrak{L}}$ projecting $\mathfrak{B}$ onto $\mathfrak{L}$, then the invariance of $\mathfrak{L}$ under A is equivalent to the relation $P_{\mathfrak{L}} A P_{\mathfrak{L}} = A P_{\mathfrak{L}}$. And A commutes with $P_{\mathfrak{L}}$ if and only if $\mathfrak{L}$ *reduces* A, i.e. both $\mathfrak{L}$ and its complement $\mathfrak{M} = (I - P_{\mathfrak{L}})\mathfrak{B}$ are invariant under A.

From what has been said it is evident that the validity of the commutativity relations (2.8) is equivalent to the assertion that each of the subspaces $\mathfrak{B}_k = P_k \mathfrak{B}$ is invariant under A.

It can be verified that the spectrum $\sigma(A \mid \mathfrak{B}_k)$ of the restriction of A to the subspace $\mathfrak{B}_k$ coincides with the spectral set $\sigma_k(A)$.

In the formula

$$P_k = -\frac{1}{2\pi i} \oint_{\Gamma_k} R_\lambda \, d\lambda$$

the right side is known as a *Riesz integral.*

The space $\mathfrak{B}$ is the direct sum $\mathfrak{B}_1 \dotplus \cdots \dotplus \mathfrak{B}_m$, as follows from the last relation in (2.7).

Let us agree to call the projection P_k and the subspace $\mathfrak{B}_k = P_k\mathfrak{B}$ *the spectral projection and invariant subspace corresponding to each other and the spectral set* $\sigma_k(A)$.

In the sequel we will often consider an operator A whose spectrum does not intersect the imaginary axis. In this case we will always let $\sigma_-(A)$ and $\sigma_+(A)$ denote the spectral sets of A lying in the interiors of the left and right halfplanes respectively. This situation will be briefly expressed by the equality $\sigma(A) = \sigma_+(A) \cup \sigma_-(A)$.

In particular, the equality $\sigma(A) = \sigma_-(A)$ will mean that $\sigma(A)$ lies in the interior of the left halfplane. The equality $\sigma(A) = \sigma_+(A)$ has an analogous meaning.

The corresponding spectral projections will be denoted by $P_- = P_-(A)$ and $P_+ = P_+(A)$.

If each of the sets $\sigma_\pm(A)$ is nonempty, then for reasons to be given in Chapters II and IV we will sometimes say that A *is exponentially dichotomic* (*e-dichotomic*).

5. *Theorem on the continuous dependence of a spectral projection on the operator.* Suppose $A \in [\mathfrak{B}]$ and Γ is a closed contour not intersecting $\sigma(A)$ and surrounding a spectral set $\sigma_0(A)$. We will briefly say that Γ *distinguishes* the set $\sigma_0(A)$. As follows from subsection 2, the contour Γ will then not intersect the spectrum $\sigma(X)$ of any

operator $X \in U_\delta(A)$ for δ sufficiently small (the role of the open set G can be played, for example, by the complement of Γ in a sufficiently large open disk $|\lambda| < R$). Moreover, a $\delta > 0$ can be chosen so small that the resolvents $R_\lambda(X)$ and $R_\lambda(A)$ will differ arbitrarily little from each other on Γ. But then the projections

$$P_A = -\frac{1}{2\pi i}\oint_\Gamma R_\lambda(A)\,d\lambda \quad \text{and} \quad P_X = -\frac{1}{2\pi i}\oint_\Gamma R_\lambda(X)\,d\lambda$$

will differ arbitrarily little from each other.

Thus the following proposition holds.

THEOREM 2.2. *Suppose $A \in [\mathfrak{B}]$ and Γ is a contour distinguishing a spectral set $\sigma_0(A)$ corresponding to a spectral projection P_A. Then for any $\varepsilon > 0$ there exists a δ-neighborhood $U_\delta(A)$ such that for all $X \in U_\delta(A)$ the contour Γ will distinguish a spectral set $\sigma_0(X)$ corresponding to a spectral projection P_X in the ε-neighborhood of P_A: $\| P_A - P_X \| < \varepsilon$.*

§ 3. Solution of some linear operator equations

1. *General formula for the solutions.* In certain problems connected with differential equations it is necessary to investigate linear operator equations similar to those of the form $AX + XB = Y$.

We consider the more general equation

$$\sum_{j,k=0}^{n} c_{jk} A^j X B^k = Y, \tag{3.1}$$

where the c_{jk} are complex numbers, X is the desired operator and A, B and Y are given operators.

In order to consider equation (3.1) in the most general setting we will assume that $B \in [\mathfrak{B}_1]$, $A \in [\mathfrak{B}_2]$, $Y \in [\mathfrak{B}_1, \mathfrak{B}_2]$ and the desired operator $X \in [\mathfrak{B}_1, \mathfrak{B}_2]$.

Let A_l and B_r denote the linear operators acting on any operator $X \in [\mathfrak{B}_1, \mathfrak{B}_2]$ that are induced by multiplying X on the left by A and on the right by B respectively:

$$A_l X \overset{\text{def}}{=} AX, \quad B_r X \overset{\text{def}}{=} XB \qquad (X \in [\mathfrak{B}_1\, \mathfrak{B}_2]).$$

It is easily seen that the operators A_l and B_r commute for any $A \in [\mathfrak{B}_2]$ and $B \in [\mathfrak{B}_1]$.

With the use of the polynomial

$$P(\lambda, \mu) = \sum_{j,k=0}^{n} c_{jk} \lambda^j \mu^k$$

we form the operator

$$P_{A,B} \overset{\text{def}}{=} P(A_l, B_r) = \sum_{j,k=0}^{n} c_{jk} A_l^j B_r^k \quad (\in [\mathfrak{B}_1, \mathfrak{B}_2]). \tag{3.2}$$

Equation (3.1) now takes the form $P_{A,B}X = Y$. We will first determine a condition for the existence of an inverse of $P_{A,B}$.

Suppose $\lambda \notin \sigma(A)$. Then the operator

$$(A_l - \lambda I)^{-1} = (A - \lambda I)_l^{-1}$$

clearly exists. This means in particular that $\sigma(A_l) \subset \sigma(A)$.[6] In exactly the same way $\sigma(B_r) \subset \sigma(B)$.

We can now explicitly express the operator $P(A_l, B_r)$ in terms of the polynomial $P(\lambda, \mu)$. In fact, by substituting expressions obtained from (2.3) for the powers of the operators A_l and B_r in (3.2), we have

$$P(A_l, B_r) = -\frac{1}{4\pi^2}\oint_{\Gamma_A}\oint_{\Gamma_B} P(\lambda, \mu)(A_l - \lambda I)^{-1}(B_r - \mu I)^{-1}\,d\lambda\,d\mu. \tag{3.3}$$

Formula (3.3) suggests the following generalization of the problem under consideration. Let $K_{A,B}$ be the set of all single valued functions $\phi(\lambda, \mu)$ that are defined in a neighborhood of the set $\sigma(A) \times \sigma(B)$ and analytic in λ (μ) in a neighborhood of each point of $\sigma(A)$ ($\sigma(B)$) for fixed $\mu \in \sigma(B)$ ($\lambda \in \sigma(A)$).

For each function $\phi(\lambda, \mu) \in K_{A,B}$ we define an operator acting in the space $[\mathfrak{B}_1, \mathfrak{B}_2]$:

$$\phi_{A,B} \stackrel{\text{def}}{=} \phi(A_l, B_r) \stackrel{\text{def}}{=} -\frac{1}{4\pi^2}\oint_{\Gamma_A}\oint_{\Gamma_B} \phi(\lambda, \mu)(A_l - \lambda I)^{-1}(B_r - \mu I)^{-1}\,d\lambda\,d\mu. \tag{3.4}$$

Formula (3.4) establishes a correspondence, between the functions of class $K_{A,B}$ and a certain commutative set of operators acting in $[\mathfrak{B}_1, \mathfrak{B}_2]$, that has the following properties:

a) If $\phi(\lambda, \mu) \equiv 1$ then $\phi(A_l, B_r) = I$.

b) If $\phi(\lambda, \mu) = \alpha_1\phi_1(\lambda, \mu) + \alpha_2\phi_2(\lambda, \mu)$, where α_1 and α_2 are constants, then

$$\phi(A_l, B_r) = \alpha_1\phi_1(A_l, B_r) + \alpha_2\phi_2(A_l, B_r).$$

c) If $\phi(\lambda, \mu) = \phi_1(\lambda, \mu)\phi_2(\lambda, \mu)$ then, as follows from Lemma 2.1,

$$\phi(A_l, B_r) = \phi_1(A_l, B_r)\,\phi_2(A_l, B_r).$$

d) If $\lim_{n\to\infty}\phi_n(\lambda, \mu) = \phi(\mu, \lambda)$ uniformly in a neighborhood of the Cartesian product $\sigma(A) \times \sigma(B)$ then

$$\lim_{n\to\infty}\phi_n(A_l, B_r) = \phi(A_l, B_r).$$

We note that if $X \in [\mathfrak{B}_1, \mathfrak{B}_2]$ then

$$\phi(A_l, B_r)\,X = -\frac{1}{4\pi^2}\oint_{\Gamma_A}\oint_{\Gamma_B}\phi(\lambda, \mu)(A - \lambda I)^{-1}X(B - \mu I)^{-1}\,dd\mu.$$

The above remarks imply the following proposition.

[6] It can in fact be shown that $\sigma(A_l) = \sigma(A_r) = \sigma(A)$.

THEOREM 3.1. *Suppose a function $\phi(\lambda, \mu)$ of class $K_{A,B}$ does not vanish for $(\lambda, \mu) \in \sigma(A) \times \sigma(B)$.*

Then the operator $\phi(A_l, B_r)$ has an inverse $\psi(A_l, B_r)$ where $\psi(\lambda, \mu) = 1/\phi(\lambda, \mu)$.

In other words, the equation

$$\phi(A_l, B_r)\, X = Y \tag{3.5}$$

has in this case a unique solution $X \in [\mathfrak{B}_1, \mathfrak{B}_2]$ for each $Y \in [\mathfrak{B}_1, \mathfrak{B}_2]$. It is representable in the form

$$X = -\frac{1}{4\pi^2} \oint_{\Gamma_A} \oint_{\Gamma_B} \frac{(A - \lambda I)^{-1}\, Y (B - \mu I)^{-1}}{\phi(\lambda, \mu)}\, d\lambda\, d\mu. \tag{3.6}$$

We formulate this result for equation (3.1) in particular.

THEOREM 3.2. *If the condition*

$$P(\lambda, \mu) = \sum_{j,k=0}^{n} c_{jk}\, \lambda^j \mu^k \neq 0 \qquad ((\lambda, \mu) \in \sigma(A) \times \sigma(B))$$

is satisfied, equation (3.1) *has for each $Y \in [\mathfrak{B}_1, \mathfrak{B}_2]$ a unique solution $X \in [\mathfrak{B}_1, \mathfrak{B}_2]$, which is representable in the form*

$$X = -\frac{1}{4\pi^2} \oint_{\Gamma_A} \oint_{\Gamma_B} \frac{(A - \lambda I)^{-1}\, Y (B - \mu I)^{-1}}{P(\lambda, \mu)}\, d\lambda\, d\mu. \tag{3.7}$$

REMARK 3.1. Suppose $A, B \in [\mathfrak{B}]$ and a function $\phi(\lambda, \mu) \in K_{A,B}$ is different from zero only on a set $\sigma_1(A) \times \sigma_2(B)$, where $\sigma_1(A)$ and $\sigma_2(B)$ are spectral sets of the operators A and B respectively.

Let $\mathfrak{B}_1$ denote the invariant subspace of A that corresponds to the spectral set $\sigma_1(A)$, let P_1 denote the corresponding spectral projection and let A_1 denote the restriction of A to $\mathfrak{B}_1$. In exactly the same way we define for the operator B the invariant subspace $\mathfrak{B}_2$ corresponding to the spectral set $\delta_2(B)$, the spectral projection P_2 and the restriction B_2.

If the operator Y in (3.1) satisfies the condition $Y = P_1 Y P_2$, equation (3.1) can be reduced to the form

$$\sum_{j,k=0}^{n} c_{jk}\, A_1^j \hat{X} B_2^k = \hat{Y},$$

by considering $\hat{X}$ and $\hat{Y}$ as operators from $\mathfrak{B}_2$ into $\mathfrak{B}_1$. Formula (3.7) then permits one to find a solution $\hat{X} \in [\mathfrak{B}_2, \mathfrak{B}_1]$. This operator can be extended to the whole space $\mathfrak{B}$ by putting $X = P_1 \hat{X} P_2$. The operator X will be the unique solution of equation (3.1) satisfying the condition $X = P_1 X P_2$. Clearly, all of what has been said also applies to the case of the general equation (3.5).

2. *Special cases.* a) We consider the equation

$$AX - XB = Y. \tag{3.8}$$

Here $P(\lambda, \mu) = \lambda - \mu$. Thus under the condition $\lambda - \mu \neq 0\,((\lambda, \mu) \in \sigma(A) \times \sigma(B))$, i.e. under the condition

$$\sigma(A) \cap \sigma(B) = \varnothing, \tag{3.9}$$

this equation has a unique solution, and it is given by the formula

$$X = -\frac{1}{4\pi^2} \oint_{\Gamma_A} \oint_{\Gamma_B} \frac{(A - \lambda I)^{-1} Y (B - \mu I)^{-1}}{\lambda - \mu} \, d\lambda \, d\mu. \tag{3.10}$$

Suppose now

$$\sigma(A) = \sigma_+(A) \cup \sigma_-(A), \qquad \sigma(B) = \sigma_+(B) \cup \sigma_-(B).$$

According to Remark 3.1 we can determine the operator

$$X = P_+(A) X P_-(B) = -\frac{1}{4\pi^2} \oint_{\Gamma_{A+}} \oint_{\Gamma_{A-}} \frac{(A - \lambda I)^{-1} Y (B - \mu I)^{-1}}{\lambda - \mu} d\lambda \, d\mu, \tag{3.11}$$

which is the unique solution of equation (3.8) satisfying the condition $P_+(A)XP_-(B) = X$, provided the condition

$$P_+(A)\, Y P_-(B) = Y \tag{3.12}$$

is satisfied.

An analogous result is obtained under the condition $P_-(A)\, Y P_+(B) = Y$.

b) We now consider the equation

$$AX - XA = Y. \tag{3.13}$$

In this case $P(\lambda, \mu) = \lambda - \mu$ and condition (3.9) is obviously not satisfied.

We assume, however, that the spectrum $\sigma(A)$ decomposes into two spectral sets: $\sigma(A) = \sigma_1(A) \cup \sigma_2(A)$. When $\lambda \in \sigma_1(A)$ and $\mu \in \sigma_2(A)$, we have $\lambda - \mu \neq 0$, and hence under the condition $Y = P_1 Y P_2$ the operator

$$X = -\frac{1}{4\pi^2} \oint_{\Gamma_1} \oint_{\Gamma_2} \frac{(A - \lambda I)^{-1} Y (A - \mu I)^{-1}}{\lambda - \mu} d\lambda \, d\mu \tag{3.14}$$

is the unique solution of equation (3.13) satisfying the condition $X = P_1 X P_2$.

c) We also consider the equation

$$\int_a^b e^{At} X e^{Bt} \, dt = Y. \tag{3.15}$$

It can be reduced to the form (3.5) if one makes the substitutions

$$e^{At} = -\frac{1}{2\pi i} \oint_{\Gamma_A} e^{\lambda t} (A - \lambda I)^{-1} \, d\lambda, \qquad e^{Bt} = -\frac{1}{2\pi i} \oint_{\Gamma_B} e^{\mu t} (B - \mu I)^{-1} \, d\mu.$$

In fact, upon making these substitutions and changing the order of integration, we get from (3.15)

$$-\frac{1}{4\pi^2}\oint_{\Gamma_A}\oint_{\Gamma_B}\phi(\lambda,\mu)\,(A-\lambda I)^{-1}X(B-\mu I)^{-1}\,d\lambda\,d\mu=Y,$$

where

$$\phi(\lambda,\mu)=\int_a^b e^{(\lambda+\mu)t}\,dt=\frac{e^{(\lambda+\mu)b}-e^{(\lambda+\mu)a}}{\lambda+\mu}.$$

This function does not vanish for $(\lambda,\mu)\in\sigma(A)\times\sigma(B)$ if, for example, the condition

$$|\operatorname{Im}(\lambda+\mu)|<2\pi/(b-a)\qquad((\lambda,\mu)\in\sigma(A)\times\sigma(B))\tag{3.16}$$

is satisfied.

Consequently, under condition (3.16) equation (3.15) has a unique solution, which is representable in the form

$$X=-\frac{1}{4\pi^2}\oint_{\Gamma_A}\oint_{\Gamma_B}\frac{(A-\lambda I)^{-1}Y(B-\mu I)^{-1}(\lambda+\mu)}{e^{(\lambda+\mu)b}-e^{(\lambda+\mu)a}}\,d\lambda\,d\mu.\tag{3.17}$$

§ 4. Exponential operator function

1. *Definition of the exponential operator function.* In the theory of differential equations an especially important role is played by the operator function e^{At}, which can be introduced by means of either one of the two relations

$$e^{At}=-\frac{1}{2\pi i}\oint_{\Gamma_A}e^{\lambda t}(A-\lambda I)^{-1}\,d\lambda,\tag{4.1}$$

$$e^{At}=\sum_{k=0}^{\infty}\frac{A^kt^k}{k!}.\tag{4.2}$$

From the multiplicativeness of the correspondence $\phi(\lambda)\leftrightarrow\phi(A)$ it follows that the operators e^{At} form a one parameter group:

$$\begin{cases} e^{At}e^{A\tau}=e^{A(t+\tau)} & (-\infty<t;\tau<\infty),\\ e^{At}\big|_{t=0}=I. \end{cases}$$

We note, incidentally, that in general $e^{A+B}\neq e^Ae^B$. The equality sign holds here whenever $AB=BA$.

Further, by differentiating relation (4.1) with respect to t under the integral sign, we obtain the formula

$$de^{At}/dt=Ae^{At}.\tag{4.3}$$

In fact,

$$\begin{aligned}\frac{de^{At}}{dt}&=-\frac{1}{2\pi i}\oint_{\Gamma_A}\lambda e^{\lambda t}(A-\lambda I)^{-1}\,d\lambda\\ &=\left\{-\frac{1}{2\pi i}\oint_{\Gamma_A}\lambda(A-\lambda I)^{-1}\,d\lambda\right\}\left\{-\frac{1}{2\pi i}\oint_{\Gamma_A}e^{\lambda t}(A-\lambda I)^{-1}\,d\lambda\right\}=Ae^{At}.\end{aligned}$$

By making use of the series (4.2) it is not difficult to obtain the estimate

$$\| e^{At} \| \leqq \sum_{k=0}^{\infty} \frac{\| A^k \| t^k}{k!} = e^{\|A\|t} \qquad (t \geqq 0).$$

This estimate, however, is too rough; it can be replaced by a more exact one if the location of the spectrum of the operator A is known.

We first introduce an important notion.

2. *Estimate of the growth of the exponential function.* Let $\phi(t)$ $(0 \leqq t < \infty)$ be a positive function. As is well known, the (*upper*) *Ljapunov exponent* of $\phi(t)$ is the quantity

$$\kappa = \overline{\lim_{t\to\infty}} \frac{\ln \phi(t)}{t}.$$

This quantity coincides with the greatest lower bound of real numbers ρ for which there exists a constant N_ρ such that $\phi(t) \leqq N_\rho e^{\rho t}$ for all $t \geqq 0$.

If the limit $\lim_{t\to\infty} \ln \phi(t)/t$ exists, κ is called the *strict Ljapunov exponent* of $\phi(t)$.

The strict Ljapunov exponent (if it exists) also coincides with the least upper bound of real numbers ρ' such that

$$\phi(t) \geqq e^{\rho' t}$$

for sufficiently large t.

THEOREM 4.1. *For any $A \in [\mathfrak{B}]$ the function* $\| \exp(At) \|$ *has a strict Ljapunov exponent* κ, *and*

$$\kappa = \lim_{t\to\infty} \frac{\ln \| e^{At} \|}{t} = \max \{\operatorname{Re} \lambda \mid \lambda \in \sigma(A)\}. \tag{4.4}$$

PROOF. We put $\phi(t) = \| \exp(At) \|$. Clearly, $\phi(t + s) \leqq \phi(t)\phi(s)$, i.e.

$$\ln \phi(t + s) \leqq \ln \phi(t) + \ln \phi(s).$$

For any $\varepsilon > 0$ there exists an h_ε such that

$$\frac{\ln \phi(h)}{h} \leqq \overline{\lim_{t\to\infty}} \frac{\ln \phi(t)}{t} + \varepsilon \qquad (h \geqq h_\varepsilon). \tag{4.5}$$

For any $t > 0$ there exists an integer $n \geqq 0$ such that $t = nh + r$ $(0 \leqq r \leqq h)$. Then

$$\frac{\ln \phi(t)}{t} \leqq \frac{n \ln \phi(h) + \ln \phi(r)}{nh + r} \leqq \frac{n \ln \phi(h) + c}{nh + r} \left(c = \max_{0 \leqq r \leqq h} \phi(r) \right),$$

and hence

$$\overline{\lim_{t\to\infty}} \frac{\ln \phi(t)}{t} \leqq \frac{\ln \phi(h)}{h}. \tag{4.6}$$

Comparing (4.5) with (4.6), we conclude that the limit $\kappa = \lim_{t\to\infty} (\phi(t)/t)$ exists.

To prove equality (4.4) we note that it is equivalent to the following assertion: *in order for a real number ρ to have a positive number N_ρ corresponding to it such that*

$$\| e^{At} \| \leqq N_\rho e^{\rho t} \qquad (t \geqq 0), \tag{4.7}$$

it is necessary that $\operatorname{Re} \lambda \leqq \rho$ *for all* $\lambda \in \sigma(A)$, *and sufficient that* $\operatorname{Re} \lambda < \rho$ *for all* $\lambda \in \sigma(A)$.

Suppose estimate (4.7) is satisfied. Then according to Lemma 2.2 the spectrum $\sigma(At)$ ($= t\,\sigma(A)$) lies in the halfplane $\operatorname{Re} \lambda \leqq \ln N_\rho + \rho t$ $(t \geqq 0)$. Hence for any $t > 0$ the spectrum $\sigma(A)$ lies in the halfplane

$$\operatorname{Re} \lambda \leqq \ln N_\rho \,/\, t + \rho.$$

In view of the arbitrariness of $t > 0$ this implies the first part of the assertion.

For a proof of the second part we write e^{At} in the form

$$e^{At} = -\frac{1}{2\pi i} \oint_{\Gamma_A} e^{\lambda t} R_\lambda \, d\lambda,$$

assuming that the contour Γ_A lies completely inside the halfplane $\operatorname{Re} \lambda < \rho$. Then

$$\| e^{At} \| \leqq \frac{1}{2\pi} \int_{\Gamma_A} | e^{\lambda t} | \, \| R_\lambda \| \, | d\lambda | \leqq N_\rho e^{\rho t},$$

where $N_\rho = (l/2\pi) \max_{\lambda \in \Gamma_A} \| R_\lambda \|$, l being the length of the contour Γ_A. The theorem is proved.

REMARK 4.1. Estimate (4.7) does not hold in general if the spectrum $\sigma(A)$ lies in the closed halfplane $\operatorname{Re} \lambda \leqq \rho$.

In order to see this it suffices to consider a finite-dimensional space $\mathfrak{B}$ $(0 < \dim \mathfrak{B} < \infty)$ and an operator A in it consisting of a single Jordan cell (see Exercises 4 and 16).

An application of equality (4.4) to the operator $-A$ gives

$$\lim_{t \to -\infty} \frac{\ln \| e^{At} \|}{t} = \min \{ \operatorname{Re} \lambda \mid \lambda \in \sigma(A) \}.$$

COROLLARY 4.1. *If the estimate* $\| e^{At} \| \leqq c$ *holds for all* $t \in (-\infty, \infty)$, *the spectrum* $\sigma(A)$ *lies on the imaginary axis.*

The converse of this assertion is not true (see Remark 4.1).

3. *Estimate connected with the theorem on a boundary point of the spectrum.* We will need below another estimate of the behavior of e^{At} that is connected with the location of the spectrum of A. We first note that if λ is an eigenvalue of A and x is a corresponding normalized eigenvector: $(A - \lambda I)x = 0$, then, as readily follows from (4.2), for example, $e^{At}x = e^{\lambda t}x$, and hence $\| e^{At}x \| = e^{t \operatorname{Re} \lambda}$.

In the case of an arbitrary point of the spectrum the situation is more complicated, although certain estimates can still be given when λ is a boundary point of $\sigma(A)$. Their derivation is based on the following important proposition.

THEOREM ON A BOUNDARY POINT OF THE SPECTRUM. *For any boundary point* λ *of a spectrum* $\sigma(A)$ $(A \in [\mathfrak{B}])$ *and any* $\varepsilon > 0$ *there exists a normalized vector* x $(\|x\| = 1)$ *such that*

$$\|(A - \lambda I)x\| < \varepsilon. \tag{4.8}$$

PROOF. Suppose $\lambda \in \sigma(A)$ and $\mu \in \rho(A)$. Then $\|R_\mu\| \geqq 1/|\mu - \lambda|$. For we would otherwise get from the obvious identity

$$(A - \lambda I)R_\mu - I = (\mu - \lambda)R_\mu$$

that $\|(A - \lambda I)R_\mu - I\| < 1$, and this would imply (see the special rule mentioned at the end of § 2.1) that $(A - \lambda I)R_\mu$ and hence $A - \lambda I$ are invertible, which is impossible.

If λ is boundary point of $\sigma(A)$, there exists for any $\varepsilon > 0$ a $\mu \in \rho(A)$ such that $|\mu - \lambda| < \varepsilon/2$. Then $\|R_\mu\| > 2/\varepsilon$ and there accordingly exists an element y $(\|y\| = 1)$ such that $\|R_\mu y\| > 2/\varepsilon$ for $|\mu - \lambda| < \varepsilon/2$.

Therefore, setting $x = R_\mu y / \|R_\mu y\|$, the above identity implies

$$\|(A - \lambda I)x\| = \|(A - \lambda I)R_\mu y / \|R_\mu y\|\| \leqq 1/\|R_\mu y\| + |\mu - \lambda| < \varepsilon.$$

A proposition on the estimates referred to above can be formulated as follows.

LEMMA 4.1. *Let* λ *be a boundary point of a spectrum* $\sigma(A)$. *For any* $\delta > 0$ *and* $T > 0$ *there exists a normalized vector* x *for which*

$$e^{t\,\mathrm{Re}\,\lambda}(1 - \delta) \leqq \|e^{At}x\| \leqq e^{t\,\mathrm{Re}\,\lambda}(1 + \delta) \qquad (0 \leqq t \leqq T).$$

PROOF. The relations

$$(e^{At} - e^{\lambda t}I)x = e^{\lambda t}[e^{(A-\lambda I)t} - I]x = e^{\lambda t}\int_0^t e^{(A-\lambda I)\tau}\,d\tau \cdot (A - \lambda I)x$$

imply the estimate

$$\|e^{At}x - e^{\lambda t}x\| \leqq e^{t\,\mathrm{Re}\,\lambda}\int_0^t \|e^{(A-\lambda I)\tau}\|\,d\tau \cdot \|(A - \lambda I)x\|.$$

Let $M = \int_0^T \|e^{(A-\lambda I)\tau}\|\,d\tau$. It but remains to choose a normalized vector x that ensures the fulfilment of (4.8) when $\varepsilon = \delta/M$.

4. *Expression for the solutions of the operator equations in terms of the exponential function.* We will make use of the properties of the operator function e^{At} studied above in order to restate some of the results of § 3 in a new form.

We consider the equation

$$AX + XB = Y, \tag{4.9}$$

which is obtained from equation (3.8) by substituting $-B$ for B. Suppose that the spectra $\sigma(A)$ and $\sigma(B)$ lie in the interior of the left halfplane. Then the condition

$\lambda + \mu \neq 0$ for $(\lambda, \mu) \in \sigma(A) \times \sigma(B)$ is satisfied, and hence equation (4.9) has the unique solution

$$X = -\frac{1}{4\pi^2} \oint_{\Gamma_A} \oint_{\Gamma_B} \frac{(A - \lambda I)^{-1} Y (B - \mu I)^{-1}}{\lambda + \mu} \, d\lambda \, d\mu. \tag{4.10}$$

We take advantage of the equality

$$\frac{1}{\lambda + \mu} = -\int_0^\infty e^{(\lambda+\mu)t} \, dt,$$

which is valid for Re $\lambda < 0$ and Re $\mu < 0$. Substituting this expression in (4.10) and changing the order of integration, we find that

$$X = \frac{1}{4\pi^2} \int_0^\infty \oint_{\Gamma_A} e^{\lambda t}(A - \lambda I)^{-1} \, d\lambda \cdot Y \cdot \oint_{\Gamma_B} e^{\mu t}(B - \mu I)^{-1} \, d\mu$$

$$= -\int_0^\infty e^{At} Y e^{Bt} \, dt. \tag{4.11}$$

It can also be verified directly that the integral (4.11),which exists by virtue of estimate (4.7), satisfies equation (4.9). In fact,

$$-\left[A \int_0^\infty e^{At} Y e^{Bt} \, dt + \int_0^\infty e^{At} Y e^{Bt} \, dt \cdot B \right] = -\int_0^\infty d(e^{At} Y e^{Bt})$$

$$= -e^{At} \, Y e^{Bt} \Big|_0^\infty = Y.$$

Analogously, if the sets $\sigma(A)$ and $\sigma(B)$ lie in the interior of the right halfplane, the solution of equation (4.9) is given by the formula

$$X = \int_0^\infty e^{-At} Y e^{-Bt} \, dt. \tag{4.12}$$

5. *Some important renormings of* $\mathfrak{B}$. Certain norms defined by the exponential operator function play a critical role in stability questions.

a) We first assume that the spectrum $\sigma(A)$ lies in the interior of the left halfplane. Then for any ν such that

$$\text{Re } \lambda < -\nu \quad \text{for all } \lambda \in \sigma(A)$$

there exists an $N > 0$ for which

$$\| e^{At} \| < N e^{-\nu t} \qquad (t \geqq 0). \tag{4.13}$$

This estimate permits one to put for any $x \in \mathfrak{B}$

$$\| x \|_{A,r} = \left\{ \int_0^\infty \| e^{At} x \|^r \, dt \right\}^{1/r} \qquad (r \geqq 1). \tag{4.14}$$

From the Minkowski inequality it at once follows that $\| x \|_{A,r}$, can serve as a norm. Moreover, we have the following fact.

THEOREM 4.2. *The norm $\| x \|_{A,r}$, is topologically equivalent to the original norm $\| x \|$.*

PROOF. The assertion of the theorem is equivalent to the existence of positive constants m_r and M_r such that $m_r \| x \| \leqq \| x \|_{A,r} \leqq M_r \| x \|$.

The right inequality immediately follows from (4.13) and the relations

$$\| x \|_{A,r}^r \leqq \| x \|^r \int_0^\infty \| e^{At} \|^r dt \leqq \| x \|^r N^r \int_0^\infty e^{-\nu r t} dt = \frac{N^r}{\nu r} \| x \|^r,$$

On the other hand, since

$$\| x \| = \| e^{-At} e^{At} x \| \leqq \| e^{-At} \| \| e^{At} x \| \leqq e^{t\|A\|} \| e^{At} x \|,$$

we have

$$\| x \|_{A,r} = \left\{ \int_0^\infty \| e^{At} x \|^r dt \right\}^{1/r} \geqq \| x \| \left\{ \int_0^\infty e^{-rt\|A\|} dt \right\}^{1/r} = \frac{\| x \|}{\sqrt[r]{r \| A \|}}.$$

Thus the theorem is proved, it being possible to put

$$M_r = N / \sqrt[r]{\nu r}; \qquad m_r = 1 / \sqrt[r]{r \| A \|}.$$

The norm $\| x \|_{A,r}$ is interesting in that for any $x \in \mathfrak{B}$ the function of t

$$\| e^{At} x \|_{A,r} = \left\{ \int_t^\infty \| e^{As} x \|^r ds \right\}^{1/r}$$

monotonically decreases with increasing t and, moreover, has a negative (when $x \neq 0$) continuous derivative. This fact will be used in Chapters II and VII.

When $\mathfrak{B}$ is a Hilbert space $\mathfrak{H}$ the most natural of the $\| x \|_{A,r}$ is $\| x \|_{A,2}$. This norm is defined by the scalar product

$$(x, y)_{A,2} = \int_0^\infty (e^{At} x, e^{At} y) \, dt.$$

b) Theorem 4.2 is easily carried over to a more general case. We assume that the spectrum of the operator A does not intersect the imaginary axis and is separated by it into two parts, so that

$$\sigma(A) = \sigma_+(A) \cup \sigma_-(A).$$

Let $\mathfrak{B}_+$ and $\mathfrak{B}_-$ be the invariant subspaces of A corresponding to these parts of the spectrum, and let P_+ and P_- be the corresponding spectral projections. A norm can be introduced in each of these spaces by using formula (4.14) (with the replacement of A by $-A$ in $\mathfrak{B}_+$):

$$\| x \|_{A_-,r} = \left\{ \int_0^\infty \| e^{At} x \|^r dt \right\}^{1/r} \qquad (x \in \mathfrak{B}_-),$$

$$\| x \|_{-A_+,r} = \left\{ \int_0^\infty \| e^{-At} x \|^r dt \right\}^{1/r} \qquad (x \in \mathfrak{B}_+).$$

We now introduce a new norm in $\mathfrak{B}$ by putting

$$\| x \|_{A,r} = \| P_+ x \|_{-A_+,r} + \| P_- x \|_{A_-,r}. \tag{4.15}$$

Since in each of the subspaces $\mathfrak{B}_\pm$ the norms $\| x \|_{-A_+,r}$, $\| x \|_{A_-,r}$ are topologically equivalent to $\| x \|$, the norm $\| x \|_{A,r}$ is topologically equivalent to the norm $\| P_+ x \| + \| P_- x \|$, which in turn is equivalent to the norm $\| x \|$.

When $\mathfrak{B}$ is a Hilbert space $\mathfrak{H}$ it is natural to consider the norm $\| x \|_A$ defined by the scalar product

$$(x, y)_A = (P_+ x, P_+ y)_{-A_+,2} + (P_- x, P_- y)_{A_-,2},$$

which is topologically equivalent to the original scalar product (see § 1.5).

The new norms that have been introduced are interesting in that the function $\| e^{At}x \|_{A,r}$ ($\| e^{At}x \|_A$ when $\mathfrak{B} = \mathfrak{H}$) monotonically decreases for $x \in \mathfrak{B}_+$ but monotonically increases for $x \in \mathfrak{B}_-$.

§ 5. Generalized Ljapunov theorem

In the next four sections we will be dealing with a Hilbert space: $\mathfrak{B} = \mathfrak{H}$.

We assume that the reader is familiar with the fundamentals of the geometry of Hilbert spaces and the theory of operators in them. The scalar product of a pair of elements $x, y \in \mathfrak{H}$ will be denoted by (x, y). In order for a linear operator $A: \mathfrak{H} \to \mathfrak{H}$ to be bounded ($A \in \{\mathfrak{H}]$) it is necessary and sufficient that there correspond to it a linear operator $A^*: \mathfrak{H} \to \mathfrak{H}$ such that $(Ax, y) = (x, A^*y)$ for any $x, y \in \mathfrak{H}$.

If $A \in [\mathfrak{H}]$ then $A^* \in [\mathfrak{H}]$ and $(A^*)^* = A$. The operator A^* is called the *adjoint* of A.

We note the following simple properties of adjoints:

1) $(A + B)^* = A^* + B^*$ for $A, B \in [\mathfrak{H}]$;

2) $(\alpha A)^* = \bar{\alpha} A^*$, where α is a scalar;

3) $(AB)^* = B^* A^*$;

4) $\| A \| = \| A^* \|$;

5) the spectra $\sigma(A)$ and $\sigma(A^*)$ are distributed symmetrically with respect to the real axis.

An operator $H \in \{\mathfrak{H}]$ is asid to be *Hermitian* if $H = H^*$. A Hermitian operator H is characterized by the fact that its Hermitian form (Hx, x) $(x \in \mathfrak{H})$ takes only real values. The spectrum $\sigma(H)$ of a Hermitian operator H is a bounded closed set on the real axis. The least segment that contains $\sigma(H)$ will be denoted by $[\lambda_m(H), \lambda_M(H)]$. As is well known,

$$\lambda_m(H) = \inf \{(Hx, x) \mid \| x \| = 1\}; \lambda_M(H) = \sup \{(Hx, x) \mid \| x \| = 1\};$$
$$\| H \| = \max \{\lambda_M(H), -\lambda_m(H)\}.$$

An operator $H \in [\mathfrak{H}]$ is said to be *positive* (*nonnegative*) if its form (Hx, x) is positive (nonnegative) for any $x \neq 0$. Whenever H is nonnegative one has $\| H \| = \lambda_M(H)$.

An operator H is said to be *uniformly positive*, and one writes $H \gg 0$, if its form (Hx, x) is uniformly positive on the unit sphere $S = \{x \mid \| x \| = 1\}$ in $\mathfrak{H}$, i.e. if $\lambda_m(H) > 0$.

Negative, nonnegative and uniformly negative operators (and the meaning of the relation $H \ll 0$) are defined analogously.

Clearly, in order for a nonnegative operator to be invertible it is necessary and sufficient that it be uniformly positive.

It is significant that every uniformly positive operator H permits one to introduce in $\mathfrak{H}$ a new scalar product $(x, y)_H = (Hx, y)$ with respect to which $\mathfrak{H}$ remains a complete Hilbert space. This occurs because the new norm $\| x \|_H = (Hx, x)^{1/2}$ is topologically equivalent to the original one by virtue of the estimates

$$\lambda_m(H) \| x \|^2 \leqq \| x \|_H^2 \leqq \| H \| \, \| x \|^2 \qquad (\lambda_m(H) > 0).$$

We will say below that two different scalar products defined on one and the same set are *topologically equivalent* if the norms defined by them are topologically equivalent.

It can be asserted that any scalar product $(x, y)_1$ that is topologically equivalent to the original one can be obtained by means of a formula of the form

$$(x, y)_1 = (x, y)_H \qquad (= (Hx, y)),$$

where H is a uniformly positive operator.

We recall that the *real part* of an operator $A \in [\mathfrak{H}]$ is the Hermitian operator

$$A_{\mathfrak{R}} \equiv \operatorname{Re} A \equiv \tfrac{1}{2}(A + A^*),$$

while its *imaginary part* is the Hermitian operator

$$A_{\mathfrak{J}} \equiv \operatorname{Im} A \equiv (1/2i)(A - A^*),$$

so that $A = \operatorname{Re} A + i \operatorname{Im} A$.

THEOREM 5.1 (GENERALIZED LJAPUNOV THEOREM). *In order for the spectrum of an operator A to lie in the interior of the left halfplane it is necessary and sufficient that there exist a uniformly positive operator W such that*

$$\operatorname{Re}(WA) \ll 0. \tag{5.1}$$

Moreover, if $\sigma(A)$ lies in the interior of the left halfplane, for any $H \gg 0$ there exists an operator $W \gg 0$ such that

$$\operatorname{Re}(WA) = -H. \tag{5.2}$$

Thus, if $\operatorname{Re} A \ll 0$, the spectrum $\sigma(A)$ lies in the interior of the left halfplane. In linear algebra this proposition is a corollary of Hirsch's theorems. It does not admit a converse even in the case of an n-dimensional space $\mathfrak{H}$ with $n > 1$.

PROOF OF THE THEOREM. We will show that if the spectrum $\sigma(A)$ lies in the interior of the left halfplane, equation (5.2) has a solution $W \gg 0$ for any $H \gg 0$. The "necessary part" of the theorem will thereby be proved.

We rewrite equation (5.2) in the more explicit form

$$A^*W + WA = -2H.$$

Its solution can be represented on the basis of formula (4.11) in the form

$$W = 2\int_0^\infty e^{A^*t} He^{At}\,dt,$$

since the spectrum $\sigma(A^*)$, like $\sigma(A)$, lies in the interior of the left halfplane.

The uniform positiveness of this operator is implied by the following estimate, which is derived with the use of Theorem 4.2:

$$\begin{aligned}(Wx, x) &= 2\int_0^\infty (He^{At}x,\ e^{At}x)\ dt \geqq 2\lambda_m(H)\int_0^\infty \| e^{At}x \|^2\,dt \\ &= 2\lambda_m(H)\ \| x \|^2_{A,2} \geqq 2\lambda_m(H)m_2^2\ \| x \|^2.\end{aligned}$$

Suppose now that there exists an operator $W \gg 0$ satisfying condition (5.2) for $H \gg 0$.

According to Theorem 4.1 the sufficiency part of the theorem will follow if we obtain the estimate

$$\| e^{At} \| \leqq Ne^{-\nu t} \qquad (\nu > 0). \tag{5.3}$$

It suffices to obtain estimate (5.3) in the norm $\| x \|_W = (Wx, x)^{1/2}$, since this norm is equivalent to the original norm.

Let

$$\phi(t) = \| e^{At}x \|^2_W = (We^{At}x,\ e^{At}x).$$

Then

$$\begin{aligned}\phi'(t) &= (WAe^{At}x,\ e^{At}x) + (We^{At}x,\ Ae^{At}x) \\ &= ((WA + A^*W)\,e^{At}x,\ e^{At}x) = -2(He^{At}x,\ e^{At}x) \\ &\leqq -2\lambda_m(H)\ \| e^{At}x \|^2 \leqq \frac{2\lambda_m(H)}{\| W \|}\ \| e^{At}x \|^2_W = -2\nu\phi(t),\end{aligned}$$

where $\nu = \lambda_m(H)\,/\,\| W \| > 0$.

Integrating the obtained inequality $\phi'(t) \leqq -2\nu\phi(t)$, we get

$$\ln\frac{\phi(t)}{\phi(0)} = \int_0^t \frac{\phi'(\tau)}{\phi(\tau)}\,d\tau \leqq -2\nu t,$$

which implies $\phi(t) \leqq \phi(0)e^{-2\nu t}$ and, finally, $\|e^{At}x\|_W \leqq \| x \|_W\,e^{-\nu t}$.

REMARK 5.1. An operator A is said to be *dissipative* if

$$((\operatorname{Re} A)\, x, x) = \operatorname{Re}\, (Ax, x) \leqq 0.$$

We will call an operator A *uniformly dissipative* if $\operatorname{Re} A \ll 0$, i.e. there exists a positive constant c such that $\operatorname{Re}\, (Ax, x) \leqq c\| x \|^2$ (as c, one can take $-\lambda_M (\operatorname{Re} A)$).

Since the scalar products (x, y) and $(x, y)_W$ are topologically equivalent, condition (5.1) for the uniform dissipativeness of the operator WA can be interpreted as a condition for the uniform dissipativeness of the operator A itself with respect to the scalar product $(\cdot, \cdot)_W$. In fact, if $\operatorname{Re}\, (WA) \ll 0$,

$$\operatorname{Re}(Ax, x)_W \,(= \operatorname{Re}(WAx, x)) < - c \| x \|^2 \leqq - (c/\| W \|) \|x \|^2_W.$$

Conversely, if $\operatorname{Re}(Ax, x)_W < - c_1\| x \|^2_W \, (c_1 > 0)$,

$$\operatorname{Re}(WAx, x) < - c_1(\lambda_m (W))^2 \| x \|^2.$$

Therefore an operator A satisfying condition (5.1) is naturally said to be *uniformly W-dissipative.*

Using this terminology, the first part of Theorem 5.1 can be rephrased in the following manner.

In order for the spectrum of an operator A to lie in the interior of the left halfplane it is necessary and sufficient that it be uniformly W-dissipative for some $W \gg 0$.

§ 6. A theorem of B. von Sz.-Nagy

1. *Commutative groups of operators.* As has already been noted, the operators e^{At} $(-\infty < t < \infty)$ form a commutative group.

It can turn out that this group is bounded. In this case it can be asserted that the spectrum $\sigma(A)$ lies on the imaginary axis (see Corollary 4.1).

If $\mathfrak{B}$ is a Hilbert space $\mathfrak{H}$, it is possible to give a complete characterization of the operator A generating a bounded group.

This result (Theorem 6.3) will be obtained from a simple extension (Theorem 6.2) of a general stability criterion (Theorem 6.1) of B. von Sz.-Nagy.

An operator $G \in [\mathfrak{B}]$ is said to be *stable* if it has an inverse and its integral powers are uniformly bounded: $\| G^n \| \leqq c$ $(n = 0, \pm 1, \pm 2, \cdots)$.

THEOREM 6.1 (SZ.-NAGY). *In order for an operator G acting in a Hilbert space $\mathfrak{H}$ ($G \in [\mathfrak{H}]$) to be stable it is necessary and sufficient that it be similar to a unitary operator U, i.e. $G = SUS^{-1}$, where $U^* = U^{-1}$ and $S, S^{-1} \in [\mathfrak{H}]$.*

Since the integral powers of an invertible operator form a commutative group, Theorem 6.1 is a corollary of the following more general proposition.

THEOREM 6.2. *Suppose $\mathfrak{G} = \{G\}$ is a commutative group of operators from $[\mathfrak{H}]$. Then G is bounded:*

$$\| G \| \leqq c \qquad (G \in \mathfrak{G}), \tag{6.1}$$

if and only if there exists in $\mathfrak{H}$ *a scalar product* $(x, y)_1$, *topologically equivalent to the original one, that is invariant under the group* $\mathfrak{G}$:

$$(Gx, Gy)_1 = (x, y)_1 \qquad (G \in \mathfrak{G};\, x, y \in \mathfrak{H}). \tag{6.2}$$

PROOF. We consider on $\mathfrak{G}$ the function

$$\phi_{x,y}(X) = (Xx, Xy) \qquad (X \in \mathfrak{G}).$$

This function is bounded on $\mathfrak{G}$ since

$$|\phi_{x,y}(X)| \leqq \|X\|^2 \|x\| \|y\| \leqq c^2 \|x\| \|y\|. \tag{6.3}$$

From a well-known theorem of A. A. Markov[7] (A. A. Markov [1]; see also M. G. Kreĭn and M. A. Rutman [1]) it follows that on the space $\mathfrak{M} = \mathfrak{M}(\mathfrak{G})$ of all bounded complex valued functions on $\mathfrak{G}$ there exists an invariant mean, i.e. a linear functional $M(\phi)$ ($\phi \in \mathfrak{M}$) having the following properties:

a) $M(1) = 1$;

b) $M(\phi(X)) \geqq 0$ for $\phi(X) \geqq 0$;

c) $M(\phi(GX)) = M(\phi(X))$ for any $G \in \mathfrak{G}$.

From the first two properties it follows that[8]

$$|M(\phi(X))| \leqq \sup\{|\phi(X)| \mid X \in \mathfrak{G}\}. \tag{6.4}$$

As the desired scalar product, one can take $(x,y)_1 = M(\phi_{x,y})$.

It can be verified directly that $(x, y)_1$ has the usual properties of a scalar product. From (6.3) and (6.4) it follows that

$$|(x, y)_1| \leqq c^2 \|x\| \|y\|; \quad \|x\|_1 \leqq c\|x\|.$$

On the other hand, we have by virtue of (6.1) that

$$\phi_{x,x}(X) = \|Xx\|^2 \geqq \frac{\|X^{-1}\|^2}{c_2} \|Xx\|^2 \geqq \frac{1}{c_2} \|X^{-1}Xx\|^2 = \frac{1}{c^2} \|x\|^2,$$

which implies by virtue of a) and b) that

$$(x, x)_1 = M(\phi_{x,x}) \geqq \frac{1}{c_2} \|x\|^2.$$

Thus the scalar product $(\cdot,\cdot)_1$ is topologically equivalent to the original one. The

---

[7] We have in mind the following theorem: *Suppose Q is an arbitrary set and $\mathfrak{G}$ is a commutative family of single valued mappings of Q into itself. Then on the set $\mathfrak{M} = \mathfrak{M}(Q)$ of all bounded complex valued functions there exists an invariant mean, i.e. a linear functional having properties* a) – c).

We apply this theorem for $Q = \mathfrak{G}$ (here an element $G \in \mathfrak{G}$ is considered as the mapping $X \to GX$ ($X \in \mathfrak{G}$).

[8] In fact, if $\phi(X)$ is real, denoting the right side of (6.4) by s_ϕ, we will have $s_\phi \pm \phi(X) \geqq 0$. It follows by virtue of b) that $M(s_\phi \pm \phi) \geqq 0$, i.e. $|M(\phi)| \leqq M(s_\phi) = s_\phi$. If $\phi(X)$ is complex valued, putting $M(\phi) = |M(\phi)| e^{i\alpha}$, we will have $|M(\phi)| = e^{-i\alpha} M(\phi) = M(e^{-i\alpha}\phi) = M(\mathrm{Re}(e^{-i\alpha}\phi))$. On the other hand, by what has been proved $M(\mathrm{Re}(e^{-i\alpha}\phi)) \leqq \sup |\mathrm{Re}(e^{-i\alpha}\phi)| \leqq \sup |\phi|$.

invariance of the scalar product $(\cdot,\ \cdot)_1$ under $\mathfrak{G}$ is obtained immediately. In fact, for every $G \in \mathfrak{G}$, on the basis of c)

$$(Gx, Gy)_1 = M(\phi_{Gx,Gy}(X)) = M(\phi_{x,y}(XG))$$
$$= M(\phi_{x,y}(GX)) = M(\phi_{x,y}(X)) = (x, y)_1.$$

The direct part of the theorem is proved.

Suppose now $(x, y)_1$ is a scalar product that is topologically equivalent to the original one and invariant under the group $\mathfrak{G}$.

As we know, topological equivalence implies the existence of a uniformly positive operator H such that $(x, y)_1 = (Hx, y)$.

Since the operator H is uniformly positive, there exists a uniquely defined uniformly positive square root $S = \sqrt{H}$ of it (see Exercise 27).

From (6.2) it follows that

$$(HGx, Gy) = (Hx, y) \quad \text{or} \quad (S^2Gx, Gy) = (S^2x, y),$$

which implies

$$G^* S^2G = S^2 \quad \text{or} \quad (S^{-1}G^*S)(SGS^{-1}) = I.$$

Thus the invertible operator $U = SGS^{-1}$ is a unitary operator.

Since $G = S^{-1}US$ and $\| U \| = 1$,

$$\| G \| \leqq \| S^{-1} \| \| S \|,$$

i.e. the group $\mathfrak{G}$ is bounded. The theorem is proved.

We have simultaneously proved the following assertion, which contains Theorem 6.1.

THEOREM 6.2′. *In order for a commutative group $\mathfrak{G}$ of operators acting in a Hilbert space $\mathfrak{H}$ to be bounded it is necessary and sufficient that it be similar to a group of unitary operators.*

REMARK 6.1. We leave it to the reader to verify that the above arguments imply that the following equality holds for a bounded group $\mathfrak{G} \subset [\mathfrak{H}]$:

$$\sup \{ \| G \| \mid G \in \mathfrak{G} \} = \inf \{ \| S \| \| S^{-1} \| \mid S \in \mathbf{S}(\mathfrak{G})\},$$

where $\mathbf{S}(\mathfrak{G})$ is the set of all invertible transformations S of $\mathfrak{G}$ into the group of unitary operators.

2. *Boundedness condition for the exponential operator function in $\mathfrak{H}$.* We obtain the result referred to in subsection 1 as a simple corollary of Theorem 6.2.

THEOREM 6.3. *Let $A \in [\mathfrak{H}]$. In order for the commutative group $\mathfrak{G} = \{e^{At}\}$ $(-\infty < t < \infty)$ to be bounded it is necessary and sufficient that A be similar to a skew-Hermitian*[9] *operator B.*

[9] An operator $B \in [\mathfrak{H}]$ is said to be *skew Hermitian* if $B^* = -B$, i.e. if the operator iB is Hermitian.

PROOF. NECESSITY. If the group G is bounded, there exists a uniformly positive operator H such that

$$(He^{At}x, e^{At}y) = (Hx, y).$$

Differentiating this equality with respect to t, we get

$$(HAe^{At}x, e^{At}y) + (He^{At}x, Ae^{At}y) = 0,$$

which for $t = 0$ implies $HA + A^*H = 0$.

Putting $S = \sqrt{H}$, $H = S^2$, we find that $SAS^{-1} = -S^{-1}A^*S$.

Since $S = S^*$, the latter equality means that the operator $B = SAS^{-1}$ is skew Hermitian while $A = S^{-1}BS$.

The sufficiency follows from the fact that the equality $A = S^{-1}BS$ implies $e^{At} = S^{-1}e^{Bt}S$, where e^{Bt} is a unitary operator ($(e^{Bt})^* = e^{-Bt}$).

REMARK 6.2. The well-known theorem of Stone, asserting that the general form of a one parameter group U_t of unitary operators is given by the formula $U_t = e^{iHt}$, where H is a selfadjoint (possibly unbounded) operator, in combination with Theorem 6.2 permits one to obtain the general form of bounded one parameter semigroups in $[\mathfrak{H}]$.

Since we are confining our attention to bounded operators, we will not go into this question in the sequel.

§ 7. Hilbert space with an indefinite metric

1. *Uniformly W-dissipative operators and the direct decompositions associated with them.* A Hermitian operator $W \in [\mathfrak{H}]$ is said to be *indefinite* if the Hermitian form (Wx, x) takes values of different sign for $x \in \mathfrak{H}$.

Suppose W is an indefinite operator. We consider the *indefinite* scalar product $(x, y)_W = (Wx, y)$. The index W will often be dropped when this does not lead to confusion.

We introduce some notation.

An element $x \in \mathfrak{H}$ is said to be *W-positive* if $(Wx, x) > 0$, *W-negative* if $(Wx, x) < 0$ and *W-neutral* if $(Wx, x) = 0$.

An element x is said to be *W-nonnegative* (*W-nonpositive*) if it is either W-positive (W-negative) or W-neutral.

Two vectors $x, y \in \mathfrak{H}$ are said to be *W-orthogonal* if $(Wx, y) = 0$. A W-neutral vector is orthogonal to itself.

Two sets $E_1, E_2 \subset \mathfrak{H}$ are said to be *W-orthogonal* if $(Wx, y) = 0$ for all $x \in E_1$ and $y \in E_2$.

A subspace $\mathfrak{L} \subset \mathfrak{H}$ is said to be *W-positive* (*W-nonnegative*) if all of its nonzero elements are W-positive (W-nonnegative).

A subspace $\mathfrak{L}$ is said to be *uniformly W-positive* if there exists a constant $\alpha > 0$ such that for all $x \in \mathfrak{L}$

$$(x, x)_W \geqq \alpha (x, x). \tag{7.1}$$

W-negative, W-nonpositive and *uniformly W-negative* spaces are defined analogously.

We note that inequality (7.1) and the boundedness of W imply that in a uniformly W-positive subspace the scalar product $(x, y)_W$ is an ordinary scalar product that is topologically equivalent to the original one.

Suppose W is an arbitrary Hermitian operator. In precisely this case we call an operator A *uniformly W-dissipative* if $\mathrm{Re}\,(WA) \gg 0$.

LEMMA 7.1. *The spectrum of a uniformly W-dissipative operator A does not intersect the imaginary axis.*

PROOF. Suppose $\lambda = \mu i$ (μ a real number) is a boundary point of the spectrum $\sigma(A)$.

Then by the theorem on a boundary point of the spectrum there exists a sequence of unit vectors f_n ($\|f_n\| = 1$) such that $g_n = Af_n - \lambda f_n \to 0$ for $n \to \infty$. But this contradicts the uniform positiveness of the operator $H = -2\mathrm{Re}\,(WA) = -(WA + A^*W)$ since

$$\begin{aligned} 0 < \lambda_m(H) &\leqq (Hf_n, f_n) = -(WAf_n, f_n) - (Wf_n, Af_n) \\ &= -\mu i(Wf_n, f_n) - (Wg_n, f_n) + \mu i(Wf_n, f_n) - (Wf_n, g_n) \\ &= -2\mathrm{Re}\,(Wg_n, f_n) \leqq 2\|W\|\|g_n\| \to 0. \end{aligned}$$

An operator $T \in [\mathfrak{H}]$ is called a *uniform W-contraction* if

$$T^*WT \ll W. \tag{7.2}$$

LEMMA 7.2. *If A is a uniformly W-dissipative operator, for any positive $a \notin \sigma(A)$ the operator*

$$T = (A + aI)(A - aI)^{-1} \tag{7.3}$$

is a uniform W-contraction whose spectrum does not intersect the unit circle.

PROOF. Let $\phi(z) = (z + a)/(z - a)$. Then by the spectral mapping theorem $\sigma(T) = \phi(\sigma(A))$. But $\phi(z)$ maps only points of the imaginary axis onto the unit circle, and hence $\sigma(T)$ does not intersect the unit circle.

Further, it follows directly from (7.3) that $A = a(T + I)(T - I)^{-1}$. Therefore the equality $WA + A^*W = -H$ implies

$$W(T + I)(T - I)^{-1} + (T^* - I)^{-1}(T^* + I)W = -(1/a)H$$

or

$$(T^* - I)W(T + I) + (T^* + I)W(T - I) = -(1/a)(T^* - I)H(T - I) \ll 0.$$[10]

Multiplying out the terms in the left side of this inequality, we arrive at (7.2).

We now prove a theorem that generalizes Theorem 5.1.

[10] We have made use of the fact that $S^*HS \gg 0$ for any invertible S if $H \gg 0$.

THEOREM 7.1. *In order for an operator $A \in [\mathfrak{H}]$ to be e-dichotomic*[11] *it is necessary and sufficient that it be uniformly W-dissipative with respect to an indefinite Hermitian operator W:*

$$\operatorname{Re}(WA) \ll 0. \tag{7.4}$$

Any operator W satisfying condition (7.4) is invertible and such that the subspace $\mathfrak{H}_-$ is uniformly W-positive and the subspace $\mathfrak{H}_+$ is uniformly W-negative.

An operator W can be chosen so that these subspaces are W-orthogonal.

PROOF. SUFFICIENCY. Suppose there exist operators $H \gg 0$ and $W = W^*$ such that

$$WA + A^*W = -H. \tag{7.4'}$$

By virtue of Lemma 7.1 the spectrum of A does not intersect the imaginary axis: $\sigma(A) = \sigma_+(A) \cup \sigma_-(A)$.

We put $\mathfrak{H}_+ = P_+\mathfrak{H}$, $\mathfrak{H}_- = P_-\mathfrak{H}$, where P_+ and P_- are the spectral projections and $\mathfrak{H}_+$ and $\mathfrak{H}_-$ are the invariant subspaces corresponding to $\sigma_+(A)$ and $\sigma_-(A)$ respectively.

The operator $T = (A + aI)(A - aI)^{-1}$ dealt with in Lemma 7.2 has the same invariant subspaces and projections. Here the set $\sigma(T \mid \mathfrak{H}_-)$ lies in the interior, while $\sigma(T \mid \mathfrak{H}_+)$ lies in the exterior, of the unit disk. We therefore have the representation

$$P_- = \frac{1}{2\pi i} \oint_{|z|=1} (zI - T)^{-1}\, dz = \frac{1}{2\pi} \int_0^{2\pi} e^{i\phi} (e^{i\phi}I - T)^{-1}\, d\phi. \tag{7.5}$$

In order to prove that $\mathfrak{H}_-$ is uniformly positive we choose an arbitrary element $x \in \mathfrak{H}_-$ $(P_-x = x)$ and consider the expression

$$\begin{aligned}(Wx, x) &= (WP_-x, x) + (Wx, P_-x) - (Wx, x)\\ &= ((WP_- + P^*_-W - W)x, x).\end{aligned} \tag{7.6'}$$

It suffices to prove that the operator $WP_- + P^*_-W - W$ is uniformly positive:

$$WP_- + P^*_-W - W \gg 0. \tag{7.6}$$

Substituting the right side of (7.5) for P_- in the left side of (7.6), we get

$$\begin{aligned} &WP_- + P^*_-W - W\\ &= \frac{1}{2\pi}\int_0^{2\pi} (We^{i\phi}(e^{i\phi}I - T)^{-1} + e^{-i\phi}(e^{-i\phi}I - T^*)^{-1}W - W)\, d\phi\\ &= \frac{1}{2\pi}\int_0^{2\pi} (e^{-i\phi}I - T^*)^{-1} H (e^{i\phi}I - T)^{-1}\, d\phi.\end{aligned}$$

[11] We recall (see § 2.4) that this means that $\mathfrak{H}$ decomposes into a direct sum of invariant (under A) subspaces: $\mathfrak{H} = \mathfrak{H}_+ \dotplus \mathfrak{H}_-$, the spectra $\sigma_\pm = \sigma(A \mid \mathfrak{H}_\pm)$ of the restrictions of A to $\mathfrak{H}_\pm$ lying respectively in the interiors of the right and left halfplanes.

In view of the fact that all of the points of the unit circle are regular points of T and T^* while $H \gg 0$, the integrand here is uniformly positive.[12]

It follows from (7.6′) that $(Wx, x) \geqq c\| x \|^2$ for any $x \in \mathfrak{H}_-$.

The uniform W-negativeness of $\mathfrak{H}_+$ is established analogously.

We show, finally, that relation (7.4′) automatically implies the invertibility of W. In fact,

$$\| x \|^2 \leqq \frac{1}{\lambda_m(H)} (Hx, x)$$

$$= \frac{1}{\lambda_m(H)} \left| (Wx, Ax) + (Ax, Wx) \right| \leqq \frac{2\| A \|}{\lambda_m(H)} \| Wx \| \| x \|,$$

and thus an inequality of type $\| Wx \| \geqq c\| x \|$ is valid.

The sufficiency of the condition of the theorem is proved.

NECESSITY. Suppose now that the spectrum of A does not intersect the imaginary axis and $\mathfrak{H}$ decomposes into a direct sum $\mathfrak{H} = \mathfrak{H}_+ \dotplus \mathfrak{H}_-$ of invariant (under A) subspaces so that the set $\sigma_+ = \sigma(A \,|\, \mathfrak{H}_+)$ lies in the interior of the right, while $\sigma_- = \sigma(A \,|\, \mathfrak{H}_-)$ lies in the interior of the left, halfplane.

Starting from any operator $H \gg 0$, we form the operator $P_+^* H P_+$ acting from $\mathfrak{H}_+$ into the subspace $\mathfrak{H}_+^*$ that is invariant under A^* and corresponds to its spectral set $\sigma_+^* = \bar{\sigma}_+$.

According to the results in § 4.4 the operator

$$X = -\int_0^\infty e^{-A^*t} P_+^* H P_+ e^{-At} \, dt$$

satisfies the equation $A^*X + XA = -P_+^* H P_+$. In addition, $X = P_+^* X P_+$ and for $f \in \mathfrak{H}_+$

$$(Xf, f) = -\int_0^\infty (HP_+ e^{-At} f, P_+ e^{-At} f) \, dt < 0.$$

Analogously, the operator

$$Y = \int_0^\infty e^{A^*t} P_-^* H P_- e^{At} \, dt$$

satisfies the equation $A^*Y + YA = -P_-^* H P_-$ and the conditions $Y = P_-^* Y P_-$ and $(Yf, f) > 0$ for $f \in \mathfrak{H}_-$.

We put $W = X + Y$.

[12] We have made use of the following easily proved fact: if $F(\phi)$ is a continuous and continuously invertible operator function on $[a, b]$ while $H \gg 0$, then $\int_a^b F^*(\phi) H F(\phi) \, d\phi \gg 0$.

Clearly $(Wf, g) = 0$ for $f \in \mathfrak{H}_+$, $g \in \mathfrak{H}_-$. In addition, the operator A is uniformly W-dissipative inasmuch as

$$A^*W + WA = -H_1,$$

where $H_1 = P_+^* HP_+ + P_-^* HP_- \gg 0$, since

$$\begin{aligned}(H_1 x, x) &= (HP_+ x, P_+ x) + (HP_- x, P_- x)\\ &\geqq \lambda_m(H)\{\|P_+ x\|^2 + \|P_- x\|^2\}\\ &\geqq \tfrac{1}{2}\lambda_m(H)\{\|P_+ x\| + \|P_- x\|\}^2 \geqq \tfrac{1}{2}\lambda_m(H)\|x\|^2.\end{aligned}$$

The theorem is proved.

In connection with Theorem 7.1 it is useful to have in mind the following general fact.

THEOREM 7.2. *Suppose W is an invertible Hermitian operator and the space $\mathfrak{H}$ decomposes into a direct sum $\mathfrak{H}_1 \dotplus \mathfrak{H}_2$ of W-orthogonal subspaces. If the subspace $\mathfrak{H}_1$ is W-positive (negative), it is uniformly W-positive (negative).*

PROOF. Let P_1 denote the projection onto $\mathfrak{H}_1$ corresponding to the given direct decomposition.

Suppose $f \in \mathfrak{H}_1$ and $g = W^{-1}f$. Then, inasmuch as the form (Wx, y) is positive in $\mathfrak{H}_1$, we have

$$\begin{aligned}\|f\|^2 &= (Wg, f) = (WP_1 g + WP_2 g, f) = (WP_1 g, f)\\ &\leqq \sqrt{(WP_1 g, P_1 g)(Wf, f)} \leqq \sqrt{\|W\|}\cdot\|P_1 W^{-1}\|\cdot\|f\|\cdot\sqrt{(Wf, f)},\end{aligned}$$

which implies $(Wf, f) \geqq m\|f\|^2$.

2. *u-dichotomic operators and uniform W-contractions.* We say that an operator T is *u-dichotomic* if its spectrum does not intersect the unit circle and contains both a spectral set $\sigma_i(T)$ lying in the interior of the unit circle and a spectral set $\sigma_e(T)$ lying in its exterior: $\sigma(T) = \sigma_i(T) \cup \sigma_e(T)$.

To this decomposition there corresponds a direct decomposition of the space $\mathfrak{H} = \mathfrak{H}_i \dotplus \mathfrak{H}_e$ into invariant subspaces of the operator T so that the spectra of the restrictions of this operator are

$$\sigma_i(T) = \sigma(T \mid \mathfrak{H}_i) \quad \text{and} \quad \sigma_e(T) = \sigma(T \mid \mathfrak{H}_e).$$

In proving Theorem 7.1 we essentially used the properties of an operator T satisfying condition (7.2). It is easily seen that Theorem 7.1 is essentially equivalent to the following proposition on the operators T.

THEOREM 7.1′. *Suppose $T \in [\mathfrak{H}]$ and $\sigma(T)$ does not cover the unit circle. In order for T to be u-dichotomic it is necessary and sufficient that it be a uniform W-contraction:*

$$T^* WT \ll W \tag{7.2}$$

for some indefinite W ($= W^* \in [\mathfrak{H}]$).

Any operator W ($= W^*$) *satisfying condition* (7.2) *is invertible and such that the invariant subspace* $\mathfrak{H}_i$ *of* T *is* W*-positive while the invariant subspace* $\mathfrak{H}_e$ *of* T *is* W*-negative.*

An operator W *can be chosen so that these subspaces are* W*-orthogonal.*

In fact, by hypothesis, the operator T has at least one regular point ζ on the unit circle. It can be assumed without loss of generality that $\zeta = 1$ (otherwise we could consider the operator $\zeta^{-1}T$ in place of T).

There then exists the operator

$$A = (T + I)(T - I)^{-1} \quad (\in [\mathfrak{H}]),$$

condition (7.2) for T being equivalent to condition (7.4) for A.

COROLLARY 7.1. *A uniform* W*-contraction is either*

1) *u-dichotomic or*

2) *has a spectrum covering the whole unit circle.*

Case 1) is obviously realized. An example will be given below in which case 2) is realized.

We first establish a number of propositions.

LEMMA 7.3. *Let* T *be a uniform* W*-contraction. Then all of the points of the unit circle are points of regular type for* T, *i.e. for any point* ζ *of the unit circle* ($|\zeta| = 1$) *the operator* $T - \zeta I$ *is an isomorphism of* $\mathfrak{H}$ *onto a closed subspace.*

PROOF. Condition (7.2) means that

$$(Wx, x) - (WTx, Tx) \geqq m \| x \|^2 \qquad (x \in \mathfrak{H}), \tag{7.7}$$

where $m = \lambda_m (W - T^* WT) > 0$.

We show that this implies the existence of a constant $c_\zeta > 0$ such that

$$\| (T - \zeta I) x \| \geqq c_\zeta \| x \| \qquad (x \in \mathfrak{H}). \tag{7.8}$$

In fact, there would otherwise exist a sequence of unit vectors $\{x_n | \, \| x_n \| = 1\}$ for which $y_n = (T - \zeta I)x_n \to 0$ as $n \to \infty$. Putting $x = x_n$, $Tx_n = \zeta x_n + y_n$ in (7.7), we would then get

$$m \| x_n \|^2 \leqq - \zeta (Wx_n, y_n) - \zeta (Wy_n, x_n) - (Wy_n, y_n) \to 0,$$

which is impossible.

Inequality (7.8) implies that the mapping $T - \zeta I$ is an isomorphism. Since $\mathfrak{H}$ is closed, $\mathfrak{L}_\zeta = (T - \zeta I)\mathfrak{H}$ is also closed.

REMARK 7.1′. A comparison of Lemma 7.3 with the theorem on a boundary point of the spectrum (see § 4.3) shows that a point of the unit circle can only belong to $\sigma(T)$ as an interior point. This in turn implies that the unit circle belongs to $\sigma(T)$ if and only if it is contained in an annulus belonging to $\sigma(T)$.

THEOREM 7.3. *Let T be a uniform W-contraction with W an invertible operator ($W^{-1} \in [\mathfrak{H}]$). In order for T to be u-dichotomic it is necessary and sufficient that T^* be a uniform W^{-1}-contraction.*

PROOF. NECESSITY. If $1 \notin \sigma(T)$, the operator $A = (T + I)(T - I)^{-1}$ exists. Then $A - I = 2(T - I)^{-1}$, i.e. $T = (A + I)(A - I)^{-1}$.

Substituting the obtained expression for T in (7.2), we find that $WA + A^*W \ll 0$. Multiplying this inequality from the left and from the right by W^{-1}, we get $\operatorname{Re}(W^{-1}A^*) \ll 0$. On the basis of Lemma 7.2 we conclude that $(A^* + I)(A^* - I)^{-1} = T^*$ is a uniform W^{-1}-contraction.

SUFFICIENCY. If T^* is a uniform W^{-1}-contraction, the equality $(T^* - I)x = 0$ implies by Lemma 7.3 that $x = 0$. This in turn implies that $\mathfrak{L}_1 = (T - I)\mathfrak{H} = \mathfrak{H}$, since $(T^* - I)x = 0$ for every vector $x \perp \mathfrak{L}_1$. Thus the point $\lambda = 1$ and hence every other point of the unit circle are regular points of T.

In certain cases Theorem 7.3 permits one to verify directly whether the spectrum of an operator T covers the unit circle or does not intersect it.

3. *Example of a uniform W-contraction that is not u-dichotomic.* Let us construct an example of a uniform W-contraction whose spectrum covers the unit circle.

We construct an operator T whose spectrum covers the unit circle and to which at the same time there corresponds even an invertible Hermitian operator W converting it into a uniform W-contraction.

Let $\{e_k\}_{-\infty}^{\infty}$ be an orthonormal basis in $\mathfrak{H}$, $\mathfrak{H}_+ = \bigvee_0^\infty e_k$, $\mathfrak{H}_- = \bigvee_1^\infty e_{-k}$,[13] P_+ and P_- be the orthogonal projections on $\mathfrak{H}_+$ and $\mathfrak{H}_-$ respectively, and let $W = P_+ - P_-$ ($= W^{-1}$). We choose a pair of complex numbers q_+ and q_-, with $|q_+| < 1$ and $|q_-| > 1$, and define the operator T:

$$Te_n = \begin{cases} q_+e_{n-1}, & n = 1, 2, \cdots, \\ q_-e_{n-1}, & n = 0, -1, -2, \cdots. \end{cases}$$

For any $x = \sum_{-\infty}^{\infty} c_k e_k \in \mathfrak{H}$ we have

$$Tx = \sum_{k=1}^{\infty} q_+c_ke_{k-1} + \sum_{-\infty}^{0} q_-c_ke_{k-1},$$

$$(Wx, x) = \sum_{k=0}^{\infty} |c_k|^2 - \sum_{-\infty}^{-1} |c_k|^2,$$

$$(WTx, Tx) = \sum_{k=1}^{\infty} |q_+|^2 |c_k|^2 - \sum_{-\infty}^{0} |q_-|^2 |c_k|^2,$$

so that for $H = W - T^* WT$

[13] The symbol $\bigvee_{k \in K} e_k$ denotes the closed linear span of the system of vectors $\{e_k\}_{k \in K}$.

$$(Hx, x) = (1 - |q_+|^2) \sum_{k=1}^{\infty} |c_k|^2 + (1 + |q_-|^2) |c_0|^2$$
$$+ (|q_-|^2 - 1) \sum_{-\infty}^{-1} |c_k|^2 \geqq \delta \sum_{-\infty}^{\infty} |c_k|^2 = \delta(x, x),$$

where $\delta = \min\{1 - |q_+|^2, |q_-|^2 - 1\}$.

Thus $H \gg 0$, i.e. T is a uniform W-contraction.

It is easily seen that $T^*e_n = \bar{q}_- e_{n+1}$ $(n = -1, -2, \cdots)$. Therefore

$$(WT^*e_{-1}, T^*e_{-1}) = |q_-|^2 (We_0, e_0) = |q_-|^2 > 0.$$

On the other hand, $(We_{-1}, e_{-1}) = -1 < 0$, so that T^* is not even an ordinary W^{-1}-contraction $(W^{-1} = W)$.

By Theorem 7.3 (and Corollary 7.1) the spectrum of the constructed operator T covers the whole unit circle.

REMARK 7.1. The spectrum of the constructed operator T contains the annulus $|q_+| \leqq |\lambda| \leqq |q_-|$ (cf. Remark 7.1′). In fact, for any $r \in (|q_+|, |q_-|)$ the operator $r^{-1} T$ is of the same type as T and hence its spectrum covers the unit circle. This means that the spectrum $\sigma(T)$ covers the circle $|\lambda| = r$.

We note that the interior points of the considered annulus are points of regular type for T for which the codimension of the space $(T - \lambda I)\mathfrak{H}$ is equal to one.

4. *u-dichotomicity conditions for uniform W-contractions.* The following important assertion is almost trivial.

THEOREM 7.4. *Suppose $W = W^* \in [\mathfrak{H}]$ is an invertible operator. Then every uniform (u-dichotomic uniform) W-contraction $T \in [\mathfrak{H}]$ has a neighborhood in $[\mathfrak{H}]$ consisting of uniform (u-dichotomic uniform) W-contractions.*

The assertion follows from the fact that the operators $W - T^*WT$ and $W^{-1} - TW^{-1}T^*$ are continuous functions of T.

THEOREM 7.5. *Suppose a uniform W-contraction T is the limit of a sequence $\{T_n\}$ of u-dichotomic uniform W-contractions. Then T is also a u-dichotomic uniform W-contraction.*

Proof. Assume the contrary. Then the point $\lambda = 1$ is a point of the spectrum of T. But according to Lemma 7.3 the operator $T - I$ maps $\mathfrak{H}$ isomorphically onto a closed subspace $\mathfrak{L} = (T - I)\mathfrak{H} \neq \mathfrak{H}$. Therefore by Banach's theorem there exists an operator R in $[\mathfrak{L}, \mathfrak{H}]$ such that $(T - I)R = R(T - I) = I$.

We choose an n such that $\| T - T_n \| < 1 / \| R \|$.

Let us show[14] that $(T_n - I)\mathfrak{H} \neq \mathfrak{H}$.

[14] This assertion is a corollary of the following more general proposition of Kreĭn, Krasnosel′skiĭ and Mil′man on the stability of the deficiency of an operator $A \in [\mathfrak{B}_1, \mathfrak{B}_2]$ injecting $\mathfrak{B}_1$ onto a closed subspace $A\mathfrak{B}_1$. Each operator X in a sufficiently small neighborhood of such an operator A

We have $T_n - I = [I - (T - T_n)R]\,(T - I)$, so that

$$(T_n - I)\,\mathfrak{H} = (I - Q)\mathfrak{L},$$

where $Q = (T - T_n)R$. We note that $\| Q \| \leqq \| T - T_n \| \| R \| < 1$.

We extend the operator Q onto all of $\mathfrak{H}$ by putting $\tilde{Q}x = QP_{\mathfrak{L}}x$, where $P_{\mathfrak{L}}$ is the orthogonal projection on $\mathfrak{L}$. Clearly, $\| \tilde{Q} \| = \| Q \| < 1$.

Therefore the operator $I - \tilde{Q}$ is invertible and consequently is a one-to-one mapping of all of $\mathfrak{H}$ onto $\mathfrak{H}$; hence $(I - \tilde{Q})\mathfrak{L} = (I - Q)\mathfrak{L} \neq \mathfrak{H}$.

On the other hand, since the operator T_n is a u-dichotomic uniform W-contraction, it follows that $1 \notin \sigma(T_n)$ and hence $(T_n - I)\mathfrak{H} = \mathfrak{H}$.

We have arrived at a contradiction. The theorem is proved.

COROLLARY 7.2. *Let $T(t)$ $(a \leqq t \leqq b)$ be a continuous one parameter family of uniform W-contractions. If $T(a)$ is a u-dichotomic operator, so will be $T(t)$ for all $t \in [a, b]$.*

For suppose there exists a nonempty set of points in $[a, b]$ at which $T(t)$ is not u-dichotomic, and let t_0 denote the infimum of this set. By Theorem 7.5, $T(t_0)$ is u-dichotomic. But according to Theorem 7.4 there would then exist a neighborhood of t_0 in which $T(t)$ is u-dichotomic, which is impossible.

§ 8. Stable W-unitary operators

1. *Stability criterion for W-unitary operators.* Let W by an invertible Hermitian operator. An operator A will be said to be *W-Hermitian* if $(Ax, y)_W = (x, Ay)_W$. This condition is equivalent to the equality $WA = A^*W$ or

$$WAW^{-1} = A^*. \tag{8.1}$$

An operator U is said to be *W-unitary* if $(Ux, y)_W = (x, U^{-1}y)_W$, which is equivalent to the equality

$$WUW^{-1} = (U^{-1})^*. \tag{8.2}$$

We note that the Cayley transform[15].

$$U = (A - iaI)\,(A + iaI)^{-1}$$

takes a W-Hermitian operator into a W-unitary one with regular point $\lambda = 1$.

In fact, it is easily seen that

$$U - I = -2ia\,(A + iI)^{-1}; \quad (U^*)^{-1} - I = -2ia\,(A^* + iI)^{-1},$$

also injects $\mathfrak{B}_1$ onto a closed subspace $X\mathfrak{B}_1$ with codim $X\mathfrak{B}_1$ = codim $A\mathfrak{B}_1$, where codim $X\mathfrak{B}_1 = \dim \mathfrak{B}_2 / X\mathfrak{B}_1$.

A still more general proposition can be found, for example, in the article [3] of Gohberg and Krein.

[15] The real number $a \neq 0$ is chosen so that the operator $(A + iaI)^{-1}$ will be bounded $(-ia \notin \sigma(A))$.

and therefore

$$WUW^{-1} = I - 2iWa(A + iI)^{-1} W^{-1}$$
$$= I - 2ia(WAW^{-1} + iI)^{-1} = I - 2ia(A^* + iI)^{-1} = (U^*)^{-1}.$$

In addition, the operator $(U - I)^{-1} = (i/2a)(A + iI)$ is bounded.

One can easily show that the inverse transform $A = ai(U + I)(I - U)^{-1}$ takes each W-unitary operator with regular point $\lambda = 1$ into a W-Hermitian operator.

We discuss some of the properties of W-Hermitian and W-unitary operators.

1) The spectrum of a W-Hermitian (W-unitary) operator is symmetric with respect to the real axis (unit circle).[16]

In fact, the relation $WAW^{-1} = A^*$ implies $\sigma(A^*) = \sigma(A)$. On the other hand, $\bar{\lambda} \in \sigma(A^*)$ if $\lambda \in \sigma(A)$.

The analogous assertion for a W-unitary operator follows in the same way from the similarity of the operators U and $(U^{-1})^*$ (see (8.2)).

2) Let Λ_1 and Λ_2 be two parts of the spectrum of a W-Hermitian (W-unitary) operator A. We assume that the symmetric image $\tilde{\Lambda}_1$ of Λ_1 with respect to the real axis (unit circle) and Λ_2 are separated from each other by nonintersecting contours $\tilde{\Gamma}_1$ and Γ_2.

Then the invariant subspaces $\mathfrak{H}_1$ and $\mathfrak{H}_2$ of A corresponding to the parts Λ_1 and Λ_2 of its spectrum are W-orthogonal.

For a proof it suffices to show that $P_1^* WP_2 = 0$, where

$$P_1 = \frac{1}{2\pi i} \oint_{\Gamma_1} (\lambda I - A)^{-1} \, d\lambda$$

and

$$P_2 = \frac{1}{2\pi i} \oint_{\Gamma_2} (\lambda I - A)^{-1} \, d\lambda$$

are the projections on $\mathfrak{H}_1$ and $\mathfrak{H}_2$ respectively. Making use of the relation $(\bar{\lambda} I - A^*)^{-1} W = W(\bar{\lambda} I - A)^{-1}$, which is easily deduced from the definition of a W-Hermitian operator, we get

$$P_1^* WP_2 = \frac{1}{4\pi^2} \oint_{\Gamma_1} \oint_{\Gamma_2} (\bar{\lambda} I - A^*)^{-1} W(\mu I - A)^{-1} \, d\bar{\lambda} \, d\mu$$
$$= \frac{W}{4\pi^2} \oint_{\Gamma_1} \oint_{\Gamma_2} (\bar{\lambda} I - A)^{-1} (\mu I - A)^{-1} \, d\bar{\lambda} \, d\mu$$
$$= -\frac{W}{4\pi^2} \oint_{\Gamma_1} \oint_{\Gamma_2} (\lambda I - A)^{-1} (\mu I - A)^{-1} \, d\lambda \, d\mu$$
$$= \frac{1}{4\pi^2} \oint_{\Gamma_1} \oint_{\Gamma_2} \frac{1}{\lambda - \mu} [(\lambda I - A)^{-1} - (\mu I - A)^{-1}] \, d\lambda \, d\mu = 0,$$

[16] We recall that complex numbers z_1 and z_2 correspond to points that are symmetric with respect to the unit circle if $z_1 \bar{z}_2 = 1$.

which follows from Cauchy's theorem (cf. the proof of Lemma 2.1).

The analogous assertion for a W-unitary operator can be deduced in the same way (or obtained from what has already been proved by applying the Cayley transform if W has at least one regular point on the unit circle).

The general stability criterion (Theorem 6.1) admits the following sharpening in the case of W-unitary operators. In order to simplify the formulation let us agree to say that a subspace $\mathfrak{L} \subset \mathfrak{H}$ is *W-definite* if it is either W-positive or W-negative.

THEOREM 8.1. *In order for a W-unitary operator U to be stable it is necessary and sufficient that $\mathfrak{H}$ decompose into a direct sum of two W-orthogonal W-definite subspaces $\mathfrak{H}_1$ and $\mathfrak{H}_2$ that are invariant under U.*

PROOF. SUFFICIENCY. By Theorem 7.2., if

$$\mathfrak{H} = \mathfrak{H}_1 \dotplus \mathfrak{H}_2, \tag{8.3}$$

where $\mathfrak{H}_1$ and $\mathfrak{H}_2$ are W-orthogonal and, for the sake of definiteness, $\mathfrak{H}_1$ is W-positive while $\mathfrak{H}_2$ is W-negative, the space $\mathfrak{H}_1$ ($\mathfrak{H}_2$) is uniformly W-positive (W-negative). Thus there exists a constant $\mu > 0$ such that

$$|(x, x)_W| \geqq \mu (x, x) \qquad (x \in \mathfrak{H}_k;\ k = 1, 2).$$

The space $\mathfrak{H}_1$ ($\mathfrak{H}_2$) remains a Hilbert space while U becomes a unitary operator if the scalar product (x, y) in it is replaced by $(x, y)_W$ ($-(x, y)_W$).

Thus

$$\mu(U^n x, U^n x) \leqq |(U^n x, U^n x)_W| = |(x,x)_W| \leqq \| W \| (x,x) \qquad (x \in \mathfrak{H}_k, k = 1,2),$$

which implies $\| U^n | \mathfrak{H}_k \| \leqq \| W \| / \mu\ (k = 1,2)$.

Inasmuch as the projections P_k $(k = 1,2)$ on the $\mathfrak{H}_k$ are bounded: $\| P_k \| \leqq c$ $(k = 1,2)$, we have

$$\| U^n \| \leqq \| U^n P_1 \| + \| U^n P_2 \| = \| U^n | \mathfrak{H}_1 \| + \| U^n | \mathfrak{H}_2 \| \leqq 2c \| W \| / \mu$$
$$(n = 0, \pm 1, \pm 2, \cdots).$$

NECESSITY. By Theorem 6.1, if U is a stable operator, there exists an invertible operator S transforming U into a unitary operator: $V = S^{-1} US$ $(V^* V = VV^* = I)$. The fact that U is a W-unitary operator implies

$$W = U^* W U = S^{*-1} V^* S^* W S V S^{-1}.$$

Putting $G = S^* W S$, we find that $G = V^* G V$, so that

$$VG = GV. \tag{8.4}$$

The operator G is Hermitian, has an inverse and

$$(Gx, x) = (WSx, Sx). \tag{8.5}$$

Therefore the real spectrum $\sigma(G) = \sigma_+(G) \cup \sigma_-(G)$.

Let $\mathfrak{H} = \mathfrak{H}_+ \oplus \mathfrak{H}_-$ be the orthogonal decomposition of $\mathfrak{H}$ into the corresponding invariant subspaces of G.

The invariant (under G) subspaces $\mathfrak{H}_+$ and $\mathfrak{H}_-$ are G-orthogonal and G-definite ($(Gx, x) > 0$ for $x \in \mathfrak{H}_+$ $(x \neq 0)$ and $(Gx, x) < 0$ for $x \in \mathfrak{H}_-$ $(x \neq 0)$).

By virtue of (8.4) the subspaces $\mathfrak{H}_\pm$ are also invariant under the unitary operator V, so that $V\mathfrak{H}_\pm = \mathfrak{H}_\pm$.

We now put $S\mathfrak{H}_+ = \mathfrak{H}_1$ and $S\mathfrak{H}_- = \mathfrak{H}_2$. Then $U\mathfrak{H}_k = \mathfrak{H}_k$ $(k = 1,2)$ and according to (8.5) the subspaces $\mathfrak{H}_1$ and $\mathfrak{H}_2$ will have all of the required properties.

2. *Normally W-decomposable operators.* Let us agree to say that an operator A is *normally W-decomposable* if its spectrum $\sigma(A)$ can be divided into nonintersecting spectral sets: $\sigma(A) = \sigma_1(A) \cup \sigma_2(A)$, so that in the corresponding decomposition

$$\mathfrak{H} = \mathfrak{H}_1 \dotplus \mathfrak{H}_2 \tag{8.6}$$

the invariant (under A) subspaces $\mathfrak{H}_1$ and $\mathfrak{H}_2$ are uniformly W-definite.

Let us agree to use the indexing under which the subspace $\mathfrak{H}_1$ will be uniformly W-positive while $\mathfrak{H}_2$ will be uniformly W-negative.

We note that if A is a W-Hermitian operator (or a W-unitary operator), the subspaces $\mathfrak{H}_1$ and $\mathfrak{H}_2$ in (8.6) will be W-orthogonal, and by Theorem 7.2 they will be uniformly W-definite whenever they are W-definite.

We note some of the properties of normally W-decomposable operators.

1) If a W-unitary operator is normally W-decomposable, it is a stable operator. This assertion follows directly from Theorem 8.1.

2) If an operator A is normally W-decomposable, so will be an operator $\phi(A)$ where the scalar function $\phi \in K_A$ is such that $\phi(\sigma_1) \cap \phi(\sigma_2) = \varnothing$.

Clearly, the latter condition is always satisfied if the function $\phi(\lambda)$ is one-to-one on the spectrum $\sigma(A)$ and, in particular, if ϕ is a linear fractional function.

3) The spectrum of a W-Hermitian normally W-decomposable operator is real.

In fact, the Cayley transform of such an operator is a W-unitary operator and hence has a spectrum lying on the unit circle (see Theorem 6.1).

4) If $H \ll 0$ (or $H \gg 0$), the operator $A = WH$ is normally W-decomposable.

In fact, assuming, for example, that $H \ll 0$ (we would otherwise consider the operator $-A$), we get $AW + WA^* = 2WHW \ll 0$.

Applying Theorem 7.1, we obtain the required assertion.

We note that the operator $A = WH$ is a W^{-1}-Hermitian operator.

THEOREM 8.2. *The set of all normally W-decomposable operators is open in the Banach space $[\mathfrak{H}]$.*

In other words, to every normally W-decomposable operator A there corresponds a neighborhood $\| X - A \| < \delta$ in $[\mathfrak{H}]$ consisting of normally W-decomposable operators.

Thus the property of being normally W-decomposable is stable.

PROOF. Let Γ_1 and Γ_2 be smooth closed contours separating the parts $\sigma_1(A)$ and $\sigma_2(A)$ of the spectrum of A from each other.

From Theorem 2.2 it follows that for a sufficiently small $\delta > 0$ the spectrum of an operator X satisfying the condition $\| X - A \| < \delta$ also decomposes into parts $\sigma_1(X)$ and $\sigma_2(X)$ separated by the same contours while the spectral projections $P_1' = P_1(X)$ and $P_2' = P_2(X)$ differ sufficiently little from the corresponding spectral projections $P_1 = P_1(A)$ and $P_2 = P_2(A)$.

Consequently, for any $\varepsilon > 0$ there exists a $\delta > 0$ such that the inequalities

$$\| P_1^* P_1 - P_1'^* P_1' \| < \varepsilon, \quad \| P_2^* P_2 - P_2'^* P_2' \| < \varepsilon,$$
$$\| P_1^* WP_1 - P_1'^* WP_1' \| < \varepsilon, \quad \| P_2^* WP_2 - P_2'^* WP_2' \| < \varepsilon$$

are satisfied for $\| X - A \| < \delta$. But then

$$\begin{aligned}(WP_1' x, P_1'x) &= (WP_1x, P_1 x) + ([P_1'^* WP_1' - P_1^* WP_1] x, x) \\ &\geqq m(P_1x, P_1 x) - \varepsilon \| x \|^2 \geqq m(P_1'x, P_1'x) \\ &\quad + m([P_1^*P_1 - P_1'^*P_1'] x, x) - \varepsilon \| x \|^2 \geqq m \| P_1'x \|^2 - 2\varepsilon \| x \|^2\end{aligned}$$

and we get $(Wx, x) \geqq (m - 2\varepsilon) \| x \|^2$ when $x = P_1'x$.

The case when $x = P_2'x$ is treated analogously.

3. *Strongly stable W-unitary operators.* A stable W-unitary operator U is said to be *strongly stable* if all of the W-unitary operators U' in some δ-neighborhood $\| U' - U \| < \delta$ of it are stable.

THEOREM 8.3. *In order for a W-unitary operator U to be strongly stable it is necessary and sufficient that it be normally W-decomposable.*

The sufficiency of the condition follows from the preceding theorem and Theorem 8.1. Let us prove its necessity.

By Theorem 8.1, if the operator U is strongly stable (and hence stable), the decomposition (8.3) with the known properties will hold.

The theorem will be proved if we prove that the strong stability of U implies

$$\sigma(U \mid \mathfrak{H}_1) \cap \sigma(U \mid \mathfrak{H}_2) = \varnothing.$$

Let P_1 and P_2 denote the projections corresponding to the decomposition (8.3). We introduce in $\mathfrak{H}$ the new scalar product

$$(x, y)_1 = (Wx_1, y_1) - (Wx_2, y_2) \qquad (x, y \in \mathfrak{H}),$$

where $x_k = P_k x$, $y_k = P_k y$ $(k = 1, 2)$.

The new scalar product is definite:

$$(x, x)_1 = (Wx_1, x_1) - (Wx_2, x_2) > 0 \qquad (x \neq 0),$$

and, moreover, as can easily be seen, the norm $\| x \|_1 = (x, x)_1^{1/2}$ is topologically equivalent to the old one.

We put

$$W_1 = P_1^* WP_1 - P_2^* WP_2; \tag{8.7}$$

this permits us to write the new scalar product in the form

$$(x, y)_1 = (W_1x, y) = (x, y)_{W_1} \qquad (x, y \in \mathfrak{H}).$$

Since the operator U commutes with the projections P_k $(k = 1, 2)$, it is a unitary operator with respect to the new scalar product and its invariant subspaces $\mathfrak{H}_1$ and $\mathfrak{H}_2$ are orthogonal to each other.

In other words, U is also *a* W_1-unitary operator while the subspaces $\mathfrak{H}_1$ and $\mathfrak{H}_2$ are W_1-orthogonal.

Let $U = \int_0^{2\pi} e^{i\lambda}\, dE_\lambda$ $(E_0 = 0;\ E_\lambda = E_{\lambda-0}$ for $0 < \lambda < 2\pi)$ be the spectral resolution of U.

In this resolution E_λ $(0 \leqq \lambda \leqq 2\pi)$ is a W_1-orthogonal projection function, i.e. $E_\lambda E_\mu = E_{\min(\lambda,\mu)}$ and $W_1E_\lambda = E_\lambda^* W_1$ $(0 \leqq \lambda \leqq 2\pi)$.

We form the operator $H = \int_0^{2\pi} \lambda dE_\lambda$. Clearly it is W_1-Hermitian, i.e.

$$W_1H = H^* W_1. \tag{8.8}$$

We show that it is also W-Hermitian, i.e.

$$WH = H^* W. \tag{8.9}$$

In fact, since the operators P_k $(k = 1, 2)$ commute with U, they commute with the projection function E_λ and H, and therefore the projections P_k^* commute with E_λ^* and H^*.

It therefore follows from (8.8), after multiplying it by P_k^* from the left and by P_k from the right, that

$$P_k^* W_1 P_k H = H^* P_k^* W_1 P_k.$$

Nothing that $W = P_1^* WP_1 + P_2^* WP_2$, we arrive at (8.9).

The fact that the operators P_k commute with H means that the subspaces $\mathfrak{H}_k$ are invariant under H. Consequently, to the decomposition (8.3) there corresponds the following representation of H in matrix form:

$$H = \begin{pmatrix} H_{11} & 0 \\ 0 & H_{22} \end{pmatrix}.$$

Then U can be represented in the form

$$U = \begin{pmatrix} \exp\,(iH_{11}) & 0 \\ 0 & \exp\,(iH_{22}) \end{pmatrix};$$

$$\sigma(U \mid \mathfrak{H}_1) = \sigma(\exp(iH_{11})); \qquad \sigma(U \mid \mathfrak{H}_2) = \sigma(\exp\,(iH_{22})).$$

The theorem will be proved (by contradiction) if we show that when the spectra $\sigma(H \mid \mathfrak{H}_1)$ and $\sigma(U \mid \mathfrak{H}_2)$ have a common point λ there exists in any neighborhood of H a W-Hermitian operator H' whose spectrum contains nonreal points. In fact, this will imply that there exists in any neighborhood of U a W-unitary operator $U' = \exp(iH')$ whose spectrum contains nonunitary points (points not lying on the unit circle) and which are thus not stable.

It can be assumed without loss of generality that the common point $\lambda \neq 1$, since we could otherwise achieve this by replacing U by a W-unitary operator ρU where $|\rho| = 1$ and $\rho \neq 1$.

If $\lambda = \exp(i\omega)$ $(0 < \omega < 2\pi)$ is a common point of the spectra $\sigma(U \mid \mathfrak{H}_1)$ and $\sigma(U \mid \mathfrak{H}_2)$ then ω will be a common point of the spectra $\sigma(H_{11})$ and $\sigma(H_{22})$. We first consider the simpler case when ω is a common eigenvalue of the operators H_{11} and H_{22}. Let ϕ_1 and ϕ_2 denote corresponding unit eigenvectors: $\phi_k \in \mathfrak{H}_k$, $\|\phi_k\|_1 = 1$, $H_{kk}\phi_k = \omega\phi_k$ $(k = 1, 2)$.

For an arbitrary $\varepsilon > 0$ we form the operator

$$H_\varepsilon = \begin{pmatrix} H_{11} & H^{22} \\ -G_\varepsilon^* & G_\varepsilon \end{pmatrix},$$

where G_ε is the operator acting from $\mathfrak{H}_2$ into $\mathfrak{H}_1$ according to the formula

$$G_\varepsilon x = i\varepsilon(x, \phi_2)_1 \phi_1 = -i\varepsilon(Wx, \phi_2)\phi_1 \qquad (x \in \mathfrak{H}_2),$$

and G_ε^* is the W_1-adjoint of G_ε acting from $\mathfrak{H}_1$ into $\mathfrak{H}_2$ according to the formula

$$G_\varepsilon^* x = -i\varepsilon(x, \phi_1)_1 \phi_2 = -i\varepsilon(Wx, \phi_1)\phi_2 \qquad (x \in \mathfrak{H}_1).$$

It is easily seen that together with H the operator H_ε $(\varepsilon > 0)$ will be W-Hermitian. A simple calculation shows that

$$H_\varepsilon(\phi_1 + \phi_2) = (\omega + i\varepsilon)(\phi_1 + \phi_2).$$

Thus the W-Hermitian operator H_ε $(\varepsilon > 0)$ always has a nonreal eigenvalue. On the other hand, H_ε will be arbitrarily close in norm to H for sufficiently small $\varepsilon > 0$.

The case when $\omega \in \sigma(H_{11}) \cap \sigma(H_{22})$ but is not necessarily an eigenvalue for each of the operators H_{11} and H_{22} reduces to the preceding one.

In fact, using the spectral resolutions of the operators H_{kk}:

$$H_{kk} = \int_0^{2\pi} \lambda dE_{k\lambda} \qquad (E_{k\lambda} = E_\lambda P_k = P_k E_\lambda;\ k = 1, 2),$$

we form the operators

$$H_{kk}^{(\delta)} = \left(\int_0^{\omega-\delta} + \int_{\omega+\delta}^{2\pi} \right) \lambda dE_{k\lambda} + \omega \hat{P}_k,$$

where $\hat{P}_k = \int_{\omega-\delta}^{\omega+\delta} dE_{k\lambda}$ $(k = 1, 2;\ 0 < \delta < \omega)$.

Since, by assumption, $\omega \in \sigma(H_{kk})$ $(k = 1, 2)$, it follows that $\hat{P}_k \neq 0$ and ω will be an eigenvalue of $H_{kk}^{(\delta)}$ with eigenspace $\hat{P}_k\mathfrak{H}$.

We have

$$\| H_{kk} - H_{kk}^{(\delta)} \| = \left\| \int_{\omega-\delta}^{\omega+\delta} (\lambda - \omega)\, dE_{k\lambda} \right\| < \delta,$$

and hence $\| H - H_\delta \| < \delta$, where H_δ is the W-Hermitian operator defined by the equality

$$H_\delta = \begin{pmatrix} H_{11}^{(\delta)} & 0 \\ 0 & H_{22}^{(\delta)} \end{pmatrix}$$

By what has been proved above there exists in any neighborhood of H_δ a W-Hermitian operator with nonreal spectrum; therefore such an operator also exists in any neighborhood of H.

The theorem is proved.

§ 9. Elements of nonlinear analysis

1. *Contraction principle.* In what follows we will assume that $\mathfrak{M}$ is a closed subset of a Banach space $\mathfrak{B}$, although the basic Theorem 9.1 can be formulated for any complete metric space.

Let S be a (not necessarily linear) operator mapping $\mathfrak{M}$ into itself. We call it a *contraction* if there exists a constant $q < 1$ such that

$$\| Sx - Sy \| \leqq q \| x - y \| \tag{9.1}$$

for any $x, y \in \mathfrak{M}$.

THEOREM 9.1. *Suppose the operator S^ν is a contraction for some natural number $\nu \geqq 1$. Then there exists in $\mathfrak{M}$ one and only one fixed point x_Φ of S:*

$$Sx_\Phi = x_\Phi. \tag{9.2}$$

This point can be obtained from any point $x_0 \in \mathfrak{M}$ as the following limit:

$$x_\Phi = \lim_{n\to\infty} S^n x_0. \tag{9.3}$$

PROOF. Suppose first $\nu = 1$. Then for the sequence

$$x_1 = Sx_0, \quad x_2 = Sx_1, \cdots, x_k = Sx_{k-1}, \cdots$$

we have according to (9.1)

$$\| x_{k+1} - x_k \| = \| Sx_k - Sx_{k-1} \| \leqq q \| x_k - x_{k-1} \| \quad (k = 1, 2, \cdots),$$

so that $\| x_{k+1} - x_k \| \leqq q^k \| x_1 - x_0 \|$.

Therefore

$$\| x_{m+p} - x_m \| \leqq \sum_{k=m}^{m+p-1} \| x_{k+1} - x_k \| \leqq \| x_1 - x_0 \| \sum_{k+m}^{m+p-1} q^k \leqq \frac{q_m}{1-q} \| x_1 - x_0 \|,$$

i.e. $\{x_n\}$ is a Cauchy sequence. The fact that $\mathfrak{M}$ is closed implies the existence of the limit $\lim_{n\to\infty} x_n = x_\Phi$.

This limit is a fixed point of S since

$$\begin{aligned}\| x_\Phi - Sx_\Phi \| &\leqq \| x_\Phi - x_{m+1} \| + \| Sx_m - Sx_\Phi \| \\ &\leqq \| x_\Phi - x_{m+1} \| + q \| x_m - x_\Phi \| \to 0\end{aligned}$$

for $m \to \infty$.

The uniqueness of the fixed point follows at once from (9.1). Thus the theorem is completely proved for the case $\nu = 1$.

Suppose now $\nu > 1$ and consider the contraction $\tilde{S} = S^\nu$. By what has been proved above $\tilde{S}$ has one and only one fixed point x_Φ. On the other hand, since $\tilde{S}$ and S commute,

$$\tilde{S}(Sx_\Phi) = S(\tilde{S}x_\Phi) = Sx_\Phi,$$

i.e. Sx_Φ is also a fixed point of $\tilde{S}$. Hence by virtue of the uniqueness (9.2) holds.

Taking into account that every fixed point of S is a fixed point of $\tilde{S}$, we conclude that S has no other fixed points.

It remains to prove that equality (9.3) holds for any $x_0 \in \mathfrak{M}$.

To this end we note that every n can be represented in the form $n = m\nu + r$, where $0 \leqq r \leqq \nu - 1$. Then $S_n x_0 = S^r \tilde{S}^m x_0$ and, since $\lim_{m\to\infty} S^r \tilde{S}^m x_0 = S^r x_\Phi = x_\Phi$ for fixed r, equality (9.3) is proved.

REMARK 9.1. If S itself is a contraction, the difference $\| x_\Phi - S^n x_0 \|$ monotonically decreases at the rate of $O(q^n)$. We leave it to the reader to formulate what happens when $\nu > 1$.

2. *An application of the contraction principle.* Let $A \in [\mathfrak{B}]$ and let $r_A = \lim_{n\to\infty} \| A^n \|^{1/n}$ be the spectral radius of A (see § 2.1).

We consider the equation

$$x = Ax + f, \tag{9.4}$$

where $f \in \mathfrak{B}$ is given. If we set $Sx = Ax + f$, we will have

$$S^n x = A^n x + \sum_{k=0}^{n-1} A^k f,$$

which implies

$$\| S^n x_2 - S^n x_1 \| = \| A^n x_2 - A^n x_1 \| \leqq \| A^n \| \| x_2 - x_1 \|.$$

When $r_A < 1$ and ν is sufficiently large we have $\| A^\nu \| < 1$, and hence S^ν is a contraction.

Thus, *when $r_A < 1$, equation* (9.4) *has for any $f \in \mathfrak{B}$ exactly one solution, which is given by the formula*

$$x = \sum_{k=0}^{\infty} A^k f \; (= \lim_{n\to\infty} S^n x_0). \tag{9.5}$$

3. *Differentiation of functions with a vector argument.* Let $y = f(x)$ be a function defined in a neighborhood $U(x_0)$ of a point x_0 of a Banach space $\mathfrak{B}_0$ and taking values in a Banach space $\mathfrak{B}_1$.

The function $f(x)$ is said to be *differentiable* (*according to Fréchet*) *at* x_0 if in the neighborhood $U(x_0)$ it can be represented in the form

$$f(x) = f(x_0) + A(x - x_0) + \eta_1(x, x_0) \| x - x_0 \|, \tag{9.6}$$

where $A \in [\mathfrak{B}_0, \mathfrak{B}_1]$ and the function $\eta_1(x, x_0)$ satisfies the condition

$$\lim_{x \to x_0} \eta_1(x, x_0) = 0.$$

The operator A is called the (*Fréchet*) *derivative of* $f(x)$ *at* x_0, and one writes $A = f'(x_0)$.

As usual, we say that $f(x)$ is *differentiable on an open set* G if it is differentiable at each point of this set.

We will say that $f(x)$ is *continuously differentiable* in G if its derivative $f'(x)$ is a continuous function of x with values in $[\mathfrak{B}_0, \mathfrak{B}_1]$.

All of the usual formal rules of differential calculus, which we will not cite here, are retained in the case under consideration. We formulate only the simplest variant of the rule for differentiating a composite function, which will be needed by us in the sequel.

If $x(t)$ is a differentiable function of a scalar argument t whose values lie in a domain G of the Banach space $\mathfrak{B}_0$ in which $f(x)$ is differentiable, then

$$(d/dt) f(x(t)) = f'(x(t))\, dx/dt.$$

In particular, for $x \in G$ and sufficiently small t

$$(d/dt) f(x + ty) = f'(x + ty)y \qquad (y \in \mathfrak{B}).$$

Integrating the latter equality from 0 to 1 and changing the notation, we obtain the equality

$$f(x_2) - f(x_1) = \int_0^1 f'(x_1 + t(x_2 - x_1))\,(x_2 - x_1)\, dt, \tag{9.7}$$

which is valid when the interval $x_1 + t(x_2 - x_1)$ $(0 \leqq t \leqq 1)$ lies entirely in the set G.

REMARK 9.2. It follows from formula (9.7) that if $f'(x)$ is bounded on a convex set G: $\| f'(x) \| \leqq M$ $(x \in G)$, then

$$\| f(x_2) - f(x_1) \| \leqq M \| x_2 - x_1 \| \qquad (x_1, x_2 \in G).$$

If $f(x)$ is differentiable in a neighborhood of x_0 then the derivative $f'(x)$ itself is a function in this neighborhood with values in the Banach space $[\mathfrak{B}_0, \mathfrak{B}_1]$. If it too is differentiable at x_0, its derivative is denoted by the symbol $f''(x_0)$ and called the *second derivative of* $f(x)$ *at* x_0. By definition, $f''(x_0) \in [\mathfrak{B}_0, [\mathfrak{B}_0, \mathfrak{B}_1]]$. In other words, it can be said that $f''(x_0)$ is a bilinear operator taking each pair (h_1, h_2) of elements of $\mathfrak{B}_0$ into an element $f''(x_0)(h_1, h_2) \in \mathfrak{B}_1$.

It is not difficult to obtain the representation

$$f(x) = f(x_0) + f'(x_0)(x - x_0), + \frac{1}{2} f''(x_0)(x - x_0, x - x_0) + \eta_2(x, x_0) \| x - x_0 \|^2, \tag{9.8}$$

where $\lim_{x \to x_0} \eta_2(x, x_0) = 0$.

The derivatives of higher order are defined analogously.

4. *Inverse function theorem.* We now use the contraction principle to obtain an important result of analysis.

THEOREM 9.2. *Suppose a function $y = f(x)$ is defined and continuously differentiable in a neighborhood $U(x_0)$ of a point x_0 of a Banach space $\mathfrak{B}_0$ and takes values in a Banach space $\mathfrak{B}_1$.*

If the operator $T = f'(x_0) \in [\mathfrak{B}_0, \mathfrak{B}_1]$ has a bounded inverse $T^{-1} \in [\mathfrak{B}_1, \mathfrak{B}_0]$, the function $f(x)$ effects a homeomorphic mapping of a neighborhood $V(x_0) \subset U(x_0)$ onto a neighborhood of the point $y_0 = f(x_0)$.

The inverse mapping $x = g(y)$ has in this neighborhood the continuous derivative

$$g'(y) = [f'(g(y))]^{-1}. \tag{9.9}$$

If $f(x)$ has continuous derivatives up to order n in a neighborhood of x_0, so will $g(y)$ in a neighborhood of $y_0 = f(x_0)$.

PROOF. We can assume without loss of generality that $x_0 = y_0 = 0$. The equation $f(x) = y$ is equivalent to the equation

$$x = x + T^{-1}[y - f(x)] \overset{\text{def}}{=} \Phi(x). \tag{9.10}$$

For fixed y we have

$$\Phi'(x) = I - T^{-1}f'(x) = T^{-1}[f'(0) - f'(x)].$$

Therefore by virtue of the continuity of $f'(x)$ there exists a $\delta > 0$ such that in the closure of the neighborhood $S_{\delta,\mathfrak{B}_0} = \{ x \mid \| x \| < \delta \}$ we have the estimate $\| \Phi'(x) \| \leqq q < 1$, and hence (see Remark 9.2) also the estimate

$$\| \Phi(x_2) - \Phi(x_1) \| \leqq q \| x_2 - x_1 \| \qquad (x_k \in \bar{S}_{\delta,\mathfrak{B}_0}; k = 1,2).$$

Suppose now

$$\| y \| < \varepsilon = (1 - q)\delta / \| T^{-1} \|. \tag{9.11}$$

Then the estimate

$$\| \Phi(x) \| \leqq \| \Phi(x) - \Phi(0) \| + \| \Phi(0) \| \leqq q \| x \| + \| T^{-1} \| \| y \| \tag{9.12}$$

implies $\| \Phi(x) \| \leqq \delta$ provided $\| x \| \leqq \delta$. Thus under condition (9.11) the mapping $\Phi(x)$ is a contraction in the closed ball $\bar{S}_{\delta,\mathfrak{B}_0}$. From Theorem 9.1 we deduce that for each $y \in S_{\varepsilon,\mathfrak{B}_1} = \{ y \in \mathfrak{B}_1 \mid \| y \| < \varepsilon \}$ there exists a unique solution $x \in S_{\delta,\mathfrak{B}_0}$ of the equation $x = \Phi(x)$ and hence of the equation $f(x) = y$.

The inverse function $x = g(y)$ is defined in $S_{\varepsilon,\mathfrak{B}_1}$ and, since δ can be chosen arbitrarily small, is continuous at $y = 0$.

In particular, by taking y sufficiently small, we can obtain an arbitrarily small $x = g(y)$ and hence an operator $f'(x)$ that is close enough to $f'(0) = T$ to have a bounded inverse. Thus all of the above arguments can still be applied if the pair (x_0, y_0) is replaced by the pair $(g(y), y)$, and hence $g(y)$ is also continuous at this point y.

In exactly the same way it also suffices to establish the differentiability of $g(y)$ only at $y = 0$. We have $y = f(x) = Tx + [f(x) - Tx]$, where, by virtue of (9.6), $f(x) - Tx = \eta\|x\|$ and $\eta = o(1)$ for $x \to 0$.

Hence

$$x = T^{-1}y - T^{-1}\eta \| x \| = T^{-1}y - T^{-1}\eta \| g(y) \|. \tag{9.13}$$

We now note that (9.12) implies the estimate

$$\| g(y) \| = \| \Phi(g(y)) \| \leqq q \| g(y) \| + \| T^{-1} \| \| y \|,$$

from which we get

$$\| g(y) \| \leqq \| T^{-1} \| \| y \| / (1 - q).$$

This shows that for $y \to 0$

$$\tilde{\eta} = T^{-1}\eta \ \| g(y) \| / \| y \| = o(1).$$

In this way we obtain the representation (see (9.13))

$$g(y) = T^{-1}y - \tilde{\eta} \| y \|,$$

showing that $g(y)$ is differentiable at $y = 0$ and $g'(0) = T^{-1}y = [f'(0)]^{-1}$. This proves equality (9.9).

The proof of the last assertion of the theorem can be obtained by induction, for which one successively differentiates (9.9).

5. *Cone inequality theorem.* In this subsection we present a simple assertion, which will be systematically used in more special cases in subsequent chapters. Its proof will be developed for linear operators, although the assertion admits certain generalizations to the nonlinear case.

We first introduce the following notion.

A closed subset $\mathfrak{K}$ of a Banach space $\mathfrak{B}$ is called a *cone* if it has the following properties:

a) $x_0 \in \mathfrak{K}$ implies $\lambda x_0 \in \mathfrak{K}$ for $\lambda \geqq 0$;

b) $x_1, x_2 \in \mathfrak{K}$ implies $x_1 + x_2 \in \mathfrak{K}$;

c) $\pm x_0 \in \mathfrak{K}$ implies $x_0 = 0$.

Suppose a cone $\mathfrak{K}$ is given in a Banach space $\mathfrak{B}$. We will write

$$x \leqslant y, \tag{9.14}$$

if $y - x \in \mathfrak{K}$. It is not difficult to see that relation (9.14) has the properties of an ordinary inequality.[17]

If the cone $\mathfrak{K}$ is invariant under a linear operator A: $A\mathfrak{K} \subset \mathfrak{K}$, then A preserves the inequalities: $x \leqslant y$ implies $Ax \leqslant Ay$.

THEOREM 9.3. *Suppose the cone $\mathfrak{K}$ is invariant under an operator $A \in [\mathfrak{B}]$ with spectral radius $r_A < 1$. If a vector $x \in \mathfrak{B}$ satisfies the inequality*

$$x \leqslant Ax + f \qquad (Ax + f \leqslant x), \tag{9.15}$$

it also satisfies the estimate $x \leqslant y$ $(y \leqslant x)$, where y is a solution of the equation

$$y = Ay + f. \tag{9.16}$$

We give a proof of the first variant of the theorem. Since, in addition to A, the operator S: $Sx = Ax + f$, also preserves inequalities, the inequality $x \leqslant Sx$ implies $x \leqslant Sx \leqslant S^2x \leqslant \cdots \leqslant S^nx$, so that

$$x \leqslant S^n x = A^n x + \sum_{k=0}^{n-1} A^k f. \tag{9.17}$$

We now note that the condition $r_A < 1$ implies (see (2.1)) the convergence of the series $\sum_0^\infty A^k$ and the equality $(I - A)^{-1} = \sum_0^\infty A^k$.

This means that the element $y = \sum_0^\infty A^k f$ is defined and is the unique solution of equation (9.16).

By virtue of the fact that the cone $\mathfrak{K}$ is closed we can pass to the limit in inequality (9.17). Since the condition $r_A < 1$ implies $\|A^n\| \to 0$, we then obtain the required estimate

$$x \leqslant \sum_{k=0}^{\infty} A^k f = y.$$

REMARK 9.3. It is not difficult to verify that the assertion of the theorem continues to hold if A is a nonlinear monotone contraction.

EXERCISES

1. An operator $A \in [\mathfrak{B}]$ is called an operator of *simplest type* if its spectrum consists of a finite number of eigenvalues and $\mathfrak{B}$ is the direct sum of its eigenspaces.

In other words, an operator A is an operator of simplest type if and only if it is representable in the form

$$A = \sum_{j=1}^{n} \lambda_j P_j, \tag{0.1}$$

where $\lambda_1, \cdots, \lambda_n$ are mutually distinct complex numbers and the P_j $(j = 1, \cdots, n)$ form a complete system $(\sum P_j = I)$ of pairwise disjoint projections.

Show that

[17] In particular, we will consider inequalities between operators in a Hilbert space $\mathfrak{H}$ defined by the cone of nonnegative operators (see § 5). In this case the relation $A \leqslant B$ $(A, B \in [\mathfrak{H}])$ means that $(Ax, x) \leqq (Bx, x)$ for all $x \in \mathfrak{H}$.

a) An operator A is an operator of simplest type if and only if there exists a polynomial $P(\lambda)$ with simple roots such that $P(A) = 0$.

Hint. Let $p(\lambda) = (\lambda - \lambda_1)\cdots(\lambda - \lambda_n)$. Use the identities

$$1 = \sum_{j=1}^{n} s_j(\lambda); \quad \lambda = \sum_{j=1}^{n} \lambda_j s_j(\lambda), \qquad \text{where} \quad s_j(\lambda) = \frac{p(\lambda)}{p'(\lambda_j)(\lambda - \lambda_j)}$$

and the relations issuing from them

$$I = \sum S_j(A); \qquad A = \sum \lambda_j S_j(A).$$

b) If $\mathfrak{B} = \mathfrak{H}$, an operator A is an operator of simplest type if and only if it is similar to a normal operator B $(B^*B = BB^*)$ and its spectrum consists of a finite number of points.

Hint. To prove the necessity of the condition, introduce on the basis of the decomposition (0.1) the new scalar product

$$(x, y)' = (P_1 x, P_1 y) + (P_2 x, P_2 y) + \cdots + (P_n x, P_n y).$$

2. Let $A = \sum_1^n \lambda_j P_j$ and $B = \sum_1^m \mu_k Q_k$ be two operators of simplest type. Condider the operator $\mathfrak{A}$ acting on the elements X of $[\mathfrak{B}]$ according to the formula $\mathfrak{A}X = AX - XB$. As is well known (cf. § 3), its spectrum consists of the numbers $\lambda_j - \mu_k$ $(j = 1,\cdots, n$ and $k = 1,\cdots,m)$.

Show that the operator $\mathfrak{A}$ is also an operator of simplest type and, if $\alpha \in \sigma(\mathfrak{A})$, the spectral projection $\mathfrak{P}_\alpha$ corresponding to α is given by the formula

$$\mathfrak{P}_\alpha X = \sum_{\lambda_j - \mu_k = \alpha} P_j X Q_k.$$

Generalize this result to the case of an operator $\mathfrak{A}$ of the form

$$\mathfrak{A}X = \sum c_{pq} A^p X B^q \quad (\text{the } c_{pq} \text{ are scalars}).$$

3. Let A be an operator acting in the finite-dimensional space $\mathfrak{B} = C^n$, let $\lambda_1,\cdots, \lambda_k$ be the different points of its spectrum, let P_j denote the spectral projection corresponding to λ_j and let n_j denote the dimension of the subspace $\mathfrak{B}_j = P_j\mathfrak{B}$.

Show that one has the representation

$$A = \sum_{j=1}^{k} (\lambda_j P_j + Q_j), \tag{0.2}$$

in which $Q_j = Q_j P_j = P_j Q_j$ and $Q_j^{m_j} = 0$, where m_j is an integer not exceeding n_j.

Hint. Make use of the Jordan normal form of the matrix A and put $Q_j = (A - \lambda_j I)P_j$.

4. Retaining the notation of the preceding exercise, verify that the resolvent R_λ of A is representable in the form

$$R_\lambda = \sum_{j=1}^{k} P_j \sum_{r=0}^{m_j - 1} (-1)^r \frac{Q_j^r}{(\lambda_j - \lambda)^{r+1}}. \tag{0.3}$$

Use formula (2.3) to obtain the representation

$$\phi(A) = \sum_{j=1}^{k} P_j \sum_{r=0}^{m_j - 1} \frac{\phi^{(r)}(\lambda_j)}{r!} Q_j^r. \tag{0.4}$$

Write out, in particular, the representation for e^{At}.

5. Using the results of Exercise 4, obtain the following representation for a function $\phi(\lambda, \mu) \in K_{A,B}$:

$$\phi(A_l, B_r)Y = \sum_{j_1=1}^{k_1} \sum_{j_2=1}^{k_2} \sum_{r_1=0}^{m'_{j_1}-1} \sum_{r_2=0}^{m''_{j_2}-1} P'_{j_1} Q'^{r_1}_{j_1} Y P''_{j_2} Q''^{r_2}_{j_2} \frac{1}{r_1!\, r_2!} \frac{\partial^{r_1+r_2} \phi(\lambda_{j_1}, \lambda_{j_2})}{\partial \lambda_{j_1}^{r_1} \partial \lambda_{j_2}^{r_2}} \tag{0.5}$$

where

$$A = \sum_{j_1=1}^{k_1} (\lambda_{j_1} P'_{j_1} + Q'_{j_1}), \qquad B = \sum_{j_2=1}^{k_2} (\lambda_{j_2} P''_{j_2} + Q''_{j_2}).$$

6. Show that in the finite-dimensional case the solution of the equation

$$\sum_{s_1, s_2} c_{s_1, s_2} A^{s_1} X B^{s_2} = Y \tag{3.1}$$

can be written in the form

$$X = \sum_{j_1=1}^{k_1} \sum_{j_2=1}^{k_2} \sum_{r_1=0}^{m'_{j_1}-1} \sum_{r_2=0}^{m''_{j_2}-1} P'_{j_1} Q'^{r_1}_{j_1} Y P''_{j_2} Q''^{r_2}_{j_2} \frac{1}{r_1!\, r_2!} \frac{\partial^{r_1+r_2}}{\partial \lambda^{r_1}_{j_1} \partial \lambda^{r_2}_{j_2}} \left\{ \frac{1}{\sum_{s_1, s_2} c_{s_1 s_2} \lambda^{s_1}_{j_1} \lambda^{s_2}_{j_2}} \right\}. \tag{0.6}$$

7. When the operators A and B are Hermitian, the operators Q'_{j_1} and Q''_{j_2} obviously vanish and the formulas given in Exercises 5 and 6 can be simplified. In particular, the latter of them takes the form

$$X = \sum_{j_1=1}^{k_1} \sum_{j_2=1}^{k_2} P'_{j_1} Y P''_{j_2} \frac{1}{\sum_{s_1, s_2} c_{s_1 s_2} \lambda^{s_1}_{j_1} \lambda^{s_2}_{j_2}}. \tag{0.7}$$

Generalize this result to the case when $\mathfrak{B}$ is an infinite-dimensional Hilbert space $\mathfrak{H}$ and show that the solution of equation (3.1) for Hermitian operators A and B is representable in the form

$$X = \int_{\sigma(A)} \int_{\sigma(B)} \frac{dE'_\lambda Y dE''_\mu}{\sum_{s_1, s_2} c_{s_1 s_2} \lambda^{s_1} \mu^{s_2}} d\lambda\, d\mu,$$

where E'_λ and E''_μ are the resolutions of the identity corresponding to the operators A and B and $\sum c_{s_1 s_2} \lambda^{s_1} \mu^{s_2} \neq 0$ (for $(\lambda, \mu) \in \sigma(A) \times \sigma(B)$).

8. Let λ_0 be an eigenvalue of an operator $A \in [\mathfrak{B}]$. The linear manifold $\mathfrak{L}_{\lambda_0}$ consisting of all of the vectors x which satisfy the condition $(A - \lambda_0 I)^k x = 0$ for some integer $k \geqq 0$ is called the *root manifold* corresponding to λ_0 (the *root space* if $\mathfrak{L}_{\lambda_0}$ is closed, in particular, finite dimensional).

The dimension of $\mathfrak{L}_{\lambda_0}$ is called the *algebraic multiplicity* of the eigenvalue λ_0.

If for some fixed natural number k

$$(A - \lambda_0 I)^k \mathfrak{L}_{\lambda_0} = 0, \tag{0.8}$$

one says that λ_0 is an eigenvalue of finite index, and the least k satisfying this condition is called the *index* of λ_0.

If (0.8) is not satisfied for any k, one says that λ_0 is of infinite index.

The index of an eigenvalue does not exceed its algebraic multiplicity.

a) Show that an isolated eigenvalue λ_0 of a linear operator $A \in [\mathfrak{B}]$ is of finite index r if and only if the resolvent $R_\lambda(A)$ has a pole of order r at λ_0.

Hint. Let $R_\lambda(A) = \sum_{-\infty}^{\infty} A_n (\lambda - \lambda_0)^n$. Make use of the relation $A_{-(n+1)} = (A - \lambda_0 I) A_{-1}$ which follows from the definition of the resolvent, and the fact that $A_{-1} = (1/2\pi i) \oint_\Gamma R_\lambda(A) d\lambda$ is a projection on $\mathfrak{L}_{\lambda_0}$.

An isolated eigenvalue is said to be *normal* if it is of finite index.

b) Show that an eigenvalue λ_0 is normal precisely when the algebraic multiplicity of λ_0 is finite and $\mathfrak{B}$ decomposes into a direct sum $\mathfrak{B} = \mathfrak{L}_{\lambda_0} \dotplus \mathfrak{N}_{\lambda_0}$, where $\mathfrak{N}_{\lambda_0}$ is an invariant subspace of A in which the operator $A - \lambda_0 I$ is invertible.

9. An operator $A \in [\mathfrak{B}]$ is called an *operator of algebraic type* if its spectrum consists of a finite number of eigenvalues $\lambda_1, \cdots, \lambda_n$ of finite index. Show that

a) An operator $A \in [\mathfrak{B}]$ is an operator of algebraic type precisely when the decomposition (0.2) is valid. In particular, every linear operator in a finite-dimensional space is an operator of algebraic type (see Exercise 3).

b) An operator A is an operator of algebraic type precisely when there exists a polynomial $P(\lambda)$ such that $P(A) = 0$.

c) As a generalization of Exercise 2, if A and B are operators of algebraic type in $\mathfrak{B}$, the operator $\mathfrak{A}X = AX - XB$ in $[\mathfrak{B}]$ is also an operator of algebraic type.

d) The results of Exercises 4 and 5 are retained for operators of algebraic type in an infinite-dimensional space.

10. a) Suppose the spectrum of an operator T lies in the interior of the unit circle. Show that the equation $W - T^*WT = S$ has for any $S \in [\mathfrak{B}]$ a unique solution, which is representable in the form of the series

$$W = \sum_{k=0}^{\infty} (T^*)^k ST^k.$$

b) Suppose T is a u-dichotomic operator. Show that for any $S \in [\mathfrak{B}]$ satisfying the condition $S = P_i^*SP_i + P_e^*SP_e$, where P_i and P_e are the spectral projections of T on the subspaces $\mathfrak{H}_i$ and $\mathfrak{H}_e$, the above equation has the solution

$$W = \sum_{k=0}^{\infty} (T^*)^k P_i^*SP_iT^k - \sum_{k=-\infty}^{-1} (P_i^* + T^*P_e^*)\, P_eSP_e(P_i + TP_e)^k.$$

This solution is unique under the additional condition

$$W = P_i^*WP_i + P_e^*WP_e.$$

The above assertion for $S \gg 0$ independently implies part of Theorem 7.1′.

11. Suppose the spectrum of an operator A lies in the interior of the left halfplane. Show without using the integral formulas of § 3 that the equation $WA + A^*W = -H$ has a unique solution $W \in [\mathfrak{B}]$. This solution satisfies the condition $W \gg 0$ if $H \gg 0$.

Hint. Make use of the results of Exercise 10 for $T = (A + aI)(A - aI)^{-1}$, where a is sufficiently large.

12. Let G be a triangular operator: $G = \begin{pmatrix} A & B \\ 0 & C \end{pmatrix}$. If $\sigma(A) \cap \sigma(C) = \varnothing$ then G is similar to the operator $H = \begin{pmatrix} A & 0 \\ 0 & C \end{pmatrix}$. If on the other hand $\sigma(A) \cap \sigma(C) \neq \varnothing$ and $\lambda \in \sigma(A) \cap \sigma(C)$ is a normal eigenvalue of both A and C (of index r_A for A and of index r_C for C), then λ is a normal eigenvalue of G of index $r_G \leqq r_A + r_C$.

Hint. According to Theorem 3.2, if $\sigma(A) \cap \sigma(C) = \varnothing$, the equation $AX - XC = B$ has a solution. A check shows that $H = TGT^{-1}$, where $T = \begin{pmatrix} I_1 & X \\ 0 & I_2 \end{pmatrix}$.

Suppose $\sigma(A) \cap \sigma(C) \neq \varnothing$, $G^{p-1}x \neq 0$, $G^px = 0$ and $x = \begin{pmatrix} x_1 \\ x_2 \end{pmatrix} \neq 0$. If $x_2 = 0$ then x_1 is a root vector of A and $p \leqq r_A$. On the other hand, if $x_2 \neq 0$ and $p > r_B$ then

$$A^{r_C}x_1 + (A^{r_B-1}B + A^{r_B-2}BC \cdots + BC^{r_B-1})\, x_2 \neq 0$$

is a root vector of A of index $p - r_B$.

13. Suppose A and B are Hermitian operators in a Hilbert space $\mathfrak{H}$ and let

$$\lambda_M(A) = \sup_{\|x\|=1} (Ax, x), \qquad \lambda_m(A) = \inf_{\|x\|=1} (Ax, x).$$

Obtain the estimate

$$e^{\lambda_m(A)} \leqq \| e^{A+iB} \| \leqq e^{\lambda_M(A)} = \| e^A \|. \tag{0.9}$$

Hint. Differentiate the expression $\phi(t) = \| e^{(A+iB)t} x \|^2$ and obtain an estimate of the logarithmic derivative of $\phi(t)$.

14. Suppose $\mathfrak{H}$ is a Hilbert space and $A \in [\mathfrak{H}]$. Show that if the spectrum $\sigma(\mathrm{Re}\ A)$, where $\mathrm{Re}\ A = (A + A^*)/2$, lies in the closed halfplane $\mathrm{Re}\ \lambda \leqq \alpha$, then the spectrum $\sigma(A)$ satisfies the same condition.

Hint. Make use of inequalities (0.9) and Theorem 5.1.

15. Prove the relation

$$e^{A+B} = \lim_{n\to\infty} [e^{A/n}e^{B/n}]^n, \tag{0.10}$$

where $A, B \in [\mathfrak{B}]$ and the limit is taken in the sense of uniform convergence.

Hint. Make use of the relation $\|e^{(A+B)t} - e^{At}e^{Bt}\| \leqq ct^2$ (c = const), which is easily deduced from Taylor's formula.

Obtain from (0.10) the upper estimate of (0.9).

16. Show that the following estimate is valid for any operator A acting in C^n.

$$\|e^{At}\| \leqq e^{\kappa t} \sum_{k=0}^{n-1} \frac{1}{k!}(2t\|A\|)^k \qquad (t \geqq 0), \tag{0.11}$$

where $\kappa = \max\{\operatorname{Re}\lambda \mid \lambda \in \sigma(A)\}$ (see B. F. Bylov, R. È. Vinograd, D. M. Grobman and V. V. Nemyckiĭ [1], page 131).

Hint. The following equality holds for any analytic function $f(\lambda)$ in a domain containing the spectrum $\sigma(A) = \{\lambda_j\}$:

$$f(A) = a_1 + a_2(A - \lambda_1 I) + a_3(A - \lambda_1 I)(A - \lambda_2 I) + \cdots + a_n(A - \lambda_1 I)(A - \lambda_2 I)\cdots(A - \lambda_{n-1} I),$$

where

$$a_{k+1} = \int_0^1 ds_1 \cdots \int_0^{s_{k-1}} f^{(k)}[\lambda_1 + (\lambda_2 - \lambda_1)s_1 + \cdots + (\lambda_k - \lambda_{k-1})s_k]\, ds_k.$$

Hence $|a_{k+1}| \leqq (1/k!)\max_{\mathscr{B}} |f^{(k)}(\lambda)|$, where $\mathscr{B}$ is the convex hull of the numbers $\lambda_1, \cdots, \lambda_k$ in the complex plane.

Put $f(A) = e^{At}$.

(In the book of I. M. Gel'fand and G. E. Šilov [1] the number $1/k!$ in the estimate for $|a_{k+1}|$ has been replaced by 1 and therefore the factors $1/k!$ in the estimate for $\|e^{At}\|$ are missing.)

An "exact" estimate for $\|e^{At}\|$ in terms of $\|A\|$ and κ is apparently not known.

The following Exercises 17–24 contain assertions permitting one to generalize the inequalities of Exercise 13 to the case of operators acting in any Banach space.

17. Prove the existence for any $x, y \in \mathfrak{B}$ of the limit

$$(\partial_y \|x\| \stackrel{\text{def}}{=}) \lim_{h\downarrow 0} \frac{\|x + hy\| - \|x\|}{h},$$

to which the quantity under the limit sign tends montonically (is nonincreasing as h (>0) decreases).

Hint. The assertion follows from the convexity of the function $\varphi_{x,y}(h) = \|x + hy\|$ ($-\infty < h < \infty$).

18. Let $x(t)$ be a vector function defined in a right halfneighborhood of a point t_0 and having at this point the strong right derivative

$$u = \lim_{h\downarrow 0} \frac{x(t_0 + h) - x(t_0)}{h}.$$

Prove that the right derivative of the norm $\|x(t)\|$ exists at t_0 and, in addition,

$$\lim_{h\downarrow 0} \frac{\|x(t_0 + h)\| - \|x(t_0)\|}{h} = \partial_u \|x(t_0)\|.$$

19. According to Exercise 17 the following limit exists for any $A \in [\mathfrak{B}]$:

$$\Lambda(A) = \lim_{h\downarrow 0} \frac{\|I + hA\| - 1}{h}.$$

Clearly,

$$\Lambda(\alpha A) = \alpha\Lambda(A), \text{ if } \alpha \geqq 0; \quad |\Lambda(A)| \leqq \|A\|;$$
$$\Lambda(A+B) \leqq \Lambda(A) + \Lambda(B); \qquad [\Lambda(A) - \Lambda(B)] \leqq \|A - B\|.$$

At the same time the "measure" Λ is not a norm since it can take negative values. But one always has

$$\Lambda(A) + \Lambda(-A) \geqq 0 \quad (= \Lambda(0)).$$

a) Show that, if $\mathfrak{B} = \mathfrak{H}$,

$$\Lambda(A) = \lambda_M(A_{\mathfrak{R}}) \qquad \left(A_{\mathfrak{R}} = \operatorname{Re} A = \frac{A + A^*}{2}\right),$$

and thus $-\Lambda(-A) = \lambda_m(A_{\mathfrak{R}})$.

Let $\mathfrak{B}$ be a Banach space of sequences of complex numbers: $x \in \mathfrak{B} \Leftrightarrow x = \{\xi_j\}_1^\infty$, the $\xi_j = \xi_j(x)$ $(j = 1, 2, \cdots)$ being continuous functionals of $x \in \mathfrak{B}$. Then to each operator $A \in [\mathfrak{B}]$ there corresponds a matrix $\|a_{jk}\|_1^\infty$ such that the equality $x' = Ax$ is equivalent to the equalities

$$\xi_j' = \sum_{k=1}^{\infty} a_{jk}\xi_k \qquad (j = 1, 2, \cdots).$$

Show that

b) If $\mathfrak{B} = c$ (c is the space of all convergent sequences with norm $\|x\| = \sup|\xi_j|$),

$$\Lambda(A) = \sup_j \left(\operatorname{Re} a_{jj} + \sum_{k \neq j} |a_{jk}|\right),$$

whereas $\|A\| = \sup_j \sum_k |a_{jk}|$.

c) If $\mathfrak{B} = l_1$ (l_1 is the space of all absolutely summable sequences with norm $\|x\| = \sum_{j=1}^\infty |\xi_j|$), then

$$\Lambda(A) = \sup_j \left(\operatorname{Re} a_{jj} + \sum_{k \neq j} |a_{jk}|\right),$$

whereas $\|A\| = \sup_k \sum_j |a_{jk}|$.

20. Extend the result of Exercise 13 by proving that for any $A \in [\mathfrak{B}]$

$$e^{-\Lambda(-A)} \leqq \|e^A\| \leqq e^{\Lambda(A)}. \tag{0.12}$$

Estimate (0.12) implies the following estimate for a strip containing the spectrum $\sigma(A)$ (see (2.5) and Lemma 2.2):

$$-\Lambda(-A) \leq \operatorname{Re}\lambda \leq \Lambda(A) \quad \text{for all} \quad \lambda \in \sigma(A). \tag{0.13}$$

Hint. Using the result of Exercise 18, estimate the derivative with respect to t of the norm $\|\exp(At)x_0\|$ for any $x_0 \in \mathfrak{B}$.

Estimate (0.9) is a special case of a more general estimate of Wintner (see Corollary III.4.3), while estimate (0.12) is a special case of one of the estimates of S. M. Lozinskiĭ referred to in Exercise III.19.

The notion of the functional $\Lambda(A)$ as well as the other results indicated in these exercises are due to S. M. Lozinskiĭ [1] (see also the paper of G. Dahlquist [1]; in the papers of these authors $\mathfrak{B}$ is a finite-dimensional space). S. M. Lozinskiĭ called the measure $\Lambda(A)$ the *logarithmic norm* of A.

21. Estimate (0.13) implies that the spectrum of an operator A lies in the closed convex domain K_A of the complex ζ plane obtained by taking the intersection of the halfplanes

$$\operatorname{Re}(e^{i\theta}\zeta) \leq \Lambda(e^{i\theta}A) \qquad (0 \leq \theta \leq 2\pi).$$

A further restriction of this convex domain can be obtained by applying the formulated result to all possible operators of the form $A_\alpha = A - \alpha I$, where α is a complex number.

For $\mathfrak{B} = \mathfrak{H}$ the assertion concerning K_A yields the well-known *Hausdorff theorem*

$$K_A = \{(Ax, x) \mid x \in \mathfrak{H},\ \|x\| \leqq 1\}.$$

When applied to operators acting in C^n with this or that norm the assertion on K_A yields one of the well-known theorems of A. Hirsch (see, for example, Marcus and Minc [1], page 140).

22. An entire function $x(\zeta)$ of a complex argument ζ with values in $\mathfrak{B}$ is called an *entire function of exponential type* σ $(0 \leqq \sigma < \infty)$ (or, according to S. N. Bernšteĭn, *of finite degree* σ) if

$$\overline{\lim_{\zeta\to\infty}} \frac{\ln\|x(\zeta)\|}{|\zeta|} = \sigma.$$

Let

$$x(\zeta) = x_0 + \frac{\zeta}{1!}x_1 + \frac{\zeta^2}{2!}x_2 + \cdots + \frac{\zeta^n}{n!}x_n + \cdots.$$

Verify that, as in the scalar case (see B. Ja. Levin [1]), a function will be an entire function of exponential type σ if $\limsup_{n\to\infty}\|x_n\|^{1/n} = \sigma$.

23. According to a theorem of S. N. Bernšteĭn, if an entire scalar function $x(\zeta)$ $(\mathfrak{B} = C^1)$ of exponential type is bounded on the real axis, then

$$|x'(\zeta)| \leqq \sigma \sup_{-\infty<t<\infty} |x(t)| \qquad (-\infty < \zeta < \infty).$$

Verify that the theorem is also valid in the case of an entire function $x(\zeta)$ of exponential type with values in any Banach space $\mathfrak{B}$, i.e.

$$\|x'(\zeta)\| \leqq \sigma \sup_{-\infty<t<\infty} \|x(t)\| \qquad (-\infty < \zeta < \infty).$$

From estimate (0.13) it incidentally follows that if an entire function of exponential type σ is bounded on the real axis then $\sigma > 0$.

Hint. Consider the scalar function $f(x(t))$, where f is an element of the conjugate space, and apply the theorem of S. N. Bernšteĭn.

24. Prove that for every nonzero operator $A \in [\mathfrak{B}]$ with spectrum concentrated at the origin (in particular, for a Volterra operator[18]) at least one of the quantities $\Lambda(A)$ or $\Lambda(-A)$ is positive (while the other is nonnegative).

Hint. Construct a proof by contradiction. Consider the entire function $\exp(\zeta A)$ and make use of (0.12), (0.13), the theorem of S. N. Bernšteĭn (see the preceding exercise) and the fact that A is a quasi-nilpotent operator: $r_A = \lim_{n\to\infty} \|A^n\|^{1/n} = 0$.

25. Let $\{A_k\}$ $(k = 1, \cdots, n)$ denote a family of pairwise commutative operators in Hilbert space with spectra lying in the interior of the left halfplane. Show that there exists a uniformly positive operator W such that all of the operators $\mathrm{Re}\,(A_k W)$ are uniformly negative (S. G. Kreĭn and S. D. Èĭdel'man; see S. G. Kreĭn [1], page 167).

Hint. Put $W = \int_0^\infty \cdots \int_0^\infty \exp\,(\sum_1^n s_k A_k)\, ds_1 \cdots ds_n$ (when $n = 1$ the assertion is contained in Theorem 5.1).

It is assumed in the following exercises that W is a Hermitian operator in a Hilbert space $\mathfrak{H}$.

26. An operator $A \in [\mathfrak{H}]$ is said to be *W-nonnegative* if $(Ax, x)_W \geqq 0$ for all $x \in \mathfrak{H}$ or, in other words, if $WA \gg 0$.

a) Show that the spectrum of a W-nonnegative operator A lies on the real axis.

---

[18] A *Volterra operator* is a completely continuous operator with spectrum concentrated at the origin.

Hint. Assume the contrary and make use of the fact that for a boundary point λ of $\sigma(A)$ not lying on the real axis there exists a sequence of $f_n \in \mathfrak{H}$ ($\|f_n\| = 1$) such that $Af_n - \lambda f_n \to 0$ for $n \to \infty$.

b) Suppose $A \in [\mathfrak{H}]$ and $AW \gg 0$ (in this case the operator A is W^{-1} nonnegative). Prove that $\mathfrak{H}$ decomposes into a W-orthogonal sum $\mathfrak{H} = \mathfrak{H}_+ \dotplus \mathfrak{H}_-$ of subspaces invariant under A such that the restrictions $A_+ = A|\mathfrak{H}_+$ and $A_- = A|\mathfrak{H}_-$ are similar to a uniformly positive and a uniformly negative operator respectively.

Hint. Make use of Theorem 7.1.

27. Suppose A is a W-Hermitian operator with positive spectrum.

Show that there exists a unique operator B with positive spectrum such that $B_2 = A$.

This operator is W-Hermitian (V. P. Potapov [1]) and Ju. P. Ginzburg [1]).

Hint. To construct the operator B make use of the Poincaré-Riesz formula (2.3).

28. An operator Y is called a *W-nonenlargement* if

$$(WYf, Yf) \leqq (Wf, f),$$

or, in other words, if $Y^*WY \leqslant W$.

Show that if Y is a W-nonenlargement having at least one regular point on the unit circle, its adjoint Y^* is a W^{-1}-nonenlargement (see, for example, V. P. Potapov [1] and Ju. P. Ginzburg [1]).

Hint. Apply the Cayley transform to reduce the problem to one of considering W-dissipative operators.

29. Let U be a W-unitary u-dichotomic operator (the spectrum $\sigma(U)$ does not intersect the unit circle).

Show that $\mathfrak{H}$ decomposes into a direct sum of two W-neutral invariant (under U) subspaces in which the spectrum $\sigma(U)$ is found respectively inside and outside the unit circle (I. S. Iohvidov and M. G. Kreĭn [1]).

Hint. Make use of the arguments employed in the proof of Theorem 7.1.

The results noted in the following exercises have been taken from an article of M. G. Kreĭn [11]; they will be used in Chapter V.

30. One calls θ ($0 \leqq \theta \leqq \pi$) the *angle* between nonzero vectors $x, y \in \mathfrak{H}$, and writes $\theta = \sphericalangle(x, y)$ ($= \sphericalangle(y, x)$) or $\theta(\widehat{x,y})$ ($\theta(\widehat{y,x})$), if

$$\cos\theta = \operatorname{Re}(x, y)/\|x\|\,\|y\|.$$

An equivalent and perhaps more natural definition of the angle $\theta = \sphericalangle(x, y)$ is given by the equality

$$2\sin(\theta/2) = \|e_x - e_y\|,$$

where $e_x = x/\|x\|$, $e_y = y/\|y\|$ are the unit vectors in the directions of the vectors x, y.

Prove that for any $x_j \in \mathfrak{H}$ ($x_j \neq 0$; $j = 1, 2, 3$)

$$\sphericalangle(x_1, x_3) \leqq \sphericalangle(x_1, x_2) + \sphericalangle(x_2, x_3), \tag{0.14}$$

the equality sign holding if and only if the vector x_2 lies "between" the vectors x_1 and x_3, i.e. $x_2 = c_1x_1 + c_3x_3$ ($c_1, c_3 > 0$).

Hint. Using the unit vectors $e_j = x_j/\|x_j\|$ and arbitrary real numbers ξ_j ($j = 1, 2, 3$), construct the nonnegative form

$$\left\|\sum_{j=1}^{3}\xi_j e_j\right\|^2 = \sum_{j,k=1}^{3}\cos\theta_{jk}\xi_j\xi_k \qquad (\theta_{jk} = \sphericalangle(e_j, e_k)).$$

Calculate the determinant of this form and explain why it is nonnegative.

31. a) Let $e(t)$ ($a \leqq t \leqq b$) be a smooth (i.e. continuously differentiable) arc on the unit sphere S of H. Prove that

$$\int_a^b \left\| \frac{de}{dt} \right\| dt \geqq \measuredangle\,(e(b), e(a)), \tag{0.15}$$

the equality sign holding if and only if $e(t)$ lies between $e(a)$ and $e(b)$ for all $t \in [a, b]$.

If the unit sphere of $\boldsymbol{R}^n$ (n-dimensional Euclidean space) is taken as S, inequality (0.15) expresses the fact that the geodesics on S are great circle arcs.

Hint. Make use of the result of the preceding exercise.

b) Generalize (0.15) to the case when the smoothness condition for $e(t)$ is replaced by the requirement that it be absolutely continuous.

Hint. Approximate the spherical curve $e(t)$ by the curves

$$e_h(t) = \frac{1}{h}\int_{t-h}^{t+h} e(\tau)d\tau \Big/ \left\| \frac{1}{h}\int_{t-h}^{t+h} e(\tau)d\tau \right\| \qquad (a \leqq t \leqq b),$$

putting $e_h(t) = e_h(a)$ for $t < a$ and $e_h(t) = e_h(b)$ for $t > b$.

32. Suppose $A \in [\mathfrak{H}]$ has an inverse $A^{-1} \in [\mathfrak{H}]$. The quantity

$$\operatorname{dev}(A) \stackrel{\text{def}}{=} \sup\{\measuredangle(Ax, x) \,|\, x \in \mathfrak{H}, x \neq 0\}$$

is called the *angular deviation* of A. Thus $0 \leqq \operatorname{dev}(A) \leqq \pi$ and

$$\cos \operatorname{dev}(A) = \inf\left\{ \frac{\operatorname{Re}(Ax, x)}{\|Ax\|\ \|x\|} \,\middle|\, x \in \mathfrak{H}, x \neq 0 \right\}.$$

Clearly, $\operatorname{dev}(cA) = \operatorname{dev}(A)$ for $c > 0$.

a) Verify that $\operatorname{dev}(A)$ is a continuous function of $A \in [\mathfrak{H}]$.

Suppose A, $B \in [\mathfrak{H}]$ are invertible operators. Since $\measuredangle(ABx, x) \leqq \measuredangle(ABx, Bx) + \measuredangle(Bx, x)$, we have

$$\operatorname{dev}(AB) \leqq \operatorname{dev}(A) + \operatorname{dev}(B).$$

b) Prove that if $\operatorname{dev}(A) < \pi$, the spectrum $\sigma(A)$ of A lies in the sector $|\arg \lambda| \leqq \operatorname{dev}(A)$.

Hint. Take a ray $\zeta = \rho e^{i\phi}$ $(0 < \rho < \infty)$ intersecting $\sigma(A)$, and take the point $\zeta_0 \in \sigma(A)$ on it with maximum modulus. There will exist a sequence $\{x_n\}$ such that $Ax_n - \zeta_0 x_n \to 0$, $\|x_n\| = 1$. Consider $\lim \measuredangle(Ax_n, x_n)$.

33. Prove that for a uniformly positive operator H

$$\cos \operatorname{dev}(H) = \frac{2\sqrt{\lambda_M(H)\lambda_m(H)}}{\lambda_M(H) + \lambda_m(H)},$$

where $\lambda_m(H) \leqq \lambda_M(H)$ are the boundaries of the spectrum $\sigma(H)$ (see Exercise 13) (L. V. Kantorovič [1]).

34. Suppose $A \in [\mathfrak{H}]$ is an invertible operator. The quantity

$$\operatorname{am}(A) \stackrel{\text{def}}{=} \min\{\operatorname{dev}(\zeta^{-1}A) \,|\, |\zeta| = 1\} \tag{0.16}$$

is called the *angular amplitude* of A. If $\operatorname{am}(A) < \pi$ then $2\operatorname{am}(A)$ is the angle of the least sector of the ζ plane containing the spectrum $\sigma(A)$.

Let $\zeta = \zeta_0 = e^{i\alpha}$ be the unique point of the circle $|\zeta| = 1$ at which the minimum in (0.16) is achieved. Then the indicated sector will be the sector $\alpha - \operatorname{am}(A) \leqq \arg \lambda \leqq \alpha + \operatorname{am}(A)$. Clearly, for any scalar $c \neq 0$ we have $\operatorname{am}(cA) = \operatorname{am}(A)$. It readily follows from (0.16) that

$$\operatorname{am}(AB) \leqq \operatorname{am}(A) + \operatorname{am}(B). \tag{0.17}$$

Prove that

a) The inequality $\operatorname{am}(U) < \pi$ holds for a unitary operator U if and only if the spectrum $\sigma(U)$ does not cover the whole unit circle. When this condition is satisfied, $2\operatorname{am}(U)$ is the length of the least arc of the unit circle containing the spectrum $\sigma(U)$.

Hint. Use the fact that if $\sigma(U)$ lies on the arc $\gamma = e^{i\phi}$ $(-\alpha \leqq \phi \leqq \alpha, 0 < \alpha < \pi)$ and consequently admits the spectral resolution

$$U = \int_{-\alpha}^{\alpha} e^{i\phi}\, dE(\phi),$$

then $\operatorname{Re}(Ux, x) \geqq (x, x) \cos \alpha$.

b) If the spectra of the unitary operators U_1 and U_2 lie on arcs γ_1 and γ_2 of the unit circle the sum of the lengths of which is less than 2π, the spectrum $\sigma(U_1U_2)$ lies on the product of these arcs $\gamma = \{\zeta_1\zeta_2 \mid \zeta_1 \in \gamma_1, \zeta_2 \in \gamma_2\}$.

For a further development of this result see Chapter V.

Hint. Use assertion a) and inequality (0.17).

35. Let T be a u-dichotomic *uniform W-contraction with coefficient q*:

$$W - T^*WT \gg qI. \tag{0.18}$$

Show that the annulus of the complex plane

$$\sqrt{1 - \frac{q}{\lambda_M(W)}} < |\lambda| < \sqrt{1 + \frac{q}{\lambda_M(-W)}} \tag{0.19}$$

does not intersect the spectrum $\sigma(T)$.

We note that the annulus (0.19) contains the annulus

$$\sqrt{1 - \frac{q}{\|W\|}} < \|\lambda\| < \sqrt{1 + \frac{q}{\|W\|}}.$$

Hint. Consider the boundary points λ of $\sigma(T)$ and the values of the form $((W - T^*WT)x, x)$ on the corresponding sequences $\{x_n\}$ of unit vectors $(\|x_n\| = 1)$ such that $\|(T - \lambda I)x_n\| \to 0$ (see the theorem on a boundary point of the spectrum in § 4.3).

36. Let W be an invertible bounded Hermitian operator in $\mathfrak{H}$ and let A be a W-Hermitian operator. Show that an equivalent norm can be introduced in $\mathfrak{H}$ in such a way that the operator A becomes $\mathcal{J}$-Hermitian with $\mathcal{J}^* = \mathcal{J}$ and $\mathcal{J}^2 = I$.

Hint. Put $(x, y)_1 = (|W|x, y)$ and $\mathcal{J} = \operatorname{sign} W$, where $|W| = (W^*W)^{1/2}$ and $\operatorname{sign} W = W|W|^{-1}$.

37. An indefinite operator $W = W^* \in [\mathfrak{H}]$ is said to be *balanced* if under some decomposition of $\mathfrak{H}$ into a direct sum $\mathfrak{H} = \mathfrak{H}_+ \dotplus \mathfrak{H}_-$ of two W-orthogonal W-definite subspaces one has the equality $\dim \mathfrak{H}_+ = \dim \mathfrak{H}_-$.

a) Show that the latter equality will hold under any such decomposition (a special case of the general *law of inertia*).

b) Show that for every balanced operator $W \in [\mathfrak{H}]$ there exists an orthogonal decomposition $\mathfrak{H} = \mathfrak{H}_1 \oplus \mathfrak{H}_2$ and an invertible operator $S \in [\mathfrak{H}]$ such that

$$S^*WS = \begin{pmatrix} 0 & I_{12} \\ I_{21} & 0 \end{pmatrix},$$

where I_{12} is an operator of identification of $\mathfrak{H}_2$ with $\mathfrak{H}_1$ (a unitary mapping of $\mathfrak{H}_2$ onto $\mathfrak{H}_1$) while $I_{21} = I_{12}^{-1}$.

Hint. Make use of the result of Exercise 36 and then make use of the equality

$$E^* \begin{pmatrix} I_{11} & 0 \\ 0 & -I_{22} \end{pmatrix}, \quad E = \begin{pmatrix} 0 & I_{12} \\ I_{21} & 0 \end{pmatrix},$$

where

$$E = \frac{1}{\sqrt{2}} \begin{pmatrix} I_{11} & I_{12} \\ I_{21} & -I_{22} \end{pmatrix}.$$

38. Suppose $\mathfrak{H} = \mathfrak{H}_1 \oplus \mathfrak{H}_2$ and $\mathfrak{L}$ $(\subset \mathfrak{H})$ is a closed subspace. Prove that there corresponds to $\mathfrak{L}$ a unique operator $K \in [\mathfrak{H}_1, \mathfrak{H}_2]$ such that

$$\mathfrak{L} = \{x + Kx \mid x \in \mathfrak{H}_1\} \tag{0.20}$$

if and only if the restriction to $\mathfrak{L}$ of the orthogonal projection on $\mathfrak{H}_1$ is a one-to-one mapping of $\mathfrak{L}$ onto $\mathfrak{H}_1$.

If $\mathfrak{L}$ admits the representation (0.20), the operator K is called the *equation operator* for $\mathfrak{L}$ with respect to $(\mathfrak{H}_1, \mathfrak{H}_2)$ (or *angular operator* of inclination of $\mathfrak{L}$ to $\mathfrak{H}_1$; see M. G. Kreĭn [10]), and one writes $K = \tan(\mathfrak{L}, \mathfrak{H}_1)$.

Hint. Make use of Banach's theorem.

39. Let $\mathcal{J} = P_+ - P_-$, where the $P_\pm$ $(\in [\mathfrak{H}])$ are complementary orthogonal projections, and let $\mathfrak{H}_\pm = P_\pm\mathfrak{H}$. A $\mathcal{J}$-nonnegative subspace $\mathfrak{L}$ $(\subset \mathfrak{H})$ is said to be *maximal* if it is not a proper part of any other $\mathcal{J}$-nonnegative subspace.

Prove (Ju. P. Ginzburg [1, 4] and R. Phillips [2]) that

a) In order for a subspace $\mathfrak{L} \subset \mathfrak{H}$ to be a maximal $\mathcal{J}$-nonnegative subspace it is necessary and sufficient that the equation operator $K_\mathfrak{L} = \tan(\mathfrak{L}, \mathfrak{H}_+)$ exist for it.

b) A maximal $\mathcal{J}$-nonnegative subspace will be (i) $\mathcal{J}$-neutral precisely when $\|K_\mathfrak{L}x\| = \|x\|$ $(x \in \mathfrak{L})$, (ii) $\mathcal{J}$-positive precisely when $\|K_\mathfrak{L}x\| < \|x\|$ $(x \in \mathfrak{H},\ x \neq 0)$ and (iii) uniformly $\mathcal{J}$-positive precisely when $\|K_\mathfrak{L}\| < 1$.

Hint. The condition $(\mathcal{J}x, x) \geq 0$ for $x \in \mathfrak{L}$ means that $\|P_+x\| \geq \|P_-x\|$ for $x \in \mathfrak{L}$. This implies the existence of an operator $K \in [\mathfrak{H}_+', \mathfrak{H}_-]$ such that $KP_+x = P_-x$ for $x \in \mathfrak{L}$ and $\|K\| \leq 1$; here $\mathfrak{H}_+' = P_+\mathfrak{L} = \overline{P_+\mathfrak{L}}$ (the closure of $P_+\mathfrak{L}$). The maximality of $\mathfrak{L}$ implies the equality $\mathfrak{H}_+' = P_+\mathfrak{L} = \mathfrak{H}_+$.

40. Suppose $\mathcal{J} = P_+ - P_-$, the $\mathfrak{H}_\pm$ are the same as in Exercise 39 and $\mathfrak{K} = \mathfrak{K}(\mathfrak{H}_+, \mathfrak{H}_-)$ is the unit ball of operators in $[\mathfrak{H}_+, \mathfrak{H}_-]$, i.e. the set of all $K \in [\mathfrak{H}_+, \mathfrak{H}_-]$ with $\|K\| \leq 1$.

Suppose, further, T is a u-dichotomic uniform $\mathcal{J}$-contraction and, by virtue of the decomposition $\mathfrak{H} = \mathfrak{H}_+ \oplus \mathfrak{H}_-$,

$$T = \begin{pmatrix} T_{11} & T_{12} \\ T_{21} & T_{22} \end{pmatrix} \qquad (T_{jk} \in [\mathfrak{H}_j, \mathfrak{H}_k],\ \mathfrak{H}_1 = \mathfrak{H}_+,\ \mathfrak{H}_2 = \mathfrak{H}_-).$$

Show that the linear fractional mapping

$$\phi_T(K) = (T_{21} + T_{22}K)(T_{11} + T_{12}K)^{-1} \qquad (K \in \mathfrak{K}(\mathfrak{H}_+, \mathfrak{H}_-))$$

is meaningful on the ball $\mathfrak{K}$ and maps this ball into a smaller ball $\mathfrak{K}_r = r\mathfrak{K}$ $(0 < r = r_T < 1)$.

Hint. Use the fact that if K is the equation operator for a maximal $\mathcal{J}$-nonnegative subspace $\mathfrak{L}$, the operator $\phi_T(K)$ will be the equation operator for the uniformly $\mathcal{J}$-positive subspace $T\mathfrak{L}$.

41. Suppose $\mathfrak{H}^{(2)} = \mathfrak{H} \oplus \mathfrak{H}$ and $W \in [\mathfrak{H}^{(2)}]$ has the form

$$W = \begin{pmatrix} 0 & I \\ I & 0 \end{pmatrix}.$$

Let $\Pi_r = \Pi_r(\mathfrak{H})$ denote the *right operator halfplane*, i.e. the set of all $X \in [\mathfrak{H}]$ such that $\operatorname{Re} X \geqslant 0$, and let Π_r^0 denote the interior of Π_r (i.e. the set of all $X \in [\mathfrak{H}]$ such that $\operatorname{Re} X \gg 0$).

Suppose

$$T = \begin{pmatrix} T_{11} & T_{12} \\ T_{21} & T_{22} \end{pmatrix} (\in [\mathfrak{H}^{(2)}])$$

is a u-dichotomic uniform W-contraction with coefficient q $(0 < q < 1)$:

$$W - T^*WT \geqslant qI. \tag{0.18}$$

Prove that the linear fractional tranformation

$$X \to (T_{21} + T_{22}X)(T_{11} + T_{12}X)^{-1} (= \phi_T(X))$$

maps Π_r into a bounded closed part of Π_r^0.

Hint. The transformation $Z = (I - X)(I + X)^{-1}$ maps Π_r into the ball $\mathfrak{K} = \{Z \mid Z \in [\mathfrak{H}], \|Z\| \leq 1\}$ and Π_r^0 onto the open ball $\mathfrak{K}^0 = \{Z \mid Z \in [\mathfrak{H}], \|Z\| < 1\}$.

If one makes the substitution $X = (I - Z)(I + Z)^{-1}$ in the mapping $X \to \phi_T(X)$, one obtains the mapping $Z \to \phi_{T'}(Z)$, where $T' = ETE$ with

$$E = \frac{1}{\sqrt{2}} \begin{pmatrix} I & I \\ I & -I \end{pmatrix} = (E^*)^{-1} = E^*.$$

Then (0.18) goes over into $W' - T'^*W'T' \gg qI$, where

$$W' = \begin{pmatrix} I & 0 \\ 0 & -I \end{pmatrix}.$$

After this it remains to make use of the assertion of the preceding Exercise 40.

For a number of other general theorems on linear fractional transformations of operator balls into halfplanes see the article [1] of M. G. Kreĭn and Ju. L. Šmul'jan.

NOTES

§ 3. Formula (3.7) for the solution of equation (3.1) was found by M. G. Kreĭn in 1948 and reported on by him at seminars at Odessa and at the Mathematics Institute of the Academy of Sciences of the Ukrainian Soviet Socialist Republic in Kiev. It was later independently obtained by Ju. L. Daleckiĭ and first published in his paper [1]. Still later an analogous formula for the equation $AX + XB = C$ was obtained by M. Rosenblum [1]. The more general approach presented in § 3 was taken from the paper [4] of Ju. L. Daleckiĭ.

Formula (3.17) for equation (3.15) was obtained by Ju. L. Daleckiĭ and first published in the paper [1] of I. Kovtun.

§ 4. The results of § 4 are due to M. G. Kreĭn and were first presented in the "Lectures."

§ 5. The algebraic case of Theorem 5.1 was obtained by A. M. Ljapunov [1]. The general case was considered by M. G. Kreĭn in the "Lectures."

§ 6. Theorem 6.1 was proved by B. von Sz.-Nagy [1]. Theorems 6.2 and 6.3 are natural generalizations of it and were presented in the original lectures of M. G. Kreĭn (see the "Lectures" [9]).

§ 7. The main results of this section are due to M. G. Kreĭn. These results are related to a number of earlier investigations of M. G. Kreĭn [4, 5, 8], I. S. Iohvidov and M. G. Kreĭn [1], M. L. Brodskiĭ [1], H. Langer [1] and others.

Later, results similar to Theorem 7.1 were restated in more precise form in the algebraic case and under other more special assumptions by A. Ostrowski and H. Schneider [1], and by O. Taussky [1].

Theorem 7.2 is part of an assertion proved by R. Phillips [1] and Ju. P. Ginzburg [2], but our proof is simpler.

It should be noted in connection with Theorem 7.3 that Ju. P. Ginzburg [1] was the first to discover in an infinite-dimensional space the existence of $\mathcal{J}$-contractions T whose adjoint T^* is not also a $\mathcal{J}$-contraction, and to investigate their structure.

§ 8. Theorem 8.1 is due to R. Phillips [1]. In the finite-dimensional case the sufficiency of the condition of Theorem 8.3 was established by M. G. Kreĭn [4], and its necessity, by I. M. Gel'fand and V. B. Lidskiĭ [1].

The generalization of these results to the case of an infinite-dimensional space was obtained by V. I. Derguzov [1, 2] under a less complete formulation (and independently by P. Jonas [1]). The proof presented here is due to M. G. Kreĭn. It is apparently simpler than the others even in the case of a finite-dimensional space.

CHAPTER II

THE LINEAR EQUATION WITH A CONSTANT OPERATOR

In this chapter, which is propaedeutic in the sense that many of the facts and properties discussed will reappear in the more complicated cases considered in later chapters, we study the simplest equation, viz. the equation $\dot{x} = Ax + f(t)$ with a constant operator A and a continuous vector function $f(t)$.

In § 1 we deduce formulas permitting one to express the solutions of the homogeneous and inhomogeneous equations in terms of the exponential operator function e^{At}. We also consider some linear equations in a space of operators.

In § 2 we study the effect of the location of the spectrum of A on the behavior of the solutions of the homogeneous equation at infinity. The geometric meaning of the considered facts is discussed in connection with the well-known second method of Ljapunov.

In § 3 we use the results of § 6 of Chapter I to obtain boundedness conditions for solutions on the real line of the homogeneous equation. Here we also consider a second order equation which reduces to a first order equation with an operator acting in the direct sum of two Banach spaces.

The longest section, § 4, is devoted to a study of the inhomogeneous equation. Here we introduce the important notion of a Green operator function, which is used to consider a number of problems concerning existence conditions for solutions bounded on the real line and on a halfline as well as periodic and almost periodic solutions.

§ 1. Solution of the homogeneous and inhomogeneous equations

1. *Vector equations.* In this chapter we will consider the simplest equation[1)]

$$dx/dt = Ax + f(t) \tag{1.1}$$

in a Banach space $\mathfrak{B}$ with a constant operator $A \in [\mathfrak{B}]$ and a continuous vector function $f(t)$. The space $\mathfrak{B}$ will often be called the *phase space* of the equation.

All of the following results are valid for a much wider class of functions, which we introduce in the next chapter. In order to attach an elementary character to this chapter we confine ourselves here to the consideration of continuous functions.

We first turn to the homogeneous equation

$$dx/dt = Ax. \tag{1.2}$$

The solution of the Cauchy problem for this equation with the condition

$$x(t_0) = x_0 \tag{1.3}$$

[1)]Here and below, when the interval of variation of t is not indicated, we have in mind an arbitrary finite or infinite interval on which $f(t)$ is defined.

is readily obtained with the use of the operator function e^{At}.

In fact it immediately follows from (I.4.3) that the vector function

$$x(t)=e^{A(t-t_0)}x_0 \tag{1.4}$$

has a continuous derivative and is a solution of problem (1.2)–(1.3).

One can easily verify that this solution is unique in the class of differentiable functions. It suffices to show that if a continuous function $x(t)$ satisfying equation (1.2) vanishes at $t = t_0$, it also vanishes in a neighborhood of this point.

In fact, such a function must satisfy the equation

$$x(t) = \int_{t_0}^{t} Ax(s)ds,$$

which implies for $|t-t_0| \leqq \delta$ the estimate

$$\| x(t) \| \leqq \delta \| A \| \sup_{|s-t_0| \leqq \delta} \| x(s) \|,$$

leading to a contradiction for $\delta \| A \| < 1$ if

$$\sup_{|s-t_0|<\delta} \| x(s) \| \neq 0.$$

Applying the method of variation of constants, we seek a solution of the Cauchy problem for the inhomogeneous equation (1.1) in the form

$$x(t) = e^{A(t-t_0)}y(t).$$

After such a substitution equation (1.1) takes the form

$$e^{A(t-t_0)}y'(t) = f(t),$$

from which we get

$$y(t) = x_0 + \int_{t_0}^{t} e^{-A(s-t_0)}f(s)ds,$$

and, finally,

$$x(t) = e^{A(t-t_0)}x_0 + \int_{t_0}^{t} e^{A(t-s)}f(s)ds. \tag{1.5}$$

Expression (1.5) is clearly a differentiable function.

The uniqueness of solution (1.5) of the Cauchy problem (1.1)–(1.3) follows from the uniqueness of the solution of the Cauchy problem for the homogeneous equation.

2. *Operator equations.* The operator function $U(t) = e^{A(t-t_0)}C$, where $C \in [\mathfrak{B}]$, satisfies the equation

$$dU / dt = AU \tag{1.6}$$

and the condition $U(t_0) = C$.

We consider the more general operator equation

$$dU/dt = AU + UB \qquad (A, B \in [\mathfrak{B}]) \tag{1.7}$$

in the phase space $[\mathfrak{B}]$. Using the arguments in § 3 of Chapter I, we rewrite equation (1.7) in the form

$$dU/dt = (A_l + B_r)U. \tag{1.8}$$

Since the operators A_l and B_r commute, a solution of equation (1.8) satisfying the condition $U(t_0) = C$ has the form

$$U(t) = e^{(A_l+B_r)(t-t_0)}C = e^{A_l(t-t_0)}e^{B_r(t-t_0)}C = e^{A(t-t_0)}Ce^{B(t-t_0)}. \tag{1.9}$$

In exactly the same way we can represent the solution of the equation

$$dU/dt = AU + UB + F(t), \tag{1.10}$$

where $F(t)$ is a continuous function with values in $[\mathfrak{B}]$, in the form

$$U(t) = e^{A(t-t_0)}U(t_0)e^{B(t-t_0)} + \int_{t_0}^{t} e^{A(t-s)}F(s)e^{B(t-s)}ds. \tag{1.11}$$

§ 2. The behavior of the solutions of the homogeneous equation at infinity

1. *Renormings of the space.* The behavior of the solutions of problem (1.2)–(1.3) at infinity essentially depends on the location of the spectrum of A.

Suppose that the spectrum $\sigma(A)$ lies in the interior of the left halfplane. Then it follows from (1.4) on the basis of Theorem I.4.1 that

$$\| x(t) \| \leqq Ne^{-\nu(t-t_0)} \| x(t_0) \| \tag{2.1}$$

for any $t \geqq t_0$ and certain positive constants N and ν.

Conversely, if estimate (2.1) holds for every solution $x(t)$ of equation (1.2), the spectrum $\sigma(A)$ lies in the interior of the left halfplane. For it follows from (2.1) that $\| e^{A(t-t_0)} \| \leqq Ne^{-\nu(t-t_0)}$, and it remains to apply Theorem I.4.1.

The behavior of the solutions of equation (1.2) under the assumption that $\sigma(A)$ lies in the interior of the left halfplane can be characterized more precisely if one introduces a new equivalent norm in $\mathfrak{B}$ with the use of the formula

$$\| x \|_A = \int_0^\infty \| e^{As}x \| ds. \tag{2.2}$$

It turns out that in this norm the solutions of equation (1.2) tend monotonically to zero as $t \to \infty$.[2)]

In fact

$$\| x(t) \|_A = \int_0^\infty \| e^{As}x(t) \| ds = \int_0^\infty \| e^{A(t+s)}x_0 \| ds = \int_t^\infty \| e^{As}x_0 \| ds$$

[2)]It is assumed without loss of generality in §§ 2 and 3 below that $t_0 = 0$.

and thus

$$(d/dt)\,\| x(t) \|_A = - \| e^{At} x_0 \| < 0.$$

We now pass to case when $\sigma(A) = \sigma_+(A) \cup \sigma_-(A)$, the spectral set $\sigma_+(A)$ being nonempty.

Let P_+ and P_- be the spectral projections corresponding to this decomposition of the spectrum and let $\mathfrak{B} = \mathfrak{B}_+ \dot{+} \mathfrak{B}_-$ be the corresponding direct decomposition of $\mathfrak{B}$ into invariant subspaces of A. Since $\mathfrak{B}_+$ and $\mathfrak{B}_-$ are also invariant under the operators e^{At} $(0 \leqq t < \infty)$, a solution $x(t) = e^{At} x_0$ of equation (1.2) remains in the subspace containing the initial vector x_0.

We introduce in $\mathfrak{B}$ the indefinite norm

$$\| x \|_A = \int_0^\infty \{ \| e^{At} P_- x \| - \| e^{-At} P_+ x \| \} dt. \tag{2.3}$$

From calculations analogous to those carried out above, we get

$$\begin{aligned} \| x(t) \|_A &= \int_0^\infty \{ \| e^{As} e^{At} P_- x_0 \| - \| e^{-As} e^{At} P_+ x_0 \| \} ds \\ &= \int_t^\infty \| e^{As} P_- x_0 \| \, ds - \int_{-t}^\infty \| e^{-As} P_+ x_0 \| \, ds. \end{aligned}$$

Therefore

$$(d/dt) \| x(t) \|_A = - \{ \| e^{At} P_- x_0 \| + \| e^{At} P_+ x_0 \| \} < 0 \tag{2.4}$$

and hence the indefinite norm of any solution of equation (1.2) decreases.

We consider two special cases.

a) Suppose $P_+ x_0 = 0$, i.e. $x_0 \in \mathfrak{B}_-$ $(= P_- \mathfrak{B})$. In this subspace $\| x \|_A \geqq 0$ and the norms $\| x \|$ and $\| x \|_A$ are equivalent. It follows from (2.4) that $\|x(t)\|_A$ tends monotonically to zero.

Thus the solutions $x(t) = e^{At} x_0$ of equation (1.2) with initial vector x_0 in $\mathfrak{B}_-$ tend to zero.

b) Suppose $P_- x_0 = 0$, i.e. $x_0 \in \mathfrak{B}_+$ $(= P_+ \mathfrak{B})$. In this subspace the quantity $- \| x \|_A \geqq 0$ is an ordinary norm that is equivalent to $\| x \|$. It follows from (2.4) that

$$-\frac{d}{dt} \| x(t) \|_A = \| x(t) \| \geqq -\frac{1}{M} \| x(t) \|_A.$$

Integrating this inequality, we see that

$$- \| x(t) \|_A \geqq - \| x_0 \|_A \, e^{(1/M)(t - t_0)},$$

i.e. the solutions with initial values in $\mathfrak{B}_+$ increase unboundedly as $t \to +\infty$.

We note that the decomposition

$$x(t) = e^{At} x_0 = e^{At} P_- x_0 + e^{At} P_+ x_0 \qquad (= P_- e^{At} x_0 + P_+ e^{At} x_0)$$

implies that any solution for which $P_+x_0 \neq 0$ increases unboundedly. In particular, if the whole spectrum lies in the interior of the right halfplane: $\sigma(A) = \sigma_+(A)$, a nonzero solution of equation (1.2) goes to infinity as $t \to +\infty$, the norm $-\|x(t)\|_A$ monotonically increasing. Moreover, this case readily reduces to the original one if we replace A by $-A$ and reverse the time.

We recall that an operator A with a spectrum decomposing into two nonempty spectral sets lying in the interiors of the right and left halfplanes respectively: $\sigma(A) = \sigma_+(A) \cup \sigma_-(A)$, was said to be *e-dichotomic*. In this case we will also say that the differential equation (1.2) is *e-dichotomic*.

The above arguments clearly imply that the phase space $\mathfrak{B}$ of an e-dichotomic equation decomposes into a direct sum $\mathfrak{B} = \mathfrak{B}_+ + \mathfrak{B}_-$, the solutions initially in $\mathfrak{B}_-$ remaining in $\mathfrak{B}_-$ and exponentially decreasing and the solutions initially in $\mathfrak{B}_+$ remaining in $\mathfrak{B}_+$ and exponentially increasing. This accounts for the term *e-dichotomy*. We will see later (in Chapter IV) that an analogous situation is encountered in the more general case.

2. *Geometric meaning of the renormings*. If the phase space $\mathfrak{B}$ is a Hilbert space $\mathfrak{H}$, it is natural to develop all of the arguments with the norms $\|x\|_A$ replaced by the Hilbert norms $\|x\|_{A,2}$ (see § I.4.5).

In this case the renorming carried out above has a simple geometric meaning.

Suppose first $A \ll 0$. Then a solution of the equation

$$dx/dt = Ax \tag{1.2}$$

satisfies the relation

$$(x'(t), x(t)) = (Ax(t), x(t)) \leqq -c\|x(t)\|^2,$$

which implies that the angle α between the directions of the radius vector $x(t)$ and the vector $x'(t)$ tangent to an integral curve of equation (1.2) satisfies the estimate

$$\cos\alpha = \frac{(x'(t), x(t))}{\|x'(t)\|\,\|x(t)\|} \leqq -\frac{c\|x(t)\|^2}{\|A\|\,\|x(t)\|^2} = -\frac{c}{\|A\|}.$$

Thus $\alpha \geqq \pi/2 + \arcsin(c/\|A\|)$ and the vector field of tangents to the integral curves of equation (1.2) is at each point essentially directed toward the interior of the sphere with center at the origin passing through this point.

We now consider the more general situation when the spectrum $\sigma(A)$ lies in the interior of the left halfplane. In this case, by Theorem I.5.1 there exists a bounded uniformly positive operator W such that $\mathrm{Re}(WA) \ll 0$. The arguments presented earlier remain valid if the estimates are carried out in the new metric $(x, y)_W = (Wx, y)$, which is equivalent to the old one.

In fact the boundedness and uniform positiveness of W ensure the topological equivalence of the norms $\|\cdot\|$ and $\|\cdot\|_W$. On the other hand, if $\mathrm{Re}(WA) \leqslant -cI$ $(c > 0)$, we will have

$$\mathrm{Re}(x'(t), x(t))_W = ([A^*W + WA]x, x) \leqq -c(x, x) \leqq -c_1(x, x)_W$$

for a solution $x(t)$ of equation (1.2).

We recall that the required operator W can be obtained as a solution of the equation $A^*W + WA = -H$, where H is an arbitrary uniformly positive operator. Putting, for example, $H = I$, we get (see the proof of Theorem I.5.1)

$$W = \int_0^\infty e^{A^*t}e^{At}\,dt,$$

i.e.

$$(x, x)_W = (Wx, x) = \int_0^\infty \| e^{At}x \|^2\,dt. \tag{2.5}$$

In this case the system of spheres into which the integral curves enter is replaced by the system of ellipsoids $(Wx, x) = \text{const}$ with center at the origin.

The consideration in a Banach phase space of the norm (2.2), which is analogous to (2.5), generalizes the above mentioned geometric arguments connected with the so-called second method of Ljapunov.

Here the role of the ellipsoids is played by the family of centrally symmetric bodies bounded by the surfaces $\| x \|_A = \text{const}$.

A somewhat more complicated but quite clear geometric picture is obtained when the spectrum of A also has a component in the interior of the right halfplane, i.e. when equation (1.2) is e-dichotomic.

We first consider the case when the phase space $\mathfrak{B}$ is a Hilbert space $\mathfrak{H}$. The following simple test is a rephrasing of Theorem I.7.1.

THEOREM 2.1. *In order for equation* (1.2) *to be e-dichotomic it is necessary and sufficient that the operator A be uniformly W-dissipative with respect to an invertible indefinite Hermitian operator $W \in [\mathfrak{H}]$:*

$$\operatorname{Re} WA \ll 0.$$

Under any choice of an operator W satisfying this relation the invariant subspace $\mathfrak{H}_+$ ($\mathfrak{H}_-$) of A corresponding to the part of the spectrum $\sigma_+(A)$ ($\sigma_-(A)$) lying in the interior of the right (left) halfplane is uniformly W-negative (W-positive).

The indefinite form (Wx, x) defines two systems of hyperboloids in $\mathfrak{H}$ by the equation $(Wx, x) = c$: *plus hyperboloids* when $c > 0$ and *minus hyperboloids* when $c < 0$.

It is not difficult to verify that the uniform W-dissipativeness of A implies the relation

$$\frac{d(Wx(t), x(t))}{dt} = 2\operatorname{Re}(WAx(t), x(t)) < 0 \qquad (x(t) \neq 0),$$

which shows that the indefinite form (Wx, x) decreases, for any solution $x(t)$ of equation (1.2).

If the point x_0 is on a minus hyperboloid, the subsequent diminution of (Wx, x) with increasing t means that the trajectory $x(t)$ intersects all of the hyperboloids with larger-in-absolute-value negative values of c and hence goes to infinity.

When x_0 is on a plus hyperboloid in $\mathfrak{H}_-$, the diminution of (Wx, x), which on $\mathfrak{H}_-$ is equivalent to an ordinary norm, leads to an arbitrarily close approach of the trajectory to the center of the hyperboloid.

But if x_0 is on a plus hyperboloid outside $\mathfrak{H}_-$, the trajectory, after intersecting the hyperboloids $(Wx, x) = c$ with diminishing values of c, enters the cone $(Wx, x) = 0$ and goes over to a system of minus hyperboloids, receding then to infinity.

This follows from the fact that any solution for which $P_+x_0 \neq 0$ increases unboundedly in norm.

We now note that by choosing for the e-dichotomic operator A a Hermitian operator W from the relation

$$2\mathrm{Re}(WA) = WA + A^*W = -(P_+^*HP_+ + P_-^*HP_-),$$

where the $P_\pm$ are the spectral projections of A corresponding to the invariant subspaces $\mathfrak{H}_\pm$, we arrive, as follows from the formulas in § I.4.4, at the form

$$(Wx, x) = \int_0^\infty \{\| He^{At}P_-x \|^2 - \| He^{-At}P_+x \|^2\}\, dt.$$

This form is analogous (when $H = I$) to the indefinite norm (2. 3) in a Banach space $\mathfrak{B}$.

It is easily seen that analogous geometric arguments can be developed in a Banach space by considering in place of the hyperboloids the surfaces $\| x \|_A = c$.

§ 3. Boundedness of the solutions of the homogeneous equation

1. *First order equation.* We wish to determine conditions under which the solutions of the homogeneous equation $dx/dt = Ax$ are bounded on the real line.

Since the solutions of this equation are described by the formula $x(t) = e^{At}x_0$ $(x_0 = x(0))$, the condition that the solutions be bounded implies the estimate

$$\| e^{At}x_0 \| \leqq c_{x_0} \qquad (-\infty < t < \infty),$$

where the constant c_{x_0} depends only on x_0.

Thus the set of operators e^{At} $(-\infty < t < \infty)$ is bounded at each element $x_0 \in \mathfrak{B}$. By the uniform boundedness principle (Theorem I.1.2) these operators are uniformly bounded:

$$\| e^{At} \| \leqq c \qquad (-\infty < t < \infty). \tag{3.1}$$

The latter estimate, as was shown in Chapter I (Corollary I.4.1), implies that the spectrum of A lies on the imaginary axis.

A more exact result can be obtained if the phase space $\mathfrak{B}$ is a Hilbert space $\mathfrak{H}$.

In fact, by virtue of Theorem I.6.3 condition (3.1) is satisfied if and only if the operator A is similar to a Hermitian operator multiplied by i (a skew-Hermitian operator): $A = S^{-1}(iB)S$ $(B^* = B)$.

Thus the following result holds.

THEOREM 3.1. *If each solution of the equation $dx/dt = Ax$ is bounded on the real line, the spectrum $\sigma(A)$ lies on the imaginary axis.*

If the phase space $\mathfrak{B}$ is a Hilbert space, all of the solutions are bounded if and only if the operator A is similar to a skew-Hermitian operator.

2. *Second order equation in a Banach space.* We consider the second order differential equation

$$\frac{d^2y}{dt^2} + Ty = 0, \tag{3.2}$$

where $T \in [\mathfrak{B}]$.

Its investigation can be reduced to the investigation of a first order equation in the doubled phase space $\mathfrak{B}^{(2)} = \mathfrak{B} + \mathfrak{B}$ whose elements are the pairs $x = (x_1, x_2)$ $(x_1, x_2 \in \mathfrak{B})$ and whose norm is given by the formula $\| x \|_2^2 = \| x_1 \|^2 + \| x_2 \|^2$.

Setting $y = x_1$ and $dy/dt = x_2$, we replace equation (3. 2) by the system

$$\frac{dx_1}{dt} = x_2; \qquad \frac{dx_2}{dt} = -Tx_1,$$

or by the equivalent equation for a vector $x = (x_1, x_2)$ in $\mathfrak{B}^{(2)}$

$$dx/dt = \mathscr{A}x, \tag{3.3}$$

where the operator $\mathscr{A} \in [\mathfrak{B}^{(2)}]$ is defined by the operator matrix

$$\mathscr{A} = \begin{pmatrix} 0 & 1 \\ -T & 0 \end{pmatrix}.$$

It is not difficult to calculate that

$$\mathscr{A}^{2k} = (-1)^k \begin{pmatrix} T^k & 0 \\ 0 & T^k \end{pmatrix}; \quad \mathscr{A}^{2k+1} = (-1)^k \begin{pmatrix} 0 & T^k \\ -T^{k+1} & 0 \end{pmatrix}. \tag{3.4}$$

The operator function $e^{\mathscr{A}t}$ defining the solutions of equation (3. 3) takes the form

$$e^{\mathscr{A}t} = \sum_{n=0}^{\infty} t^n \frac{\mathscr{A}^n}{n!} = \sum_{k=0}^{\infty} t^{2k} \frac{\mathscr{A}^{2k}}{(2k)!} + \sum_{k=0}^{\infty} t^{2k+1} \frac{\mathscr{A}^{2k+1}}{(2k+1)!}. \tag{3.5}$$

Making use of the correspondence between scalar and operator functions, we put

$$\cos T^{1/2}t = \sum_{k=0}^{\infty} (-1)^k \frac{T^k t^{2k}}{(2k)!}; \quad T^{-1/2} \sin T^{1/2}t = \sum_{k=0}^{\infty} (-1)^k \frac{T^k t^{2k+1}}{(2k+1)!}.$$

In this notation it follows from (3. 4) and (3. 5) that

$$e^{\mathscr{A}t} = \begin{pmatrix} \cos T^{1/2}t & T^{-1/2}\sin T^{1/2}t \\ -T^{-1/2}\sin T^{1/2}t & \cos T^{1/2}t \end{pmatrix}.$$

The formula $x(t) = e^{\mathscr{A}t}x_0$, where $x_0 = (y_0, y_0')$, now leads us to a representation of the solution of equation (3. 2) satisfying the conditions

$$y(0) = y_0, \quad y'(0) = y_0', \tag{3.6}$$

in the form

$$y(t) = (\cos T^{1/2}t)\,y_0 + (T^{-1/2}\sin T^{1/2}t)\,y_0'. \tag{3.7}$$

It can be verified by direct substitution that the vector function (3. 7) satisfies equation (3. 2) and the conditions (3. 6).

From formula (3. 7) it follows that the boundedness for $t \in (-\infty, \infty)$ of each solution of equation (3. 2) is equivalent to the boundedness of the operator functions $\cos T^{1/2}t$ and $T^{-1/2}\sin T^{1/2}t$.

We now show that it suffices to require the boundedness of the operator function $T^{-1/2}\sin T^{1/2}t\ (-\infty < t < \infty)$.

We consider the vector function

$$y(t) = (T^{-1/2}\sin T^{1/2}t)y_0.$$

Its derivatives are the vector functions

$$y'(t) = (\cos T^{1/2}t)y_0 \quad \text{and} \quad y''(t) = -(T^{1/2}\sin T^{1/2}t)y_0 = -Ty(t).$$

By hypothesis, for each fixed $y_0 \in \mathfrak{B}$ the vector function $y(t)$, and hence also $y''(t) = -Ty(t)$, is bounded.

If we prove that $y'(t)$ is also bounded, we can conclude that the set of operators $\cos T^{1/2}t\ (-\infty < t < \infty)$ is bounded at each element $y_0 \in \mathfrak{B}$, and hence is bounded in norm by virtue of the uniform boundedness principle (Theorem I.1.2).

Thus it remains to show that the boundedness of $y(t)$ and $y''(t)$ implies the boundedness of $y'(t)$.

We put

$$y''(t) - y(t) = f(t). \tag{3.8}$$

The function $f(t)$ is bounded on the real line together with $y(t)$ and $y''(t)$. If we consider (3. 8) as a differential equation, we can express the solution of it that is bounded on the real line by means of the easily verified (and well known) formula[3)]

$$y(t) = \frac{1}{2}\int_{-\infty}^{\infty} e^{-|t-s|}f(s)\,ds.$$

By differentiating this expression with respect to t, it is easily shown that

[3)] This formula is a consequence of a general formula involving the Green function of a stationary linear equation (see § 4).

$$\sup_t \| y'(t) \| \leqq \sup_t \| f(t) \| < \infty.$$

Thus we get that *the boundedness of each solution of the second order equation* (3.2) *on the real line is equivalent to the boundedness of the operator function* $T^{-1/2} \sin T^{1/2}t$.

We have simultaneously proved the following assertion: *in order for all of the solutions of equation* (3.2) *to be bounded it is necessary and sufficient that the solutions satisfying the condition* $y(0) = 0$ *be bounded.*

3. *Second order equation in Hilbert space.* A more precise investigation can be carried out when the space $\mathfrak{B}$ is a Hilbert space $\mathfrak{H}$.

THEOREM 3.2. *In order for each solution of the equation*

$$\frac{d^2y}{dt^2} + Ty = 0 \qquad (-\infty < t < \infty) \tag{3.2}$$

in a Hilbert space $\mathfrak{H}$ *to be bounded on the real line it is necessary and sufficient that the operator* T *be similar to a uniformly positive operator.*

PROOF. For a proof of the sufficiency of the condition we note that if $T = S^{-1}T_1S$, where T_1 is uniformly positive, the substitution $y = Sx$ reduces equation (3.2) to the form

$$d^2x/dt^2 + T_1x = 0.$$

Thus we can assume from the very beginning without loss of generality that $T \gg 0$.

We consider equation (3.3), which is equivalent to (3.2), in the doubled Hilbert space $\mathfrak{H}^{(2)} = \mathfrak{H} \oplus \mathfrak{H}$. A direct calculation shows that

$$\mathscr{A} = i\mathscr{Q}\mathscr{B}\mathscr{Q}^{-1},$$

where

$$\mathscr{A} = \begin{pmatrix} 0 & 1 \\ -T & 0 \end{pmatrix}, \quad \mathscr{B} = \begin{pmatrix} T^{1/2} & 0 \\ 0 & -T^{1/2} \end{pmatrix}$$

is a Hermitian operator in $\mathfrak{H}^{(2)}$, and

$$\mathscr{Q} - \begin{pmatrix} I & I \\ -iT^{1/2} & iT^{1/2} \end{pmatrix} \quad \text{and} \quad \mathscr{Q}^{-1} = \frac{1}{2}\begin{pmatrix} I & iT^{-1/2} \\ I & -iT^{-1/2} \end{pmatrix}$$

are bounded operators in $\mathfrak{H}^{(2)}$.

Thus the operator $\mathscr{A}$ is similar to a skew-Hermitian operator and therefore, by virtue of Theorem 3.1, each solution of equation (3.3) is bounded on the real line. This implies the boundedness of each solution together with its derivative of equation (3.2).

We now prove the necessity of the condition of the theorem.

We first note that if a solution of equation (3.2) is bounded, its derivative is also bounded. In fact, it follows from the equation that

$$\sup_{-\infty<t<\infty} \| y''(t) \| \leqq \| T \| \sup_{-\infty<t<\infty} \| y(t) \|,$$

while it has already been shown above that the boundedness of a function and its second derivative on the real line implies the boundedness of its first derivative. Thus the boundedness of the solutions of equation (3.2) implies the boundedness of the solutions of equation (3.3). By virtue of Theorem 3.1 this means that the operator $\mathscr{A}$ is similar to a skew-Hermitian operator. The latter condition can be written in the form $\mathscr{H}\mathscr{A} + \mathscr{A}^*\mathscr{H} = 0$ (see the proof of Theorem I.6.3), where $\mathscr{H}$ is a uniformly positive operator in $\mathfrak{H}^{(2)}$.

Substituting the matrices

$$\mathscr{H} = \begin{pmatrix} H_{11} & H_{12} \\ H_{21} & H_{22} \end{pmatrix}, \quad \mathscr{A} = \begin{pmatrix} 0 & I \\ -T & 0 \end{pmatrix}$$

in this equality and multiplying them out, we obtain from a consideration of the element in the second row and first column the equality $-H_{22}T + H_{11} = 0$.

We now note that the uniform positiveness of H in $\mathfrak{H}^{(2)}$ implies the uniform positiveness of H_{11} and H_{22} in $\mathfrak{H}$. For when $x = (x_1, 0)$ we have

$$(\mathscr{H}x, x)_2 = (H_{11}x_1, x_1) \geqq m(\mathscr{H})(x, x)_2 = m(\mathscr{H})(x_1, x_1).$$

The assertion concerning H_{22} is proved analogously.

This permits us to write

$$T = H_{22}^{-1} H_{11} = H_{22}^{-1/2} D H_{22}^{1/2},$$

where $D = H_{22}^{-1/2} H_{11} H_{22}^{-1/2}$ is a uniformly positive operator.

The theorem is proved.

COROLLARY. *In a Hilbert space $\mathfrak{H}$ the boundedness of the operator function $T^{-1/2} \sin T^{1/2}t$ holds precisely when the operator T is similar to a uniformly positive operator.*

§ 4. Conditions for the existence of a bounded solution of the inhomogeneous equation

1. *Green functions.* We consider the inhomogeneous equation

$$dx/dt = Ax + f(t) \tag{4.1}$$

with a continuous function $f(t)$.

We assume that the spectrum of the operator A decomposes into two spectral sets: $\sigma(A) = \sigma_1(A) \cup \sigma_2(A)$. Let $\mathfrak{B}_1$ and $\mathfrak{B}_2$ denote the invariant subspaces of A corresponding to these sets and let P_1 and P_2 denote the corresponding spectral projections.

We recall (see I.2.4) that

$$P_k = -\frac{1}{2\pi i}\oint_{\Gamma_k} R_\lambda d\lambda \qquad (k = 1, 2).$$

We introduce the *Green operator function*

$$G(t) = \begin{cases} e^{At}P_1 = -\dfrac{1}{2\pi i}\oint_{\Gamma_1} e^{\lambda t}R_\lambda d\lambda; & t > 0; \\ -e^{At}P_2 = \dfrac{1}{2\pi i}\oint_{\Gamma_2} e^{\lambda t}R_\lambda d\lambda; & t < 0. \end{cases} \tag{4.2}$$

It has the following properties.

1) When $t \neq 0$ it is continuously differentiable and satisfies the homogeneous equation

$$dG(t)/dt = AG(t).$$

This fact follows directly from (4.2).

2) Its jump at the origin is equal to the identity operator.

In fact,

$$G(+0) = P_1,\ G(-0) = -P_2;.\ G(+0) - G(-0) = P_1 + P_2 = I.$$

3) The vector function

$$x(t) = \int_a^b G(t-s)f(s)ds, \tag{4.3}$$

where $f(t)$ is continuous, satisfies the inhomogeneous equation (4.1) when $a \leqq t \leqq b$.

For a proof we differentiate the equality

$$x(t) = \int_t^b G(t-s)f(s)ds + \int_a^t G(t-s)f(s)ds.$$

We get

$$\begin{aligned}\frac{dx}{dt} &= -G(-0)f(t) \\ &\quad + \int_t^b \frac{dG(t-s)}{dt}f(s)ds + G(+0)f(t) + \int_a^t \frac{dG(t-s)}{dt}f(s)ds \\ &= \int_a^b AG(t+s)f(s)ds + [G(+0) - G(-0)]f(t) = Ax + f(t).\end{aligned}$$

As a rule, in what follows we will consider the case when the spectrum $\sigma(A)$ does not intersect the imaginary axis (in particular, when A is e-dichotomic): $\sigma(A) = \sigma_+(A) \cup \sigma_-(A)$.

The Green function defined by the formula

$$G_A(t) = \begin{cases} e^{At}P_- & \text{for } t > 0, \\ -e^{At}P_+ & \text{for } t < 0, \end{cases} \tag{4.4}$$

is called the *principal Green function* for equation (4.1).

We do not assume that both of the sets $\sigma_{\pm}(A)$ are nonempty: in formula (4.4) we take $P_+ = I$, $P_- = 0$ when $\sigma(A) = \sigma_+(A)$, and $P_+ = 0$, $P_- = I$ when $\sigma(A) = \sigma_-(A)$.

Since the spectrum $\sigma(A)$ does not intersect the imaginary axis, there exist numbers $\nu > 0$ and $N > 0$ for which (see (I.4.7))

$$\| G_A(t) \| \leqq Ne^{-\nu|t|}. \tag{4.5}$$

2. *Solutions bounded on the real line.* The principal Green function (4.4) plays an important role in the determination of conditions for the existence of solutions of equation (4.1) that are bounded on the real line.

THEOREM 4.1. *In order for there to correspond to any bounded-on-the-real-line continuous vector function $f(t)$ one and only one bounded-on-the-real-line solution of equation* (4.1) *it is necessary and sufficient that the spectrum $\sigma(A)$ not intersect the imaginary axis.*

This solution is given by the formula

$$x(t) = \int_{-\infty}^{\infty} G_A(t - s) f(s) ds, \tag{4.6}$$

where $G_A(t)$ is the principal Green function for equation (4.1).

PROOF. Suppose there exists a unique bounded solution for any bounded continuous $f(t)$. We put $f(t) = y$, where y is a constant vector, and let $x(t)$ be the unique bounded solution of the equation

$$dx/dt = Ax + y.$$

The vector $x(t + \tau)$ is also a solution of this equation for any τ, and by virtue of the uniqueness $x(t + \tau) = x(t)$, i.e. $x(t) = x =$ const, which implies $Ax = -y$.

From the arbitrariness of y it follows that the continuous linear operator A maps $\mathfrak{B}$ onto itself. By Banach's theorem (I.1.1) such an operator has a continuous inverse A^{-1}, i.e. the point $\lambda = 0$ is a regular point of A.

Suppose now ρi is an arbitrary imaginary number. We consider the equation

$$dx/dt = Ax + ye^{\rho it}.$$

The substitution $x = \xi e^{\rho it}$ gives

$$d\xi/dt = (A - \rho iI)\xi + y.$$

Repeating the above arguments, we get that the operator $A - \rho iI$ has a continuous inverse, i.e. ρi is a regular point.

Thus the necessity of the condition of the theorem is proved. For a proof of the sufficiency we make use of estimate (4.5). It follows from this estimate that the vector function (4. 6) is bounded:

$$\| x(t) \| \leqq N \int_{-\infty}^{\infty} e^{-\nu|t-s|} \| f(s) \| \, ds \leqq \frac{2N}{\nu} \sup_{-\infty < t < \infty} \| f(s) \|.$$

The fact that this function satisfies equation (4.1) for $t \in (-\infty, \infty)$ was established earlier (property 3) of subsection 1).

It remains for us to prove the uniqueness of a bounded solution. To this end it suffices to verify that the homogeneous equation $dx/dt = Ax$ does not have any nontrivial solutions that are bounded on the real line.

Let us assume that such a solution $x(t) = e^{At}x_0$ exists. If we let $A_- = P_-A$ and $A_+ = P_+A$, we can write it in the form

$$x(t) = e^{A_-t}P_-x_0 + e^{A_+t}P^+x_0.$$

Since the spectrum of the operator A_- in $\mathfrak{B}_-$ is the set $\sigma_-(A)$ lying in the interior of the left halfplane, the first term, and hence also the second term, is bounded for $t > 0$:

$$\| e^{A_+t}P_+x_0 \| \leqq c \qquad (t > 0).$$

But then, taking into account the fact that the spectrum $\sigma_+(A)$ of the operator A_+ in $\mathfrak{B}_+$ (if it is not empty) lies in the interior of the right halfplane, we get

$$\| P_+x_0 \| = \| e^{-A_+t}(e^{A_+t} P_+x_0) \| \leqq Ne^{-\nu t}c \qquad (t > 0).$$

This inequality for $t \to \infty$ shows that $P_+x_0 = 0$. It is shown analogously that $P_-x_0 = 0$.

The theorem is proved.

3. *Solutions bounded on a halfline.* Under the assumption that the spectrum of A does not intersect the imaginary axis: $\sigma(A) = \sigma_+(A) \cup \sigma_-(A)$, it is possible to completely describe the solutions of equation (4.1) that are bounded on a halfline $[t_0, \infty)$.

THEOREM 4.2. *Suppose the spectrum $\sigma(A)$ does not intersect the imaginary axis and $f(t)$ is a continuous function that is bounded on a halfline $[t_0, \infty)$. To each element $x_0^- \in \mathfrak{B}_-$ there corresponds a unique solution $x(t)$ of equation* (4.1) *that is bounded on $[t_0, \infty)$ and satisfies the condition $P_-x(t_0) = x_0^-$. This solution is given by the formula*

$$x(t) = e^{A(t-t_0)}x_0^- + \int_{t_0}^{\infty} G_A(t-s) f(s)\, ds. \tag{4.7}$$

PROOF. Equation (4.1) is equivalent to the system of two independent equations

$$\begin{cases} dx_+/dt = A_+x_+ + f_+(t), & (4.8) \\ dx_-/dt = A_-x_- + f_-(t), & (4.9) \end{cases}$$

where $x_\pm = P_\pm x$, $f_\pm = P_\pm f$, $A_\pm = P_\pm A$. The first of these equations is considered in the subspace $\mathfrak{B}_+ = P_+\mathfrak{B}$, and the second, in the subspace $\mathfrak{B}_- = P_-\mathfrak{B}$. Here the spectrum of A_+ is the set $\sigma_+(A)$ lying in the interior of the right halfplane while the spectrum $\sigma_-(A)$ of A_- lies in the interior of the left halfplane.

Equation (4. 9) has a solution that is bounded for $t \geqq t_0$ for any $x_-(t_0) = P_-x_0$. This solution is given by the formula

$$x_-(t) = e^{A_-(t-t_0)} P_-x_0 + \int_{t_0}^{t} e^{A_-(t-s)} f_-(s)\,ds \qquad (4.10)$$
$$= e^{A(t-t_0)} P_-x_0 + \int_{t_0}^{t} e^{A(t-s)} P_-f(s)\,ds.$$

Equation (4. 8) can have no more than one solution that is bounded for $t \geqq t_0$, inasmuch as the corresponding homogeneous equation cannot have such solutions. It is easily verified that this unique solution is the function

$$x_+(t) = -\int_{t}^{\infty} e^{A_+(t-s)} f_+(s)\,ds = -\int_{t}^{\infty} e^{A(t-s)} P_+f(s)\,ds. \qquad (4.11)$$

Comparing (4. 10) with (4. 11), we obtain the required formula (4. 7). The uniqueness of the solution follows from the above arguments.

REMARK 4.1. Each bounded solution on $[t_0, \infty)$ of equation (4.1) is uniquely determined by the element $x_0^- = P_-x(t_0)$.

Thus there exists an initial manifold $\mathfrak{L}_{t_0}^-$ of "the same dimension" as $\mathfrak{B}_-$ having the property that those (and only those) trajectories intersecting it at time t_0 remain bounded for $t \to \infty$.

We note that all of these trajectories converge together as $t \to \infty$. In fact it follows from (4.7) that for any two such trajectories

$$\| x_2(t) - x_1(t) \| \leqq \| e^{A(t-t_0)} P_-[x_2(t_0) - x_1(t_0)] \|$$
$$\leqq Ne^{-\nu(t-t_0)} \| x_2(t_0) - x_1(t_0) \|.$$

It can be said in particular that all of these trajectories approach the unique trajectory that is bounded on the real line (if the vector function $f(t)$ is bounded on the real line).

We note further that if the spectral set $\sigma_-(A)$ is empty: $\sigma(A) = \sigma_+(A)$, there exists only one trajectory that is bounded on a halfline and for it $x_0^- = 0$.

An analogous situation holds for the trajectories that are bounded for $t \to -\infty$. They are described by the formula

$$x(t) = e^{A(t-t_0)} x_0^+ + \int_{-\infty}^{t_0} G_A(t-s) f(s)\,ds \qquad (4.12)$$

for $x_0^+ \in \mathfrak{B}_+$.

The corresponding initial manifold $\mathfrak{L}_{t_0}^+$ has "the same dimension" as $\mathfrak{B}_+$. It is obvious that the intersection of the manifolds $\mathfrak{L}_{t_0}^+$ and $\mathfrak{L}_{t_0}^-$ consists of the single point $\int_{-\infty}^{\infty} G_A(t_0 - s) f(s)\,ds$.

REMARK 4.2. In a finite-dimensional space one can easily deduce from the existence of a bounded solution on a halfline of equation (4.1) for each bounded $f(t)$ that the spectrum $\sigma(A)$ does not have any points on the imaginary axis. It is not known whether or not the analogous assertion in the infinite case is true.

Under certain additional assumptions on the operator A this assertion follows from the more general results of Chapter IV.

4. *Periodic solutions.* We consider the equation

$$dx/dt = Ax + f(t) \qquad (-\infty < t < \infty), \tag{4.1}$$

assuming that the function $f(t)$ is continuous and periodic:

$$f(t + T) = f(t). \tag{4.13}$$

We first adduce some formal heuristic arguments permitting one to determine conditions for the existence of periodic solutions of equation (4. 1) and to find these solutions.

We consider the expansion of $f(t)$ in a Fourier series

$$f(t) \sim \sum_{k=-\infty}^{\infty} f_k e^{2k\pi it/T}. \tag{4.14}$$

We will seek a periodic solution of equation (4.1) also in the form of a sum of a Fourier series:

$$x(t) = \sum_{k=-\infty}^{\infty} x_k e^{2k\pi it/T}. \tag{4.15}$$

Formally[4)] substituting expansions (4. 15) and (4. 14) in equation (4. 1) and comparing the coefficients of $e^{2k\pi it/T}$, we obtain the system of equations

$$\left(A - \frac{2k\pi i}{T} I\right) x_k = - f_k. \tag{4.16}$$

We first assume that the following condition is satisfied:

$$2k\pi i/T \in \rho(A) \qquad (k = 0, \pm 1, \pm 2, \cdots), \tag{4.17}$$

i.e. that the spectrum of A does not contain the points $2k\pi i/T$, where $k = 0, \pm 1, \pm 2, \cdots$, of the imaginary axis. Then (4. 16) implies

$$x_k = -\left(A - \frac{2k\pi i}{T} I\right)^{-1} f_k, \tag{4.18}$$

so that

$$x(t) = -\sum_{k=-\infty}^{\infty} \left(A - \frac{2k\pi i}{T} I\right)^{-1} f_k e^{2k\pi it/T}. \tag{4.19}$$

4) Formula (4.16) can also be obtained formally, by multiplying (4.1) by $\exp(2\pi ki/T)$ and integrating from 0 to T.

Substituting in the right side the expressions

$$f_k = \frac{1}{T}\int_0^T f(s)\,e^{-k\pi is/T}\,ds$$

for the Fourier coefficients, we obtain the relation

$$x(t) = -\frac{1}{T}\int_0^T \sum_{k=-\infty}^{\infty}\left(A - \frac{2k\pi i}{T}I\right)^{-1} e^{2k\pi i(t-s)/T} f(s)\,ds \tag{4.20}$$

$$= \int_0^T \Gamma_T(t-s) f(s)\,ds.$$

We consider in more detail the integrand function

$$\Gamma_T(t) = -\frac{1}{T}\sum_{k=-\infty}^{\infty}\left(A - \frac{2k\pi i}{T}I\right)^{-1} e^{2k\pi it/T}. \tag{4.21}$$

Subtracting the T-periodic function

$$\chi(t)\,I = \frac{1}{2\pi i}\sum_{k\neq 0}\frac{1}{k}e^{2k\pi it/T}I \qquad \left(\chi(t) = \frac{1}{2} - \frac{t}{T} \text{ for } 0<t<T\right)$$

from both sides of this equality, we obtain the regularized representation

$$\Gamma_T(t) = \chi(t)\,I - \frac{1}{T}\cdot\sum_{k\neq 0}\left[\left(A - \frac{2k\pi i}{T}I\right)^{-1} + \frac{IT}{2k\pi i}\right]e^{2k\pi it/T} - \frac{1}{T}A^{-1} \tag{4.22}$$

$$= \chi(t)\,I - \frac{1}{T}A^{-1} + A\sum_{k\neq 0}\frac{1}{2k\pi i}e^{2k\pi it/T}\left(\frac{2k\pi i}{T}I - A\right)^{-1}.$$

We now note that for k sufficiently large in absolute value

$$\left\|\left(\frac{2k\pi i}{T}I - A\right)^{-1}\right\| \leqq \frac{c}{|k|}.$$

Therefore the series in formula (4.22) converges abolutely and uniformly, and hence its sum is a continuous periodic function. Thus the operator function $\Gamma_T(t)$ $(-\infty < t < \infty)$ obtains a precise meaning if it is defined by equality (4. 22). This operator function will be called the *T-periodic Green function* for equation (4. 1). It is characterized by the following properties.

1) It is a periodic function: $\Gamma_T(t+T) = \Gamma_T(t)$.

2) It is continuous in the operator norm for all t with the exception of the points $t = Tk$ $(k = 0, \pm 1, \pm 2,\cdots)$, where $\Gamma_T(+0) - \Gamma_T(-0) = I$.

3) At points of continuity it is differentiable in norm and satisfies the differential equation

$$\Gamma_T'(t) = A\Gamma_T(t). \tag{4.23}$$

The latter equality is readily obtained by differentiating (4. 22) (after first removing another singularity in order to improve the convergence).

Properties 1) – 3) uniquely determine the function $\Gamma_T(t)$. Since a continuous solution of equation (4.23) has the form $e^{At}C$, where C is a constant operator, we easily find that

$$\Gamma_T(t) = e^{At}(I - e^{AT})^{-1} \qquad (0 < t < T).$$

The operator $(I - e^{AT})^{-1}$ exists, since by the spectral mapping theorem (see I.2.5) the spectrum $\sigma(A)$ would otherwise contain at least one point of the form $2k\pi i/T$.

It remains to extend the obtained expression periodically onto all of the real line.

Using the periodic Green function $\Gamma_T(t)$, we at once obtain the following result.

THEOREM 4.3. *If the spectrum* $\sigma(A)$ *does not contain the points* $2k\pi i/T$ $(k = 0, \pm 1, \pm 2, \cdots)$ *of the imaginary axis, equation* (4.1) *for any continuous T-periodic function* $f(t)$ *has one and only one T-periodic solution* $x(t)$.

This solution is given by the formula

$$x(t) = \int_0^T \Gamma_T(t - s) f(s)\, ds. \tag{4.24}$$

PROOF. The integral (4.24) exists inasmuch as the integrand is continuous. The periodicity of $\Gamma_T(t)$ implies that $x(t)$ is also periodic.

Rewriting (4.24) in the form

$$x(t) = \int_0^t \Gamma_T(t - s) f(s)\, ds + \int_t^T \Gamma_T(t - s) f(s)\, ds$$

and differentiating with respect to t, we get

$$x'(t) = \int_0^t A\Gamma_T(t - s) f(s)\, ds + \int_t^T A\Gamma_T(t - s) f(s)\, ds$$
$$+ [\Gamma_T(+0) - \Gamma_T(-0)] f(t) = Ax(t) + f(t).$$

The uniqueness of the solution follows from the fact that under the conditions of the theorem the homogeneous equation cannot have nontrivial solutions that are continuous and T-periodic.

For suppose that such a solution $x(t) = e^{At}x_0$ exists. We would then have the relation $e^{TA}x_0 = x_0$, which contradicts the existence of the operator $(I - e^{AT})^{-1}$.

The theorem is proved.

We now drop condition (4.17) and consider the general case. Let $K_{A;T}$ be the set of those indices k for which $2k\pi i/T \in \sigma(A)$. Since the spectrum $\sigma(A)$ is a bounded set, the set $K_{A;T}$ is finite.

The following assertion holds: *in order for equation* (4.1) *to have a T-periodic solution* $x(t)$ *it is necessary and sufficient that each of the equations*

$$\left(A - \frac{2k\pi i}{T} I\right) x_k = -f_k \qquad (k \in K_{A;T}) \tag{4.25}$$

have at least one solution $x_k \in \mathfrak{B}$.

For, if equation (4.1) has a T-periodic solution $x(t)$, its Fourier coefficients x_k $(k = 0, \pm 1, \pm 2, \cdots)$ satisfy conditions (4.16). This proves the necessity of the assertion.

Suppose now x_k^0 $(k \in K_{A;T})$ is a solution of equation (4.25). We put

$$x(t) = y(t) + \sum_{k \in K_{A;T}} x_k^0 e^{(2k\pi i/T)t}.$$

Substituting this expression in (4.1), we obtain for $y(t)$ the equation

$$y'(t) = Ay(t) + \phi(t), \tag{4.26}$$

where $\phi(t) = f(t) - \sum_{k \in K_{A;T}} f_k e^{-2k\pi i/T}$ is a function whose Fourier coefficients with index in $K_{A;T}$ are equal to zero:

$$\phi_k = 0 \qquad (k \in K_{A;T}). \tag{4.27}$$

It remains to note that a solution of equation (4.26) under condition (4.27) can be obtained by making use of formula (4.24), since the terms in (4.21) corresponding to the indices $k \in K_{A;T}$ drop out thanks to (4.27).

If we introduce the "incomplete" Green function

$$\tilde{\Gamma}_T(t) = -\frac{1}{T} \sum_{k \bar{\in} K_{A;T}} \left(A - \frac{2k\pi i}{T} I\right)^{-1} e^{(2k\pi i/T)t} \tag{4.28}$$

we can write an arbitrary T-periodic solution of equation (4.1) in the form

$$x(t) = \int_0^T \tilde{\Gamma}_T(t-s) f(s)\, ds + \sum_{k \in K_{A;T}} x_k^0 e^{(2k\pi i/T)t}, \tag{4.29}$$

where the x_k^0 $(k \in K_{A;T})$ are solutions of equations (4.16).

5. *Almost periodic solutions.* In this subsection we consider the differential equation

$$dx/dt = Ax + f(t) \qquad (-\infty < t < \infty), \tag{4.1}$$

under the assumption that the function $f(t)$ is continuous and almost periodic.

We recall that a continuous function $f(t)$ on the real line with values in a Banach space $\mathfrak{B}$ is said to be *almost periodic* (according to Bohr) if for each $\varepsilon > 0$ there exists an $L_\varepsilon > 0$ such that each interval of the real line of length not less than L_ε contains a point $\tau = \tau(\varepsilon)$ (an ε-translation number) for which

$$\| f(t) - f(t+\tau) \| < \varepsilon \qquad (-\infty < t < \infty).$$

According to a well-known theorem of S. Bochner a necessary and sufficient condition for the almost periodicity of a continuous function is the precompactness of its family of translates $f_\tau(t) = f(t+\tau)$ $(-\infty < \tau < \infty)$ in the topology of uniform (on the real line) convergence in norm.

It follows in particular from this proposition that the set of all values of an almost periodic function is precompact and hence bounded: $\sup_{-\infty<t<\infty} \|f(t)\| < \infty$.

Suppose that the spectrum $\sigma(A)$ does not intersect the imaginary axis: $\sigma(A) = \sigma_+(A) \cup \sigma_-(A)$. Then, as follows from Theorem 4.1, equation (4.1) has the bounded solution

$$x(t) = \int_{-\infty}^{\infty} G_A(t-s) f(s)\, ds, \tag{4.30}$$

where $G_A(t)$ is the principal Green function for equation (4.1).

THEOREM 4.4. *If the spectrum $\sigma(A)$ does not intersect the imaginary axis, the differential equation* (4.1) *with an almost periodic function $f(t)$ has one and only one almost periodic solution. This solution is given by formula* (4.30).

PROOF. Since equation (4.1) has only one bounded solution, we need only prove that the function (4.30) is also almost periodic. To this end it suffices to establish that the set of functions $x_\tau(t) = x(t+\tau)$ $(-\infty < \tau < \infty)$ is precompact in the topology of uniform convergence.

We consider a sequence $x_{\tau_k}(t)$. By virtue of the almost periodicity of $f(t)$ we can select from the sequence of functions $f_{-\tau_k}(t) = f(t - \tau_k)$ a Cauchy subsequence $f_{\tau_{k_l}}(t)$. In this connection

$$x_{\tau_{k_l}}(t) = x(t + \tau_{k_l}) = \int_{-\infty}^{\infty} G_A(t-s) f(s - \tau_{k_l})\, ds$$

$$= \int_{-\infty}^{\infty} G_A(t-s) f_{-\tau_{k_l}}(s)\, ds.$$

This relation implies that $x_{\tau_{k_l}}(t)$ is a Cauchy sequence. In fact,

$$\| x_{\tau_{k_i}}(t) - x_{\tau_{k_l}}(t) \| \leqq \int_{-\infty}^{\infty} \| G_A(t-s) \| \| f_{-\tau_{k_i}}(s) - f_{-\tau_{k_l}}(s) \| \, ds$$

$$\leqq \sup_s \| f_{-\tau_{k_i}}(s) - f_{-\tau_{k_l}}(s) \| \int_{-\infty}^{\infty} \| G_A(s) \| \, ds.$$

We have thus proved the precompactness of the set of functions $x_\tau(t) = x(t+\tau)$, and hence the almost periodicity of $x(t)$.

REMARK 4.3. For the readers familiar with the theory of almost periodic functions we note that the arguments presented in the proof of Theorem 4.4 imply the coincidence of the spectra of $x(t)$ and $f(t)$ as well as a simple connection between the Fourier coefficients of these functions (see Exercise 11).

REMARK 4.4. If $f(t)$ is a periodic function $(f(t+T) = f(t))$, formula (4.30) clearly gives a unique periodic solution $x(t)$ $(\equiv x(t+T))$.

One can easily obtain an expression for the periodic Green function $\Gamma_T(t)$ in terms of the principal Green function $G_A(t)$.

In fact, if $f(t)$ is a T-periodic function, (4.30) implies

$$x(t) = \sum_{k=-\infty}^{\infty} \int_{kT}^{(k+1)T} G_A(t-s) f(s)\, ds$$
$$= \sum_{k=-\infty}^{\infty} \int_0^T G_A(t-s-kT) f(s)\, ds$$
$$= \int_0^T \left[\sum_{k=-\infty}^{\infty} G_A(t-s-kT) \right] f(s)\, ds.$$

From a comparison of the latter integral with formula (4.24) we get

$$\Gamma_T(t) = \sum_{k=-\infty}^{\infty} G_A(t+kT). \tag{4.31}$$

The latter series converges uniformly and "rapidly" since its terms decrease exponentially (see (4.5)).

Equality (4.31) can also be obtained directly by making use of the well-known Poisson summation formula (see Courant and Hilbert [1]).

Despite Remark 4.4, Theorem 4.4 does not contain Theorem 4.3 since its conditions involve the rigid requirement that there be no points of the spectrum of the operator A on the imaginary axis.

EXERCISES

1. Let $A, B \in [\mathfrak{B}]$. Derive the formula

$$e^{(A+B)t} = e^{At} + \sum_{n=1}^{\infty} \int_0^t \int_0^{s_1} \cdots \int_0^{s_{n-1}} e^{A(t-s_1)} B e^{A(s_1-s_2)} \cdots B e^{A(s_{n-1}-s_n)} B e^{As_n}\, ds_n \cdots ds_1.$$

Hint. Consider the differential equation $d\psi/dt = A\psi + f$, where $f = B\psi$, and make use of formula (1.5).

2. Give an example of an equation $dx/dt = Ax$ all of whose solutions tend to zero as $t \to +\infty$ whereas $\| e^{At} \|$ does not tend to zero.

Hint. Consider a nonpositive Hermitian operator in Hilbert space $\mathfrak{H}$ having $\lambda = 0$ as a point of its continuous spectrum.

3. Show that the situation described in Exercise 2 is impossible in the finite case.

4. Let $\{A_k\}$ $(k = 1, \cdots, n)$ be a commutative collection of operators from $[\mathfrak{B}]$, the spectrum of each of them lying in the interior of the left halfplane. Show that a new norm can be introduced in $\mathfrak{B}$ in such a way that all of the solutions of each of the equations $dx/dt = A_k x$ will monotonically tend to zero (S. G. Kreĭn and S. D. Èĭdel'man; see S. G. Kreĭn [1], page 167).

Hint. Put

$$|||x||| = \int_0^\infty \cdots \int_0^\infty \| \exp (\textstyle\sum_{k=1}^n s_k A_k) x \| \, ds_1 \cdots ds_n$$

(see Exercise I.25).

5. Suppose $W, H \in [\mathfrak{H}]$, $H \gg 0$ and $W = W^*$ is an invertible Hermitian operator ($W^{-1} \in [\mathfrak{H}]$). Show that

a) Independently of whether W is definite or indefinite, all of the solutions of the equation

$$dx/dt = iWHx \tag{0.1}$$

are bounded on the real line.

b) If W is definite, the Ljapunov exponents of all of the solutions of the equation

$$dx/dt = WHx \tag{0.2}$$

lie in the interval (a, b), where $a = \min\{\lambda \mid \lambda \in \sigma(WH)\}$ and $b = \max\{\lambda \mid \lambda \in \sigma(WH)\}$. Here $a > 0$ if $W \gg 0$ while $b < 0$ if $W \ll 0$.

c) If W is indefinite, equation (0.2) is e-dichotomic and $\mathfrak{H}$ decomposes into a W-orthogonal sum of subspaces $\mathfrak{H} = \mathfrak{H}_+ \dotplus \mathfrak{H}_-$ such that every solution of equation (0.2) initially in $\mathfrak{H}_+$ ($\mathfrak{H}_-$) remains in that subspace and its Ljapunov exponent lies on a certain segment of the positive (negative) halfline.

Hint. Make use of the result of Exercise I.26b).

6. Suppose $A, B, C \in [\mathfrak{B}]$ and the inverse $A^{-1} \in [\mathfrak{B}]$ exists. Show that

a) For any $y_0, y_0' \in \mathfrak{B}$ the equation

$$Ad^2y/dt^2 + Bdy/dt + Cy = 0 \tag{0.3}$$

has one and only one solution $y(t) \in C_2(\mathfrak{B})$ (i.e. a twice continuously differentiable solution with values in $\mathfrak{B}$) such that $y(0) = y_0$ and $y'(0) = y_0'$.

Hint. Reduce equation (0.3) to a stationary equation in the phase space $\mathfrak{B}^{(2)} = \mathfrak{B} \dotplus \mathfrak{B}$ by setting, for example, $dy/dt = z$ and $x = y \dotplus z$.

b) For any solution $y(t) \in C_2(\mathfrak{B})$ the Ljapunov exponents of the functions $y(t)$ and $y(t) + dy(t)/dt$ coincide.

Hint. Make use of the arguments in § 3.2.

By the *(upper) Ljapunov exponent* κ_L *of equation* (0.3) is meant the upper bound of the set of Ljapunov exponents of all of the solutions $y(t) \in C_2(\mathfrak{B})$ of this equation (see § 4 of Chapter III).

By the *λ spectrum* of equation (0.3) is meant the set σ of those complex λ for which the operator $\lambda^2 A + \lambda B + C$ does not have an inverse in $[\mathfrak{B}]$.

c) For equation (0.3) the upper Ljapunov exponent coincides with the supremum of the set of real parts of the points of the λ spectrum: $\kappa_L = \sup\{\mathrm{Re}\,\lambda \mid \lambda \in \sigma\}$.

d) Derive a formula for the solution of the equation

$$d^2y/dt^2 + Ty = f(t) \qquad (T \in [\mathfrak{B}]),$$

satisfying the condition $y(t_0) = y_0$, $y'(t_0) = y_0'$.

7. Suppose the state of a certain linear autonomous mechanical system S is specified by a vector $y \in \mathfrak{H}$, its potential energy is given by the form $(Cy, y)/2$ and its kinetic energy is given by the form $(A\dot{y}, \dot{y})/2$ $(\dot{y} = dy/dt)$. Then its equation of motion is written in the form $A\ddot{y} + Cy = -B\dot{y}$ $(\ddot{y} = d^2y/dt^2)$, where $-B\dot{y}$ gives the effect of the external forces, which are assumed to depend only on the velocity $\dot{y}$. For the total mechanical energy $E = (Cy, y)/2 + (Ay, y)/2$ we will have $dE/dt = -(\mathrm{Re}\,B\dot{y}, \dot{y})$.

Thus, if $B = -B^*$, the force $-B\dot{y}$ does not cause a dissipation of energy and is said to be *gyroscopic*. If $B = B^* \geqq 0$, the force $-B\dot{y}$ is called a *resistance* (or *Rayleigh*) *force*, and such a force causes a dissipation of energy.

In problems of mechanics one frequently encounters the case when $B = B_{\mathfrak{R}} + iB_{\mathfrak{J}}$ $(B_{\mathfrak{R}}^* = B_{\mathfrak{R}}$, $B_{\mathfrak{J}}^* = B_{\mathfrak{J}})$, i.e. the force $-B\dot{y}$ contains both a Rayleigh and a gyroscopic component.

Since the kinetic energy of a motion is always positive, the operator A must be positive. We assume that $A \gg 0$. If the system oscillates about the state $y = 0$ of stable equilibrium (minimum potential energy), the potential energy will be positive for $y \neq 0$, i.e. $C \gg 0$. If the motion of the system began near a state of maximum (a saddle point of the) potential energy, the operator C must be negative (indefinite).

These remarks lead to a mechanical interpretation of the results cited in this and the next two exercises (8 and 9).

a) Suppose $A \gg 0$ and $C \in [\mathfrak{H}]$. Show that all of the solutions of the equation $A\ddot{y} + Cy = 0$ are bounded on the real line precisely when the operator C is uniformly positive: $C \gg 0$.

Hint. Make use of Theorem 3.2.

b) Suppose $A \gg 0$, $C \gg 0$ and $B^* = -B$. Show that all of the solutions of the equation

$$A\ddot{y} + B\dot{y} + Cy = 0 \tag{0.4}$$

in $\mathfrak{H}$ are bounded on the real line.

Hint. Reduce the equation to a stationary equation in the phase space $\mathfrak{H}^{(2)} = \mathfrak{H} \oplus \mathfrak{H}$ by putting $x = y \oplus z$ and

$$z = A\, dy/dt + \tfrac{1}{2} By, \tag{0.5}$$

and then make use of the assertion of Exercise 5a).

8. a) Suppose $A \gg 0$, $\operatorname{Re} B \gg 0$ and $C \gg 0$. Show that the upper Ljapunov exponent κ_L of equation (0.4) is negative and, moreover,

$$\kappa_L \leqq \sup \left\{ \operatorname{Re} \frac{-(Bf, f) + \sqrt{(Bf, f)^2 - 4(Af, f)(Cf, f)}}{(Af, f)} \,\middle|\, f \in \mathfrak{H},\ f \neq 0 \right\}.$$

Hint. Consider a boundary point of the spectrum of the coefficient of the first order equation in $\mathfrak{H}^{(2)}$ into which (0.4) is transformed, and make use of the theorem on a boundary point of the spectrum (see § I.4.3).

b) Suppose $A = A^*$ is invertible, $B = B^*$, $C = C^*$ and for any $x \in \mathfrak{H}$

$$(Bx, x)^2 \geqq 4(Ax, x)(Cx, x).$$

Show that the λ spectrum of equation (0.4) is real.

9. Suppose $A \gg 0$, $B = B^*$ and $\operatorname{Re} C \ll 0$. Show that

a) The differential equation

$$A\ddot{Y} + B\dot{Y} + CY = 0 \tag{0.6}$$

in $[\mathfrak{H}]$ has the two solutions $Y_\pm = \exp(tZ_\pm)$, where the $Z_\pm$ ($\in [\mathfrak{H}]$) are the unique roots of the quadratic "algebraic" equation

$$AZ^2 + BZ + C = 0 \tag{0.7}$$

with the properties $\operatorname{Re}(AZ_+ + B/2) \gg 0$ and $\operatorname{Re}(AZ_- + B/2) \ll 0$. The spectrum $\sigma(Z_+)$ lies in the interior of the right halfplane: $\operatorname{Re} \lambda > 0$, while the spectrum $\sigma(Z_-)$ lies in the interior of the left halfplane: $\operatorname{Re} \lambda < 0$.

b) The general solution of equation (0.6) is given by the formula

$$Y(t) = \exp(tZ_+)\, C_+ + \exp(tZ_-)\, C_-,$$

where C_+ and C_- are arbitrary operators from $[\mathfrak{H}]$.

c) The general solution of equation (0.4) is given by the formula

$$y(t) = \exp(tZ_+)\, y_+ + \exp(tZ_-)\, y_-,$$

where y_+ and y_- are arbitrary vectors from $\mathfrak{H}$.

d) If $C = C^*$ ($\ll 0$), then $AZ_+ + B/2 = -(AZ_- + B/2)^*$. The operator Z_+ is similar to a uniformly positive operator while Z_- is similar to a uniformly negative operator.

Hint. Make the substitution (0.5), verify that the coefficient $\mathscr{A}$ of the corresponding equation in $\mathfrak{H}^{(2)}$ satisfies the condition $\operatorname{Re}(W\mathscr{A}) \gg 0$ for $W = \begin{pmatrix} 0 & I \\ I & 0 \end{pmatrix}$ and make use of Theorem 2.1. Under the decomposition $\mathfrak{H}^{(2)} = \mathfrak{H}_+ \dotplus \mathfrak{H}_-$ corresponding to the decomposition $\sigma(\mathscr{A}) = \sigma_+(\mathscr{A}) \cup \sigma_-(\mathscr{A})$ of the spectrum there will correspond to the W-definite spaces $\mathfrak{H}_\pm$ uniformly W-definite operators $K_\pm$ such that

$$\mathfrak{H}_\pm = \{ y \oplus K_\pm y \mid y \in \mathfrak{H} \}$$

(cf. the proof of Theorem V.2.4).

The desired roots $Z_\pm$ are found from the equations $AZ_\pm + B/2 = K_\pm$.

When $C = C^*$ ($\ll 0$) we have $W\mathscr{A} \gg 0$, and the restrictions $\mathscr{A} \mid \mathfrak{H}_+$ and $\mathscr{A} \mid \mathfrak{H}_-$ are similar to a uniformly positive and a uniformly negative operator respectively (see Exercise I.26b)).

Assertions a) – d) can also be proved by making use of the usual substitution $\dot{y} = z$ and setting $W = \begin{pmatrix} B & A \\ A & 0 \end{pmatrix}$.

We conclude this group of exercises by noting that more sophisticated investigations of the differential equation (0.4) and the corresponding quadratic equation (0.7) can be found in the papers of M. G. Kreĭn and H. Langer [1, 2], H. Langer [1, 2] and R. Kühne [1, 2] (see also M. G. Kreĭn [10]). Some of these investigations were generalizations and extensions of the investigations of R. Duffin [1, 2] of equation (0.4) in the finite-dimensional real space $\mathfrak{H} = R^n$.

10. Show that the principal Green function $G_A(t)$ for an operator $A \in [\mathfrak{B}]$ whose spectrum does not intersect the imaginary axis is representable in the form of a Fourier transform:

$$G_A(t) = \frac{1}{2\pi} \int_{-\infty}^{\infty} (A - i\lambda I)^{-1} e^{i\lambda t}\, d\lambda \qquad (t \neq 0),$$

where the integral is understood in the sense of the Cauchy principal value.

11. Let $f(t)$ be an almost periodic function. It is known that there exists (uniformly with respect to t_0) the limit

$$\mathfrak{M}(f) = \lim_{T\to\infty} \frac{1}{T} \int_{t_0}^{t_0+T} f(t)\, dt,$$

which is called the *mean value of an almost periodic function* $f(t)$. Since along with $f(t)$ the function $f(t)e^{-i\lambda t}$ is also almost periodic for any real λ, we can speak of the quantity $c_\lambda(f) = \mathfrak{M}(fe^{-i\lambda t})$, the nonzero values of which are called the *Fourier coefficients of an almost periodic function* $f(t)$.

The set $\Lambda(f)$ of those values of λ for which $c_\lambda(f) \neq 0$ is called the *spectrum of an almost periodic function* $f(t)$.

It is known that the spectrum of an almost periodic function is at most a countable set. This permits one to associate with each almost periodic function $f(t)$ the Fourier series $f(t) \sim \sum_\lambda c_\lambda(f) e^{i\lambda t}$,

Show that under the conditions of Theorem 4.4 the Fourier coefficients of an almost periodic solution $x(t)$ of equation (4.1) are connected with the Fourier coefficients of the function $f(t)$ by the relation

$$c_\lambda(x) = -(A - i\lambda I)^{-1} c_\lambda(f).$$

12. Let $A \in [\mathfrak{H}]$. Prove that in order for the vector function $e^{iAt}x$ to be almost periodic for all $x \in [\mathfrak{H}]$ it is necessary and sufficient that the operator A have a complete system of eigenvectors that is orthogonal with respect to a scalar product that is topologically equivalent to the original one.

Hint. Make use of Theorem I.6.1.

For a completely continuous operator $A \in [\mathfrak{H}]$ the condition that all of the vector functions $e^{iAt}x$ ($x \in \mathfrak{H}$) be almost periodic is equivalent to the simpler condition that they be bounded on the real line.

In this case the spectrum of the almost periodic operator function e^{iAt} is concentrated near $\lambda = 0$.

The following question can be asked. Suppose A is a completely continuous operator in a Banach space $\mathfrak{B}$ ($A \in [\mathfrak{B}]$) and the operator function e^{iAt} ($-\infty < t < \infty$) is bounded. What can be said about the spectral properties of A?

This question has been fully and elegantly investigated for a weakly complete (in particular, reflexive) space $\mathfrak{B}$ by Ju. I. Ljubič [1].

13. (*A. I. Perov; personal communication*). Suppose $A \in [\mathfrak{B}]$. Prove that in order for the operator function e^{iAt} to be almost periodic (in the topology of the space $[\mathfrak{B}]$) it is necessary and sufficient that the operator A be an operator of simplest type with a real spectrum.

NOTES

In the finite-dimensional case the arguments given in § 2 are due to A. M. Ljapunov [1]. In his paper presented at a meeting of the Moscow Mathematical Society in 1948 M. G. Kreĭn noted that the use by Ljapunov of different quadratic forms can be regarded as a renorming, possibly indefinite, of the space, and indicated how to carry out such a renorming in a Banach space. The simplest variant is considered in this section; the essential importance of similar renormings is brought out in proving theorems on the stability and instability of solutions of nonlinear equations (see Chapter VII).

The results of § 3 are closely connected with Theorem I.6.2 and have been taken from the "Lectures" of M. G. Kreĭn. The boundedness conditions for solutions of the second order equation were also indicated in the above mentioned paper of M. G. Kreĭn (see M. G. Kreĭn [2]).

The construction of Green functions by means of contour integrals goes back to Cauchy. The principal Green function was used in latent form in the investigations of P. Bohl [1].

CHAPTER III
THE NONSTATIONARY LINEAR EQUATION. BOHL EXPONENTS

In this chapter we consider for the first time homogeneous and inhomogeneous linear equations with a variable operator coefficient $A(t)$. We begin by constructing the evolution operator $U(t,\tau)$, which replaces the exponential operator function $e^{A(t-\tau)}$ and reduces to it when $A(t)$ is a constant operator: $A(t) \equiv A$ (§ 1). In § 2 we give some important estimates which are systematically used in this and subsequent chapters. Some elementary notions connected with the behavior of the solutions of a nonstationary equation are considered in § 3.

The central notion of this chapter, which is studied in § 4, is that of an (upper) Bohl exponent. The solutions of a nonstationary equation of negative Bohl exponent behave at infinity in the same way as the solutions of a stationary equation with an operator whose spectrum lies in the interior of the left halfplane. In § 5 we discuss the role of the negativeness of the Bohl exponent in the study of bounded solutions of the inhomogeneous equation with a bounded free term.

Finally, in § 6 we consider a class of equations with variable coefficients, viz. equations with precompactly valued operator functions, for which, as in the stationary case, one can estimate and sometimes also calculate the Bohl exponent in terms of the limiting characteristics of the operator function $A(t)$ for $t \to \infty$.

§ 1. Evolution operator and formulas for the solution of linear equations

1. *Bochner integrable functions.* In this and several subsequent chapters we will study on a finite or infinite interval $\mathscr{J}$ the linear differential equation

$$dx/dt = A(t)x + f(t) \qquad (t \in \mathscr{J}) \tag{1.1}$$

with variable operator functions $A(t)$ and vector functions $f(t)$ depending on a real parameter t. We would too greatly restrict the class of equations being studied if we were to consider only strongly continuous functions.

In order to avoid complicated theories of integration we will not strive for the greatest generality but rather confine our attention to the sufficiently wide class of equations whose coefficients are strongly measurable and locally Bochner integrable.

We recall some definitions and facts connected with these notions (the reader who is unfamiliar with them, however, can as before regard all of the functions encountered as continuous).

A vector function $x(t)$ on a finite or infinite interval $\mathscr{J} = [a,b]$ with values in a Banach space $\mathfrak{B}$ is said to be *countably valued* if it takes on $[a, b]$ no more than a countable number of nonzero values x_k $(k = 1, 2, \cdots)$, the sets $E_k = \{t \mid x(t) = x_k\}$ $(k = 1, 2, \cdots)$ being Lebesgue measurable.

A countably valued function is Bochner integrable on $[a, b]$ if and only if the numerical function $\|x(t)\|$ is Lebesgue integrable on $[a, b]$.

The Bochner integral of a countably valued function is by definition

$$\int_a^b x(t)dt = \sum_{k=1}^{\infty} x_k mE_k$$

(mE_k is the Lebesgue measure of the set E_k).

A vector function $x(t)$ is said to be *strongly measurable* on $[a, b]$ if it is the limit on this interval of an almost everywhere convergent sequence of countably valued functions $x_n(t)$. If a function $x(t)$ is strongly measurable, the function $\|x(t)\|$ is Lebesgue measurable. If $\|x(t)\|$ is also integrable, $x(t)$ is said to be *Bochner integrable* (*strongly integrable*) and by definition

$$\int_a^b x(t)dt = \lim_{n\to\infty} \int_a^b x_n(t)dt.$$

It can be shown that such an integral does not depend on the choice of the sequence $\{x_n(t)\}$ converging to $x(t)$ and is subject to the estimate

$$\left\| \int_a^b x(t)dt \right\| \leqq \int_a^b \|x(t)\| dt,$$

from which one can deduce various theorems on convergence under the integral sign.

We will not enumerate the various properties of the Bochner integral that are completely analogous to properties of the Lebesgue integral (linearity, countable additivity, etc.).

We note the relation

$$A \int_a^b x(t)dt = \int_a^b Ax(t)dt \qquad (A \in [\mathfrak{B}, \mathfrak{B}_1]).$$

By considering in place of the space $\mathfrak{B}$ a space $[\mathfrak{B}, \mathfrak{B}_1]$ of operators, we can transfer all of the above definitions and notions to operator functions. In this connection the product of two integrable functions (operator functions or an operator function and a vector function) of which one is bounded is also an integrable function.

If $x(t)$ is Bochner integrable, the relation

$$\lim_{\Delta t\to 0} \frac{1}{\Delta t} \int_t^{t+\Delta t} \|x(\tau) - x(t)\| d\tau = 0$$

is valid for almost all values of $t \in [a, b]$ and, in particular, the function

$$y(t) = \int_a^t x(\tau)d\tau \tag{1.1a}$$

is continuous and almost everywhere differentiable in $[a, b]$.

Further, wherever it is not otherwise stipulated, we will understand by the differentiability of a vector or operator function its representability in the form of an integral of type (1.1a) of a strongly integrable function. It is not difficult to verify that all of the basic properties of the derivative (in particular, the rules for differentiating a sum and a product) remain valid in this case.

2. *Existence of solutions.* We consider on a finite or infinite interval $\mathscr{J}$ the differential equation

$$dx/dt = A(t)x + f(t) \qquad (t \in \mathscr{J}) \tag{1.1}$$

in a Banach space $\mathfrak{B}$.

In this and the next two chapters, wherever it is not otherwise stipulated, we will assume that the functions $f(t)$ and $A(t)$ with values in $\mathfrak{B}$ and $[\mathfrak{B}]$ respectively are strongly measurable and Bochner integrable on the finite subintervals of $\mathscr{J}$.

By a solution of equation (1.1) we will understand a continuous function $x(t)$ that is differentiable in the sense described above and satisfies (1.1) almost everywhere.

Thus a solution of the integral equation

$$x(t) = x_0 + \int_{t_0}^{t} A(\tau)x(\tau)d\tau + \int_{t_0}^{t} f(\tau)d\tau, \tag{1.2}$$

where $x_0 = x(t_0)$, is by definition a solution of equation (1.1).

If, in particular, $f(t)$ is continuous while $A(t)$ is strongly continuous,[1] a solution of equation (1.2) is continuously differentiable at each point $t \in \mathscr{J}$ and relation (1.1) is satisfied everywhere in $\mathscr{J}$.

We consider in place of (1.2) the more general equation

$$x(t) = g(t) + \int_{t_0}^{t} A(\tau)x(\tau)d\tau \tag{1.3}$$

with a continuous vector function $g(t)$ on $\mathscr{J}$, and show that it has a continuous solution on any finite interval $[a, b] \subset \mathscr{J}$.

Let $C(\mathfrak{B}; [a, b])$ denote[2] the Banach space of continuous functions on $[a, b]$ with values in $\mathfrak{B}$ and norm

$$|||x||| = \max_{t \in [a, b]} \|x(t)\|.$$

We consider in this space the operator

$$(Sx)(t) = g(t) + \int_{t_0}^{t} A(\tau)x(\tau)d\tau,$$

[1] This means that the vector function $A(t)x$ is continuous for any $x \in \mathfrak{B}$.

[2] If it is clear from the context what the interval is, we will use the less cumbersome notation $C(\mathfrak{B})$.

defined by the right side of equation (1.3). This operator takes $C(\mathfrak{B})$ into itself since the function $(Sx)(t)$ is obviously continuous.

It is not difficult to verify by induction the formula

$$\begin{aligned}(S^n x)(t) = g(t) &+ \int_{t_0}^{t} A(t_1)g(t_1)dt_1 \\ &+ \int_{t_0}^{t}\int_{t_0}^{t_2} A(t_2)A(t_1)g(t_1)dt_1\,dt_2 + \cdots \\ &+ \int_{t_0}^{t}\int_{t_0}^{t_{n-1}}\cdots\int_{t_0}^{t_2} A(t_{n-1})A(t_{n-2})\cdots A(t_1)g(t_1)dt_1\cdots dt_{n-1} \\ &+ \int_{t_0}^{t}\int_{t_0}^{t_n}\cdots\int_{t_0}^{t_2} A(t_n)A(t_{n-1})\cdots A(t_1)x(t_1)dt_1\cdots dt_n,\end{aligned} \tag{1.4}$$

which implies the relation

$$\begin{aligned}&(S^n x_2)(t) - (S^n x_1)(t) \\ &\quad = \int_{t_0}^{t}\int_{t_0}^{t_n}\cdots\int_{t_0}^{t_2} A(t_n)A(t_{n-1})\cdots A(t_1)[x_2(t) - x_1(t)]dt_1\cdots dt_n\end{aligned}$$

and the estimate

$$\begin{aligned}&\|(S^n x_2)(t) - (S^n x_1)(t)\| \\ &\quad \leqq |||x_2 - x_1||| \int_{t_0}^{t}\int_{t_0}^{t_n}\cdots\int_{t_0}^{t_2} \|A(t_n)\|\,\|A(t_{n-1})\|\cdots\|A(t_1)\|dt_1\cdots dt_n.\end{aligned}$$

Since the integrand is invariant under any permutation of the variables $t_1,\cdots, t_n$, we get the equality

$$\begin{aligned}&\int_{t_0}^{t}\int_{t_0}^{t_n}\cdots\int_{t_0}^{t_2} \|A(t_n)\|\,\|A(t_{n-1})\|\cdots\|A(t_1)\|dt_1\cdots dt_n \\ &\quad = \frac{1}{n!}\int_{t_0}^{t}\int_{t_0}^{t}\cdots\int_{t_0}^{t} \|A(t_n)\|\,\|A(t_{n-1})\|\cdots\|A(t_1)\|dt_1\cdots dt_n \\ &\quad = \frac{1}{n!}\left[\int_{t_0}^{t}\|A(\tau)\|d\tau\right]^n.\end{aligned} \tag{1.5}$$

We finally obtain the estimate

$$|||S^n x_2 - S^n x_1||| \leqq \frac{1}{n!}\left[\int_a^b \|A(\tau)\|d\tau\right]^n |||x_2 - x_1|||,$$

showing that when n is sufficiently large, S^n is a contraction in $C(\mathfrak{B})$.

On the basis of Theorem I.9.1 we conclude that equation (1.3) has exactly one continuous solution in $[a, b]$.

We note that this solution can be obtained by means of the relation

$$x(t) = \lim_{n\to\infty} S^n x_0(t)$$

for any $x_0(t) \in C(\mathfrak{B})$ and hence, as follows from (1.4), can be represented by the series

$$\begin{aligned} x(t) &= g(t) + \int_{t_0}^{t} A(t_1)g(t_1)dt_1 \\ &\quad + \sum_{n=2}^{\infty} \int_{t_0}^{t}\int_{t_0}^{t_n}\cdots\int_{t_0}^{t_2} A(t_n)A(t_{n-1})\cdots A(t_1)g(t_1)dt_1\cdots dt_n \qquad (1.6) \\ &= g(t) + \sum_{k=1}^{\infty} g_k(t), \end{aligned}$$

where

$$g_k(t) = \int_{t_0}^{t} A(\tau)g_{k-1}(\tau)d\tau, \quad g_0(t) = g(t).$$

It follows from (1.5) that this series is majorized in norm by the series

$$|||g|||\left\{1+\sum_{n=1}^{\infty}\frac{1}{n!}\left[\int_{t_0}^{t}\|A(\tau)\|d\tau\right]^n\right\}, \tag{1.7}$$

which implies the estimate

$$|||x||| \leq |||g|||\exp\left\{\int_a^b \|A(\tau)\|d\tau\right\}. \tag{1.8}$$

We consider in particular the integral equation

$$x(t) = x_0 + \int_{t_0}^{t} A(\tau)x(\tau)d\tau,$$

which is equivalent to the differential equation

$$dx/dt = A(t)x \tag{1.9}$$

with the initial condition

$$x(t_0) = x_0. \tag{1.10}$$

The solution of the Cauchy problem (1.9)–(1.10) is obtained from (1.6) by setting $g(t) \equiv x_0$. In this connection one obtains the formula

$$\begin{aligned} x(t) &= x_0 + \int_{t_0}^{t} A(t_1)x_0dt_1 \\ &\quad + \sum_{n=2}^{\infty} \int_{t_0}^{t}\int_{t_0}^{t_n}\cdots\int_{t_0}^{t_2} A(t_n)A(t_{n-1})\cdots A(t_1)x_0dt_1\cdots dt_n. \end{aligned}$$

We introduce the linear operator

$$U(t) = I + \int_{t_0}^{t} A(t_1)dt_1 + \sum_{n=2}^{\infty} \int_{t_0}^{t} \int_{t_0}^{t_n} \cdots \int_{t_0}^{t_2} A(t_n)A(t_{n-1})\cdots A(t_1)dt_1\cdots dt_n. \tag{1.11}$$

The series (1.11) is majorized by the series in braces in (1.7) and hence converges uniformly in the operator norm (i.e. in $[\mathfrak{B}]$) on the interval $[a, b]$.

Using the operator $U(t)$, the solution of the Cauchy problem (1.9)–(1.10) can be expressed by the formula

$$x(t) = U(t)x_0. \tag{1.12}$$

We note that (1.8) implies the estimate

$$\|U(t)\| \leqq \exp\left\{\int_{t_0}^{t} \|A(\tau)\| d\tau\right\}.$$

3. *Some operator equations.* It is not difficult to verify that the operator $U(t)$, which by definition is continuous in $[\mathfrak{B}]$ and almost everywhere differentiable, satisfies the equation

$$dU/dt = A(t)U \tag{1.13}$$

and the condition $U(t_0) = I$.

We consider the more general equation

$$dZ/dt = A(t)Z - ZB(t). \tag{1.14}$$

It can be written in the form

$$dZ/dt = \mathfrak{A}(t)Z, \tag{1.15}$$

where $\mathfrak{A}(t)\colon Z \to A_lZ - B_rZ = A(t)Z - ZB(t)$ is a linear operator (transformer) acting in $[\mathfrak{B}]$ (see § I.3.1).

All of the results obtained in the preceding subsection can be applied to equation (1.15).

It consequently has a unique continuous differentiable solution $Z(t)$ which satisfies the condition $Z(t_0) = Z_0$ and which furthermore can be represented in the form

$$Z(t) = \sum_{k=0}^{\infty} Z_k(t), \tag{1.16}$$

where

$$Z_n(t) = \int_{t_0}^{t} \mathfrak{A}(\tau)Z_{n-1}(\tau)d\tau = \int_{t_0}^{t} \{A(\tau)Z_{n-1}(\tau) - Z_{n-1}(\tau)B(\tau)\}d\tau.$$

We note in this connection that the continuity of $Z(t)$ as a vector function with values in [𝔅] means its continuity in the operator norm.

These results apply in particular to equation (1.13), which results from (1.14) when $B = 0$, as well as to the equation

$$dZ/dt = -ZB(t), \tag{1.17}$$

resulting from (1.14) when $A(t) = 0$.

A special case of the latter equation is the equation

$$dV/dt = -VA(t), \tag{1.18}$$

which is called *adjoint associate equation* of (1.13).

We consider the solution of equation (1.18) satisfying the condition $V(t_0) = I$. It is not difficult to verify that $U^{-1}(t)$ exists and $V(t) = U^{-1}(t)$.

For suppose $Z_1(t) = V(t)U(t)$ and $Z_2(t) = U(t)V(t)$. Then

$$\frac{dZ_1(t)}{dt} = \frac{dV}{dt}U + V\frac{dU}{dt} = -VAU + VAU = 0,$$

which implies that $Z_1(t) \equiv Z_1(t_0) = I$.

On the other hand, $Z_2(t)$ satisfies the system

$$\begin{cases} \dfrac{dZ_2(t)}{dt} = \dfrac{dU}{dt}V + U\dfrac{dV}{dt} = AZ_2 - Z_2A, \\ Z_2(t_0) = I, \end{cases}$$

the unique solution of which is $Z_2(t) \equiv I$.

4. *Formula for the solutions of the inhomogeneous equation.* We return to a consideration of the Cauchy problem for the inhomogeneous equation

$$\begin{cases} dx/dt = A(t)x + f(t), & (1.1) \\ x(t_0) = x_0. & (1.10) \end{cases}$$

We will seek its solution, the existence and uniqueness of which have already been proved, in the form $x = U(t)y$, where $U(t)$ is the operator function (1.11).

Making this substitution, we find

$$dy/dt = U^{-1}(t)f(t) \quad \text{and} \quad y(t_0) = x_0.$$

Integrating from t_0 to t, we get

$$y = x_0 + \int_{t_0}^{t} U^{-1}(\tau)f(\tau)d\tau.$$

Thus

$$x(t) = U(t)x_0 + \int_{t_0}^{t} U(t)U^{-1}(\tau)f(\tau)d\tau. \tag{1.19}$$

5. *Evolution* (*solving*) *operator of the equation.* Let $U(t, \tau) = U(t)U^{-1}(\tau)$. The operator $U(t, \tau)$ will be called the *evolution* (*solving*) *operator* of the differential equation (1.1) (or equation (1.9)).

Since

$$\frac{dU(t, \tau)}{dt} = \frac{dU(t)}{dt} U^{-1}(\tau) = A(t)U(t)U^{-1}(\tau) = A(t)U(t, \tau),$$

this operator satisfies the system

$$dX/dt = A(t)X, \quad X(\tau) = I \tag{1.20}$$

and does not depend on the choice of the value t_0 with respect to which the operator $U(t)$ was constructed. The notation $U(t)$ will be employed in the sequel for the operator $U(t) = U(t, 0)$. This operator will usually be called the *Cauchy operator* of the equation. An arbitrary invertible solution $U(t)C$ of the operator equation (1.13) is called a *fundamental operator* of equations (1.1) and (1.9).

With the use of the evolution operator the solution of the Cauchy problem for the homogeneous equation $dx/dt = A(t)x$, $x(\tau) = x_\tau$ can be written in the form

$$x(t) = U(t, \tau)x_\tau,$$

and, for the inhomogeneous equation, in the form

$$x(t) = U(t, t_0)x_0 + \int_{t_0}^{t} U(t, \tau)f(\tau)d\tau. \tag{1.21}$$

The following fundamental properties of the evolution operator follow directly from its definition.

a) $U(t, t) = I$,

b) $U(t, s)U(s, \tau) = U(t, \tau)$,

c) $U(t, \tau) = [U(\tau, t)]^{-1}$.

In addition, it follows from the foregoing that

d) $\|U(t, \tau)\| \leqq \exp[\int_\tau^t \|A(\tau)\| d\tau]\ (\tau \leqq t)$.

We note that if the family of operators $U(t, \tau)$ $(t, \tau \in \mathcal{J})$ has properties a) and b), it also has property c). For a proof it suffices to put $\tau = t$ in b).

We leave it to the reader to verify that if the evolution operator $U(t, s)$ corresponds to the operator function $A(t)$, the evolution operator $U_\alpha(t, s) = e^{\alpha(t-s)}U(t, s)$ corresponds to the operator function $A_\alpha(t) = A(t) + \alpha I$ (where α is a complex number).

When $\mathfrak{B} = \mathfrak{H}$ we encounter an interesting special case.

LEMMA 1.1. *Let $\mathfrak{H}$ be a Hilbert space in which there has been introduced a (possibly indefinite) scalar product $(x, y)_W = (Wx, y)$.*

If the operator $A(t)$ is W-skew-Hermitian:

$$(A(t)x, y)_W = -(x, A(t)y)_W,$$

the evolution operator $U(t, \tau)$ corresponding to it is a W-unitary operator.

PROOF. Using the differential equation (1.20) for $U(t, \tau)$, we get

$$\frac{d(U(t,\tau)\phi, U(t,\tau)\psi)_W}{dt} = (AU(t,\tau)\phi, U(t,\tau)\psi)_W + (U(t,\tau)\phi, AU(t,\tau)\psi)_W \equiv 0.$$

Since, in addition,

$$(U(t,\tau)\phi, U(t,\tau)\psi)_W = (\phi, \psi)_W \qquad \text{for } t = \tau,$$

this equality is satisfied for all t, which proves the lemma.

It is easily seen that when $A(t) \equiv A$, i. e. when $A(t)$ does not depend on t, all of the formulas obtained here go over into the corresponding formulas of Chapter II. In this connection $U(t, \tau) = e^{A(t-\tau)}$.

We note that by virtue of property b) the evolution operator can be written in the form

$$U(t,\tau) = U(t, t_n)U(t_n, t_{n-1})\cdots U(t_1, \tau) = \lim_{n\to\infty} \prod_{k=0}^{n} U(t_{k+1}^{(n)}, t_k^{(n)}),$$

where $t_k^{(n)} = \tau + k(t - \tau)/(n + 1)$. From (1.13) it follows that for a continuous operator function $A(t)$

$$U(t_{k+1}^{(n)}, t_k^{(n)}) = I + A(t_k^{(n)})\cdot\frac{t-\tau}{n} + o\left(\frac{1}{n}\right) = \exp\left[A(t_k^{(n)})\cdot\frac{t-\tau}{n}\right] + o\left(\frac{1}{n}\right),$$

and, since the latter term turns out to be unessential in passing to the limit,

$$U(t,\tau) = \lim_{n\to\infty} \prod_{k=0}^{n} \exp\left[A(t_k^{(n)})\cdot\frac{t-\tau}{n}\right].$$

The limit on the right bears the name of *multiplicative integral* and is denoted by the symbol

$$U(t,\tau) = \widehat{\int_\tau^t} \exp A(\theta)d\theta.$$

This construction can be generalized so as to make it suitable for discontinuous operator functions $A(t)$.

To this end we note that the differential equation for the operator $U(t, \tau)$ corresponding to an integrable operator function $A(t)$ is equivalent to the equation

$$U(t,\tau) = I + \int_\tau^t A(\theta)U(\theta,\tau)d\theta, \tag{1.22}$$

which can also be written in the form

$$U(t,\tau) = I + \int_\tau^t [dF(\theta)]U(\theta,\tau),$$

where $F(t) = \int A(t)dt$.

The latter equation can be generalized by dropping the requirement of differentiability of the operator function $F(t)$ and rewriting it in the more general form

$$U(t, \tau) = I + \int_{\tau}^{t} [S_1(\theta)dF(\theta)S_2(\theta)]U(\theta, \tau), \tag{1.23}$$

where the $S_k(\theta)$ $(k = 1, 2)$ are continuous operator functions, $F(\theta)$ is of bounded variation and the integral is understood in the sense of Stieltjes.

The solution of equation (1.23) can be written in the form

$$U(t, \tau) = \widehat{\int_{\tau}^{t}} \exp [S_1(\theta)dF(\theta)S_2(\theta)], \tag{1.24}$$

if this expression (which is called a *multiplicative Stieltjes integral*) is understood as the limit of the corresponding product of operators; in particular, the solution of equation (1.22) can be written in the form

$$U(t, \tau) = \widehat{\int_{\tau}^{t}} \exp A(\theta)d\theta.$$

A substantial part of the results of this chapter can be generalized along these same lines (see Exercises 1–4).

6. *Formulas for the solutions of the operator equations.* Here we express the solutions of equation (1.14) and the inhomogeneous equation (1.26) corresponding to it in terms of the evolution operator $U_A(t, \tau)$ of equation (1.9) and the evolution operator $U_B(t, \tau)$ of the equation $dx/dt = B(t)x$.

We consider the transformer defined by the formula

$$\mathfrak{U}(t, \tau)X = U_A(t, \tau)XU_B(\tau, t). \tag{1.25}$$

Differentiating with respect to t, we get

$$(d/dt)\mathfrak{U}(t, \tau)X = A(t)U_A(t, \tau)XU_B(\tau, t) - U_A(t, \tau)XU_B(\tau, t)B(t) = A(t)\mathfrak{U}(t, \tau)X - \mathfrak{U}(t, \tau)XB(t).$$

Since we also have $\mathfrak{U}(\tau, \tau)X = X$, the transformer (1.25) is the evolution operator for equation (1.15). Therefore the solution of the equation

$$dX/dt = A(t)X - XB(t) + F(t) \tag{1.26}$$

can be written in the form

$$X(t) = \mathfrak{U}(t, t_0)X(t_0) + \int_{t_0}^{t} \mathfrak{U}(t, \tau)F(\tau)d\tau$$

$$= U_A(t, t_0)X(t_0)U_B(t_0, t) + \int_{t_0}^{t} U_A(t, \tau)F(\tau)U_B(\tau, t)d\tau. \tag{1.27}$$

In particular, for the adjoint equation (1.18) of (1.13)

$$X(t) = X(t_0)U_A(t_0, t) + \int_{t_0}^{t} F(\tau)U_A(\tau, t)d\tau. \qquad (1.28)$$

7. *Examples.* We consider some examples of the equations of type (1.1), (1.9). Suppose the space $\mathfrak{B}$ is finite dimensional ($\mathfrak{B} = C^n$). We choose an orthonormal basis $e_1, \cdots, e_n$ in it and let $x_k = (x, e_k)$ $(k = 1, \cdots, n)$ be the coordinates of a vector $x \in C^n$ and $a_{jk} = (Ae_k, e_j)$ be the matrix elements of the operator A in this basis.

In this case equation (1.1) is equivalent to the system of ordinary differential equations

$$\frac{dx_j}{dt} = \sum_{k=1}^{n} a_{jk}(t)x_k \qquad (j = 1, 2, \cdots, n). \qquad (1.29)$$

The evolution operator $U(t, \tau)$ is given by a matrix $\|u_{jk}(t, \tau)\|$ satisfying the system

$$\begin{cases} \dfrac{du_{jk}}{dt} = \sum_{s=1}^{n} a_{js}(t)u_{sk}, \\ u_{jk}(\tau, \tau) = \delta_{jk}. \end{cases}$$

It is easily seen that the columns of this matrix form a fundamental system of solutions of system (1.29).

The general solution is a linear combination of these columns:

$$x_j(t) = \sum_{k=1}^{n} u_{jk}(t, \tau)x_k(\tau).$$

Finally, the solution of the inhomogeneous system

$$\frac{dx_j}{dt} = \sum_{k=1}^{n} a_{jk}(t)x_k + f_j(t) \qquad (1.30)$$

is given by the formula

$$x_j(t) = \sum_{k=1}^{n} u_{jk}(t, t_0)x_k(t_0) + \sum_{k=1}^{n} \int_{t_0}^{t} u_{jk}(t, \tau)f_k(\tau)d\tau,$$

which follows directly from (1.21).

A natural generalization of system (1.30) is a countable system of differential equations

$$\frac{dx_j}{dt} = \sum_{k=1}^{\infty} a_{jk}(t)x_k + f_j(t) \qquad (j = 1, 2, \cdots).$$

This system can be considered as a vector equation in various sequence spaces, thereby obtaining various conditions for the existence of solutions.

The natural continuous analog of a countable system of differential equations is an integrodifferential equation of, for example, the form

$$\frac{dx(t,s)}{dt} = \int_a^b \mathcal{K}(t,s,\sigma)x(t,\sigma)\,d\sigma + f(t,s).$$

By considering this equation in various function spaces, we can obtain various conditions for the existence of its solutions.

§ 2. Integral inequalities. Comparison of evolution operators

1. *Integral inequalities.* The following lemma plays a decisive role in obtaining various estimates for the solutions of linear (and, as we will see in Chapter VII, nonlinear) differential equations.

LEMMA 2.1. *Suppose $\mathcal{K}(t,\tau)$ is a nonnegative kernel on a finite or infinite interval $\mathcal{J}$ such that the integral operator*

$$(Kx)(t) = \int_{\mathcal{J}} \mathcal{K}(t,\tau)x(\tau)d\tau$$

leaves the space $\mathbf{C}(\mathcal{J})$ of bounded continuous functions on $\mathcal{J}$ invariant and has a spectral radius less than one in this space.

Then a continuous function $\phi(t)$ satisfying the inequality

$$\phi(t) \leqq f(t) + \int_{\mathcal{J}} \mathcal{K}(t,\tau)\phi(\tau)d\tau \qquad (t \in \mathcal{J}), \tag{2.1}$$

where $f(t)$ is a continuous function on $\mathcal{J}$, satisfies the inequality $\phi(t) \leqq \psi(t)$ $(t \in \mathcal{J})$, where $\psi(t)$ is the solution of the integral equation

$$\psi(t) = f(t) + \int_{\mathcal{J}} \mathcal{K}(t,s)\psi(s)ds. \tag{2.2}$$

The assertion is also true when the signs $\leqq$ are replaced by $\geqq$.

We note that we could formulate an analogous assertion for vector functions.

PROOF. The lemma is obtained by applying Theorem I.9.3 to the space $\mathbf{C}(\mathbf{R}^1, \mathcal{J})$, the cone $\mathfrak{K}$ of nonnegative functions and the operator K.

We will give below (see Remarks 2.1 and 2.2 and the corollaries following each of them) a number of concrete applications of Lemma 2.1. The concrete estimates obtained in this connection will be systematically used in this and the following chapters of the book.

REMARK 2.1. Suppose $\mathcal{J} = [a, b]$ is an arbitrary finite interval while the kernel $\mathcal{K}(t,\tau)$ is nonnegative on $[a, b]$, continuous for $\tau \leqq t$ and vanishes for $\tau > t$. The integral operator K:

$$(Kx)(t) = \int_{\mathcal{J}} \mathcal{K}(t,\tau)x(\tau)d\tau = \int_a^t \mathcal{K}(t,\tau)x(\tau)d\tau$$

is a Volterra operator and, as is well known, has a spectral radius equal to zero.

Therefore Lemma 2.1 is applicable and if a function $\phi(t)$ satisfies the inequality

$$\phi(t) \leqq f(t) + \int_a^t \mathscr{K}(t, \tau)\phi(\tau)d\tau \qquad (t \geqq a), \tag{2.1'}$$

then $\phi(t) \leqq \psi(t)$, where

$$\psi(t) = f(t) + \int_a^t \mathscr{K}(t, s)\psi(s)ds. \tag{2.2'}$$

The converse estimates are obtained in exactly the same way. We note that the interval $[a, b]$ can be chosen arbitrarily large and hence the estimates are valid for all $t \geqq a$ provided the kernel $\mathscr{K}(t, \tau)$ and the function $f(t)$ are defined for $a \leqq \tau \leqq t < \infty$.

In the same way one can consider the inequality

$$\phi(t) \leqq f(t) + \int_t^b \mathscr{K}(t, \tau)\phi(\tau)d\tau. \tag{2.1''}$$

for $t \leqq b$ and obtain the estimate $\phi(t) \leqq \psi(t)$, where

$$\psi(t) = f(t) + \int_t^b \mathscr{K}(t, s)\psi(s)ds. \tag{2.2''}$$

COROLLARY 2.1 *Suppose*

$$\phi(t) \leqq c + \int_{t_0}^t h(\tau)\phi(\tau)d\tau \qquad (t \geqq t_0), \tag{2.3}$$

where $h(t)$ is a continuous nonnegative function. Then

$$\phi(t) \leqq ce^{\int_{t_0}^t h(\tau)d\tau}. \tag{2.4}$$

For if we differentiate the equality

$$\psi(t) = c + \int_{t_0}^t h(\tau)\psi(\tau)d\tau \qquad (t \geqq t_0)$$

with respect to t, we obtain the differential equation $\psi'(t) = h(t)\psi(t)$ with the condition $\psi(t_0) = c$, which implies

$$\psi(t) = ce^{\int_{t_0}^t h(\tau)d\tau}$$

COROLLARY 2.2. *Suppose*

$$\phi(t) \leqq \alpha e^{-\nu(t-t_0)} + \beta \int_{t_0}^t e^{-\nu(t-\tau)}p(\tau)\phi(\tau)d\tau \qquad (t \geqq t_0), \tag{2.5}$$

where $\beta p(t)$ is a continuous nonnegative function. Then

$$\phi(t) \leqq \alpha e^{-\nu(t-t_0)+\beta\int_{t_0}^{t} p(\tau)d\tau} \tag{2.6}$$

For if we set $\phi(t) = \phi_1(t)e^{-\nu t}$, we obtain the following inequality for ϕ_1:

$$\phi_1(t) \leqq \alpha e^{\nu t_0} + \beta \int_{t_0}^{t} p(\tau)\phi_1(\tau)d\tau.$$

Therefore by virtue of (2.4) when $c = \alpha e^{\nu t_0}$

$$\phi_1(t) \leqq \alpha e^{\nu t_0} e^{\beta\int_{t_0}^{t} p(\tau)d\tau} \qquad (t \geqq t_0),$$

from which the required estimate follows.

If we reverse the inequality signs in (2.3) and (2.5), we obtain estimates (2.4) and (2.6) with their inequality signs reversed.

REMARK 2.2. Suppose $\mathscr{K}(t, \tau)$ is a continuous nonnegative kernel on $\mathscr{I}$ and

$$\sup_t \int_{\mathscr{I}} \mathscr{K}(t, \tau)\, d\tau = q < 1. \tag{2.7}$$

Then in $C(\boldsymbol{R}^1, \mathscr{I})$ the spectral radius $r(K) \leqq \|K\| = q < 1$, and hence Lemma 2.1 is applicable.

COROLLARY 2.3. *Suppose for* $0 < \beta < \nu/2$

$$\phi(t) \leqq \alpha e^{-\nu(t-t_0)} + \beta \int_{t_0}^{\infty} e^{-\nu|t-\tau|}\phi(\tau)d\tau. \tag{2.8}$$

Then

$$\phi(t) \leqq \frac{2\alpha\nu}{\nu + \sqrt{\nu^2 - 2\beta\nu}} e^{-\sqrt{\nu^2-2\beta\nu}(t-t_0)}. \tag{2.9}$$

PROOF. Here $\mathscr{K}(t, \tau) = \beta e^{-\nu|t-\tau|}$ and

$$\int_{t_0}^{\infty} \mathscr{K}(t, \tau)\, d\tau = \beta\left\{\int_{t_0}^{t} e^{-\nu(t-\tau)}d\tau + \int_{t}^{\infty} e^{-\nu(\tau-t)}d\tau\right\} \leqq \frac{2\beta}{\nu}.$$

Therefore when $\beta < \nu/2$ condition (2.7) is satisfied and Lemma 2.1 is applicable. We consider the equation

$$\psi(t) = \alpha e^{-\nu(t-t_0)} + \beta \int_{t_0}^{\infty} e^{-\nu|t-\tau|}\psi(\tau)d\tau. \tag{2.10}$$

Differentiating it twice with respect to t, we obtain the relations

$$\psi'(t) = -\alpha\nu e^{-\nu(t-t_0)} - \beta\nu \int_{t_0}^{t} e^{-\nu(t-\tau)}\psi(\tau)d\tau + \beta\nu \int_{t}^{\infty} e^{-\nu(\tau-t)}\psi(\tau)d\tau,$$

$$\psi''(t) = \alpha\nu^2 e^{-\nu(t-t_0)} + \beta\nu^2 \int_{t_0}^{\infty} e^{-\nu|t-\tau|}\psi(\tau)d\tau - 2\beta\nu\psi(t),$$

which together with (2.10) give the differential equation

$$\psi''(t) - (\nu^2 - 2\beta\nu)\psi(t) = 0.$$

Since we are interested in the solution of equation (2.10) that is bounded on $[t_0, \infty)$, we have

$$\psi(t) = ce^{-\sqrt{\nu^2-2\beta\nu}(t-t_0)} \left(\nu^2 - 2\beta\nu \geqq 0 \quad \text{for} \quad \beta \leqq \frac{\nu}{2}\right).$$

The constant c can be found by substituting this expression in (2.10). Carrying out the calculations, we obtain the equality $c = 2\alpha\nu/(\nu + \sqrt{\nu^2 - 2\beta\nu})$ from which (2.9) follows.

Let us consider the more complicated inequality

$$\phi(t) \leqq \alpha e^{-\nu(t-t_0)} + \beta \int_{t_0}^{\infty} e^{-\nu|t-\tau|}p(\tau)\phi(\tau)d\tau \qquad (\nu > 0, t \geqq t_0). \tag{2.11}$$

It immediately reduces to the previous inequality if $p(t)$ is a bounded function on $[t_0, \infty)$. We will need an estimate for $\phi(t)$, however, under the more general assumption that

$$\sup_t \frac{1}{\tau_0} \int_t^{t+\tau_0} p(\tau)d\tau = M_{\tau_0} < \infty \tag{2.12}$$

for some fixed τ_0.

In the present case equation (2.2) has the form

$$\psi(t) = \alpha e^{-\nu(t-t_0)} + \beta \int_{t_0}^{\infty} e^{-\nu|t-\tau|}p(\tau)\psi(\tau)d\tau. \tag{2.13}$$

It can be shown in the same way as above that the solution of this equation reduces to finding a bounded solution of the differential equation $\psi''(t) - (\nu^2 - 2\nu\beta p(t))\psi(t) = 0$.

In order to avoid the task of estimating the solutions of this equation, we proceed differently. We put $\psi(t) = u(t)e^{-\mu(t-t_0)}$, where the number $\mu > 0$ will be chosen later. The function $u(t)$ must satisfy the equation

$$u(t) = \alpha e^{-(\nu-\mu)(t-t_0)} + \beta \int_{t_0}^{\infty} e^{-\nu|t-\tau|+\mu(t-\tau)}p(\tau)u(\tau)d\tau, \tag{2.14}$$

which we consider in the space $\boldsymbol{C}(\boldsymbol{R}^1, [t_0, \infty))$ of continuous bounded functions on $[t_0, \infty)$.

We consider in this space the operator

$$(Au)(t) = \beta \int_{t_0}^{\infty} e^{-\nu|t-\tau|+\mu(t-\tau)}p(\tau)u(\tau)d\tau.$$

It is easily seen that

$$\|A\| = \beta \sup \int_{t_0}^{\infty} e^{-\nu|t-\tau|+\mu(t-\tau)}p(\tau)d\tau.$$

We estimate the integral in this inequality. Let $n = [t/\tau_0]$. Then

$$\int_{t_0}^{\infty} e^{-\nu|t-\tau|+\mu(t-\tau)}p(\tau)d\tau$$

$$= \sum_{k=1}^{\infty} \int_{t_0+(k-1)\tau_0}^{t_0+k\tau_0} e^{-\nu|t-\tau|+\mu(t-\tau)}p(\tau)d\tau$$

$$\leqq \sum_{k=1}^{\infty} \max_{[t_0+(k-1)\tau_0, t_0+k\tau_0]} e^{-\nu|t-\tau|+\mu(t-\tau)} \int_{t_0+(k-1)\tau_0}^{t_0+k\tau_0} p(\tau)d\tau$$

$$\leqq \tau_0 M_{\tau_0}\left\{\sum_{k=1}^{n} e^{-(\nu-\mu)(n-k)\tau_0} + 1 + \sum_{k=n+2}^{\infty} e^{-(\nu+\mu)(k-n-2)\tau_0}\right\}$$

$$\leqq M_{\tau_0}C(\mu, \nu, \tau_0),$$

where

$$C(\mu, \nu, \tau_0) = \tau_0 + \frac{\tau_0}{1-e^{-(\nu-\mu)\tau_0}} + \frac{\tau_0}{1-e^{-(\nu+\mu)\tau_0}} < \infty$$

for $\mu < \nu$. Thus

$$\|A\| \leqq \beta M_{\tau_0}C(\mu, \nu, \tau_0). \tag{2.15}$$

We require that

$$\beta M_{\tau_0}C(\mu, \nu, \tau_0) \leqq q < 1. \tag{2.16}$$

Under this requirement the equation

$$u - Au = \alpha e^{-(\nu-\mu)(t-t_0)} \tag{2.14}$$

will be solvable in $C(\boldsymbol{R}^1, [t_0, \infty))$, i. e. the function $u(t)$ is bounded:

$$\sup_t u(t) = c < \infty. \tag{2.17}$$

This gives the following estimate for $\psi(t)$:

$$\psi(t) \leqq ce^{\mu-(t-t_0)}. \tag{2.18}$$

We will assume that τ_0 is sufficiently small. Then

$$C(\mu, \nu, \tau_0) = \frac{1}{\nu-\mu} + \frac{1}{\nu+\mu} + O(\tau_0) = \frac{2\nu}{\nu_2-\mu_2} + O(\tau_0),$$

and condition (2.16) reduces to the relation

$$\beta M_{\tau_0}\left[\frac{2\nu}{\nu_2-\mu_2} + O(\tau_0)\right] < 1$$

or

$$\mu < \sqrt{\nu_2 - \frac{2\nu\beta M_{\tau_0}}{1 - O(\tau_0 M_{\tau_0})}}. \tag{2.19}$$

Thus an estimate of form (2.18) holds under condition (2.19), and it only remains to find the constant c. This can be done by considering (2.14), which implies

$$c \leqq \alpha + \beta c \sup_t \int_{t_0}^{\infty} e^{-\nu|t-\tau|+\mu(t-\tau)} p(\tau)d\tau \leqq \alpha + \beta c M_{\tau_0} C(\mu, \nu, \tau_0)$$

and, finally,

$$c \leqq \frac{\alpha}{1 - \beta M_{\tau_0} C(\mu, \nu, \tau_0)} \leqq \frac{\alpha}{1-q}.$$

The above arguments imply the following result.

LEMMA 2.2. *Suppose for* $\nu > 0$ *and* $\beta > 0$

$$\phi(t) \leqq \alpha e^{-\nu(t-t_0)} + \beta \int_{t_0}^{\infty} e^{-\nu|t-\tau|} p(\tau)\phi(\tau)d\tau, \tag{2.11}$$

where $p(t)$ *is a continuous nonnegative function.*

For any $\mu < \nu$, $\tau_0 > 0$ *and* $q < 1$ *there exists a* $\delta > 0$ *such that if the condition*

$$M_{\tau_0} = \sup_t \frac{1}{\tau_0} \int_t^{t+\tau_0} p(\tau)d\tau < \delta$$

is satisfied, then $\phi(t) \leqq (\alpha/(1-q))e^{-\mu(t-t_0)}$ $(t \geqq t_0)$.

REMARK 2.3. If τ_0 is sufficiently small, relation (2.19) indicates the values of μ for which the desired estimate is valid for a given M_{τ_0}.

If the condition $\sup p(t) \leqq M$ is satisfied, inequality (2.8) with β replaced by βM will hold and Corollary 2.3 can be applied.

2. *Comparison of evolution operators.*

LEMMA 2.3. *Let* $U_k(t, s)$ $(k = 1, 2; a \leqq s, t \leqq b)$ *be the evolution operators of the equations*

$$dx/dt = A_k(t)x \qquad (k = 1,2).$$

If the estimate

$$\|U_1(t, s)\| \leqq Ne^{-\nu_1(t-s)} \qquad (a \leqq s, t \leqq b) \tag{2.20}$$

is satisfied for some $N > 0$ *and real* ν_1, *then for* $s \leqq t$

$$\|U_2(t, s) - U_1(t, s)\| \leqq Ne^{-\nu_1(t-s)}\left(e^{N\int_s^t \|A_2(\tau)-A_1(\tau)\|d\tau} - 1\right), \tag{2.21}$$

and consequently

$$\|U_2(t, s)\| \leqq Ne^{-\nu_1(t-s)} e^{N\int_s^t \|A_2(\tau)-A_1(\tau)\|d\tau}. \tag{2.22}$$

Analogous estimates hold for $t \leqq s$.

PROOF. It clearly suffices to prove the assertion of the lemma for $s = 0$. The operator $U_2(t) = U_2(t, 0)$ is a solution of the system

$$\begin{cases} dU_2/dt - A_1U_2 = (A_2 - A_1)U_2, \\ U_2(0) = I. \end{cases}$$

Using formula (1.21) for the solution of the operator equation

$$dX/dt - A_1X = F(t), \qquad X(0) = I,$$

where $F(t) = [A_2(t) - A_1(t)]U_2(t)$, we find that

$$U_2(t) = U_1(t, 0) + \int_0^t U_1(t, \tau)[A_2(\tau) - A_1(\tau)]U_2(\tau)d\tau. \tag{2.23}$$

Setting $\phi(t) = \|U_2(t)\|$ and using estimate (2.20), we get

$$\phi(t) \leqq Ne^{-\nu_1 t} + N\int_0^t e^{-\nu_1(t-\tau)}p(\tau)\phi(\tau)d\tau,$$

where $p(t) = \|A_2(t) - A_1(t)\|$.

Thus $\phi(t)$ is subject to estimate (2.6) when $\alpha = \beta = N$, which implies estimate (2.22).

To obtain estimate (2.21) we note that (2.23) and estimate (2.22) imply

$$\begin{aligned} \|U_2(t) - U_1(t)\| &\leqq \int_0^t \|U_1(t, \tau)\| \|A_2(\tau) - A_1(\tau)\| \|U_2(\tau)\| d\tau \\ &\leqq N\int_0^t e^{-\nu_1(t-\tau)}p(\tau)Ne^{-\nu_1\tau}e^{N\int_0^\tau p(s)ds}\, d\tau \\ &= Ne^{-\nu_1 t}\int_0^t Np(\tau)e^{N\int_0^\tau p(s)ds}\, d\tau = Ne^{-\nu_1 t}(e^{N\int_0^t p(s)ds} - 1). \end{aligned}$$

Setting $A_1(t) = 0$ and $A_2(t) = A(t)$ in Lemma 2.3, we obtain some important estimates.

LEMMA 2.4. *The following estimate holds for the evolution operator* $U(t, s)$ *of equation* (1.1) *when* $s \leqq t$*:*

$$\|U(t, s) - I\| \leqq e^{\int_s^t |A(\tau)| d\tau} - 1; \tag{2.24}$$

in addition,

$$e^{-\int_s^t |A(\tau)| d\tau} \leqq \|U^{\pm 1}(t, s)\| \leqq e^{\int_s^t |A(\tau)| d\tau}. \tag{2.25}$$

§ 3. Stability and bistability. Comparable equations

1. *Right stability.* We introduce some notions connected with the behavior of the solutions of the equation

$$dx/dt = A(t)x \tag{3.1}$$

on the halfline $[0, \infty)$.

Equation (3.1) is said to be *stable* (more precisely, *right stable*) if every solution of it is bounded on $[0, \infty)$.

LEMMA 3.1. *A necessary and sufficient condition for the stability of equation* (3.1) *is the uniform boundedness of its Cauchy operator*:

$$\sup_{t \geqq 0} \|U(t)\| < \infty.$$

PROOF. The sufficiency follows at once from the formula $x(t) = U(t)x_0$; the necessity, from the uniform boundedness principle.

REMARK 3.1. Rephrasing the condition of the lemma, we say that a necessary and sufficient condition for right stability is the existence of a constant $q > 0$ such that any solution $x(t)$ of equation (3.1) satisfies the estimate

$$\|x(t)\| \leqq q\|x(0)\|. \tag{3.2}$$

The least value q_0 of this constant is clearly given by the equality

$$q_0 = \sup_{t \geqq 0} \|U(t)\|.$$

REMARK 3.2. The right stability of equation (3.1) is equivalent to the right stability of the operator equation $dX/dt = A(t)X$.

It is easily seen that the choice of the interval $[0, \infty)$ is not essential for the definition of stability. It can be replaced by any fixed interval $[t_0, \infty)$ on which the operator $A(t)$ is defined and satisfies the conditions adopted by us. The role of the operator $U(t)$ in the above formulas will be played by the operator $U(t, t_0)$ and estimate (3.2) is replaced by the estimate

$$\|x(t)\| \leqq q_{t_0}\|x(t_0)\| \qquad (t \geqq t_0), \tag{3.3}$$

where $q_{t_0} = \sup_{t \geqq t_0}\|U(t, t_0)\|$. The constant q_{t_0} here generally depends on the choice of t_0.

Equation (3.1) is said to be *uniformly* (*right*) *stable* if there exists a constant $N > 0$ such that any solution $x(t)$ of this equation satisfies the following estimate for all $t \geqq s \geqq 0$:

$$\|x(t)\| \leqq N\|x(s)\|. \tag{3.4}$$

In exactly the same way as above it can be shown that uniform right stability is equivalent to the condition

$$N = \sup_{t \geqq s \geqq 0} \|U(t, s)\| < \infty, \tag{3.5}$$

the indicated value of N being the best possible for estimate (3.4).

REMARK 3.3. Suppose $A(t) = A$ is a constant operator. In this case $U(t, s) =$

$e^{A(t-s)}$ and, as is easily seen, the stability condition

$$\sup_{t \geqq 0} \|e^{At}\| < \infty \tag{3.6}$$

coincides with the uniform stability condition.

As follows from the results of Chapter I, for the fulfilment of (3.6) it is necessary that the spectrum $\sigma(A)$ lie in the closed left halfplane and sufficient that it lie in the interior of this halfplane.

2. *Left stability.* We say that equation (3.1) is *left stable* if there exists a constant $q' > 0$ such that any solution $x(t)$ of this equation satisfies the following estimate for all $t \geqq 0$:

$$\|x(0)\| \leqq q' \|x(t)\| \qquad (t \geqq 0).$$

Left stability is clearly equivalent to the uniform boundedness of the operator $U^{-1}(t)$:

$$q' = \sup_{t \geqq 0} \|U^{-1}(t)\| < \infty,$$

the indicated value of q' being the best possible.

We note that since the solutions of the adjoint equation $dX/dt = -A(t)X$ are expressed by the formula $X(t) = X(0)U^{-1}(t)$, the left stability of equation (3.1) is equivalent to the right stability of the adjoint equation.

In exactly the same way as above we can speak of *uniform left stability* if we understand by this the validity of the estimate $\|x(s)\| \leqq N\|x(t)\|$ for all $0 \leqq s \leqq t$ and N not depending on s and t. Clearly, uniform left stability is equivalent to the condition

$$\sup_{t \geqq s \geqq 0} \|U(s, t)\| < \infty. \tag{3.7}$$

In regard to the left stability of an equation with a constant operator A it is possible to make the same statements concerning the spectrum $\sigma(A)$ as in Remark 3.3, but with the left halfplane replaced by the right halfplane.

3. *Bistability.* We say that equation (3.1) is *bistable* on the halfline $[0, \infty)$ if it is both right and left stable. Thus equation (3.1) is bistable if and only if

$$\sup_{0 \leqq t < \infty} \{\|U^{\pm 1}(t)\|\} < \infty,$$

or, equivalently, if and only if

$$\sup_{0 \leqq s, t < \infty} \{\|U(t, s)\|\} < \infty. \tag{3.8}$$

A comparison of this condition with (3.5) and (3.7) shows that a bistable equation is uniformly right and left stable.

In other words, it can be said that an equation is stable if there exists a constant $q > 0$ such that any solution $x(t)$ satisfies the estimate

$$\|x(t)\| \leqq q\|x(s)\| \qquad (0 \leqq s, t < \infty).$$

From the above remarks it follows that in order for a stationary equation to be bistable it is necessary (but not sufficient) that the spectrum $\sigma(A)$ lie on the imaginary axis. It is not difficult to see that the bistability of a stationary equation is equivalent to the boundedness of its solutions on the real line. As we know, when $\mathfrak{B}$ is a Hilbert space $\mathfrak{H}$, the equation is bistable precisely when A is similar to a skew-Hermitian operator.

4. *Integrally comparable equations.* Suppose given on the halfline $[0, \infty)$ the two equations

$$dx/dt = A_k(t)x \qquad (k = 1,2). \tag{3.9}$$

We say that these equations are *integrally comparable* if

$$|||A_2 - A_1||| = \int_0^\infty \|A_2(t) - A_1(t)\| dt < \infty.$$

LEMMA 3.2. *If one of the two integrally comparable equations* (3.9) *is uniformly right or left stable or bistable, so is the other equation.*

PROOF. The first assertion of the lemma follows at once from estimates (3.5) and (2.22) when $\nu_1 = 0$. The second assertion is proved analogously. The third assertion is a consequence of the first two.

We consider in particular the equation

$$dx/dt = A(t)x \tag{3.10}$$

with an integrable coefficient:

$$|||A||| = \int_0^\infty \|A(\tau)\| d\tau < \infty. \tag{3.11}$$

This equation is integrally comparable with the bistable equation $dx/dt = 0$ and hence is bistable itself.

Moreover, it can be asserted in this case that the limit $x(\infty) = \lim_{t\to\infty} x(t)$ exists for every solution $x(t)$ of equation (3.10), and for every $y \in \mathfrak{B}$ there exists one and only one trajectory $x(t)$ such that $x(\infty) = y$.

This fact follows from

THEOREM 3.1. *Suppose relation* (3.11) *is satisfied. Then the limit* $U(\infty) = \lim_{t\to\infty} U(t)$ *exists and is an invertible operator.*

PROOF. By virtue of Lemma 2.4 we have

$$\|U^{\pm 1}(t)\| \leq M = \exp|||A|||$$

and, further,

$$\|U(t) - U(s)\| = \|(U(t, s) - I)U(s)\| \leq M\left(\exp \int_s^t \|A(\tau)\| d\tau - 1\right) < \varepsilon$$

for sufficiently large t and s. The latter estimate implies the existence of the limit

$$\lim_{t\to\infty} U(t) = U(\infty).$$

The existence of the limit

$$V(\infty) = \lim_{t\to\infty} U^{-1}(t)$$

is proved analogously.

Since, clearly, $U(\infty)V(\infty) = V(\infty)U(\infty) = I$, the theorem is proved.

REMARK. The theorem can also be proved by making the change of variable $s = \int_0^t \|A(\tau)\|d\tau$, which transforms equation (3.10) into the equation

$$\frac{dx}{ds} = \frac{A(t(s))}{\|A(t(s))\|}x \qquad (0 \leqq s \leqq |||A|||)$$

with a bounded operator function on a finite interval.

5. *Asymptotically equivalent equations.* If equation (3.1) is left stable, then, as follows from the definition, a solution of this equation that tends to zero at infinity is identically equal to zero. For such equations the following definition is meaningful.

We say that equations (3.9) are *asymptotically equivalent* if it is possible to establish a one-to-one attraction correspondence $x_1(t) \leftrightarrow x_2(t)$ in such a way that

$$\lim_{t\to\infty} [x_2(t) - x_1(t)] = 0.$$

THEOREM 3.2. *Suppose that equations* (3.9) *are integrally comparable and one of them* (*and hence also the other*) *is bistable. Then these equations are asymptotically equivalent, the attraction correspondence* $x_1(t) \leftrightarrow x_2(t)$ *between their solutions being uniquely established and the estimate*

$$\|x_2(t) - x_1(t)\| = O\left(\int_t^\infty \|A_2(s) - A_1(s)\|ds\right) \tag{3.12}$$

being satisfied by corresponding solutions.

In addition, the initial values of a pair of corresponding solutions are connected by a continuous invertible operator from $[\mathfrak{B}]$.

PROOF. The fact that if one of the integrally comparable equations (3.9) is bistable, so is the other, is indicated in Lemma 3.2. We note that the solutions of these equations are bounded.

Let $x_1(t)$ be a solution of the first equation. We consider the function

$$x_2(t) = x_1(t) + \int_t^\infty U_2(t, s)[A_1(s) - A_2(s)]x_1(s)ds. \tag{3.13}$$

This expression is meaningful since

$$\left\| \int_t^\infty U_2(t, s)\,[A_1(s) - A_2(s)]x_1(s)ds \right\| \leqq |||x_1(s)|||\; |||A_1 - A_2|||\sup_{t,s} \| U_2(t, s)\|.$$

We verify that $x_2(t)$ is a solution of the second equation of (3.9). In fact,

$$\begin{aligned} x_2'(t) &= x_1'(t) - [A_1(t) - A_2(t)]x_1(t) \\ &\quad + \int_t^\infty A_2(t)U_2(t, s)[A_1(s) - A_2(s)]x_1(s)ds \\ &= A_2(t)\left\{x_1(t) + \int_t^\infty U_2(t, s)[A_1(s) - A_2(s)]x_1(s)ds\right\} \\ &= A_2(t)x_2(t). \end{aligned}$$

From (3.13) it clearly follows that $\lim_{t\to\infty}[x_2(t) - x_1(t)] = 0$.

Suppose now $\tilde{x}_2(t)$ is another solution of the second equation of (3.9) such that $\lim_{t\to\infty}[\tilde{x}_2(t) - x_1(t)] = 0$. Then

$$\lim_{t\to\infty} [\tilde{x}_2(t) - x_2(t)] = 0,$$

and since, by virtue of the bistability of the equation for any τ,

$$\|\tilde{x}_2(\tau) - x_2(\tau)\| \leqq \|\tilde{x}_2(t) - x_2(t)\| \sup_{0\leqq t,\, \tau\leqq\infty} \|U(\tau, t)\|,$$

we get $\tilde{x}_2(t) \equiv x_2(t)$.

Equations (3.9) are equivalent, and therefore equation (3.13) admits the inversion

$$x_1(t) = x_2(t) + \int_t^\infty U_1(t, s)[A_2(s) - A_1(s)]x_2(s)ds.$$

It remains to note that

$$x_2(0) = \left\{I + \int_0^\infty U_2(0, s)[A_1(s) - A_2(s)]\, U_1(s, 0)ds\right\} x_1(0). \tag{3.14}$$

An analogous expression exists for $x_1(0)$ in terms of $x_2(0)$.

Estimate (3.12) immediately follows from (3.13).

The theorem is proved.

§ 4. Ljapunov and Bohl exponents of the homogeneous equation on $[0, \infty)$

1. *Ljapunov exponents.* We consider the homogeneous differential equation

$$dx/dt = A(t)x \qquad (0 \leqq t < \infty). \tag{4.1}$$

If the operator function $A(t)$ is bounded: $\|A(t)\| \leqq M$ $(0 \leqq t < \infty)$, or, more generally, *integrally bounded*:

$$\sup_t \int_t^{t+1} \|A(\tau)\|d\tau \leqq M \qquad (0 \leqq t < \infty), \tag{4.2}$$

then, as follows from (2.25), its solutions are subject to the estimate $\|x(t)\| \leqq e^{M(t+1)}\|x(0)\|$, and thus the *(upper) Ljapunov exponent*

$$\kappa = \overline{\lim_{t\to\infty}} \frac{\ln\|x(t)\|}{t}$$

of each solution $x(t)$ of equation (4.1) satisfies the condition $\kappa \leqq M$.

It is not at all necessary that condition (4.2) be fulfilled in order for all of the solutions of equation (4.1) to have finite Ljapunov exponents (or even to be bounded) (see Exercise 19).

Without assuming the fulfilment of condition (4.2), we consider the set Σ of not necessarily finite Ljapunov exponents of all of the possible solutions of equation (4.1). This set is called the *(upper) Ljapunov spectrum* of equation (4.1).

We leave it to the reader to prove that the Ljapunov spectrum Σ of the homogeneous equation $dx/dt = Ax (A = \text{const})$ in a finite-dimensional space $\mathfrak{B}$ consists of the real parts of the eigenvalues of A.

By the *(upper) Ljapunov exponent of equation* (4.1) is meant the quantity

$$\kappa_L = \sup_{\kappa\in\Sigma} \kappa.$$

THEOREM 4.1. *The Ljapunov exponent of equation* (4.1) *coincides with the (upper) Ljapunov exponent of the operator function* $U(t) = U(t, 0)$:

$$\kappa_L = \overline{\lim_{t\to\infty}} \frac{\ln\|U(t)\|}{t}. \tag{4.3}$$

PROOF. Let μ denote the right side of formula (4.3). From the equality $x(t) = U(t)x_0$ it follows that

$$\frac{\ln\|x(t)\|}{t} \leqq \frac{\ln\|U(t)\|}{t} + \frac{\ln\|x_0\|}{t},$$

which implies

$$\overline{\lim_{t\to\infty}} \frac{\ln\|x(t)\|}{t} \leqq \overline{\lim_{t\to\infty}} \frac{\ln\|U(t)\|}{t},$$

i. e. $\kappa \leqq \mu$ for $\kappa \in \Sigma$, and hence $\kappa_L \leqq \mu$.

It remains to prove the converse inequality for the case $\kappa_L < \infty$. For any solution $x(t) = U(t)x_0$ of equation (4.1)

$$\|U(t)x_0\| \leqq N_{\varepsilon, x_0} e^{(\kappa_L+\varepsilon)t},$$

where ε is an arbitrary positive number and N_{ε,x_0} is a constant generally depending on ε and x_0.

The latter inequality can be written in the form

$$\|U(t)x_0 e^{-(\kappa_L+\varepsilon)t}\| \leqq N_{\varepsilon,x_0}.$$

The family of operators $U(t)e^{-(\kappa_L+\varepsilon)t}$ $(0 \leqq t < \infty)$ is bounded at each element $x_0 \in \mathfrak{B}$. It follows from the uniform boundedness principle that

$$\|U(t)e^{-(\kappa_L+\varepsilon)t}\| \leqq N_\varepsilon \qquad (t \geqq 0).$$

This means that $\|U(t)\| \leqq N_\varepsilon e^{(\kappa_L+\varepsilon)t}$, which implies $\mu \leqq \kappa_L$.

The theorem is proved.

REMARK 4.1. One could analogously introduce the *lower Ljapunov exponent* $\kappa' = \underline{\lim} \ln\|x(t)\|/t$ *of a solution* $x(t)$ and then the lower Ljapunov exponent κ_L' *of an equation* as the greatest lower bound of lower Ljapunov exponents of its solutions.

In this connection $\kappa_L' = -\overline{\lim}_{t\to\infty} \ln\|U^{-1}(t)\|/t$.

2. *Bohl exponents.* Here we introduce another characteristic of the behavior of the solutions of equation (4.1) which, as will be seen, has more natural properties.

Let $x(t) = U(t)x_0$ be a solution of this equation. By the *(upper) Bohl exponent* $\kappa_B(x_0)$ *of this solution* is meant the greatest lower bound of all those numbers ρ for which there exist numbers N_ρ such that

$$\|x(t)\| \leqq N_\rho e^{\rho(t-\tau)}\|x(\tau)\| \tag{4.4}$$

for any τ, $t \in [0, \infty)$ such that $\tau \leqq t$. If such numbers ρ do not exist, we put $\kappa_B(x_0) = +\infty$.

In exactly the same way the *lower Bohl exponent* $\kappa_B'(x_0)$ *of a solution* $x(t)$ is the least upper bound of those numbers ρ' for which there exists a number $N_\rho' > 0$ such that

$$\|x(t)\| \geqq N_{\rho'} e^{\rho'(t-\tau)}\|x(\tau)\| \qquad (0 \leqq \tau \leqq t < \infty). \tag{4.4'}$$

If $\kappa(x_0)$ is a Ljapunov exponent of a solution, clearly,

$$-\infty \leqq \kappa_B'(x_0) \leqq \kappa(x_0) \leqq \kappa_B(x_0) \leqq +\infty.$$

The interval $[\kappa_B'(x_0), \kappa_B(x_0)]$ is called the *Bohl interval of the solution* in question. The following formulas are easily verified:

$$\kappa_B(x_0) = \overline{\lim_{\substack{\tau\to\infty \\ t-\tau\to\infty}}} \frac{\ln\|x(t)\| - \ln\|x(\tau)\|}{t-\tau},$$

$$\kappa_B'(x_0) = \underline{\lim_{\substack{\tau\to\infty \\ t-\tau\to\infty}}} \frac{\ln\|x(t)\| - \ln\|x(\tau)\|}{t-\tau}.$$

Suppose now P is a projection in $\mathfrak{B}$ and $\mathfrak{B}_P = P\mathfrak{B}$ is the corresponding subspace. We consider the totality of solutions $x(t) = U(t)x_0$ of equation (4.1) that are initially in $\mathfrak{B}_P$: $x_0 \in \mathfrak{B}_P$.

By the *upper (lower) Bohl exponent* $\kappa_B(P)$ $(\kappa_B'(P))$ *of this totality of solutions* is meant the greatest lower (least upper) bound of the exponents ρ for which formula (4.4) ((4.4′)) is valid for all of the solutions $x(t) = U(t)x_0$ with $x_0 \in \mathfrak{B}_P$ and a number $N_\rho > 0$ not depending on x_0.

We will call the exponents $\kappa_B(P)$ and $\kappa_B'(P)$ and the interval $[\kappa_B'(P), \kappa_B(P)]$ *the upper and lower Bohl exponents and Bohl interval of equation* (4.1) *corresponding to the projection* P (*subspace* $\mathfrak{B}_P$).

In particular, when $P = I$ we will speak simply of the *Bohl exponents of equation* (4.1) and use the notation

$$\kappa_B(I) = \kappa_B, \qquad \kappa_B'(I) = \kappa_B'.$$

It is clear that an extension of the subspace $\mathfrak{B}_p$ can only widen the corresponding Bohl interval and, in particular,

$$\kappa_B' \leqq \kappa_B'(P) \leqq \kappa_B(P) \leqq \kappa_B. \tag{4.5}$$

We cite the following finiteness criterion for Bohl exponents.

THEOREM 4.2. *In order for the upper* (*lower*) *Bohl exponent of equation* (4.1) *to be finite*:

$$\kappa_B < +\infty \qquad (\kappa_B' > -\infty),$$

it is necessary and sufficient that

$$\left.\begin{aligned} K &= \sup_{0 \leqq t-\tau \leqq 1} \|U(t,\tau)\| < \infty \\ (K' &= \sup_{0 \leqq t-\tau \leqq 1} \|U(\tau,t)\| < \infty). \end{aligned}\right\} \tag{4.6}$$

We will prove below the more general Theorem 4.2′, which is closely connected with the results presented in Chapter IV.

We consider the subspace $\mathfrak{B}_{P(t)} = U(t)\mathfrak{B}_P$. It is the range of the projection $P(t) = U(t)PU^{-1}(t)$.

An important role will be played in the sequel by the condition

$$m_P = \sup_{0 \leqq t < \infty} \|U(t)PU^{-1}(t)\| < \infty. \tag{4.7}$$

If it is satisfied, we will say that *the projection P is uniformly conjugated by the Cauchy operator of equation* (4.1). The geometrical significance of this definition will be made clear in Chapter IV. This condition is obviously satisfied if P commutes with $U(t)$ and, in particular, if $P = I$.

THEOREM 4.2′. *In order for the estimate $\kappa_B(P) < \infty$ ($\kappa_B'(P) > -\infty$) to be satisfied it is sufficient and, for the projections P that are uniformly conjugated by $U(t)$, also necessary that*

$$\left.\begin{aligned} K_P &= \sup_{0 \leqq t-\tau \leqq 1} \|U(t)PU^{-1}(\tau)\| < \infty \\ (K_P' &= \sup_{0 \leqq t-\tau \leqq 1} \|U(\tau)PU^{-1}(t)\| < \infty). \end{aligned}\right\} \tag{4.6′}$$

PROOF. Suppose estimate (4.4) is satisfied for every solution $x(t) = U(t)x_0$ with $x_0 \in \mathfrak{B}_P$. We write it in the form

$$\|U(t)Px\| \leqq N_\rho e^{\rho(t-\tau)} \|U(\tau)Px\| \qquad (t \geqq \tau),$$

where x is an arbitrary element of $\mathfrak{B}$. Setting $x = U^{-1}(\tau)y$ and using the arbitrariness of y and estimate (4.5), we obtain the inequality

$$\|U(t)PU^{-1}(\tau)\| \leqq N_\rho m_P e^{\rho(t-\tau)} \qquad (t \geqq \tau), \tag{4.8}$$

which implies that $K_P < \infty$. The necessity of the condition $K'_P < \infty$ can be shown in the same way if we rewrite (4.4') in the form

$$\|U(\tau)Px\| \leqq N_{\rho'} e^{-\rho'(t-\tau)} \|U(t)Px\| \qquad (\tau \leqq t).$$

Suppose now $K_P < \infty$. For $t - \tau > 0$ and $n = [t - \tau]$ we put $\tau_k = \tau + k$ ($k = 1, 2, \cdots, n$) and $\tau_{n+1} = t$. Since

$$U(t)PU^{-1}(\tau) = \prod_{k=1}^{n+1} U(\tau_k)PU^{-1}(\tau_{k-1}),$$

we get

$$\begin{aligned}\|U(t)PU^{-1}(\tau)\| &\leqq \prod_{k=1}^{n+1} \|U(\tau_k)PU^{-1}(\tau_{k-1})\| \\ &\leqq K_P^{n+1} \leqq K_P e^{n \ln K_P} \leqq K_P e^{(t-\tau)\ln K_P},\end{aligned}$$

and, consequently, for $x_0 \in \mathfrak{B}_P$

$$\|U(t)x_0\| = \|U(t)PU^{-1}(\tau)U(\tau)x_0\| \leqq K_P e^{(t-\tau)\ln K_P} \|U(\tau)x_0\|,$$

i. e. $\kappa(P) \leqq \ln K_P$.

The sufficiency of the condition $K'_P < \infty$ can be shown analogously.

REMARK 4.2. We have incidentally obtained the estimates

$$-\ln K'_P \leqq \kappa'_B(P) \leqq \kappa_B(P) \leqq \ln K_P.$$

In particular, the Bohl exponents of equation (4.1) ($P = I$) satisfy the estimates

$$\begin{aligned}\inf \ln \frac{1}{\|U(t,\tau)\|} &= -\sup_{0 \leqq t-\tau \leqq 1} \ln \|U(t,\tau)\| \\ &\leqq \kappa'_B \leqq \kappa_B \leqq \sup_{0 \leqq t-\tau \leqq 1} \ln \|U(t,\tau)\|.\end{aligned} \tag{4.9}$$

REMARK 4.3. Since

$$[U(t)PU^{-1}(s)][U(s)PU^{-1}(\tau)] = U(t)PU^{-1}(\tau),$$

the first of the conditions (4.6') is equivalent to the condition

$$\sup_{0 \leqq t-\tau \leqq T} \|U(t)PU^{-1}(\tau)\| < \infty$$

for any $T > 0$. An analogous remark can be made concerning the second condition of (4.6′).

As a corollary of Theorem 4.2 we obtain the following result.

THEOREM 4.3. *If the operator function $A(t)$ is integrally bounded, the Bohl exponents of equation* (4.1) *are finite.*

The proof immediately follows from Lemma 2.4 and formulas (4.7) and (4.9).

We now find expressions for the Bohl exponents corresponding to a projection P.

THEOREM 4.4. *If condition* (4.7) *is satisfied and the Bohl exponents are finite, they are representable by the formulas*

$$\left.\begin{aligned} \kappa_B(P) &= \overline{\lim_{\tau,\, s\to\infty}} \frac{\ln\|U(\tau+s)PU^{-1}(\tau)\|}{s}; \\ -\kappa_B'(P) &= \overline{\lim_{\tau,\, s\to\infty}} \frac{\ln\|U(\tau)PU^{-1}(\tau+s)\|}{s}; \end{aligned}\right\} \tag{4.10}$$

in particular,

$$\left.\begin{aligned} \kappa_B &= \overline{\lim_{\tau,s\to\infty}} \frac{\ln\|U(\tau+s,\tau)\|}{s}; \\ -\kappa_B' &= \overline{\lim_{\tau,s\to\infty}} \frac{\ln\|U(\tau,\tau+s)\|}{s}. \end{aligned}\right\} \tag{4.11}$$

PROOF. Suppose $\rho > \kappa_B(P)$. Then inequality (4.8) is satisfied. It follows that

$$\frac{\ln\|U(\tau+s)PU^{-1}(\tau)\|}{s} \leqq \frac{\ln N_\rho m_P}{s} + \rho,$$

and, further,

$$\lambda = \overline{\lim_{\tau,s\to\infty}} \frac{\ln\|U(\tau+s)PU^{-1}(\tau)\|}{s} \leqq \rho,$$

which implies $\lambda \leqq \kappa_B(P) < \infty$.

Suppose now $\rho > \lambda$. By the definition of the upper limit there exists a number $T = T_\rho$ such that for $\tau, s \geqq T_\rho$

$$\frac{\ln\|U(\tau+s)PU^{-1}(\tau)\|}{s} \leqq \rho,$$

i. e. $\|U(t)PU^{-1}(\tau)\| \leqq e^{\rho(t-\tau)}$ for $\tau,\ t-\tau \geqq T_\rho$.

By Remark 4.3 the quantity $K_T = \sup_{0\leqq t-\tau\leqq T}\|U(t)PU^{-1}(\tau)\|$ is finite.

Hence for all $t \geqq \tau \geqq T$

$$\|U(t)PU^{-1}(\tau)\| \leqq Ne^{\rho(t-\tau)},$$

where $N = \max(1, K_T e^{|\rho| T})$. This same estimate is also valid for $0 \leqq \tau \leqq t \leqq Y$. Finally, when $0 \leqq \tau \leqq T < t$, we have

$$\|U(t)PU^{-1}(\tau)\| \leqq \|U(t)PU^{-1}(T)\| \, \|U(T)PU^{-1}(\tau)\|$$
$$\leqq Ne^{\rho(t-T)}Ne^{\rho(T-\tau)} = N^2 e^{\rho(t-\tau)}.$$

Setting $N_\rho = \max(N, N^2)$, we obtain the inequality

$$\|U(t)PU^{-1}(\tau)\| \leqq N_\rho e^{\rho(t-\tau)} \qquad (t \geqq \tau \geqq 0),$$

which implies for

$$x(\tau) = U(\tau)Px_0 \quad \text{and} \quad x(t) = U(t)PU^{-1}(\tau)x(\tau) = U(t)Px_0$$

the estimate

$$\|x(t)\| \leqq N_\rho e^{\rho|t-\tau|} \|x(\tau)\|,$$

showing that $\kappa_B(P) \leqq \rho$.

Thus $\kappa_B \leqq \lambda$ and the first formula is proved. The second is established analogously.

3. *Stationary equation case.* In this case we have

THEOREM 4.5. *The upper (lower) Ljapunov and upper (lower) Bohl exponents of the homogeneous stationary equation $dx/dt = Cx$ ($C =$ const) coincide with each other and are equal to the supremum (infimum) of the real parts of the numbers $\lambda \in \sigma(C)$.*

PROOF. For the stationary equation

$$U(t) = e^{Ct}, \quad U(t, \tau) = e^{C(t-\tau)}$$

and according to Theorem I.4.1

$$\kappa_L = \lim_{t\to\infty} \frac{\ln\|e^{Ct}\|}{t} = \lim_{\substack{\tau\to\infty \\ t-\tau\to\infty}} \frac{\ln\|e^{C(t-\tau)}\|}{t-\tau} = \kappa_B = \max\{\operatorname{Re}\lambda \,|\, \lambda \in \sigma(C)\}.$$

The assertion in parentheses is proved analogously.

DEFINITION. By the *strict Ljapunov exponent of equation* (4.1) is meant the limit

$$\lim_{t\to\infty} \frac{\ln\|U(t)\|}{t},$$

if it exists.

Analogously, by the *strict Bohl exponent of equation* (4.1) is meant

$$\lim_{\tau,\, s\to\infty} \frac{\ln\|U(\tau + s, \tau)\|}{s},$$

if this limit exists.

We have just convinced ourselves that a stationary equation has strict exponents.

4. *Perron's example.* We will show by an example that the case $\kappa_L < \kappa_B$ is possible.

We consider the scalar Perron equation

$$dx/dt = (\sin \ln t + \cos \ln t)x. \tag{4.12}$$

Here $\mathfrak{B}$ is a one-dimensional space and $A(t)$ is the operator of multiplication by the function $\sqrt{2} \sin (\ln t + \pi/4)$.

The solution of the equation has the form $x(t) = e^{t \sin \ln t}x(0)$, which implies $U(t) = e^{t \sin \ln t}$.

We calculate the upper Ljapunov exponent of the equation:

$$\kappa_L = \overline{\lim_{t\to\infty}} \frac{\ln|U(t)|}{t} = \overline{\lim_{t\to\infty}} \sin \ln t = 1.$$

Let us now show that $\kappa_B = \sqrt{2}$. We have

$$\kappa_B = \overline{\lim_{\substack{\tau\to\infty \\ t-\tau\to\infty}}} \frac{\ln\|U(t)U^{-1}(\tau)\|}{t-\tau} = \overline{\lim_{\substack{\tau\to\infty \\ t-\tau\to\infty}}} \frac{t \sin \ln t - \tau \sin \ln \tau}{t - \tau}$$

$$= \overline{\lim_{\substack{\tau\to\infty \\ t-\tau\to\infty}}} (\cos \ln t_{av} + \sin \ln t_{av}),$$

where t_{av} is a point lying between τ and t (according to the mean value theorem). We now choose a sequence of pairs of points τ_n, t_n as follows:

$$\ln t_n = 2n\pi + \pi/4 + \varepsilon; \quad \ln \tau_n = 2n\pi + \pi/4.$$

Then for $\tau_n < \bar{t}_n < t_n$ we have $\ln \bar{t}_n = 2n\pi + \pi/4 + \eta_n$, where $0 < \eta_n < \varepsilon$. Under this choice of τ_n, t_n for $n \to \infty$ we have $\tau_n \to \infty$ and

$$t_n - \tau_n = e^{2n\pi+\pi/4}(e^\varepsilon - 1) \to \infty,$$

and hence

$$\kappa_B \geqq \overline{\lim_{n\to\infty}} (\cos \ln \bar{t}_n + \sin \ln \bar{t}_n)$$

$$= \overline{\lim_{n\to\infty}} [\cos (\pi/4 + \eta_n) + \sin (\pi/4 + \eta_n)] = \sqrt{2} + \lambda_n,$$

where $\lambda_n \to 0$ as $\varepsilon \to 0$.

Thus $\kappa_B \geqq \sqrt{2}$. But on the other hand, $\kappa_B \leqq M = \sqrt{2}$.

5. *Stability of Bohl exponents.* We now discuss the variation of the Bohl exponents of equation (4.1) under certain transformations of this equation (see also in this connection § IV.2).

We will consider only upper exponents, since the properties of the lower exponents are established analogously. It will be convenient for us to make use of the following definition.

We will say that *equation* (4.1) *has property* $\mathscr{B}(\nu, N)$,[3] where ν is real and N

[3] This property was denoted by $L(\nu, N)$ in the "Lectures." The new notation has been introduced in recognition of the fact that this property was first used by P. Bohl (see the Notes).

is positive, if all of its solutions are subject to the estimate

$$\|x(t)\| \leqq Ne^{-\nu(t-\tau)}\|x(\tau)\| \qquad (t \geqq \tau)$$

or, equivalently,

$$\|U(t, \tau)\| \leqq Ne^{-\nu(t-\tau)} \qquad (t \geqq \tau). \tag{4.13}$$

We note that the possession of property $\mathscr{B}(0, N)$ with some $N > 0$ is equivalent to the uniform right stability of equation (4.1).

Clearly, $-\kappa_B$ coincides with the least upper bound of those ν for which equation (4.1) has property $\mathscr{B}(\nu, N)$ for $N = N_\nu$.

We note further that if an equation has property $\mathscr{B}(\nu, N)$ for $t, \tau \geqq t_0 > 0$, it has property $\mathscr{B}(\nu, N')$ with the same exponent ν and some constant N' for $t, \tau \geqq 0$. For, the evolution operator $U(t, \tau)$ is bounded on a finite interval by virtue of its continuity and it is always possible (by increasing the constant N if necessary) to preserve the validity of (4.13).

The following two properties of exponents immediately follow from the definitions.

1) All of the exponents of the equation

$$dy/dt = [A(t) + \alpha I]y$$

are obtained from the corresponding exponents of equation (4.1) by means of a shift to the right of magnitude $\operatorname{Re}\alpha$.

In particular, we note that the upper Ljapunov and upper Bohl exponents of the equation

$$dx/dt = (\sin \ln t + \cos \ln t + \alpha)x$$

are the numbers $1 + \alpha$ and $\sqrt{2} + \alpha$ respectively. For $-\sqrt{2} < \alpha < -1$ we get $\kappa_L < 0$ and $\kappa_B > 0$. Thus all of the solutions of the equation in question decrease exponentially as $t \to \infty$, although the upper Bohl exponent of this equation is positive.

2) For any $l > 0$ the upper Ljapunov and upper Bohl exponents of the equation

$$dx/dt = A(t + l)x \qquad (0 \leqq t < \infty)$$

coincide with the corresponding exponents of equation (4.1).

We now consider along with equation (4.1) the equation

$$dx/dt = A(t)x + B(t)x. \tag{4.14}$$

LEMMA 4.1. *Suppose equation* (4.1) *has property* $\mathscr{B}(\nu, N)$ *and for some* $s_0 \geqq 0$

$$\sup_{t\geqq 0} \frac{1}{s_0} \int_t^{t+s_0} \|B(\tau)\| d\tau = M_{s_0} < \infty. \tag{4.15}$$

Then equation (4.14) *has property* $\mathscr{B}(\nu', N')$, *where* $\nu' = \nu - NM_{s_0}$ *and* $N' = Ne^{NM_{s_0}s_0}$.

PROOF. From estimate (2.22) we deduce that for $s \geqq 0$

$$\|U_2(t+s,t)\| \leqq Ne^{-\nu s}e^{N\int_t^{t+s}\|B(\tau)\|d\tau}$$
$$\leqq Ne^{-\nu s}e^{NM_{s_0}(s+s_0)} = Ne^{NM_{s_0}s_0}e^{-(\nu-NM_{s_0})s}.$$

Lemma 4.1 implies the following important result.

THEOREM 4.6. *The upper Bohl exponent κ_B of equation* (4.1) *has the following stability property: for any $\varepsilon > 0$ there exists a number $\delta > 0$ depending only on ε and equation* (4.1) *such that if*

$$\overline{\lim_{t,s\to\infty}} \frac{1}{s}\int_t^{t+s}\|B(\tau)\|d\tau < \delta, \tag{4.16}$$

then

$$\tilde{\kappa}_B \leqq \kappa_B + \varepsilon, \tag{4.17}$$

where $\tilde{\kappa}_B$ is the Bohl exponent of equation (4.14).

The lower Bohl exponent has an analogous property.

PROOF. Equation (4.1) has property $\mathscr{B}(-\kappa_B - \varepsilon/2, N_{\varepsilon/2})$. On the other hand, for sufficiently large t_0 and some $s_0 > 0$

$$\sup_{t\geqq t_0} \frac{1}{s_0}\int_t^{t+s_0}\|B(\tau)\|d\tau < 2\delta.$$

Therefore, by virtue of Lemma 4.1, equation (4.14) has property $\mathscr{B}(-\kappa_B - \varepsilon/2 - N_{\varepsilon/2}\cdot 2\delta, N')$ with some constant $N' > 0$, which implies $\tilde{\kappa}_B \leqq \kappa_B + \varepsilon/2 + 2\delta N_{\varepsilon/2}$. It remains to choose $\delta < \varepsilon/4N_{\varepsilon/2}$.

REMARK 4.4. Condition (4.16) is satisfied if, beginning with some t_0, either $\|B(t)\| \leqq \delta$ or $\int_t^{t+1}\|B(\tau)\|d\tau \leqq \delta$.

COROLLARY 4.1. *The upper Bohl exponent of the equation $dx/dt = Ax + B(t)x$ is negative if the spectrum $\sigma(A)$ lies in the interior of the left halfplane and condition* (4.16) *is satisfied with a sufficiently small $\delta > 0$.*

COROLLARY 4.2. *If* $\lim_{t,s\to\infty}(1/s)\int_t^{t+s}\|B(\tau)\|\,d\tau = 0$, *the upper (lower) Bohl exponents of equations* (4.1) *and* (4.14) *coincide.*

This is true in particular when either $\int^\infty\|B(\tau)\|d\tau < \infty$ *or* $\lim_{\tau\to\infty}\|B(\tau)\| = 0$.

Equations whose coefficients differ by a term tending to zero at infinity are said to be *asymptotically comparable.*

Thus *the upper (lower) Bohl exponents of integrally or asymptotically comparable equations coincide.*

6. *Estimates of exponents in Hilbert space.* Suppose the space $\mathfrak{B}$ is a Hilbert space $\mathfrak{H}$. We recall that if H is a Hermitian operator then $\lambda_m(H)$ and $\lambda_M(H)$ are respectively the least and greatest numbers of the spectrum $\sigma(H)$, so that

$$\lambda_m(H)(x, x) \leqq (Hx, x) \leqq \lambda_M(H)(x, x) \qquad (x \in \mathfrak{H}). \tag{4.18}$$

THEOREM 4.7. *For any solution $x(t)$ of equation* (4.1) *in a Hilbert space $\mathfrak{H}$ the function*

$$\phi_M(t) = \|x(t)\| \exp\left\{- \int_{t_0}^{t} \lambda_M(A_{\mathfrak{R}}(\tau))d\tau\right\}$$

is nonincreasing while the function

$$\phi_m(t) = \|x(t)\| \exp\left\{- \int_{t_0}^{t} \lambda_m[A_{\mathfrak{R}}(\tau)]d\tau\right\}$$

is nondecreasing. In particular, for $t \geqq s$

$$\|x(s)\| \exp\left\{\int_{s}^{t} \lambda_m[A_{\mathfrak{R}}(\tau)]d\tau\right\} \leqq \|x(t)\| \tag{4.19}$$

$$\leqq \|x(s)\| \exp\left\{\int_{s}^{t} \lambda_M[A_{\mathfrak{R}}(\tau)]d\tau\right\}.$$

where $A_{\mathfrak{R}} = (A + A^*)/2$ is the real part of the operator A.

PROOF. Differentiating the function

$$\phi_M^2(t) = (x(t), x(t)) \exp\left\{- 2\int_{t_0}^{t} \lambda_M[A_{\mathfrak{R}}(\tau)]d\tau\right\},$$

we obtain the inequality

$$\frac{d}{dt}\phi_M^2(t) = [(x'(t), x(t)) + (x(t), x'(t))] \exp\left\{- 2\int_{t_0}^{t} \lambda_M[A_{\mathfrak{R}}(\tau)]d\tau\right\}$$

$$- 2(x(t), x(t))\lambda_M[A_{\mathfrak{R}}(t)] \exp\left\{- 2\int_{t_0}^{t} \lambda_M[A_{\mathfrak{R}}(\tau)]d\tau\right\}$$

$$\leqq 2\{(A_{\mathfrak{R}}(t)x(t), x(t)) - \lambda_M[A_{\mathfrak{R}}(t)](x(t), x(t))\} \exp\left\{- 2\int_{t_0}^{t} \lambda_M[A_{\mathfrak{R}}(\tau)]d\tau\right\},$$

which by virtue of (4.18) implies the estimate $(d/dt)\,\phi_M^2(t) \leqq 0$, proving the first assertion of the theorem. The second assertion is proved analogously.

COROLLARY 4.3. *The Bohl exponents of equation* (4.1) *satisfy the estimates*

$$\varliminf_{t,s\to\infty} \frac{1}{s}\int_{t}^{t+s} \lambda_m[A_{\mathfrak{R}}(\tau)]d\tau \leqq \kappa_B \leqq \varlimsup_{t,s\to\infty} \frac{1}{s}\int_{t}^{t+s} \lambda_M[A_{\mathfrak{R}}(\tau)]d\tau;$$

$$- \varlimsup_{t,s\to\infty} \frac{1}{s}\int_{t}^{t+s} \lambda_M[A_{\mathfrak{R}}(\tau)]d\tau \leqq \kappa_B' \leqq - \varliminf_{t,s\to\infty} \frac{1}{s}\int_{t}^{t+s} \lambda_m[A_{\mathfrak{R}}(\tau)]d\tau,$$

obtained by comparing formulas (4.11) *with inequalities* (4.19).

The Ljapunov spectrum is included in the interval

$$\left[\overline{\lim_{t\to\infty}}\frac{1}{t}\int_0^t \lambda_m[A_{\Re}(\tau)]d\tau,\ \overline{\lim_{t\to\infty}}\frac{1}{t}\int_0^t \lambda_M[A_{\Re}(\tau)]d\tau\right].$$

These results can be extended to any Banach space $\mathfrak{B}$ (see Exercise 19).

§ 5. Condition for boundedness on $[0, \infty)$ of the solutions of the inhomogeneous equation

1. *Necessity of negativeness of the upper Ljapunov exponent.*[4)] In this section we discuss the role of negativeness of the upper Bohl exponent of equation (4.1).

We consider the equation

$$\frac{dx}{dt} = A(t)x + f(t) \qquad (t \in [0, \infty)). \tag{5.1}$$

THEOREM 5.1. *If on the halfline* $[0, \infty)$ *the solution* $x(t)$ *of the Cauchy problem for equation* (5.1) *with the initial condition*

$$x(0) = 0 \tag{5.2}$$

is bounded whenever the vector function $f(t)$ *is bounded and continuous, then there exist positive constants* N *and* ν *such that*

$$\|U(t)\| \leqq Ne^{-\nu t} \qquad (t \geqq 0) \tag{5.3}$$

and every solution of equation (5.1) *with a bounded continuous vector function* $f(t)$:

$$|||f||| = \sup_{0\leqq t<\infty} \|f(t)\| < \infty, \tag{5.4}$$

is bounded.

PROOF. We consider the complete space $C(\mathfrak{B})$ of bounded continuous vector functions $f(t)$ $(0 \leqq t < \infty)$ with values in $\mathfrak{B}$ and norm (5.4).

The solution of the Cauchy problem (5.1)–(5.2) is given by the formula

$$x(t) = \int_0^t U(t, \tau)f(\tau)d\tau \qquad (0 \leqq t < \infty). \tag{5.5}$$

The right side of expression (5.5) represents for each t a bounded linear operator V_t acting from $C(\mathfrak{B})$ into $\mathfrak{B}$ since

$$\|x(t)\| \leqq \int_0^t \|U(t, \tau)\|d\tau \cdot |||f|||.$$

By hypothesis, there corresponds to each $f \in C(\mathfrak{B})$ a bounded solution $x(t)$

[4)]By a *Ljapunov or Bohl exponent of an inhomogeneous equation* (5.1) we will mean the Ljapunov or Bohl exponent respectively of the corresponding homogeneous equation.

on $[0, \infty)$. This means that the operators V_t $(0 \leqq t < \infty)$ are uniformly bounded at each element of $C(\mathfrak{B})$. The uniform boundedness principle therefore implies that

$$\|x(t)\| = \|V_t f\| \leqq k\|\|f\|\| \qquad (0 \leqq t < \infty), \tag{5.6}$$

where $k = \text{const}$.

We put $\chi(t) = \|U(t)\|$ and consider the function $f(t) = U(t)y/\chi(t)$, where y is an arbitrary element of $\mathfrak{B}$.

It is an element of $C(\mathfrak{B})$ since $\|\|f\|\| \leqq \|y\| < \infty$. Therefore the corresponding solution of problem (5.1)–(5.2) must by virtue of (5.6) satisfy the condition

$$\|x(t)\| \leqq k\|y\|. \tag{5.7}$$

This solution is given by the formula

$$x(t) = \int_0^t U(t, \tau)\frac{U(\tau)y}{\chi(\tau)}\,d\tau = U(t)y\cdot\phi(t), \tag{5.8}$$

where $\phi(t) = \int_0^t [\chi(\tau)]^{-1} d\tau$.

From (5.7) and (5.8) we obtain the inequality $\|U(t)y\|\phi(t) \leqq k\|y\|$, which by virtue of the arbitrariness of $y \in \mathfrak{B}$ and the equality

$$\frac{1}{\phi'(t)} = \chi(t) = \sup_y \frac{\|U(t)y\|}{\|y\|}$$

reduces to the relation $\phi'(t)/\phi(t) \geqq 1/k$.

Integrating this relation from 1 to t, we get

$$\phi(t) \geqq \phi(1)e^{(t-1)/k}$$

and hence

$$\frac{1}{\chi(t)} = \phi'(t) \geqq \frac{\phi(t)}{k} \geqq \frac{\phi(1)}{k}e^{(t-1)/k}.$$

The latter relation means that

$$\|U(t)\| = \chi(t) \leqq N_1 e^{-\nu t} \qquad (t \geqq 1),$$

where $\nu = 1/k$ and $N_1 = ke^{1/k}/\phi(1)$.

Setting, finally, $N = \max(N_1, \max_{0\leqq t\leqq 1} e^{\nu t}\|U(t)\|)$, we obtain the required inequality (5.3).

The second assertion of the theorem follows upon noting that an arbitrary solution of equation (5.1) differs from the solution distinguished by condition (5.2) by a term $U(t)x_0$:

$$x(t) = U(t)x_0 + \int_0^t U(t, \tau)f(\tau)d\tau,$$

the boundedness of which follows from estimate (5.3).

The theorem is proved.

From Theorem 5.2 below and the fact that there are examples of equations for which $A(t)$ is integrally bounded and $\kappa_L < 0 < \kappa_B$ (see § 4.5) it follows that the converse of the first assertion of Theorem 5.1 does not hold, i.e. the presence of estimate (5.3) does not guarantee the boundedness of the solution of problem (5.1) $-$(5.2) for any bounded continuous vector function $f(t)$ on $[0, \infty)$.

2. *Role of negativeness of the Bohl exponent in the case of an integrally bounded* $A(t)$. A more complete result is obtained under the additional condition of *integral boundedness* on $[0, \infty)$ of the operator function $A(t)$:

$$\int_t^{t+1} \|A(\tau)\| d\tau \leqq M_1 \qquad (t \in [0, \infty)). \tag{5.9}$$

THEOREM 5.2. *Suppose condition* (5.9) *is satisfied. In order for the Cauchy problem* (5.1)$-$(5.2) *to have a bounded solution on* $[0,\infty)$ *for every bounded continuous function* $f(t)$ *on* $[0,\infty)$ *it is necessary and sufficient that the Bohl exponent of equation* (5.1) *be negative.*

PROOF. The condition is sufficient since the estimate

$$\|U(t, \tau)\| \leqq Ne^{-\nu(t-\tau)} \qquad (\nu > 0, t \geqq \tau)$$

implies

$$\left\| \int_0^t U(t, \tau) f(\tau) d\tau \right\| \leqq N \|\|f\|\| \int_0^t e^{-\nu(t-\tau)} d\tau \leqq \frac{N}{\nu} \|\|f\|\|.$$

To prove the necessity of the condition we first show that the boundedness of the solution of the Cauchy problem (5.1)$-$(5.2) implies the boundedness of the solution of the Cauchy problem

$$\begin{cases} dx(t)/dt = A(t)x + f(t) & (t \geqq t_0), \\ x(t_0) = 0. \end{cases} \tag{5.10}$$

The solution of problem (5.10) is given by the formula

$$x(t) = \int_{t_0}^t U(t, \tau) f(\tau) d\tau.$$

If the function $f(t)$ were not necessarily continuous, the solution of problem (5.10) could be regarded as the solution of problem (5.1)$-$(5.2) with free term

$$f^*(t) = \begin{cases} 0, & 0 \leqq t < t_0, \\ f(t), & t \geqq t_0. \end{cases}$$

In order to avoid working with discontinuous functions we consider the following problem of type (5.1)$-$(5.2):

$$\begin{cases} dx_\varepsilon(t)/dt = A(t)x_\varepsilon(t) + f_\varepsilon(t), \\ x_\varepsilon(0) = 0, \end{cases}$$

where

$$f_\varepsilon(t) = \begin{cases} 0 & \text{for} \quad 0 \leqq t < t_0 - \varepsilon, \\ \dfrac{1}{\varepsilon} f(t_0)(t - t_0 + \varepsilon) & \text{for} \quad t_0 - \varepsilon \leqq t < t_0, \\ f(t) & \text{for} \quad t \geqq t_0. \end{cases}$$

Its solution is representable in the form

$$x_\varepsilon(t) = \int_0^t U(t, \tau) f_\varepsilon(\tau) d\tau$$

$$= \int_{t_0 - \varepsilon}^{t_0} U(t, \tau) f_\varepsilon(\tau) d\tau + \int_{t_0}^{t} U(t, \tau) f(\tau) d\tau.$$

By assumption (see (5.6)),

$$\|x_\varepsilon(t)\| \leqq k |||f_\varepsilon||| = k |||f|||.$$

Suppose now $\varepsilon \to 0$ for fixed t. Then the norm of the first integral tends to zero and hence $x_\varepsilon(t) \to x(t)$; consequently, $\|x(t)\| \leqq k|||f|||$ and the assertion is proved. Choosing now in problem (5.10) the function $f(t) = U(t, t_0)y/\chi(t)$, where $\chi(t) = \|U(t, t_0)\|$, and repeating all of the arguments presented in the proof of Theorem 5.1, we get

$$\chi(t) = \|U(t, t_0)\| \leqq N e^{-\nu(t - t_0)},$$

where N is a constant satisfying the single condition

$$N \geqq \max\left\{ \frac{k e^{1/k}}{\phi(1)}, \max_{0 \leqq t - t_0 \leqq 1} e^{\nu(t - t_0)} \|U(t, t_0)\| \right\},$$

$$\nu = \frac{1}{k}, \quad \phi(1) = \int_{t_0}^{t_0 + 1} \frac{d\tau}{\chi(\tau)}.$$

It still remains to show that N can be chosen independently of t_0. To this end we note that

$$\chi(t) = \|U(t, t_0)\| \leqq e^{\int_{t_0}^{t} \|A(\tau)\| d\tau} \leqq e^{M_1} \qquad (t_0 \leqq t \leqq t_0 + 1).$$

But then in the first place

$$\phi(1) \geqq \int_{t_0}^{t_0 + 1} e^{-M_1} d\tau = e^{-M_1}$$

and in the second place

$$\max_{0 \leqq t - t_0 \leqq 1} \{ e^{\nu(t - t_0)} \|U(t, t_0)\| \} \leqq \max_{0 \leqq t - t_0 \leqq 1} e^{\nu(t - t_0) + M_1} = e^{\nu + M_1}.$$

Thus we can put

$$N = e^{1/k + M_1} \max\{1, k\}.$$

The theorem is proved.

3. *Essentiality of the condition of integral boundedness.* Here we give an example showing that condition (5.9) in Theorem 5.2 cannot be dropped.

To this end we consider a continuously differentiable positive scalar function $\upsilon(t)$ having the following properties:

$$\text{a)} \qquad \int_0^t \upsilon(s)ds < \upsilon(t),$$

$$\text{b)} \qquad \upsilon(n - \alpha_n) > n\upsilon(n) \qquad (n = 1,2,\cdots; \alpha_n \to 0). \tag{5.11}$$

An example of such a function is $\upsilon(t) = e^t + r(t)$, where

$$r(t) = \begin{cases} 0 & \text{for } n - 1 \leqq t < n - 2\alpha_n \\ ne^n \sin^2\left\{\dfrac{\pi}{2\alpha_n}(t - n + 2\alpha_n)\right\} & \text{for } n - 2\alpha_n \leqq t < n, \end{cases}$$

with $\alpha_n = 1/2^{n+1}ne^n$.

If we put

$$a(t) = \upsilon(t)\frac{d}{dt}\frac{1}{\upsilon(t)} = -\frac{\upsilon'(t)}{\upsilon(t)} = -[\ln \upsilon(t)]',$$

the evolution operator of the scalar differential equation $dx/dt = a(t)x$ will be $U(t, s) = \upsilon(s)/\upsilon(t)$.

The Bohl exponent of this equation is obviously nonnegative and in fact infinite since

$$U(n + m, n - \alpha_n) > ne^{-m} \qquad (m, n = 1,2,\cdots).$$

But the solution of the Cauchy problem

$$\frac{dx}{dt} = a(t)x + g(t), \qquad x(0) = 0$$

for a bounded function $g(t)$ will be bounded since

$$x(t) = \int_0^t U(t, s)g(s)ds$$

implies that

$$|x(t)| = \left|\int_0^t U(t, s)g(s)ds\right|$$

$$= \left|\int_0^t \frac{\upsilon(s)g(s)}{\upsilon(t)}ds\right| \leqq \max_s |g(s)| \frac{\int_0^t \upsilon(s)ds}{\upsilon(t)} \leqq \max_s |g(s)|.$$

The function $\int_t^{t+1} a(\tau)\,d\tau$ in this example is of course unbounded.

§ 6. Equations with precompactly valued operator functions

1. *Criterion for negativeness of the upper Bohl exponent.* In the preceding section we demonstrated the importance of the role of negativeness of the (upper[5)]) Bohl exponent of equation (4.1).

But our ability to verify that an equation has this property is presently limited to only the simplest cases, when the equation is stationary or differs only slightly from a stationary equation (see Corollary 4.1 of Theorem 4.6).

In this section we indicate another class of equations for which the negativeness of the Bohl exponent can be determined from the properties of the coefficient of the equation.

We first establish an auxiliary result.

THEOREM 6.1. *Suppose the Bohl exponent* κ_B *of the equation*

$$\frac{dx}{dt} = A(t)x \qquad (0 \leqq t < \infty) \tag{6.1}$$

is finite. In order for it to be negative it is necessary and sufficient that there exist positive numbers T *and* $q < 1$ *for which the following condition is satisfied: for every* $x \in \mathfrak{B}$ *and* $t \geqq 0$ *there exists a number* $\theta_{x,t} \in [0, T]$ *with the property that*

$$\|U(t + \theta_{x,t}, t)x\| \leqq q\|x\|. \tag{6.2}$$

PROOF. NECESSITY. If the Bohl exponent is negative, there exist positive numbers N and ν for which equation (6.1) has property $\mathscr{B}(\nu, N)$. Then, for any T such that $Ne^{-\nu T} < 1$, estimate (6.2) is satisfied when $\theta_{x,t} = T$ and $q = Ne^{-\nu T}$.

SUFFICIENCY. Suppose $0 \leqq t_0 < t < \infty$. In view of the continuity of the operator $U(\tau, \tau')$ there exists a $\theta > 0$ such that $\|U(\tau, \tau')\| < 1/q$ for τ and τ' satisfying the conditions

$$t_0 \leqq \tau, \quad \tau' < 2t, \quad |\tau' - \tau| \leqq \theta.$$

For these τ, τ' and for any $x \in \mathfrak{B}$ the inequality

$$\begin{aligned} \|x\| &= \|U(\tau, \tau')U(\tau', \tau)x\| \\ &\leqq \|U(\tau, \tau')\| \cdot \|U(\tau', \tau)x\| < (1/q)\|U(\tau', \tau)x\| \end{aligned}$$

implies that $\|U(\tau, \tau')\| > q\|x\|$, and thus $\theta_{x,t} > \theta$ provided $t_0 \leqq \tau < \tau + \theta_{x,t} \leqq 2t$.

Let x_0 be an arbitrary element of $\mathfrak{B}$. We put

$$\begin{aligned} t_1 &= t_0 + \theta_{x_0, t_0}; \qquad & x_1 &= U(t_1, t_0)x_0; \\ t_2 &= t_1 + \theta_{x_1, t_1}; \qquad & x_2 &= U(t_2, t_1)x_1 = U(t_2, t_0)x_0; \end{aligned}$$

$$\cdots\cdots\cdots\cdots\cdots\cdots\cdots\cdots\cdots$$

[5)] Later in this section the word *upper* will be dropped for the sake of brevity.

$$t_{k+1} = t_k + \theta_{x_k t_k}; \quad x_{k+1} = U(t_{k+1}, t_k)x_k = U(t_{k+1}, t_0)x_0,$$
$$(k = 1,2,\cdots).$$

After a finite number m $(m < (t - t_0)/\theta)$ of steps it turns out that $t_m < t < t_{m+1}$.

From the equality $U(t, t_0)x_0 = U(t, t_m)x_m$ and the estimate $\|x_m\| \leqq q^m\|x_0\|$ we get

$$\|U(t, t_0)x_0\| \leqq kq^m\|x_0\|, \tag{6.3}$$

where $k = \sup_{0 \leqq t-\tau \leqq T} \|U(t, \tau)\| < \infty$ by virtue of the finiteness of κ_B (see Theorem 4.2 and Remark 4.3). Since $\theta_{x_k t_k} \leqq T$ $(k = 0, 1,\cdots, m + 1)$, we have $t \leqq t_0 + (m + 1)\, T$ and consequently $m + 1 \geqq (t - t_0)T$.

This together with (6.3) implies $\|U(t, t_0)x_0\| \leqq (k/q)q^{(t-t_0)/T}\|x_0\|$ or

$$\|U(t, t_0)x_0\| \leqq Ne^{-\nu(t-t_0)}\|x_0\|,$$

where $N = k/q$ and $\nu = T^{-1} \ln q^{-1}$. Q.E.D.

REMARK 6.1. The condition of Theorem 6.1 can be verified by verifying it for sufficiently large t $(t > \bar{t}$, where $\bar{t}$ does not depend on $x)$.

In fact, whenever the condition is satisfied for $t > \bar{t}$, it can be asserted that for some $\bar{N} > 0$ and $\nu > 0$ equation (6.1) will have property $\mathscr{B}(\nu, \bar{N})$ in the interval $[\bar{t}, \infty)$. But then, as is easily seen, under a suitable choice of $N > 0$ equation (4.1) will have property $\mathscr{B}(\nu, N)$ on the whole halfline $[0, \infty)$.

The usefulness of Theorem 6.1 is illustrated in particular by the following application of it.

THEOREM 6.2. *Suppose equation* (6.1) *has a finite Bohl exponent* κ_B *and* p *is any positive number. The Bohl exponent of the equation is negative precisely when there exists a positive constant* c *for which*

$$\left\{\int_\tau^\infty \|U(t, \tau)x\|^p dt\right\}^{1/p} \leqq c \,\|\, x \,\| \qquad (t_0 \leqq \tau < \infty). \tag{6.4}$$

PROOF. The necessity of this condition is obvious. Its sufficiency can be proved by showing that it implies condition (6.2) of Theorem 6.1. For suppose the contrary, i.e. suppose that for any q $(0 < q < 1)$ and any $T > 0$ there exist x_0 and τ_0 such that

$$\|U(t, \tau_0)x_0\| > q\|x_0\| \qquad (t \in [\tau_0, \tau_0 + T]).$$

Then

$$\int_{\tau_0}^\infty \|U(t, \tau_0)x_0\|^p dt \geqq \int_{\tau_0}^{\tau_0+T} \|U(t, \tau_0)x_0\|^p dt \geqq q^p\|x_0\|^p T.$$

Choosing $q^pT > c^p$, we arrive at a contradiction with condition (6.3). The theorem is proved.

REMARK 6.2. When $p \geqq 1$ condition (6.4) can be replaced by the simpler condition

$$\sup_{0 \leqq \tau < \infty} \int_{\tau}^{\infty} \|U(t, \tau)x\|^p dt < \infty \qquad (x \in \mathfrak{B}). \tag{6.5}$$

In fact, condition (6.5) can be written in the form

$$\sup_{0 \leqq \tau < \infty} \int_{0}^{\infty} \|e_\tau(t)U(t, \tau)x\|^p dt < \infty,$$

where $e_\tau(t)$ is the characteristic function of the interval $[\tau, \infty)$. This inequality implies the boundedness at each element x of the family of operators $V_\tau: x \to e_\tau(t)U(t, \tau)x$ acting from $\mathfrak{B}$ into the space $L_p(\mathfrak{B})$ of p power integrable functions $x(t)$ $(0 \leqq t < \infty)$ with values in $\mathfrak{B}$.

When $p \geqq 1$ the space $L_p(\mathfrak{B})$ with norm $\|x\|_p = \{\int_0^\infty \|x(t)\|^p dt\}^{1/p}$ is a Banach space. Therefore, by virtue of the uniform boundedness principle, the set of operators V_t is bounded in norm, and this is condition (6.4).

REMARK 6.3. The condition that the Bohl exponent be finite encountered in Theorems 6.1 and 6.2 is obviously satisfied if the operator function $A(t)$ is integrally bounded:

$$\sup_{t \geqq 0} \int_{t}^{t+1} \|A(\tau)\| d\tau < \infty$$

(a fortiori if it is simply bounded).

We note further that condition (6.2) is satisfied if the relation

$$\lim_{\tau \to \infty} \|x(t_0 + \tau)\| = 0$$

holds uniformly in t_0 and $x(t_0)$ for any solution of equation (6.1) satisfying the condition $\|x(t_0)\| < 1$.

Thus from the latter condition and the integral boundedness of $A(t)$ we obtain the negativeness of the Bohl exponent of equation (6.1). Moreover, it is easily seen that for equations with an integrally bounded $A(t)$ the condition in question is equivalent to the negativeness of the Bohl exponent.

2. *Auxiliary propositions on compact families of operators.*

LEMMA 6.1. *Suppose Γ is a compact*[6] *set of operators of* $[\mathfrak{B}]$ *whose spectra lie in an open set G. Then the set of resolvents $R_\lambda(A) = (A - \lambda I)^{-1}$, where $\lambda \notin G$ and $A \in \Gamma$, is bounded in norm.*

In addition, there exists a bounded closed set $F \subset G$ containing the spectra of all of the operators of Γ.

PROOF. Suppose there exist a sequence of operators $A_n \in \Gamma$ and a sequence of complex numbers $\lambda_n \notin G$ such that $\|R_{\lambda_n}(A_n)\| \to \infty$. It can be assumed without

[6] This means that every sequence of operators in Γ contains a subsequence converging to an operator of Γ.

loss of generality that $A_n \to A_0$ ($\in \Gamma$). But then Theorem I.2.1 implies $\|R_\lambda(A_n)\| \to \|R_\lambda(A_0)\|$. We have obtained a contradiction, which proves the first assertion of the lemma.

Suppose now λ_0 is a boundary point of G whose every neighborhood contains a point of the spectra of the operators of Γ. We consider a sequence $\lambda_n \to \lambda_0$, where $\lambda_n \in \sigma(A_n)$, $A_n \in \Gamma$. It can be assumed that $A_n \to A_0$ ($\in \Gamma$). But the spectrum $\sigma(A_0)$ is closed and lies in G. For sufficiently large n the set $\sigma(A_n)$ is in an arbitrarily small neighborhood of $\sigma(A_0)$. This again brings us to a contradiction, which proves the second assertion of the lemma.

LEMMA 6.2. *Suppose Γ is a compact set of operators of* [𝔅] *whose spectra lie in a fixed halfplane* Re $\lambda < -\nu_0$ ($\nu_0 > 0$). *Then there exists a constant N_0 such that for all $A \in \Gamma$*

$$\|e^{At}\| \leqq N_0 e^{-\nu_0 t}.$$

PROOF. As was shown in Lemma 6.1, there exists a domain G bounded by a simple smooth closed contour γ lying in the halfplane Re $\lambda < -\nu_0$ ($\nu_0 > 0$) and containing the spectra of all of the operators of Γ.

By the same Lemma 6.1 the resolvents $R_\lambda(A)$ are uniformly bounded outside G and, in particular, on the contour γ:

$$\|R_\lambda(A)\| \leqq c \qquad (\lambda \in \gamma, A \in \Gamma).$$

From the formula

$$e^{At} = -\frac{1}{2\pi i} \oint_\gamma e^{\lambda t} R_\lambda(A)\, d\lambda$$

we obtain the required estimate

$$\|e^{At}\| \leqq (c/2\pi) l e^{-\nu_0 t} \qquad (A \in \Gamma),$$

where l is the length of the contour γ.

DEFINITION. An operator function $A(t)$ ($0 \leqq t < \infty$) will be said to be *precompactly valued* if its range is precompact in [𝔅], i.e. if every sequence $A(t_n)$ contains a subsequence converging to an operator A of [𝔅].

Clearly, the continuity of $A(t)$ implies that it is precompactly valued on each finite interval of variation of t.

An operator C will be called an *ω-limit operator* of $A(t)$ if there exists a sequence $t_n \to \infty$ such that $A(t_n) \to C$.

Let $\Gamma = \{C\}$ be the set of all ω-limit operators of $A(t)$. It is easily seen that Γ is a closed set and moreover, as a subset of the precompact set $\{A(t)\} \cup \Gamma$, is precompact.

We emphasize that the values themselves of $A(t)$ are generally not contained in Γ. If the limit $\lim_{t\to\infty} A(t) = A(\infty)$ exists, for example, the set Γ of all ω-limit operators of $A(t)$ consists only of the single operator $A(\infty)$.

LEMMA 6.3. *Suppose the spectra of the ω-limit operators of a precompactly valued operator function $A(t)$ lie in one and the same halfplane* $\operatorname{Re}\lambda < -\nu_0$ $(\nu_0 > 0)$. *Then there exists a number $T_0 > 0$ such that when $t > T_0$*

$$\|e^{A(t)\tau}\| < N_0 e^{-\nu_0\tau},$$

where N_0 and ν_0 do not depend on t.

PROOF. By virtue of the preceding lemma it suffices to prove that there exists a T_0 such that the spectra of the operators $A(t)$ lie in the halfplane $\operatorname{Re}\lambda < -\nu_0$ when $t \geqq T_0$. Reasoning by contradiction, we suppose that there exists a sequence $t_n \to \infty$ for which each operator $A(t_n)$ has a point of its spectrum lying outside this open halfplane. It can be assumed without loss of generality that $A(t_n) \to C \in \Gamma$. Since the spectrum of C lies in the halfplane in question, the spectra of the $A(t_n)$, beginning with some n, also lie in it, which contradicts the supposition. The lemma is proved.

3. *Condition of negativeness of the Bohl exponent for an equation with a precompactly valued operator function.* We will say that an operator function $A(t)$ *satisfies condition $S_{\varepsilon,L}$* for some $\varepsilon > 0$ and $L > 0$ if there exists a number $T > 0$ such that the inequality $\|A(s) - A(t)\| \leqq \varepsilon$ is satisfied when $s, t \geqq T$ and $|s - t| \leqq L$.

Clearly, if $A(t)$ satisfies condition $S_{\varepsilon,L}$, it satisfies condition $S_{\varepsilon',L'}$ for any $\varepsilon' > \varepsilon$ and $L' \leqq L$ $(L' > 0)$ as well as condition $S_{n\varepsilon,nL}$ for any natural number n.

A function $A(t)$ is said to be *stationary at infinity* if it satisfies condition $S_{\varepsilon,L}$ for any arbitrarily small $\varepsilon > 0$ and some positive L (and hence arbitrarily large L).

In particular, it is easily seen that a function $A(t)$ is stationary at infinity if one of the following two conditions is satisfied:

a) the limit $\lim_{t\to\infty} A(t)$ exists;

b) $A'(t)$ exists for sufficiently large t and $\lim_{t\to\infty} A'(t) = 0$.

Clearly, a function of the form

$$A(t) = A_1(t) + A_2(t), \tag{6.6}$$

where $A_1(t)$ satisfies condition a) and $A_2(t)$ satisfies condition b), is also stationary at infinity:

Condition $S_{\varepsilon,L}$ is satisfied if

a′) the condition $\|A(t) - A_0\| \leqq \varepsilon/2$, where A_0 is a constant operator, is satisfied for sufficiently large t;

b′) $A'(t)$ exists and the estimate $\|A'(t)\| \leqq \varepsilon/L$ is valid for sufficiently large t.

On the other hand, if we set

$$A(t) = A_1(t) + A_2(t), \tag{6.7}$$

where

$$A_2(t) = \frac{1}{L}\int_t^{t+L} A(\tau)\,d\tau; \quad A_1(t) = \frac{1}{L}\int_t^{t+L} [A(t) - A(\tau)]\,d\tau,$$

we can represent a function satisfying condition $S_{\varepsilon,L}$ in the form of a sum of two functions such that

$$\|A_1(t)\| \leqq \varepsilon; \ \|A_2'(t)\| = \left\| \frac{A(t+L) - A(t)}{L} \right\| \leqq \frac{\varepsilon}{L}$$

for sufficiently large t.

In particular, letting ε tend to zero, we get that every stationary at infinity function is representable in the form (6.6), where $A_1(t)$ satisfies condition a) and $A_2(t)$ satisfies condition b).

THEOREM 6.3. *Suppose an operator function $A(t)$ is precompactly valued and the spectra of its ω-limit operators lie in one and the same halfplane* $\operatorname{Re}\lambda < -\nu_0$ $(\nu_0 > 0)$ (then by Lemma 6.3 there exists a $T_0 > 0$ such that when $t > T_0$

$$\|e^{A(t)\tau}\| \leqq N_0 e^{-\nu_0 \tau}, \tag{6.8}$$

where N_0 and ν_0 do not depend on t).

If, in addition, $A(t)$ satisfies condition $S_{\varepsilon,L}$ for sufficiently small $\varepsilon > 0$ and sufficiently large L $(\varepsilon < \nu_0/N_0,\ L > \ln N_0/(\nu_0 - N_0\varepsilon))$ depending only on the set Γ of ω-limit operators, the Bohl exponent κ_B of equation (6.1) *is negative.*

PROOF. By assumption, for sufficiently large τ $(\tau > t_0)$ we have

$$\|A(t) - A(\tau)\| < \varepsilon \qquad (\tau \leqq t \leqq \tau + L).$$

We apply estimate (2.22) for the case when $a = \tau$, $b = \tau + L$ and

$$A_1(t) \equiv A(\tau),\ A_2(t) = A(t) \qquad (\tau \leqq t \leqq \tau + L).$$

Since in the present case $U_1(t, s) = e^{(t-s)A(\tau)}$ while

$$\|e^{(t-s)A(\tau)}\| \leqq N_0 e^{-\nu_0(t-s)},$$

it can be asserted that

$$\|U(t, s)\| \leqq N_0 e^{-\nu(t-s)} \qquad (\tau \leqq s \leqq t \leqq \tau + L),$$

where $\nu = \nu_0 - N_0\varepsilon > 0$ since $\varepsilon < \nu_0/N_0$. Thus

$$\|U(\tau + L, \tau)\| \leqq N_0 e^{-\nu L} \ (= q).$$

It is easily seen that $q < 1$ for $L > \ln N_0/(\nu_0 - N_0\varepsilon)$. Therefore condition (6.2) of Theorem 6.1 is satisfied (for $\theta_{x,t} \equiv L$ and the indicated q) when t is sufficiently large. Since the precompactly valued operator function $A(t)$ is bounded, Theorem 6.1 is applicable (see Remarks 6.1 and 6.3) and consequently $\kappa < 0$.

REMARK 6.4. From the proof of the theorem it follows that the condition of a precompact range of the operator function $A(t)$ can be replaced by the condition that estimate (6.8) exist for sufficiently large t $(t > T_0)$.

The latter condition (i.e. estimate) will hold if the spectrum of $A(t)$ lies in the halfplane Re $\lambda < -\nu_0$ for $t > T_0$ and if, in addition, the resolvents $(A(t) - \lambda I)^{-1}$ are uniformly bounded on the straight line Re $\lambda = -\nu_0$ for $t > T_0$.

It can be shown directly that when estimate (6.8) is satisfied for $t > T_0$ the Cauchy problem

$$\begin{cases} \dfrac{dx}{dt} = A(t)x + f(t) & (0 \leqq t < \infty), \\ x(0) = 0 \end{cases}$$

has a bounded solution for any continuous vector function $f(t)$.

We recall that when $A(t)$ is integrally bounded the latter property is equivalent to the negativeness of the Bohl exponent of equation (6.1).

We formulate a theorem which in a certain sense can be regarded as the converse of Theorem 6.3.

THEOREM 6.4. *Suppose an operator function* $A(t)$ *is precompactly valued and the Bohl exponent* κ_B *of the corresponding equation* (6.1) *is negative* (*i.e. the corresponding equation* (6.1) *has property* $\mathscr{B}(\nu, N)$ *for certain positive* N *and* ν).

If, in addition, $A(t)$ *satisfies condition* $S_{\varepsilon,L}$ *for sufficiently small* $\varepsilon > 0$ *and sufficiently large* L ($\varepsilon < \nu/N$, $L > \ln N/(\nu - N\varepsilon)$), *the spectra of its* ω*-limit operators lie in one and the same halfplane* Re $\lambda \leqq -\nu_0$ ($\nu_0 > 0$).

PROOF. Suppose C is an ω-limit operator of $A(t)$, i. e. there exists a sequence $t_n \to \infty$ such that $A(t_n) \to C$ in $[\mathfrak{B}]$. Then for sufficiently large n and arbitrarily small $\delta > 0$

$$\|C - A(t_n)\| < \delta.$$

On the other hand, since by assumption $A(t)$ satisfies condition $S_{\varepsilon,L}$, for sufficiently large n,

$$\|A(t_n) - A(t)\| \leqq \varepsilon \qquad (t_n \leqq t \leqq t_n + L).$$

Therefore, for sufficiently large n

$$\|C - A(t)\| \leqq \varepsilon + \delta \qquad (t_n \leqq t \leqq t_n + L)$$

and by Lemma 2.3 (for $s = t_n$ and $t = t_n + L$) we get

$$\|e^{LC}\| \leqq Ne^{-\nu' t},$$

where $\nu' = \nu - N(\varepsilon + \delta)$.

It follows from Lemma I.2.2 that the spectrum $\sigma(C)$ of C lies in (and hence, by decreasing δ, in the interior of) the closed halfplane

$$\text{Re}\lambda \leqq \ln N/L - \nu' = -[\nu - N(\varepsilon + \delta) - \ln N/L] = -\nu_0.$$

Since by assumption $\nu - N\varepsilon - L^{-1} \ln N > 0$, we can choose δ so small that $\nu_0 > 0$. In view of the fact that the obtained value of ν_0 does not depend on C the theorem is proved.

As a corollary of Theorems 6.3 and 6.4 we obtain the following proposition.

THEOREM 6.5. *Suppose $A(t)$ is a precompactly valued operator function that is stationary at infinity. In order for the Bohl exponent κ_B of equation* (6.1) *to be negative it is necessary and sufficient that the spectra of the ω-limit operators of $A(t)$ lie in one and the same halfplane* Re $\lambda < -\nu_0$ ($\nu_0 > 0$).

By the *right spectral bound* of an operator A is meant the supremum of the real parts of the points of its spectrum $\sigma(A)$. This number coincides with the Ljapunov exponent (and hence the Bohl exponent) of the stationary equation $dx/dt = Ax$. The following fact is a consequence of the results presented above.

THEOREM 6.6. *Suppose $A(t)$ is a precompactly valued operator function that is stationary at infinity. Then the upper Bohl exponent κ of equation* (6.1) *coincides with the supremum of the right spectral bounds of the ω-limit operators of $A(t)$.*

PROOF. Let κ denote the supremum mentioned in the theorem. We pass from equation (6.1) to the equation

$$dx/dt = A_\alpha(t)x, \qquad \text{where } A_\alpha(t) = A(t) + \alpha I$$

(α is a real number).

Under such a passage the numbers κ_B and κ go over into the numbers $\kappa_B + \alpha$ and $\kappa + \alpha$. By virtue of Theorem 6.5 these numbers can be negative only simultaneously. Hence $\kappa = \kappa_B$.

REMARK 6.5. An analogous property is possessed by the lower Bohl exponent.

We note that in the case of a finite-dimensional space $\mathfrak{B}$ the condition of a precompact range of $A(t)$ is equivalent to the condition of boundedness of $A(t)$.

Clearly, Theorems 6.3–6.6 remain valid if the condition of a precompact range of $A(t)$ is replaced by the condition of a precompact range of $A(t)$ beginning with a sufficiently large t, i.e. by the condition of a precompact range of the translate $A(t + l)$ of $A(t)$ with a sufficiently large l.

EXERCISES

We adopt the following definition in the exercises below (see V. P. Potapov [1] and Ju P. Ginzburg [3]). Suppose given on $[a, b]$ a scalar function $f(t)$ and an operator function $F(t)$ with values in $[\mathfrak{B}]$. We consider the product

$$\Pi_\Delta = e^{f(\tau_n)[F(t_n)-F(t_{n-1})]} \cdots e^{f(\tau_1)[F(t_1)-F(t_0)]} \quad (t_{j-1} \leqq \tau_j \leqq t_j).$$

for an arbitrary subdivision $\Delta = \{a = t_0 < t_1 < \cdots < t_n = b\}$.

The limit

$$\lim_{|\Delta|\to 0} \Pi_\Delta = \int_a^{b\,\curvearrowleft} \exp[f(t)dF(t)] \qquad (|\Delta| = \sup_k (t_k - t_{k-1})) \tag{0.1}$$

is called a *multiplicative Stieltjes integral* if it exists in the norm topology of $[\mathfrak{B}]$.

1. Show that the multiplicative integral (0.1) exists if $f(t)$ is a continuous function and $F(t)$ is a continuous function of bounded variation in $[\mathfrak{B}]$.

Derive the estimates ($v(t) = \mathrm{Var}_{[0,t]}F$)

$$\left\| \overset{b}{\underset{a}{\widehat{\int}}} \exp[f(t)dF(t)] \right\| \leq e^{\int_a^b |f(t)|dv(t)};$$

$$\left\| \overset{b}{\underset{a}{\widehat{\int}}} \exp[f(t)dF(t)] - I \right\| \leq \left(\int_a^b |f(t)|dv(t) \right) e^{\int_a^b |f(t)|dv(t)};$$

$$\left\| \overset{b}{\underset{a}{\widehat{\int}}} \exp[f(t)dF(t)] - I - \int_a^b f(t)dF(t) \right\| \leq \left(\int_a^b |f(t)|dv(t) \right)^2 e^{\int_a^b |f(t)|dv(t)}$$

2. Prove the existence of the more general multiplicative Stieltjes integral

$$\overset{b}{\underset{a}{\widehat{\int}}} \exp[S_1(t)dF(t)S_2(t)], \tag{0.2}$$

where $S_1(t)$ and $S_2(t)$ are continuous operator functions with values in $[\mathfrak{B}, \mathfrak{B}_1]$ and $[\mathfrak{B}_1, \mathfrak{B}]$ respectively.

Use this integral to obtain a representation of the solution of the integral equation (1.23).

3. Consider the integral equation

$$x(t) = \int_0^t dF(\tau)x(\tau) + g(t), \tag{0.3}$$

where $g(t)$ is a continuous vector function of bounded variation.

Obtain formulas analogous to (1.21) for the solution of equation (0.3), using as $U(t, \tau)$ the multiplicative integral

$$U(t, \tau) = \overset{t}{\underset{\tau}{\widehat{\int}}} \exp\ [dF(s)].$$

Extend the results of Exercise 11 below to this case.

4. Let $A(t)$ and $B(t)$ $(a \leq t \leq b)$ be continuous operator functions with values in $[\mathfrak{B}]$. We introduce the "*alternating*" multiplicative integral

$$\overset{b}{\underset{a}{\widehat{\int}}} \exp\ [A(t)dt]\ \exp\ [B(t)dt] \tag{0.4}$$

$$= \lim_{|\Delta| \to 0} \prod_{k=1}^{n} e^{A(t_k)(t_k - t_{k-1})} e^{B(t_{k-1})(t_k - t_{k-1})}.$$

a) Extend Exercise I. 15 by showing that this integral exists and

$$\overset{b}{\underset{a}{\widehat{\int}}} \exp\ [A(t)dt]\ \exp\ [B(t)dt] = \overset{b}{\underset{a}{\widehat{\int}}} \exp\ [A(t) + B(t)]dt. \tag{0.5}$$

Establish the formula

$$\overset{t}{\underset{t_0}{\widehat{\int}}} \exp\ [A(\tau)d\tau]\ \exp\ [B(\tau)d\tau] = U(t, t_0) \overset{t}{\underset{t_0}{\widehat{\int}}} \exp\ [U^{-1}(\tau, t_0)B(\tau)U(\tau, t_0)d\tau],$$

where $U(t, t_0)$ is the evolution operator of the equation $dx/dt = A(t)x$.

Hint. Make the substitution $V(t, t_0) = U(t, t_0)S(t)$ in the equation $\dot{V} = A(t)V + B(t)V$.

b) Extend the results described above to the case of an integral of the form

$$\int_a^b \exp [dG(\tau)] \exp [dF(\tau)],$$

where $G(t)$ is a continuous operator function and $F(t)$ is an operator function of bounded variation.

This integral includes, in particular, the case when

$$G(t) = \int A(t)dt, \quad F(t) = \int B(t)dt,$$

where $A(t)$ and $B(t)$ are Bochner integrable operator functions. In this case we agree as before to write the multiplicative integral in the form (0.4).

The integral of form (0.4) and formula (0.5) for it can also be extended in certain cases to differential equations with unbounded operator functions. The results obtained in this direction lead to a representation of the solutions of partial differential equations of parabolic and Schrödinger types in the form of functional integrals (see Ju. L. Daleckiĭ [7]).

5. The following results extend a special case of Theorem 3.2.

We consider in a Banach space $\mathfrak{B}$ the stationary equation

$$dx/dt = Ax \tag{0.6}$$

and the equation

$$dy/dt = Ay + R(t)y \tag{0.7}$$

with an integrable operator function $R(t) \in [\mathfrak{B}]$:

$$\int_0^\infty \|R(t)\| dt < \infty. \tag{0.8}$$

We assume that

$$\sigma(A) = \sigma_-(A) \cup \sigma_0(A), \tag{0.9}$$

the spectral set $\sigma_0(A)$ lying on the imaginary axis and $\sigma_-(A)$ lying in the interior of the left halfplane.

a) Show that if in addition

$$\sup_{-\infty<t<\infty} \|e^{P_0At}\| < \infty, \tag{0.10}$$

where P_0 is the spectral projection of A corresponding to $\sigma_0(A)$, then equations (0.6) and (0.7) are asymptotically equivalent.

Hint. An attraction correspondence $x(t) \leftrightarrow y(t)$ between the solutions of equations (0.6) and (0.7) is established by the relation

$$x(t_0) = y(t_0) + \int_{t_0}^\infty P_0 e^{A(t_0-\tau)} R(\tau) y(\tau) d\tau,$$

where t_0 is a sufficiently large fixed number (the correspondence depends on the choice of t_0 provided $\sigma_-(A) \neq \emptyset$ (see Theorem 3.2)).

b) The corresponding result in a Hilbert phase space can be formulated more simply: if the stationary equation (0.6) is right stable and condition (0.9) is satisfied, equations (0.6) and (0.7) are asymptotically equivalent.

Condition (0.9) is atuomatically satisfied in the finite-dimensional case, and then the result coincides with Levinson's theorem (E. A. Coddington and N. Levinson [1]).

It is not clear whether condition (0.9) can be dropped in the infinite-dimensional case.

c) Condition (0.10) is satisfied if the operator $A_0 = P_0A$ is of simplest type (see Exercise I.1).

Extend the results described above to the case when A_0 is an operator of algebraic type (see Exercise I.9). In this connection condition (0.8) must be replaced by the condition

$$\int_{t_0}^{\infty} t^n \|R(t)\| dt < \infty, \tag{0.11}$$

where n is the maximum index of the imaginary eigenvalues of A_0 (see V. A. Jakubovič [3]).

6. In this and the following exercises we consider in a doubled Hilbert space $\mathfrak{H}^{(2)} = \mathfrak{H} \oplus \mathfrak{H}$ the differential equation on a halfline

$$\mathcal{J}_\Gamma \, dX/dr = \lambda X + V(r)X \qquad (0 \leqq r < \infty). \tag{0.12}$$

Here $\mathcal{J}_\Gamma = \begin{pmatrix} 0 & I \\ -I & 0 \end{pmatrix}$, where I is the identity operator in $\mathfrak{H}$, while $V(r)$ is an operator function with values in $[\mathfrak{H}^{(2)}]$ which is called the *potential* of equation (0.12). It will be convenient to understand by X the column

$$X(r; \lambda) = \begin{pmatrix} X_1(r; \lambda) \\ X_2(r; \lambda) \end{pmatrix},$$

whose elements are not vector functions with values in $\mathfrak{H}$ but operator functions with values in $[\mathfrak{H}]$.

When equation (0.12) has a Hermitian potential $V = V^*$ it is called a *canonical equation* (in the case of real $\mathfrak{H}$ and real λ it is called a *Hamiltonian equation* (cf. § V.4)).

We note that the operator $\mathcal{J}_\Gamma$ is an operator of simplest type with its spectrum on the imaginary axis consisting of the two points $\pm i$:

$$\mathcal{J}_\Gamma = iP^{(+)} - iP^{(-)} \qquad \left(P^{(\pm)} = \frac{1}{2}(I \mp i\mathcal{J}_\Gamma)\right), \tag{0.13}$$

while the operator $\mathcal{J} = i\mathcal{J}_\Gamma$ is an indefinite Hermitian operator.

a) Show that the Cauchy operator of equation (0.12) is

$\mathcal{J}$-unitary ($U\mathcal{J}_\Gamma U^* = U^*\mathcal{J}_\Gamma U = \mathcal{J}_\Gamma$)	for $V = V^*$	and $\operatorname{Im}\lambda > 0$;
$\mathcal{J}$-nonexpansive ($U^*(i\mathcal{J}_\Gamma)U \leqq i\mathcal{J}_\Gamma$)	for $\operatorname{Im} V \geqq 0$	and $\operatorname{Im}\lambda \geqq 0$;
$\mathcal{J}$-noncontractive ($U^*(i\mathcal{J}_\Gamma)U \geqq i\mathcal{J}_\Gamma$)	for $\operatorname{Im} V \leqq 0$	and $\operatorname{Im}\lambda \leqq 0$;
unitary ($U^*U = UU^* = I$)	for $\mathcal{J}V = V\mathcal{J}$	and $\operatorname{Re}\lambda = 0$.

b) Show that a canonical equation can be reduced by means of a unitary transformation $X = Q(t)Y$ ($QQ^* = Q^*Q = I$) to a form in which the Hermitian operator V ($= V^*$) is also $\mathcal{J}$-Hermitian ($\mathcal{J}V = V\mathcal{J}$).

Hint. Represent V in the form of a sum of $\mathcal{J}$-Hermitian and $\mathcal{J}$-skew-Hermitian terms:

$$V(r) = \frac{V(r) - \mathcal{J}_\Gamma V(r)\mathcal{J}_\Gamma}{2} + \frac{V(r) + \mathcal{J}_\Gamma V(r)\mathcal{J}_\Gamma}{2}$$

and make use of the last result of exercise a) (see also § 2 of Chapter IV).

7. Suppose in addition to the conditions of the preceding exercise that the potential $V(r)$ is integrable on $[0, \infty)$:

$$\int_0^\infty \|V(r)\| dr < \infty.$$

Then equation (0.12) and the equation

$$\mathcal{J}_\Gamma \, dY/dr = \lambda Y \tag{0.14}$$

are integrally comparable.

a) Show that equations (0.12) and (0.14) are asymptotically equivalent for $\operatorname{Im}\lambda = 0$ and that there exists precisely one operator function $A_V(\lambda)$ ($\operatorname{Im}\lambda = 0$) such that when $X(0; \lambda) = A_V(\lambda)Y(0; \lambda)$

$$\lim_{r\to\infty} \|X(r; \lambda) - Y(r; \lambda)\| = 0.$$

The operator $A_V(\lambda)$ is called the *asymptotic equivalence operator* (*A-operator*) of equations (0.12) and (0.14).

Hint. Make use of Theorem 3.2.

b) Prove that an A-operator is a $\mathcal{J}$-unitary operator when $V = V^*$ and a $\mathcal{J}$-contraction when $\operatorname{Im} V \geqq 0$ ($\operatorname{Im} \lambda = 0$).

Hint. Show that

$$A_V(\lambda) = \lim_{r \to \infty} (U_1^{-1}(r; \lambda) U_2(r; \lambda),$$

where U_1 and U_2 are the Cauchy operators of equations (0.12) and (0.14), and make use of the result of Exercise 6a).

c) Show that there exist solutions $X^{(k)}(r; \lambda)$ ($k = 1,2$) of equation (0.12) satisfying the initial conditions

$$X_2^{(1)}(0; \lambda) = 0, \quad X_1^{(2)}(0; \lambda) = 0$$

and having the following asymptotic behavior at infinity for $\operatorname{Im} \lambda = 0$:

$$X^{(1)}(r; \lambda) = \begin{pmatrix} I \\ -iI \end{pmatrix} e^{i\lambda r} S_1(\lambda) + \begin{pmatrix} I \\ iI \end{pmatrix} e^{-i\lambda r} + o(1);$$

$$X^{(2)}(r; \lambda) = \begin{pmatrix} iI \\ I \end{pmatrix} e^{i\lambda r} S_2(\lambda) + \begin{pmatrix} -iI \\ I \end{pmatrix} e^{-i\lambda r} + o(1), \quad (r \to \infty; I = I_{\mathfrak{H}}),$$

where the $S_k(\lambda)$ ($k = 1,2$) are operator functions with values in $[\mathfrak{H}]$ (the *S-matrices* of equation (0.12) in the terminology of quantum mechanics).

d) Prove that the S-matrices of equation (0.12) are unitary matrices when $V = V^*$ ($\operatorname{Im} \lambda = 0$).

Hint for Exercises 7c) *and* 7d). Use the fact that the leading terms of the indicated asymptotic expansions are solutions of the unperturbed equation (0.14) and therefore

$$\begin{pmatrix} X_1^{(1)}(0; \lambda) & 0 \\ 0 & X_2^{(2)}(0; \lambda) \end{pmatrix} = \begin{pmatrix} A_{11}(\lambda) & A_{12}(\lambda) \\ A_{21}(\lambda) & A_{22}(\lambda) \end{pmatrix} \begin{pmatrix} S_1(\lambda) + I & i(S_2(\lambda) - I) \\ -i(S_1(\lambda) - I) & S_2(\lambda) + I \end{pmatrix},$$

where $\|A_{jk}(\lambda)\|_1^2 = A_V(\lambda)$. The $S_k(\lambda)$ ($k = 1,2$) are found by equating the nondiagonal terms of the product of matrices on the right side of the equality. Since an A-operator is a $\mathcal{J}$-unitary operator, the $S_k(\lambda)$ ($k = 1,2$) are unitary matrices when $V = V^*$ ($\operatorname{Im} \lambda = 0$) (see Exercise 6a)).

It should be noted that when $\dim \mathfrak{H} < \infty$ the S-matrices $S_k(\lambda)$ ($k = 1,2$) for the canonical equation (0.12) admit a completely intrinsic characterization and either of them can be used to uniquely recover the normalized potential V ($V = V^*$, $\mathcal{J}V = V\mathcal{J}$) and consequently the matrix $A_V(\lambda)$. Moreover, there exist comparatively simple formulas for directly recovering $A_V(\lambda)$ by either of the S-matrices. All of these results are in the papers of M. G. Kreĭn and F. È. Melik-Adamjan [1,2], from which Exercises 7a)—7d) have been taken. Other aspects of the S-matrix for canonical equations in Hilbert space ($\dim \mathfrak{H} \leqq \infty$) have been studied by V. M. Adamjan [1].

8. Show that Theorem 6.2 remains valid if condition (6.4) is replaced by the condition

$$\left\{\int_{t_0}^{t} \|U(t, \tau)x\|^p d\tau\right\}^{1/p} \leqq c \|x\|$$

or when $p \geqq 1$ by

$$\sup_{t_0 \leqq t < \infty} \int_{t_0}^{t} \|U(t, \tau)x\|^p d\tau < \infty.$$

9. Prove that the upper Bohl exponent of equation (4.1) is negative if the following conditions are satisfied (E. A. Barbašin [1], Lemma 5.5):

$$\sup_{t_0 \leqq t < \infty} \int_{t_0}^{t} U\|(t, \tau)\| d\tau < \infty; \quad \sup_{t_0 \leqq t < \infty} \|U(t, t_0)\| < \infty.$$

Hint. Make use of the result of the preceding exercise.

10. Prove the following theorem (D. L. Kučer [1]): *in order for the Cauchy problem* (5.1)—(5.2) *to have a bounded solution for any vector function* $f(t)$ *satisfying the condition* $\int_0^\infty \|f(t)\|^p\, dt < \infty$ *for some fixed* $p > 1$ *it is necessary that the following condition be satisfied for certain* $N > 0$, $\nu > 0$:

$$\|U(t, 0)\| \leqq Ne^{-\nu t^{1/q}} \qquad (1/p + 1/q = 1, \quad t > 0),$$

and if $\sup_t \|A(t)\| < \infty$ *it is necessary and sufficient that the upper Bohl exponent of equation* (5.1) *be negative.*

11. Prove the following theorem (R. Bellman [1] and D. L. Kučer [1]): *in order for the Cauchy problem* (5.1)—(5.2) *to have a bounded solution for any vector function* $g(t)$ *satisfying the condition* $\int_0^\infty \|g(t)\| dt < \infty$ *it is necessary that the following condition be satisfied*:

$$\sup_{t \geqq 0} \|U(t, 0)\| < \infty,$$

and if $\sup_t \|A(t)\| < \infty$ *it is necessary and sufficient that the following condition be satisfied*:

$$\sup_{t, t_0 \geqq 0} \|U(t, t_0)\| < \infty.$$

12. Verify that the proofs of the theorems formulated in Exercises 10 and 11 can be extended to the case obtained by replacing the boundedness condition $\sup_t \|A(t)\| < \infty$ by the integral boundedness condition $\sup_t \int_t^{t+1} \|A(t)\| dt < \infty$ (E. A. Barbašin [1]).

13. Consider in place of equation (5.1) the more general integral equation

$$x(t) = x_0 + \int_0^t A(\tau)x(\tau)d\tau + g(t), \tag{0.15}$$

where $g(t)$ is a continuous function of strong bounded variation.

Prove the following theorem (E. A. Barbašin [1]): *in order for equation* (0.15) *to have a bounded solution for every function* $g(t)$ *of bounded variation on the halfline* $[0, \infty)$ *it is necessary and sufficient that the following condition be satisfied*:

$$\sup_{t, \tau} \|U(t, \tau)\| < \infty.$$

14. Under the conditions of the preceding exercise prove the following theorem (E. A. Barbašin [1]): *in order for equation* (0.15) *to have a bounded solution for every function* $g(t)$ *satisfying the condition* $\sup_{t>0} \mathrm{Var}_{[t, t+1]}\, g(t) < \infty$ *it is necessary and sufficient that the upper Bohl exponent of the equation* $x'(t) = A(t)x(t)$ *be negative.*

15. We consider in a Hilbert phase space $\mathfrak{H}$ the differential equation

$$dx/dt = A(t)x \qquad (0 \leqq t < \infty) \tag{0.16}$$

the real part of the coefficient of which is integrally bounded:

$$\sup_{0 \leqq t < \infty} \int_t^{t+1} \mathrm{Re}\, A(\tau) d\tau < \infty.$$

Show that the following assertions are equivalent.

a) The upper Bohl exponent of equation (0.16) is negative.

b) The Cauchy problem

$$\begin{cases} dx/dt = A(t)x + f(t) & (0 \leqq t < \infty), \\ x(0) = 0 \end{cases} \tag{0.17}$$

has on $[0,\infty)$ a bounded solution $x(t)$ for each bounded continuous vector function $f(t)$.

c) There exists a bounded uniformly positive operator function $W(t) \gg \alpha I$ $(0 \leqq t < \infty)$ such that the following condition is satisfied for any solution $x(t)$ of equation (0.16):

$$(d/dt)(W(t)x(t), x(t)) \leqq -\beta\|x(t)\|^2 \qquad (\beta > 0, 0 \leqq t < \infty).$$

The equivalence of these assertions under the condition that $A(t)$ be bounded was proved in the finite-dimensional case by I. G. Malkin [3] with the use of Perron's technique of transforming the homogeneous equation into a triangular form [instead of b), Malkin dealt with the stronger assertion that the solutions of equation (0.17) are bounded under any initial condition $x(0) = x_0$ and not just the condition $x(0) = 0$].

Hint. Go over from equation (0.16) to the equation $\dot{x} = [\operatorname{Re} A(t)]x$ by means of a unitary kinematic similarity transformation (see § IV.2). The equivalence of a) and b) then follows from Theorem 5.2. To determine the operator $W(t)$ consider the differential equation

$$dW/dt + WA + A^*W = -H(t)$$

for a uniformly positive $H(t)$.

Show that the bounded solution of this equation is given by the formula

$$W(t) = \int_t^\infty U^*(\tau, t)H(t)U(\tau, t)d\tau.$$

16. Extend the result of the preceding exercise to the case of a Banach phase space under the assumption that $A(t)$ is integrally bounded.

Hint. Introduce in $\mathfrak{B}$ the new norm (depending on t)

$$\|x\|_t = \int_t^\infty \|U(\tau, t)x\|d\tau$$

(see § II.2).

17. Suppose that a periodic operator function $A(t) = A(t + T)$ is continuous and that for each $t \in [0, T]$ the spectrum $\sigma(A(t))$ lies in the halfplane $\operatorname{Re} \zeta \leqq -\nu < 0$.

a) Show that the equation $dx/dt = \lambda A(t)x$ has a negative Bohl exponent for all sufficiently large positive λ and, moreover, has property $\mathscr{B}(\lambda\nu_0, N)$ for some $\nu_0 > 0$.

This assertion is equivalent to the following: the Bohl exponent of the equation $dx/dt = A(\varepsilon t)x$ is negative for $0 \leqq \varepsilon < \varepsilon_0$ whenever ε_0 is sufficiently small.

Hint. Make use of Theorem 6.3 (for a direct proof see Lemma VIII.2.1).

Suppose the spectrum $\sigma(A_0)$ of an operator A_0 lies in the interior of the left halfplane and $B(t)$ is a continuous periodic operator function.

b) Show that there exist constants $q > 0$ (depending only on A_0) and $\varepsilon_0 > 0$ such that the Bohl exponent of the equation

$$dx/dt = (A_0 + B(\varepsilon t))x$$

is negative for $\|B\| \leqq q$, $0 \leqq \varepsilon < \varepsilon_0$.

Hint. Make use of the result of Exercise 15a).

c) Extend the results given in a) and b) to the case of almost periodic operator functions $A(t)$ and $B(t)$ or to the still more general case of precompactly valued uniformly continuous operator functions.

18. Consider in a Hilbert phase space $\mathfrak{H}$ the equation

$$A\frac{d^2x}{dt^2} + B\frac{dx}{dt} + [C_0 + C_1(\varepsilon t)]x = 0, \tag{0.15a}$$

where A, $\operatorname{Re} B$ and C_0 are uniformly positive operators and $C_1(\tau)$ is a periodic operator function.

a) Show that there exist constants $q > 0$ (depending only on the operators A, B and C_0) and $\varepsilon_0 > 0$ such that the upper Bohl exponent of equation (0.15a) is negative for $0 \leqq \varepsilon < \varepsilon_0$ and $0 < \|C_1(\tau)\| \leqq q$.

This assertion can be interpreted in the theory of parametric resonance as follows: if a damped system is exponentially stable (Re $B \gg 0$), it will remain exponentially stable under a sufficiently small and sufficiently slow parametric excitation.

Hint. Make use of the results of Exercises 15b) and II. 8.

b) In the scalar case it is known that for sufficiently large real B equation (0.15a) has solutions that together with their first derivatives tend to zero for any $\varepsilon > 0$. In the case considered by us the validity of this assertion for small $\varepsilon > 0$ follows from a); for large $\varepsilon > 0$ it will follow from the results of Chapter V. The conditions under which it is valid for $B = B^*$ and all $\varepsilon > 0$ are not known (in this case the equation is said to be *unlimitedly stable*). An interesting investigation of unlimited stability in the scalar case has been conducted by V. A. Jakubovič in [4], where the literature on this question is indicated.

19. Obtain for the solutions of equation (4.1) with a continuous operator function $A(t)$ in an arbitrary Banach space $\mathfrak{B}$ estimates generalizing the Wintner estimates (4.19) (S. M. Lozinskiĭ [1]). Use them to derive estimates for the Bohl exponents of the equation.

Hint. Replace the expressions $\lambda_M(A_\mathfrak{R}(\tau))$ and $\lambda_m(A_\mathfrak{R}(\tau))$ by $\Lambda[A(\tau)]$ and $-\Lambda[-A(\tau)]$ respectively (see Exercises I.18—I.20).

NOTES

The Ljapunov exponents were introduced back in 1892 under the name of characteristic numbers (differing from them in sign) for the finite-dimensional case of solutions of differential equations in the famous doctoral dissertation of A. M. Ljapunov. The results obtained by him have received further development in many papers (see B. F. Bylov et al. [1]).

The investigations presented in this book are closely connected with the notion of Bohl exponents; in particular, the main results of the present chapter are concentrated about the upper Bohl exponent. This fundamental notion was first introduced by P. Bohl in 1913 in a paper [2] published in a well-known mathematical journal; but for some reason, like many other results of this remarkable and apparently very modest mathematician, it went unnoticed. An attempt by the present authors to establish just what was done by P. Bohl in this memoir has led them to some sensational (at least for them) conclusions.[7]

This work actually contains a number of results that are presently well known from the works of other authors published up to 15 and more years later (see also the Notes to Chapter VII). The upper Bohl exponent taken with opposite sign was called the index by P. Bohl. And he considered the property $\mathscr{B}(\nu, N)$, calling the numbers N and ν/N the auxiliary and principal stability coefficients respectively. Bohl arrived at these notions by essentially studying (if one uses modern terminology) the question of stability under constantly acting perturbations (see Exercise VII. 9). He proceeded under the methodological premise that uncontrollable dissipation of energy always plays

[7] The contributions of Bohl to the theory of almost periodic functions were made long ago and are well known (he is the creator of the theory of quasiperiodic functions). His results concerning continuous flows on a torus have found a place in textbooks and monographs. A. D. Myškis and I. M. Rabinovič wrote a sensational survey (Uspehi Mat. Nauk **10** (1955), no. 3 (65), 188–192) of his results on continuous vector fields, which contain as a trivial consequence the famous fixed point theorem of Brouwer (these results were published in 1904, i.e. 5 years before the first paper of Brouwer) and were regarded by Bohl as auxiliary results of doubtful originality and of interest mainly for their applications in the theory of differential equations.

His work is now known in the Soviet Union thanks to the publication in 1961 of his selected works (P. Bohl [3]) with an introductory article by A. D. Myškis and I. M. Rabinovič, the brochure [1] of these two authors and the jubilee readings devoted to the memory of Bohl that were organized by the Academy of Sciences of the Latvian Soviet Socialist Republic in 1965.

In spite of all this, the priority of Bohl in the above mentioned matters connected with the Bohl exponent has yet to be generally recognized.

an essential role in the problems of geomechanics (Erdische Mechanik) and hence that the behavior of the solutions of correctly posed problems must be stable relative to small perturbations of the equations. It is noteworthy that Bohl's unperturbed equation was generally nonlinear. With this approach he established the stability of the upper Bohl exponent. His arguments essentially made use of an assertion similar to the assertions of Corollaries 2.1 and 2.2 and encountered in the contemporary literature under a wide variety of names. We will not cite the long and unfortunate list of papers especially devoted to the rediscovery of these and similar estimates.

The property $\mathscr{B}(\nu, N)$ appeared later in the papers of K. P. Persidskiĭ [1, 2, 3] in connection with a study of the question of asymptotic stability of the solutions of finite nonlinear systems of equations with a nonstationary principal linear part. Persidskiĭ discovered a criterion similar to the one formulated in Theorem 6.1, introduced the notion of a function stationary at infinity (with weak variation at infinity, in his terminology) and proved that a system of equations has property $\mathscr{B}(\nu, N)$ $(\nu > 0)$ if its coefficients are stationary at infinity and the spectrum of its coefficient matrix lies in the interior of the left halfplane.

All of these results are actually contained in the above mentioned paper of P. Bohl. Moreover, this paper also essentially contains (for finite-dimensional systems) the test for negativeness of the Bohl exponent formulated in Theorem 6.3, where the property of being stationary at infinity has been replaced by the condition $S_{\varepsilon, L}$ for sufficiently small (but not arbitrarily small) $\varepsilon > 0$ and sufficiently large (but not arbitrarily large) $L > 0$. This result was later rediscovered by N. Ljaščenko [1, 2] and M. A. Rutman [1]. The latter obtained this result for the infinite-dimensional case, generalizing the investigations of M. G. Kreĭn, who had generalized and sharpened Persidskiĭ's results.

The following proposition is an elementary corollary of a basic result (concerning nonlinear equations) in the paper of Bohl.

In order for an equation

$$dx/dt = A(t)x \qquad (0 \leq t < \infty)$$

with bounded continuous coefficients to have the property that

$$\sup_t |y(t) - x(t)| < KM_y, \tag{1}$$

whenever $y(t)$ is a function such that $M_y = \sup_t |dy/dt - A(t)y| < \infty$ and $x(t)$ is the solution satisfying the condition $x(0) = y(0)$, it is necessary and sufficient that the Bohl exponent of this equation be negative.

Under the indicated restrictions on $A(t)$ this result would be completely equivalent to the result of Theorem 5.2 if condition (1) were replaced by the condition

$$\sup_t |y(t) - x(t)| > \infty. \tag{2}$$

In fact, condition (2) implies condition (1) by virtue of the uniform boundedness principle, which unfortunately was not available to Bohl.

Theorem 5.2 and the closely related Theorem 5.1 were established directly for a Banach space by M. G. Kreĭn in [2], where the passage from condition (2) to condition (1) is effected with the use of the uniform boundedness principle. These theorems sharpened and strengthened the corresponding results of I. G. Malkin [3] (see Exercise 15) even for the finite-dimensional case.

It should be noted that Bohl's arguments differ from the arguments of Persidskiĭ, Malkin and others in that they do not make use of the cumbersome methods connected with the application of the (now) well-known theorem of Perron [1] (1928) on the reduction to triangular form of a system of differential equations (this theorem was "fortunately" unavailable to Bohl).

Unlike Bohl's methods, the methods connected with the application of Perron's theorem (although useful in many problems) do not appear to admit a direct extension to the infinite-dimensional case.

In 1947 M. G. Kreĭn noted (without knowing of Bohl's work) that the methods of functional analysis can be used to significantly simplify all of these results (with some sharpenings even in the finite-dimensional case) and to extend them to the case of equations in a Banach space.[8)] A number of results, in particular Theorem 5.2 and Theorem 6.5, which extends the above-mentioned result of Persidskiĭ on equations with operator functions stationary at infinity, were announced by him at a meeting of the Moscow Mathematical Society and published in [2]. They are contained in the "Lectures" along with the other results described in §§ 4—6.

In the course of writing up this material we obtained a conversion of Theorem 6.3 (Theorem 6.4) and an exact formula for the upper Bohl exponent in the case of a precompactly valued operator function. Questions similar to these investigations were also considered by V. M. Millionščikov [1].

The method of M. G. Kreĭn was subsequently extended by D. L. Kučer [1,2] (some of his results are given in the Exercises).

D. L. Kučer considered, in particular, equations in the wider class of weakly, and not just strongly, measurable functions. This class turns out to be not only more general but also more convenient in the applications, since weak measurability is more readily verified in individual Banach spaces than strong measurability. This is especially true in considering countable systems of differential equations, in application to which the investigations of D. L. Kučer have led in a natural way to significantly more general results than those obtained by K. P. Persidskiĭ in [4].

We note some other results contained in this chapter. The estimates for the solutions of the homogeneous equation in Hilbert space contained in Theorem 4.7 were established by A. Wintner [1] and later repeated by many authors. The example cited in § 4.4 is due to O. Perron [2]. The decomposition (6.7) for the special case of scalar functions stationary at infinity was obtained in a more complicated way by N. I. Gavrilov [1]. The elegant simple argument giving the general decomposition (6.7) directly for vector functions with values in a Banach space is due to M. L. Brodskiĭ.

The presentation of the majority of the results of this chapter differs from the presentation of the corresponding results in the "Lectures" by the fact that the requirement that the coefficient of the equation be bounded has been replaced by the requirement that it be integrally bounded.

A systematic consideration of differential equations with integrally bounded coefficients was apparently first carried out in the papers and book [1] of J. Massera and J. Schäffer.

[8)] The important role of the uniform boundedness principle in investigating the solutions of equations in a finite-dimensional phase space was independently discovered by R. Bellman [1]. This paper underwent some criticism by D. L. Kučer [1].

CHAPTER IV

EXPONENTIAL SPLITTING OF THE SOLUTIONS OF THE LINEAR EQUATION

In this chapter we continue our investigation of the solutions of the linear equation. Our main purpose is to conduct a study of the notion of an exponential dichotomy of the solutions and of the more general notion of an exponential splitting of order n. By this we mean, roughly speaking, the existence of a direct decomposition of the space into n subspaces with nonintersecting intervals of Bohl exponents of the solutions initially in these subspaces.

The notion of an exponential dichotomy is introduced in § 3, where we also study its role in questions concerning the existence of bounded solutions of the inhomogeneous equation. In § 4 we consider equations with an exponential splitting of higher order. In § 5 we show that the exponential splittability property is stable with respect to small perturbations of the equation. A deeper investigation of the exponential splittability of an equation can be carried out in a Hilbert phase space; in particular, we are able to establish in this case an exponential splittability test for equations with precompactly valued operator functions (§ 6).

These considerations are essentially based on a new geometric method involving the use of the so called conjugation operators for projections. The fundamentals of these operators are presented in § 1. Conjugation operators naturally arise in the study of certain questions connected with the general theory of perturbations of linear operators. In this book we also make use of them in Chapter VIII.

Of importance in the study of the various growth characteristics of the solutions of an equation are the kinematic similarity transformations under which these characteristics are invariant, as well as the related notion of reducibility of an equation. These questions are considered in § 2.

§ 1. Conjugation operators for projections

1. *A conjugation operator function.* We consider in a space $\mathfrak{B}$ two decompositions of the identity

$$\sum_{k=1}^{n} P_k = I; \quad \sum_{k=1}^{n} Q_k = I, \tag{1.1}$$

composed of pairwise disjoint projections:

$$P_kP_j = \delta_{kj}P_k; \quad Q_kQ_j = \delta_{kj}Q_k \qquad (1 \leqq k, j \leqq n). \tag{1.2}$$

Let

$$S = \sum_{j=1}^{n} Q_j P_j. \tag{1.3}$$

Clearly, this operator has the properties

$$S(P_j\mathfrak{B}) \subset Q_j\mathfrak{B}; \quad SP_j = Q_jS \qquad (j = 1,2,\cdots, n).$$

It can be written in the form

$$S = I + \sum_{j=1}^{n} Q_j(P_j - Q_j).$$

Therefore, if

$$\sum_{j=1}^{n} \|Q_j\| \, \|P_j - Q_j\| < 1, \tag{1.4}$$

then the operator S has a bounded inverse, the above inclusion relations are converted into equalities:

$$S(P_j\mathfrak{B}) = Q_j\mathfrak{B} \qquad (j = 1,2, \cdots,n),$$

and the operators P_j, Q_j are found to be similar:

$$SP_jS^{-1} = Q_j \qquad (j = 1,2,\cdots,n). \tag{1.5}$$

Suppose now

$$\sum_{j=1}^{n} P_j(t) = I \qquad (a \leqq t \leqq b) \tag{1.6}$$

is a decomposition composed of continuous pairwise disjoint projection functions on $[a, b]$:

$$P_k(t)P_j(t) = \delta_{kj}P_k(t) \qquad (a \leqq t \leqq b). \tag{1.7}$$

Setting $P_j = P_j(t)$ and $Q_j = P_j(t + \tau)$, we form the operator

$$S(t + \tau, t) = \sum_{j=1}^{n} P_j(t + \tau)P_j(t).$$

We note that the continuity implies the existence of a number $\delta > 0$ such that when $|\tau| < \delta$

$$\max_t \sum_{j=1}^{n} \| P_j(t + \tau) \| \, \| P_j(t + \tau) - P_j(t) \| < 1$$

on the finite interval $[a, b]$. The operator $S(t + \tau, t)$ therefore has a bounded inverse when $|\tau| < \delta$.

We consider a subdivision q of $[a, b]$:

$$a = t_0 < t_1 < \cdots < t_{k-1} < t_k < t_{k+1} < \cdots < t_m = b$$

such that

$$\max_k |t_k - t_{k-1}| < \delta,$$

and construct the operator function

$$\begin{aligned} \Omega_q(t) &= S(t, t_k)\, S(t_k, t_{k-1}) \cdots S(t_1, t_0) \\ &= \sum_{j=1}^{n} P_j(t)P_j(t_k)P_j(t_{k-1}) \cdots P_j(t_1)P_j(a) \end{aligned} \tag{1.8}$$

for $t_k \leqq t \leqq t_{k+1}$.

This operator function has the following properties.

1) $\Omega_q(a) = S(a, a) = I$.

2) The operators $\Omega_q(t)$ and $\Omega_q^{-1}(t)$ are bounded and continuous functions of t.

The invertibility of $\Omega_q(t)$ follows from the invertibility of each factor in (1.8), while its continuity follows from the continuity of $S(t, \tau)$ and the relation $S(t, t) = I$.

3) The operator $\Omega_q(t)$ maps the subspace $P_j(a)\mathfrak{B}$ onto $P_j(t)\mathfrak{B}$ ($j = 1,\cdots, n$), the corresponding projections being similar:

$$\Omega_q(t)P_j(a)\Omega_q^{-1}(t) = P_j(t). \tag{1.9}$$

In fact,

$$\Omega_q(t)P_j(a) = P_j(t)P_j(t_k) \cdots P_j(t_1)P_j(a) = P_j(t)\Omega_q(t).$$

4) If the projections $P_j(t)$ satisfy a Lipschitz condition

$$\| P_j(t_2) - P_j(t_1) \| \leqq c|t_2 - t_1| \qquad (j = 1,2,\cdots,n;\ t_1, t_2 \in [a, b]),$$

so does the operator $\Omega_q(t)$.

For it suffices to verify this property in each closed interval $[t_{k-1}, t_k]$, where it is obviously satisfied.

2. *Differential conjugation equation.* The constructed operator $\Omega_q(t)$ depends on the choice of the subdivision of the interval $[a, b]$. This psychologically disturbing defect can be removed by passing to the limit with respect to the filter formed from the subdivisions of $[a, b]$. The validity of such a limit passage is assured under certain conditions (see Exercise 1). In this way we obtain an operator function $\Omega(t)$, which has certain properties in addition to those of $\Omega_q(t)$.

This operator function admits a quite simple description when the projection functions $P_j(t)$ are differentiable. In fact if follows from (1.8) in this case that

$$\begin{aligned} \Omega_q'(t) &= \sum_{j=1}^{n} P_j'(t)P_j(t_k) \cdots P_j(a) \\ &= \left[\sum_{j=1}^{n} P_j'(t)P_j(t_k) \right]\left[\sum_{j=1}^{n} P_j(t_k) \cdots P_j(a) \right] \\ &= \left[\sum_{j=1}^{n} P_j'(t)P_j(t_k) \right]\Omega_q(t_k). \end{aligned}$$

If we could pass to the limit in the latter formula under the condition that $\max_k|t_k - t_{k-1}| \to 0$, we would obtain the differential equation

$$\Omega'(t) = \left[\sum_{j=1}^{n} P_j'(t)P_j(t)\right]\Omega(t) \tag{1.10}$$

Following these heuristic arguments we study the properties of the solutions of equation (1.10), which is called the *conjugation equation for the projections* $P_j(t)$.

THEOREM 1.1. *Suppose the disjoint projection functions* $P_j(t)$ ($j = 1,\cdots,n$) *giving a decomposition of the identity are differentiable. Then the evolution operator* $\Omega(t, \tau)$ ($a \leqq t, \tau \leqq b$) *of equation* (1.10) *has in addition to the usual properties* a)–d) (*see* § III.1.5) *the intertwining property*

e) $P_j(t)\Omega(t, \tau) = \Omega(t, \tau)P_j(\tau) \qquad (a \leqq t, \tau \leqq b; j = 1,2,\cdots,n)$;

and the property that

f) *it is a W-unitary operator when the space* $\mathfrak{B}$ *is a Hilbert space* $\mathfrak{H}$ *and the operators* $P_j(t)$ *are W-Hermitian with respect to some scalar product* $(x, y)_W = (Wx, y)$ (*possibly indefinite*).

We note that the intertwining property admits a simple geometric interpretation. To the decomposition (1.6) there corresponds a direct decomposition of the space $\mathfrak{B}$:

$$\mathfrak{B} = \mathfrak{B}_1(t) \dotplus \mathfrak{B}_2(t) \dotplus \cdots \dotplus \mathfrak{B}_n(t) \qquad (a \leqq t \leqq b). \tag{1.11}$$

By virtue of properties a), c) and e) the operator $\Omega(t, \tau)$ effects a linear isomorphism between the subspaces $\mathfrak{B}_j(\tau)$ and $\mathfrak{B}_j(t)$ ($j = 1,\cdots,n$; $t, \tau \in [a, b]$) and thus "tracks" the subspaces $\mathfrak{B}_j(t)$ under a variation of t. When the decompositions (1.6) and (1.11) are orthogonal (W-orthogonal), this isomorphism, as follows from property f), is isometric (W-unitary).

As in (1.9), the operators $\Omega(t, \tau)$ effect a conjugation of the projections:

$$P_j(t) = \Omega(t, \tau)P_j(\tau)\Omega^{-1}(t, \tau) \qquad (a \leqq t, \tau \leqq b; j = 1,2,\cdots,n).$$

PROOF OF THE THEOREM. We differentiate relation (1.7), multiply it from the right by P_j and sum the resulting equality over j. As a result, we get

$$P_k'(t) + \sum_{j=1}^{n} P_k(t)P_j'(t)P_j(t) = P_k'(t)P_k(t). \tag{1.12}$$

We now consider the operator

$$S_k(t, \tau) = P_k(t)\Omega(t, \tau) - \Omega(t, \tau)P_k(\tau).$$

It clearly satisfies the condition $S_k(\tau, \tau) = 0$.

If we can prove that it satisfies the differential equation

$$\frac{dS_k(t, \tau)}{dt} = \left[\sum_{j=1}^{n} P_j'(t)P_j(t) \right] S_k(t, \tau), \tag{1.13}$$

the uniqueness of the solution of the Cauchy problem for equation (1.13) will imply the identity $S_k(t, \tau) \equiv 0$ equivalent to condition e) of the theorem.

It suffices to show that equation (1.13) is satisfied by the first summand of $S_k(t, \tau)$, since it is obviously satisfied by the second. We differentiate this summand $S_{k1}(t, \tau)$ with respect to t and make use of relation (1.12) as well as the equation for the evolution operator

$$\Omega'(t, \tau) = \left[\sum_{j=1}^{n} P_j'(t)P_j(t) \right] \Omega(t, \tau).$$

This gives

$$\begin{aligned} S_{k1}'(t, \tau) &= P_k'(t)\Omega(t, \tau) + P_k(t)\Omega'(t, \tau) \\ &= \left[P_k'(t) + \sum_{j=1}^{n} P_k(t)P_j'(t)P_j(t) \right] \Omega(t, \tau) = P_k'(t)P_k(t)\Omega(t, \tau) \\ &= \left[\sum_{j=1}^{n} P_j'(t)P_j(t) \right] P_k(t)\Omega(t, \tau) = \left[\sum_{j=1}^{n} P_j'(t)P_j(t) \right] S_{k1}(t, \tau). \end{aligned}$$

For a proof of property f) it suffices to show, as follows from Lemma III.1.1, that under the considered conditions the operator $\sum_{j=1}^{n} P_j'(t)P_j(t)$ will be W-skew-Hermitian.

Since $\sum_j P_j^2(t) = I$, we get $(\sum_j P_j^2(t))' = 0$, and hence

$$\begin{aligned} &\Big(\sum_j P_j'P_j x, y\Big)_W + \Big(x, \sum_j P_j'P_j y\Big)_W \\ &= \Big(x, \sum_j P_jP_j' y\Big)_W + \Big(x, \sum_j P_j'P_j y\Big)_W = \Big(x, \Big(\sum_j P_j^2\Big)' y\Big)_W = 0. \end{aligned}$$

The theorem is proved.

3. *Tracking of the invariant subspaces of an operator functon.* We consider a differentiable operator function $F(t)$ for $t \in [a, b]$. Suppose that, for some $t_0 \in [a, b]$, the spectrum of the operator $F(t_0)$ decomposes into n spectral sets:

$$\sigma(F(t_0)) = \sigma_1(t_0) \cup \sigma_2(t_0) \cup \cdots \cup \sigma_n(t_0). \tag{1.14}$$

Let $\gamma_k(t_0)$ denote a closed curve (or system of curves) separating $\sigma_k(t_0)$ from the rest of the spectrum of $F(t_0)$. As before, we see from the continuity of $F(t)$ that its spectrum $\sigma(F(t))$ is not intersected by $\gamma_k(t_0)$ ($k = 1, \cdots, n$) for sufficiently small values of $|t - t_0|$, and we let $\sigma_k(t)$ denote the part of this spectrum lying inside $\gamma_k(t_0)$. The spectral projections

$$P_k(t) = -\frac{1}{2\pi i} \oint_{\gamma_k(t_0)} [F(t) - \lambda I]^{-1} \, d\lambda$$

corresponding to these parts of the spectrum are differentiable with respect to t and have all of the properties required by Theorem 1.1. The operator $\Omega(t, \tau)$ whose existence is guaranteed by this theorem "tracks" the invariant subspaces $\mathfrak{B}_k(t) = P_k(t)\mathfrak{B}$ of $F(t)$ as t varies in a neighborhood of t_0.

If the spectral sets $\sigma_k(t)$ $(k = 1,\cdots,n)$ remain pairwise disjoint for all $t \in [a, b]$, the operator $\Omega(t, \tau)$ "tracks" the subspaces $\mathfrak{B}_k(t)$ on the whole interval $[a, b]$.

4. *Normalization of a conjugation operator.* The above constructed operator functions $\Omega_q(t)$ and $\Omega(t, \tau)$ effecting conjugations of the projections of (1.6), although continuous, can be unbounded on an infinite interval even if the projection functions themselves are bounded.

If the space $\mathfrak{B}$ is a Hilbert space $\mathfrak{H}$, the situation can be corrected by using the following result.

THEOREM 1.2. *Suppose $P_k(\xi)$ $(k = 1,\cdots,n)$ are pairwise disjoint projections in $\mathfrak{H}$ that compose a decomposition of the identity for each ξ of some metric space Ξ and are continuous and bounded functions on Ξ.*

If there exists a continuous invertible operator function $S(\xi)$ effecting a conjugation between the projections, there also exists a continuous invertible operator function $\tilde{S}(\xi)$ that is bounded on Ξ and has the same property:

$$P_k(\xi) = \tilde{S}(\xi)P_k(\xi_0)\tilde{S}^{-1}(\xi) \qquad (\xi \in \Xi).$$

If all of the $P_k(\xi)$ are orthogonal (viz. Hermitian), the operator function $\tilde{S}(\xi)$ can be chosen so that it is unitary.

PROOF. It can be assumed without loss of generality that the initial projections $P_k(\xi_0)$ are orthogonal $(P_k^*(\xi_0) = P_k(\xi_0))$. For one can always introduce an equivalent scalar product in $\mathfrak{H}$ with respect to which this property will be possessed (see Remark I.1.1).

We put

$$R^2(\xi) = \sum_{k=1}^{n} P_k(\xi_0)\, S^*(\xi)S(\xi)P_k(\xi_0) = S^*(\xi)\Big[\sum_{k=1}^{n} P_k^*(\xi)P_k(\xi)\Big]S(\xi). \qquad (1.15)$$

Suppose $f \in \mathfrak{H}$. Then

$$\begin{aligned}(R^2(\xi)f,f) &= \sum_{k=1}^{n} (S(\xi)\, P_k(\xi_0)f,\, S(\xi)P_k(\xi_0)f) \\ &= \sum_{k=1}^{n} \| S(\xi)P_k(\xi_0)f \|^2 \geqq \frac{1}{\|S^{-1}(\xi)\|^2} \sum_{k=1}^{n} \|P_k(\xi_0)f\|^2 \\ &\geqq \frac{1}{n\|S^{-1}(\xi)\|^2} \Big\|\sum_{k=1}^{n} P_k(\xi_0)f\Big\|^2 = \frac{1}{n\|S^{-1}(\xi)\|^2} \|f\|^2\end{aligned}$$

and hence the operator $R^2(\xi)$ is uniformly positive.

Let $R(\xi)$ be the positive square root of $R^2(\xi)$.

We show that the operator $\tilde{S}(\xi) = S(\xi)R^{-1}(\xi)$ has the required properties. From (1.15) we obtain the equality

$$I = R^{-1}(\xi)S^*(\xi)\Big[\sum_k P_k^*(\xi)P_k(\xi)\Big]S(\xi)R^{-1}(\xi)$$

and hence the inequality

$$\begin{aligned}\|f\|^2 &= \sum_k (R^{-1}(\xi)S^*(\xi)P_k^*(\xi)P_k(\xi)S(\xi)R^{-1}(\xi)f,f)\\ &= \sum_k \|P_k(\xi)S(\xi)R^{-1}(\xi)f\|^2\\ &\geqq \frac{1}{n}\Big(\sum_k \|P_k(\xi)\,S(\xi)\,R^{-1}(\xi)f\|\Big)^2 \geqq \frac{1}{n}\,\|S(\xi)R^{-1}(\xi)f\|^2,\end{aligned}$$

indicating the boundedness of the operator function $\tilde{S}(\xi)$:

$$\|\tilde{S}(\xi)\| \leqq \sqrt{n}.$$

On the other hand, from (1.15) we obtain the equality

$$S^{*^{-1}}(\xi)R^2(\xi)S^{-1}(\xi) = \sum_{k=1}^{n} P_k^*(\xi)P_k(\xi),$$

which implies

$$\begin{aligned}\|\tilde{S}^{-1}(\xi)f\|^2 &= \|R(\xi)S^{-1}(\xi)f\|^2 = (S^{*^{-1}}(\xi)R^2(\xi)S^{-1}(\xi)f,f)\\ &= \sum_{k=1}^{n}(P_k^*(\xi)P_k(\xi)f,f) = \sum_{k=1}^{n}\|P_k(\xi)f\|^2 \leqq n\sup_{k,x}\|P_k(\xi)\|^2\|f\|^2.\end{aligned}$$

This proves the boundedness of the operator function $\tilde{S}^{-1}(\xi)$. The continuity of the operator function $R^2(\xi)$ and hence of $R(\xi)$ follows directly from formula (1.15).

Suppose, finally, $P_k(\xi) = P_k^*(\xi)$ $(\xi \in \Xi)$. Then (1.15) implies that $R^2(\xi) = S^*(\xi)S(\xi)$ and hence that the operator $\tilde{S}(\xi) = S(\xi)R^{-1}(\xi) = S(\xi)[S^*(\xi)S(\xi)]^{-1/2}$ is unitary.

Since the operator $R^2(\xi)$ and hence also $R(\xi)$ clearly commute with the projections $P_k(\xi_0)$, we have

$$\tilde{S}(\xi)P_k(\xi_0)\tilde{S}^{-1}(\xi) = S(\xi)R^{-1}(\xi)P_k(\xi_0)R(\xi)S^{-1}(\xi) = S(\xi)P_k(\xi_0)S^{-1}(\xi) = P_k(\xi).$$

REMARK 1.1. If the space Ξ is a segment of the real axis and the operators $P_k(\xi)$ and $S(\xi)$ are differentiable, the operator $\tilde{S}(\xi)$ is also differentiable (it suffices to note that a differentiable uniformly positive operator has a differentiable root).

5. *Angular distance between disjoint subspaces.* In the sequel we will often encounter bounded families of projections in a general Banach space $\mathfrak{B}$. Here we indicate the geometric significance of this boundedness.

Let $\mathfrak{B}_1$ and $\mathfrak{B}_2$ be a pair of nonzero subspaces of a Banach space $\mathfrak{B}$ that are *disjoint*:

$\mathfrak{B}_1 \cap \mathfrak{B}_2 = \{0\}$. We introduce the following characterization of their mutual inclination.

Let

$$\mathrm{Sn}\,(\mathfrak{B}_1, \mathfrak{B}_2) = \inf_{\substack{x_k \in \mathfrak{B}_k\ (k=1,2) \\ \| x_k \| = 1}} \| x_1 + x_2 \|, \tag{1.16}$$

where the infimum is taken over all pairs of unit vectors belonging to $\mathfrak{B}_1$ and $\mathfrak{B}_2$ respectively. One can easily verify that in the case of a Hilbert space

$$\mathrm{Sn}(\mathfrak{B}_1,\mathfrak{B}_2) = 2\sin(\theta/2),$$

where θ is the minimal angle between the subspaces $\mathfrak{B}_1$ and $\mathfrak{B}_2$ (see Exercise I.30). In the general case we call the quantity $\mathrm{Sn}(\mathfrak{B}_1, \mathfrak{B}_2)$ the *angular distance* between $\mathfrak{B}_1$ and $\mathfrak{B}_2$

LEMMA 1.1. *Suppose the space $\mathfrak{B}$ decomposes into a direct sum $\mathfrak{B} = \mathfrak{B}_1 \dotplus \mathfrak{B}_2$ of closed subspaces and P_1, $P_2 = I - P_1$ are the corresponding projections. The following estimate is valid*:

$$1/\| P_k \| \leqq \mathrm{Sn}(\mathfrak{B}_1, \mathfrak{B}_2) \leqq 2/\| P_k \| \qquad (k = 1,2). \tag{1.17}$$

PROOF. We choose an aribtrary $\delta > \mathrm{Sn}(\mathfrak{B}_1, \mathfrak{B}_2)$. There then exists a pair of unit vectors $x_k \in \mathfrak{B}_k$ $(k = 1, 2)$ for which $\| x_1 + x_2 \| < \delta$.

We put $x_1 + x_2 = x$. Then $x_k = P_k x$ and

$$1 = \|x_k\| \leqq \|P_k\|\,\|x\| < \|P_k\|\,\delta,$$

whence $1/\|P_k\| < \delta$ and consequently $1/\|P_k\| \leqq \mathrm{Sn}(\mathfrak{B}_1, \mathfrak{B}_2)$.

On the other hand, for aribtrary $x \in \mathfrak{B}$

$$\begin{aligned}\mathrm{Sn}(\mathfrak{B}_1, \mathfrak{B}_2) &\leqq \left\| \frac{P_1x}{\|P_1x\|} + \frac{P_2x}{\|P_2x\|} \right\| = \frac{1}{\|P_1x\|} \left\| P_1x + \frac{\|P_1x\|}{\|P_2x\|} P_2x \right\| \\ &= \frac{1}{\|P_1x\|} \left\| x + \frac{\|P_1x\| - \|P_2x\|}{\|P_2x\|} P_2x \right\| \\ &\leqq \frac{1}{\|P_1x\|} \left\{ \|x\| + \left\| \frac{\|P_1x + P_2x\|}{\|P_2x\|} P_2x \right\| \right\} \leqq 2\frac{\|x\|}{\|P_1x\|},\end{aligned}$$

and therefore

$$\mathrm{Sn}(\mathfrak{B}_1,\mathfrak{B}_2) \leqq 2 \inf_{x\in\mathfrak{B}} \frac{\|x\|}{\|P_1x\|} = \frac{2}{\|P_1\|}.$$

Analogously, $\mathrm{Sn}(\mathfrak{B}_1, \mathfrak{B}_2) \leqq 2/\|P_2\|$. The lemma is proved.

COROLLARY 1.1. *The boundedness from above of a set $\mathcal{P} = \{P\}$ of projections in a Banach space $\mathfrak{B}$ is equivalent to the boundedness from below of the set*

$\{\mathrm{Sn}\,(P\mathfrak{B},\ (I-P)\mathfrak{B})\,|\,P\in\mathscr{P}\}$ *of angular distances between the subspaces* $P\mathfrak{B}$ *and their complements* $(I-P)\mathfrak{B}$.

Let P be a projection in a Banach space $\mathfrak{B}$ and $U(\xi)$ $(\xi\in\Xi)$ be an invertible operator such that $U(\xi_0)=I$.

We consider the projection function

$$P(\xi)=U(\xi)PU^{-1}(\xi)\qquad(\xi\in\Xi)\tag{1.18}$$

and the corresponding subspaces $\mathfrak{B}(\xi)=U(\xi)P\mathfrak{B}$.

If the projection function (1.18) is bounded, we will say that the operator $U(\xi)$ *uniformly conjugates the projection* P *on* Ξ.

The significance of this definition is to be found in the fact that, when its condition is satisfied, the set $\{\mathrm{Sn}(P(\xi)\mathfrak{B},\ (I-P(\xi))\mathfrak{B})\,|\,\xi\in\Xi\}$ of angular distances is bounded from below.

§ 2. Kinematically similar equations

1. *Definition and tests of kinematic similarity of equations.* We consider the pair of equations

$$dx_k/dt=A_k(t)x_k\qquad(t\in\mathscr{J},\ k=1,2).\tag{2.1}$$

We say that these equations are *kinematically similar on the interval* $\mathscr{J}$ if it is possible to establish between the totalities of all solutions of these equations a one-to-one correspondence

$$x_2(t)=Q(t)x_1(t)\qquad(t\in\mathscr{J}),\tag{2.2}$$

where $Q(t)$ is a bounded linear operator function with a bounded inverse:

$$\|Q(t)\|\leqq q_1,\quad\|Q^{-1}(t)\|\leqq q_2\qquad(t\in\mathscr{J}).\tag{2.3}$$

Letting $U_k(t,\tau)$ denote the evolution operators of equations (2.1), we obtain from (2.2) the relation

$$U_2(t,\tau)x_2(\tau)=Q(t)U_1(t,\tau)x_1(\tau),$$

which by virtue of the arbitrariness of $x_1(\tau)$ and the formula $x_2(\tau)=Q(\tau)x_1(\tau)$ reduces to the equality

$$U_2(t,\tau)Q(\tau)=Q(t)U_1(t,\tau)\tag{2.4}$$

and, in particular, when $\tau=0$, to the equality

$$U_2(t)CU_1^{-1}(t)=Q(t),\tag{2.5}$$

where $C=Q(0)$.

Since the operators $\tilde{U}_2(t)=U_2(t)C$ and $\tilde{U}_1(t)=U_1(t)$ are fundamental operators of equations (2.1), the operator $Q(t)$ can be written in the form of a ratio of fundamental operators of these equations:

$$\tilde{U}_2(t)\tilde{U}_1^{-1}(t) = Q(t). \tag{2.6}$$

Equality (2.6) implies the differentiability of $Q(t)$ and the equality

$$Q'(t) = A_2(t)Q(t) - Q(t)A_1(t). \tag{2.7}$$

Suppose on the other hand that $Q(t)$ is the solution of equation (2.7) satisfying the condition $Q(0) = C$, where C is an invertible operator. This solution, as follows from (III.1.25), has the form

$$Q(t) = U_2(t)Q(0)U_1^{-1}(t),$$

and hence the operator $Q(t)$ is invertible for each $t \in \mathcal{J}$. Since (2.5) clearly implies (2.2) for $x_2(0) = Cx_1(0)$, all five of the relations (2.2), (2.4)–(2.7) are equivalent. This brings us to the following assertion.

LEMMA 2.1. *In order for equations* (2.1) *to be kinematically similar on an interval* $\mathcal{J}$ *it is necessary and sufficient that a bounded and boundedly invertible operator function* $Q(t)$ *on* $\mathcal{J}$ *satisfy one of the following conditions:*

a) *The evolution operators of the equations are connected by the relation*

$$U_2(t, \tau)Q(\tau) = Q(t)U_1(t, \tau). \tag{2.4}$$

b) *For some invertible operator* $C \in [\mathfrak{B}]$ *the Cauchy operators of the equations are connected by the relation*

$$U_2(t)CU_1^{-1}(t) = Q(t). \tag{2.5}$$

c) *There exist fundamental operators* $\tilde{U}_k(t)$ $(k = 1, 2)$ *of equations* (2.1) *for which*

$$\tilde{U}_2(t)\tilde{U}_1^{-1}(t) = Q(t). \tag{2.6}$$

d) *The operator* $Q(t)$ *satisfies the differential equation*

$$Q'(t) = A_2(t)Q(t) - Q(t)A_1(t). \tag{2.7}$$

We note that two equations are kinematically similar on a semi-infinite interval if their coefficients coincide outside of a finite interval.

For suppose for the sake of definiteness that $\mathcal{J} = [0, \infty)$ and $A_1(t) = A_2(t)$ when $t \geqq t_0 > 0$. Then when $t, \tau \geqq t_0$ the evolution operators also coincide: $U_1(t, \tau) = U_2(t, \tau)$, and hence

$$U_1(t) = U_1(t, t_0)U_1(t_0);$$
$$U_2(t) = U_2(t, t_0)U_2(t_0) = U_1(t, t_0)U_2(t_0).$$

Consequently, for $C = U_2^{-1}(t_0)U_1(t_0)$ we have

$$U_2(t)CU_1^{-1}(t) = Q(t),$$

where

$$Q(t) = \begin{cases} U_2(t, t_0)U_1^{-1}(t, t_0) & \text{when } 0 \leqq t \leqq t_0, \\ I & \text{when } t \geqq t_0, \end{cases}$$

i.e. condition (2.5) is satisfied.

THEOREM 2.1. *Equations kinematically similar on a halfline have the same upper Ljapunov and the same Bohl exponents.*

PROOF. From formulas (2.3) and (2.4) it follows that

$$(1/q_1q_2)\|U_1(t+\tau, t)\| \leqq \|U_2(t+\tau, t)\| = q_1q_2\|U_1(t+\tau, t)\|,$$

which implies

$$\frac{\ln\|U_2(t+\tau, t)\|}{\tau} - \frac{\ln(q_1q_2)}{\tau} \leqq \frac{\ln\|U_2(t+\tau, t)\|}{\tau}$$
$$\leqq \frac{\ln\|U_1(t+\tau, t)\|}{\tau} + \frac{\ln(q_1q_2)}{\tau}.$$

Passing to the upper limit as $\tau \to \infty$ and $t \to \infty$, we obtain the required relation

$$\kappa_B(U_2) = \overline{\lim_{t,\tau\to\infty}} \frac{\ln\|U_2(t+\tau, t)\|}{\tau} = \overline{\lim_{t,\tau\to\infty}} \frac{\ln\|U_1(t+\tau, t)\|}{\tau} = \kappa_B(U_1).$$

The coincidence of the lower Bohl and upper Ljapunov exponents is proved analogously.

2. *Reducible equations.* The equation $dx/dt = A(t)x$ is said to be *reducible according to Ljapunov* if it is kinematically similar to a stationary equation $dy/dt = By$ ($B = \text{const}$).

Theorems III.4.5 and 2.1 directly imply the following result.

THEOREM 2.2. *The upper Ljapunov and upper Bohl exponents of a reducible equation coincide.*

REMARK 2.1. It is not difficult to verify that the upper Ljapunov and upper Bohl exponents in the case in question are strict.

By definition, the reducibility of an equation is equivalent to the relation

$$U(t) = Q(t)e^{Bt}, \tag{2.8}$$

where $Q(t)$ and $Q^{-1}(t)$ are bounded continuous operator functions on $\mathscr{J}$ (one can easily see that it is always possible in this case to take $C = I$).

We note that in the present case $Q(t)$ satisfies the equation

$$Q' = A(t)Q - QB. \tag{2.9}$$

3. *Reduction to an equation with a Hermitian coefficient.* Suppose now that the space $\mathfrak{B}$ is a Hilbert space $\mathfrak{H}$. We represent the operator $A(t)$ by the sum of its real

and imaginary parts: $A(t) = A_{\mathfrak{R}}(t) + iA_{\mathfrak{J}}(t)$, and make the change of variables $x = V(t)y$ in the equation

$$\frac{dx}{dt} = A(t)x, \tag{2.10}$$

where $V(t)$ is the unitary operator function satisfying the relations

$$\frac{dV}{dt} = iA_{\mathfrak{J}}(t)V, \qquad V(0) = I.$$

Here $y(t)$ satisfies the equation

$$iA_{\mathfrak{J}}Vy + Vy' = A_{\mathfrak{R}}Vy + iA_{\mathfrak{J}}Vy$$

or, finally,

$$y' = V^{-1}(t)A_{\mathfrak{R}}(t)V(t)y. \tag{2.11}$$

Since the operator $V(t)$ is unitary for each t, we obtain the following assertion.

LEMMA 2.2. *In a Hilbert space equation* (2.10) *is kinematically similar to equation* (2.11), *the coefficient of which is unitarily equivalent to the real part of the coefficient of the original equation.*

4. *Theorem on the decomposition of an equation in Hilbert space.* We consider a direct decomposition of the space $\mathfrak{H}$:

$$\mathfrak{H} = \mathfrak{H}_1 \dotplus \mathfrak{H}_2 \dotplus \cdots \dotplus \mathfrak{H}_n, \tag{2.12}$$

and let $P_1, \cdots, P_n$ be the corresponding projections. Applying the Cauchy operator $U(t)$ of equation (2.10) to the decomposition (2.12), we obtain the new decomposition

$$\mathfrak{H} = \mathfrak{H}_1(t) \dotplus \mathfrak{H}_2(t) \dotplus \cdots \dotplus \mathfrak{H}_n(t), \tag{2.13}$$

where $\mathfrak{H}_k(t) = U(t)\mathfrak{H}_k$ $(k = 1, \cdots, n)$.

The projections corresponding to the decomposition (2.13) are similar to the projections P_k and are expressed by means of the formulas

$$P_k(t) = U(t)P_kU^{-1}(t) \qquad (t \in \mathscr{I}). \tag{2.14}$$

THEOREM 2.3. *Suppose the projections P_k are uniformly conjugated by the operator $U(t)$ on an interval $\mathscr{I}$. Then equation* (2.10) *is kinematically similar to an equation*

$$dy/dt = B(t)y, \tag{2.15}$$

the coefficient of which commutes with each of the projections P_k $(k = 1, \cdots, n)$ and which therefore decomposes into a system of independent equations in the phase spaces $\mathfrak{H}_k$ of the decomposition (2.12).

The kinematic similarity transformation can be chosen so that the following estimate is satisfied for some constant c depending only on the projections P_k:

$$\|B(t)\| \leqq c\|A(t)\| \qquad (t \in \mathscr{I}). \tag{2.16}$$

PROOF. The operators (2.14) satisfy the conditions of Theorem 1.2 on $\mathscr{I}$. There therefore exists a bounded invertible operator function $\tilde{S}(t) = U(t)R^{-1}(t)$ satisfying the conditions

$$P_k(t) = \tilde{S}(t)P_k\tilde{S}^{-1}(t) \qquad (k = 1,2,\cdots,n).$$

We will assume that an equivalent renorming of the space to make the operators P_k Hermitian (see the proof of Theorem 1.2) has already been carried out. It then follows from (1.15) that

$$R^2(t) = \sum_{k=1}^{n} P_kU^*(t)U(t)P_k. \tag{2.17}$$

Equation (2.10) is kinematically similar to the equation

$$dy/dt = \tilde{B}(t)y \qquad (\tilde{B}(t) = R'(t)R^{-1}(t)), \tag{2.18}$$

since $R(t)$ is clearly a fundamental operator of the latter equation and the ratio $U(t)R^{-1}(t) = \tilde{S}(t)$ satisfies the kinematic siimlarity condition (2.6).

Formula (2.17) shows that $R(t)$ and hence $\tilde{B}(t)$ commute with each of the P_k.

We must still prove the last assertion of the theorem. Differentiating (2.17), we obtain the relation

$$R(t)R'(t) + R'(t)R(t) = \sum_{k=1}^{n} P_kU^*(t)[A^*(t) + A(t)]U(t)P_k.$$

By virtue of the boundedness of the Hermitian operator $A^*(t) + A(t)$ there exist real constants α and β such that

$$\alpha(f,f) \leqq ([A^*(t) + A(t)]f,f) \leqq \beta(f,f) \qquad (f \in \mathfrak{H}).$$

Therefore the relation

$$([RR' + R'R]f,f) = \sum_{k=1}^{n} ([A^* + A]U(t)P_kf, U(t)P_kf)$$

implies the estimate

$$\alpha \sum_k (U(t)P_kf, U(t)P_kf) \leqq ([RR' + R'R]f,f)$$
$$\leqq \beta \sum_k (U(t)P_kf, U(t)P_kf),$$

which can be rewritten in the form

$$\alpha(R^2f,f) \leqq ([RR' + R'R]f,f) \leqq \beta(R^2f,f),$$

or, setting $f = R^{-1}g$,

$$\alpha\|g\|^2 \leqq ([R'R^{-1} + R^{-1}R']g, g) \leqq \beta\|g\|^2. \tag{2.19}$$

Thus we see that the operator

$$\tfrac{1}{2}(R'R^{-1} + R^{-1}R') = \tfrac{1}{2}(\tilde{B}^* + \tilde{B}) = \tilde{B}_{\mathfrak{R}}$$

is bounded, it following from (2.19) that $\|\tilde{B}_{\mathfrak{R}}(t)\| \leqq \|A(t)\|$.

It remains for us to make use of Lemma 2.2 and pass from equation (2.18) to the kinematically similar equation whose coefficient $B(t) = V(t)\tilde{B}_{\mathfrak{R}}(t)V^{-1}(t)$ is unitarily equivalent to $\tilde{B}_{\mathfrak{R}}$. It is not difficult to see that commutability with the projections P_k is not violated here since the unitary operator $V(t)$ also commutes with them.

The operator $B(t)$ which is unitarily equivalent to $\tilde{B}_{\mathfrak{R}}(t)$ satisfies the same estimate $\|B(t)\| \leqq \|A(t)\|$. The constant c in (2.16) arises in connection with the renorming of the space.

REMARK 2.2. The Cauchy operator of equation (2.15) clearly also commutes with with each of the projections P_k $(k = 1, \cdots, n)$.

§ 3. Exponentially dichotomic equations

1. *Basic definitions.* In the preceding chapter we were primarily concerned with the equation

$$dx/dt = A(t)x \tag{3.1}$$

when its upper Bohl exponent is negative and hence when all of its solutions decrease exponentially as $t \to \infty$. We recall that such a situation arises for an equation with a constant coefficient $A(t) \equiv A$ when the spectrum $\sigma(A)$ lies in the interior of the left halfplane.

A more complicated situation arises when the spectrum $\sigma(A)$ also contains a component lying in the interior of the right halfplane:

$$\sigma(A) = \sigma_+(A) \cup \sigma_-(A). \tag{3.2}$$

We recall that the operator A was then said to be *exponentially dichotomic*. It was shown in § II.2.1 that in this case the space decomposes into a direct sum

$$\mathfrak{B} = \mathfrak{B}_+ \dotplus \mathfrak{B}_- \tag{3.3}$$

of subspaces such that the solutions initially in $\mathfrak{B}_+$ exponentially increase (decrease) as $t \to +\infty$ $(t \to -\infty)$ whereas the solutions initially in $\mathfrak{B}_-$, as before, exponentially decrease (increase) as $t \to +\infty$ $(t \to -\infty)$. We note that these solutions remain in the fixed subspaces $\mathfrak{B}_+$ and $\mathfrak{B}_-$ respectively. If P_+ and P_- are the spectral projections commuting with A that correspond to the decompositions (3.2) and (3.3), there exist positive constants N and ν for which

$$\begin{aligned} \|e^{A(t-s)}P_-\| &\leqq Ne^{-\nu(t-s)} \qquad (t \geqq s), \\ \|e^{A(t-s)}P_+\| &\leqq Ne^{-\nu(s-t)} \qquad (s \geqq t). \end{aligned} \tag{3.4}$$

We will see at once that an analogous description is possible for an equation with a variable coefficient $A(t)$. And what is more, this description permits us to generalize a number of the results of the preceding chapter.

DEFINITION 3.1. We will say that *a regular*[1)] *exponential dichotomy holds on* $\mathscr{I}$ for the solutions of equation (3.1) (more briefly, *the equation is e-dichotomic*) if for some $t_0 \in \mathscr{I}$ the space $\mathfrak{B}$ decomposes into a direct sum

$$\mathfrak{B} = \mathfrak{B}_1(t_0) \dotplus \mathfrak{B}_2(t_0), \tag{3.5}$$

of closed subspaces such that the following conditions are satisfied:

a) The solutions $x_1(t) = U(t, t_0)x_1^0$ of equation (3.1) in the subspace $\mathfrak{B}_1(t_0)$ at $t = t_0$ $(x_1^0 \in \mathfrak{B}_1(t_0))$ are subject to the estimate

$$\|x_1(t)\| \leqq N_1 e^{-\nu_1(t-s)} \|x_1(s)\| \qquad (t \geqq s;\, t, s \in \mathscr{I}) \tag{3.6a}$$

with some exponent $\nu_1 > 0$.

b) The solutions $x_2(t) = U(t, t_0)x_2^0$ of equation (3.1) in the subspace $\mathfrak{B}_2(t_0)$ at $t = t_0$ $(x_2^0 \in \mathfrak{B}_2(t_0))$ are subject to the estimate

$$\|x_2(t)\| \leqq N_2 e^{-\nu_2(s-t)} \|x_2(s)\| \qquad (t \leqq s;\, t, s \in \mathscr{I}) \tag{3.6b}$$

with some exponent $\nu_2 > 0$.

c) The angular distance between the subspaces $\mathfrak{B}_1(t) = U(t,t_0)\mathfrak{B}_1(t_0)$ and $\mathfrak{B}_2(t) = U(t, t_0)\mathfrak{B}_2(t_0)$ cannot become arbitrarily small under a variation of t; more precisely, there exists a constant $\gamma > 0$ such that

$$\operatorname{Sn}(\mathfrak{B}_1(t), \mathfrak{B}_2(t)) \geqq \gamma \qquad (t \in \mathscr{I}). \tag{3.7}$$

Let us make some remarks concerning the above definition.

REMARK 3.1. The choice of t_0 is not important. In fact, a change in the value of t_0 in formulas (3.5)–(3.7) will affect only the constants N_k $(k = 1, 2)$.

For this reason, in the sequel we will always put $t_0 = 0$ (assuming that the interval $\mathscr{I}$ contains this point) and use the more concise notation $\mathfrak{B}_k(0) = \mathfrak{B}_k$ $(k = 1, 2)$.

The projections corresponding to the direct decomposition $\mathfrak{B} = \mathfrak{B}_1 \dotplus \mathfrak{B}_2$ will be denoted by P_1 and P_2.

REMARK 3.2. Let $P_k(t)$ $(k = 1, 2)$ be the mutually complementary projections on the subspaces $\mathfrak{B}_k(t)$ $(t \in \mathscr{I})$. Then

$$P_k(t) = U(t)P_k U^{-1}(t) \qquad (t \in \mathscr{I}). \tag{3.8}$$

As follows from Lemma 1.1, condition c) in Definition 3.1 is equivalent to the uniform boundedness of the projections $P_k(t)$:

$$\|P_k(t)\| \leqq M \qquad (k = 1, 2;\, t \in \mathscr{I}), \tag{3.9}$$

i.e. to the uniform conjugatability of the projections P_k.

REMARK 3.3. Suppose for the sake of definiteness that $\mathscr{I}$ contains the halfline

---

1) For the sake of simplicity we consider a less general situation than the one described in the book of J. Massera and J. Schäffer [1], and for this reason we speak of a *regular* exponential dichotomy.

The word *regular* will be dropped below.

$[0, +\infty)$. Then the solutions of equation (3.1) initially in $\mathfrak{B}_1$ remain bounded on $[0, +\infty)$ while the solutions initially in $\mathfrak{B}_2$ increase unboundedly as $t \to +\infty$. It turns out, moreover, that all of the solutions $x(t) = U(t)x_0$ for which $x_0 \in \mathfrak{B}_1$ increase unboundedly as $t \to +\infty$.

For suppose $P_2 x_0 \neq 0$. We consider the solution

$$x_2(t) = U(t)P_2x_0 = P_2(t)U(t)x_0 = P_2(t)x(t),$$

corresponding to an initial value $P_2x_0 \in \mathfrak{B}_2$.

From (3.6b) and (3.9) we obtain the inequality

$$\begin{aligned} \|x(s)\| &\geqq \frac{1}{\|P_2(s)\|}\, \|x_2(s)\| \geqq \frac{1}{MN_2}\, e^{\nu_2(s-t)}\, \|x_2(t)\| \\ &\left(= \frac{1}{MN_2}\, e^{\nu_2(s-t)}\, \|U(t)P_2x_0\| \right) \qquad (s \geqq t), \end{aligned} \tag{3.10}$$

the right side of which unboundedly increases as $s \to \infty$.

Thus the subspace $\mathfrak{B}_1$ consists of precisely the initial values $x(0)$ of those solutions $x(t)$ of equation (3.1) which remain bounded on $[0, +\infty)$.

An analogous situation in which $\mathfrak{B}_1$ is replaced by $\mathfrak{B}_2$ holds on the left halfline.

REMARK 3.4. Suppose $\mathscr{I} = [0, \infty)$. As was indicated in the preceding remark, the subspace $\mathfrak{B}_1$ is then uniquely determined by equation (3.1). On the other hand, the subspace $\mathfrak{B}_2$ can always be replaced by another subspace $\tilde{\mathfrak{B}}_2$ that preserves the direct decomposition: $\mathfrak{B} = \mathfrak{B}_1 \dotplus \tilde{\mathfrak{B}}_2$.

It turns out that if condition c) of Definition 3.1 is satisfied, this direct decomposition also determines an exponential dichotomy of the solutions of equation (3.1) in which the constants ν_k remain the same [it will be shown below that condition c) is a consequence of conditions a) and b) if $A(t)$ is an integrally bounded operator function].

We prove this assertion by verifying condition (3.6b) for the solutions initially in $\tilde{\mathfrak{B}}_2$.

Suppose $x_0 \in \tilde{\mathfrak{B}}_2$ and $x(t) = U(t)x_0$. Then by virtue of (3.6a) and (3.6b)

$$\begin{aligned} \|x(t)\| &\leqq \|U(t)P_1x_0\| + \|U(t)P_2x_0\| \\ &\leqq N_1\|P_1x_0\| + \|U(t)P_2x_0\| \leqq \left(\frac{N_1}{N_2}\frac{\|P_1x_0\|}{\|P_2x_0\|} + 1\right)\|U(t)P_2x_0\|. \end{aligned}$$

Combining this inequality with (3.10), we obtain the estimate

$$\|x(t)\| \leqq M\left(N_1\frac{\|P_1x_0\|}{\|P_2x_0\|} + N_2\right)e^{-\nu_2(s-t)}\|x(s)\|,$$

and it only remains to show that the quantity $\|P_1x_0\|/\|P_2x_0\|$ is bounded for $x_0 \in \tilde{\mathfrak{B}}_2$.

The latter fact follows directly from Banach's theorem. For the operator P_2 is bounded and, as is easily seen, is a one-to-one mapping of $\tilde{\mathfrak{B}}_2$ onto $\mathfrak{B}_2$. Therefore the inverse mapping is continuous:

$$\|P_2x_0\| \geqq q\,\|x_0\| \qquad (x_0 \in \tilde{\mathfrak{B}}_2)$$

and, finally,

$$\frac{\|P_1x_0\|}{\|P_2x_0\|} \leqq \frac{\|P_1x_0\|}{q\|x_0\|} \leqq \frac{1}{q}\,\|P_1\|.$$

We now cite another, in certain cases more convenient, e-dichotomicity condition.

LEMMA 3.1. *In order for equation* (3.1) *to be e-dichotomic on* $\mathscr{I}$ *with exponents* $\nu_1 > 0$ *and* $\nu_2 > 0$ *it is necessary and sufficient that the conditions*

$$\|U(t)P_1U^{-1}(s)\| \leqq \tilde{N}_1e^{-\nu_1(t-s)} \qquad (t \geqq s), \tag{3.11a}$$

$$\|U(t)P_2U^{-1}(s)\| \leqq \tilde{N}_2e^{-\nu_2(s-t)} \qquad (s \geqq t) \tag{3.11b}$$

with certain constants $\tilde{N}_k$ $(k = 1, 2)$ *be satisfied on this interval.*

PROOF. Inequalities (3.6) immediately follow from (3.11). For, when $x(0) = P_1x(0) \in \mathfrak{B}_1$ and $t \geqq s$, we have

$$\|x(t)\| = \|U(t)P_1x(0)\| = \|U(t)P_1U^{-1}(s)x(s)\| \leqq \tilde{N}_1e^{-\nu_1(t-s)}\|x(s)\|.$$

In addition, (3.11) also implies estimate (3.9).

Inequality (3.6b) is proved analogously.

Conversely, if in (3.6a) we consider the solution $x_1(s)$ with initial value $x_1^0 = P_1U^{-1}(s)x$ $(x \in \mathfrak{B})$ and take into account (3.9), we obtain the following estimate for $t \geqq s$:

$$\begin{aligned}\|U(t)P_1U^{-1}(s)x\| &= \|U(t)x_1^0\| \leqq N_1e^{-\nu_1(t-s)}\|U(s)x_1^0\| \\ &\leqq N_1e^{-\nu_1(t-s)}\|U(s)P_1U^{-1}(s)x\| \leqq N_1Me^{-\nu_1(t-s)}\|x\|.\end{aligned}$$

The estimate for $t \leqq s$ is obtained in exactly the same way.

We can simplify the e-dichotomicity conditions by making the additional assumption that the operator function $A(t)$ is integrally bounded:

$$\int_t^{t+1} \|A(\tau)\|d\tau \leqq M_1 \qquad (t \in \mathscr{I}), \tag{3.12}$$

LEMMA 3.2. *If the function* $A(t)$ *is integrally bounded, condition* c) *of Definition* 3.1 *is a consequence of conditions* a) *and* b).

PROOF. Estimate (3.12) implies by virtue of Lemma III.2.4 that $\|U(t + m, t)\| \leqq e^{mM_1}$ for integral m.

We consider, for some fixed t, a pair of unit vectors $x_k(t) \in \mathfrak{B}_k(t)$ $(k = 1, 2)$ and put $x_k(\tau) = U(\tau, t)x_k(t)$ $(\tau \in \mathscr{I})$. From (3.6a) and (3.6b) we obtain the estimates

$$\|x_1(t + m)\| \leqq N_1 e^{-\nu_1 m}\|x_1(t)\| = N_1 e^{-\nu_1 m},$$
$$\|x_2(t + m)\| \geqq N_2^{-1} e^{\nu_2 m}\|x_2(t)\| = N_2^{-1} e^{\nu_2 m},$$

which show that

$$\begin{aligned} &\|x_1(t) + x_2(t)\| \\ &\quad \geqq e^{-mM_1}\| U(t + m, t)x_1(t) + U(t + m, t)x_2(t)\| \\ &\quad \geqq e^{-mM_1}(\|x_2(t + m)\| - \|x_1(t + m)\|) \\ &\quad \geqq e^{-mM_1}(N_2^{-1}e^{\nu_2 m} - N_1 e^{-\nu_1 m}) = \gamma_m. \end{aligned}$$

It follows from (1.16) that

$$\mathrm{Sn}(\mathfrak{B}_1(t), \mathfrak{B}_2(t)) \geqq \gamma_m.$$

Since the constant $\gamma_m > 0$ for sufficiently large m, the assertion is proved.

2. *Preservation of e-dichotomicity under a kinematic similarity transformation.* It is quite easy to establish the following assertion.

THEOREM 3.1. *If the equations*

$$dx/dt = A_k(t)x \qquad (k = 1,2;\ t \in \mathscr{I}) \tag{3.13}$$

are kinematically similar and one of them is e-dichotomic on the interval $\mathscr{I}$, this property (*with the same exponents ν_1 and ν_2*) *is also possessed by the other equation.*

PROOF. The Cauchy operators $U_1(t)$ and $U_2(t)$ of the kinematically similar equations are connected by a relation of the form (see (2.5))

$$U_2(t) = Q(t)U_1(t)C,$$

where $Q(t)$ is a bounded operator function with a bounded inverse and C is a bounded invertible operator. Since

$$U_2(t)P_k U_2^{-1}(s) = Q(t)U_1(t)\tilde{P}_k U_1^{-1}(s)Q^{-1}(s),$$

where $\tilde{P}_k = CP_kC^{-1}$, the assertion of the theorem immediately follows from Lemma 3.1.

3. *Green functions. Bounded solutions of the inhomogeneous equation.* We now show that an e-dichotomy plays the same role in an investigation of the boundedness of the solutions of the inhomogeneous equation

$$dx/dt = A(t)x + f(t) \qquad (t \in \mathscr{I}) \tag{3.14}$$

as condition (3.2) plays for an equation with a constant operator A.

To this end we introduce a Green function analogous to the one considered in § II.4

Let P_1 and P_2 be a pair of mutually complementary projections: $P_1 + P_2 = I$. If $U(t)$ is the Cauchy operator of equation (3.14). we put

$$G(t, \tau) = \begin{cases} U(t)P_1U^{-1}(\tau) & \text{for } t > \tau, \\ -U(t)P_2U^{-1}(\tau) & \text{for } t < \tau \end{cases} \quad (t, \tau \in \mathscr{I}). \tag{3.15}$$

It immediately follows from the definition that the Green function $G(t, \tau)$ satisfies the following differential equations on $\mathscr{I}$ for $t \neq \tau$:

$$\frac{\partial G(t, \tau)}{\partial t} = A(t)G(t, \tau) \tag{3.16a}$$

and

$$\frac{\partial G(t, \tau)}{\partial \tau} = -G(t, \tau)A(\tau). \tag{3.16b}$$

At $t = \tau$ it has a discontinuity such that

$$\begin{aligned} G(\tau + 0, \tau) - G(\tau - 0, \tau) &= U(\tau)P_1U^{-1}(\tau) + U(\tau)P_2U^{-1}(\tau) \\ &= U(\tau)(P_1 + P_2)U^{-1}(\tau) = U(\tau)U^{-1}(\tau) = I, \end{aligned} \tag{3.17a}$$

and analogously

$$G(t, t + 0) - G(t, t - 0) = -I. \tag{3.17b}$$

Suppose the function $f(t)$ is continuous on $\mathscr{I}$. We consider the integral

$$g(t) = \int_{\mathscr{I}} G(t, \tau)f(\tau)d\tau \tag{3.18}$$

under the assumption that it exists and permits differentiation under the integral sign. If we set $\mathscr{I} = [a, b]$ and write

$$g(t) = \int_a^t G(t, \tau)f(\tau)d\tau + \int_t^b G(t, \tau)f(\tau)d\tau,$$

we get

$$\begin{aligned} g'(t) &= G(t, t - 0)f(t) - G(t, t + 0)f(t) \\ &\quad + \int_a^t \frac{\partial}{\partial t} G(t, \tau)f(\tau)d\tau + \int_t^b \frac{\partial}{\partial t} G(t, \tau)f(\tau)d\tau \\ &= f(t) + \int_a^b A(t)G(t, \tau)f(\tau)d\tau = A(t)g(t) + f(t). \end{aligned}$$

Thus formula (3.18) provides a solution of the inhomogeneous equation (3.14) on $\mathscr{I}$ as long as the above calculations are permissible. This is always the case when $\mathscr{I}$ is a finite interval; but special estimates, which can be carried out if equation (3.1) is e-dichotomic, are needed for an infinite or semi-infinite interval.

Suppose P_1 and P_2 are the projections corresponding to a direct decomposition $\mathfrak{B} = \mathfrak{B}_1 + \mathfrak{B}_2$ for which equation (3.1) is e-dichotomic. The Green function obtained when these projections are substituted in (3.15) is called a *principal Green function of equation* (3.14).

This is the Green function that we will normally make use of in the sequel; for this reason we will usually drop the term *principal.*

Using estimates (3.11), we easily obtain the following estimate for a principal Green function:

$$\int_{\mathscr{J}} \|G(t, \tau)\| d\tau \leqq \frac{N_1}{\nu_1} + \frac{N_2}{\nu_2}. \tag{3.19}$$

We recall that $C(\mathfrak{B})$ denotes the Banach space of bounded continuous functions on $\mathscr{J}$ with values in $\mathfrak{B}$ and norm $|||f||| = \sup_{\mathscr{J}} \|f(t)\|$.

Estimate (3.19) immediately implies the following result.

THEOREM 3.2. *If equation* (3.1) *is e-dichotomic on* $\mathscr{J}$, *the inhomogeneous equation* (3.14) *has at least one solution* $x(t) \in C(\mathfrak{B})$ *for each function* $f(t) \in C(\mathfrak{B})$. *This solution is given by the formula*

$$x(t) = \int_{\mathscr{J}} G(t, \tau) f(\tau) d\tau, \tag{3.20}$$

where $G(t, \tau)$ *is a principal Green function of equation* (3.1).

PROOF. Estimate (3.19) implies that the integral (3.20) converges and

$$|||x||| \leqq (N_1/\nu_1 + N_2/\nu_2) |||f|||. \tag{3.21}$$

It is easily seen that differentiation under the integral sign is valid in the present case and that therefore, as was shown above, the function (3.20) satisfies equation (3.14)

REMARK 3.5. Suppose $\mathscr{J} = [0, \infty)$. In this case the bounded solution

$$x(t) = \int_0^\infty G(t, \tau) f(\tau) d\tau$$

of equation (3.14) on $\mathscr{J}$ has the initial value

$$x(0) = \int_0^\infty G(0, \tau) f(\tau) d\tau = - P_2 \int_0^\infty U^{-1}(\tau) f(\tau) d\tau,$$

belonging to $\mathfrak{B}_2$.

We obtain the general form of the bounded solutions on $[0, \infty)$ of equation (3.14) by adding to the solution already obtained an arbitrary bounded solution of the homogeneous equation (3.1). These are precisely the solutions that are initially in $\mathfrak{B}_1$.

Thus all of the bounded solutions on $[0, \infty)$ of equation (3.14) are represented by the formula

$$x(t) = U(t)y + \int_0^\infty G(t, \tau)f(\tau)d\tau, \tag{3.22}$$

where $y = P_1x(0)$ is an arbitrary element of $\mathfrak{B}_1$.

REMARK 3.6. The solution (3.20) remains bounded when the boundedness condition for $f(t)$ is replaced by the more general integral boundedness condition

$$\int_t^{t+1} \|f(\tau)\|d\tau \leqq M_1.$$

For in this case we have the estimate

$$\begin{aligned}
\|x(t)\| &\leqq \int_{\mathscr{I}} \|G(t, \tau)\| \, \|f(\tau)\|d\tau \\
&= \int_{t\leqq\tau} \|G(t, \tau)\| \, \|f(\tau)\|d\tau + \int_{t\geqq\tau} \|G(t, \tau)\| \, \|f(\tau)\|d\tau \\
&\leqq N_2 \int_{t\leqq\tau} e^{-\nu_1(\tau-t)}\|f(\tau)\|d\tau + N_1 \int_{t\geqq\tau} e^{-\nu_1(t-\tau)}\|f(\tau)\|d\tau \\
&\leqq N_2 \int_{s\geqq 0} e^{-\nu_1 s}\|f(t+s)\|ds + N_1 \int_{s\leqq 0} e^{\nu_1 s}\|f(t+s)\|ds \\
&\leqq N_2M_1 \sum_{k=0}^{\infty} e^{-\nu_1 k} + N_1M_1 \sum_{k=0}^{\infty} e^{-\nu_1 k} = \frac{N_2M_1}{1-e^{-\nu_1}} + \frac{N_1M_1}{1-e^{-\nu_1}}.
\end{aligned}$$

4. *e-dichotomicity on a halfline.* As was just shown, the e-dichotomicity of equation (3.1) is a sufficient condition for the existence of bounded solutions of the inhomogeneous equation with a bounded free term.

In order to explain the extent to which this condition is necessary we must introduce some additional assumptions.

We consider the equation on the halfline $\mathscr{I} = [0, \infty)$. The linear manifold $\mathfrak{B}_1$ consisting of the initial values x_0 of the solutions of equation (3.1) that are bounded on $[0, \infty)$ is called the *(right)* $\mathscr{S}$*-set* of this equation.

We will assume that $\mathfrak{B}_1$ is a complemented subspace, i.e. that it is closed and has a direct complement: $\mathfrak{B} = \mathfrak{B}_1 + \mathfrak{B}_2$.

In the finite-dimensional case this condition is automatically satisfied. In a Hilbert space $\mathfrak{H}$ the second part of the condition is superfluous since an orthogonal complement always exists in $\mathfrak{H}$.

We note that this condition is essentially contained in the definition of e-dichotomicity of an equation. It is trivially satisfied for equations with a constant operator whose spectrum does not intersect the imaginary axis.

It is easily seen that the closedness and complementedness properties of the $\mathscr{S}$-set are not violated under a kinematic similarity transformation. We will subsequently see that they are also preserved under "small" perturbations of e-dichotomic equations. All of these transformations permit one to widen the class of equations for which the $\mathscr{S}$-set has the indicated properties.

If an equation is also defined on the left halfline, the *left $\mathscr{S}$-set* (to which all of the remarks made above for the right $\mathscr{S}$-set apply) is introduced analogously.

LEMMA 3.3. *Suppose that equation* (3.14) *has for each function $f(t) \in C(\mathfrak{B})$ at least one solution x that is bounded on* $[0, \infty)$:

$$|||x||| = \sup_{0 \leq t < \infty} \|x(t)\| < \infty.$$

Suppose further that the $\mathscr{S}$-set $\mathfrak{B}_1$ of equation (3.1) *is a complemented subspace and that $\mathfrak{B}_2$ is a complement of it.*

Then to each function $f(t) \in C(\mathfrak{B})$ there corresponds a unique solution $x(t)$ that is bounded on $[0, \infty)$ and initially in $\mathfrak{B}_2$: $x(0) \in \mathfrak{B}_2$.

This solution is subject to the estimate $|||x||| \leq K|||f|||$, where $K > 0$ is a constant not depending on f.

PROOF. Suppose $f(t) \in C(\mathfrak{B})$. By hypothesis, there exists a solution $x(t) \in C(\mathfrak{B})$ of equation (3.14). Let P_k $(k = 1,2)$ be the mutually complementary projections on the subspaces $\mathfrak{B}_k$.

We denote by $x_1(t)$ the solution of the corresponding homogeneous equation which satisfies the condition $x_1(0) = P_1x(0)$. This solution is bounded by definition of the subspace $\mathfrak{B}_1$. But then the solution $x_2(t) = x(t) - x_1(t)$ of the inhomogeneous equation for which $x_2(0) = x(0) - P_1x(0) = P_2x(0) \in \mathfrak{B}_2$ is also bounded.

The uniqueness follows from the fact that the difference of two similar solutions would be bounded by a solution initially in $\mathfrak{B}_2$ of the homogeneous equation, which is possible only for the zero solution.

It remains for us to prove the latter assertion of the lemma. To this end we consider the space C_1 of all functions $x(t)$ that are solutions of equations of the form

$$x'(t) - A(t)x(t) = f(t)$$

under the conditions $x(0) \in \mathfrak{B}_2$ and $f(t) \in C(\mathfrak{B})$. It was essentially shown above that the operator $\Phi x(t) = x'(t) - Ax(t)$ effects a one-to-one mapping of the linear space C_1 onto $C(\mathfrak{B})$. If in C_1 we introduce the norm

$$\langle x \rangle = |||x||| + |||\Phi x|||,$$

the operator Φx automatically turns out to be continuous. If, in addition, the space C_1 turns out to be complete, the inverse operator Φ^{-1} will also be continuous by Banach's theorem, and the solution $x = \Phi^{-1}f$ of equation (3.14) will then satisfy the required estimate

$$|||x||| \leq \langle x \rangle \leq \|\Phi^{-1}\| \, |||f|||.$$

Thus it remains to prove the completeness of C_1. Let $\{x_n(t)\}$ be a Cauchy sequence in it.

Such a sequence is also a Cauchy sequence in $C(\mathfrak{B})$ and hence has a limit $x(t)$ in it. In this case, clearly,

$$x(0) = \lim_{n\to\infty} x_n(0) \in \mathfrak{B}_2.$$

In exactly the same way it follows that the sequence $\{f_n(t)\} = \{\Phi x_n(t)\}$ has a limit $f(t)$ in $C(\mathfrak{B})$. Therefore, for each $t \in \mathscr{I}$,

$$x(t) - x(0) = \lim_{n\to\infty} \int_0^t x_n'(\tau)d\tau$$

$$= \lim_{n\to\infty} \int_0^t [f_n(\tau) + A(\tau)x_n(\tau)]d\tau = \int_0^t [f(\tau) + A(\tau)x(\tau)]d\tau,$$

which implies that $x(t)$ satisfies the equation $x'(t) - A(t)x(t) = f(t)$.

Thus $x(t) \in C_1$ and, as is easily seen, $\langle x - x_n\rangle \to 0$ for $n \to \infty$, i.e. C_1 is complete.

The lemma is proved.

THEOREM 3.3. *Suppose the operator function $A(t)$ is integrally bounded. In order for equation* (3.1) *to be e-dichotomic on* $[0, \infty)$ *it is necessary and sufficient that its right $\mathscr{S}$-set be a complemented subspace and that there correspond to each function $f(t) \in C(\mathfrak{B})$ at least one bounded solution on* $[0, \infty)$ *of the inhomogenous equation.*

PROOF. The necessity of the second condition follows from Theorem 3.1, while the necessity of the first was noted in defining the $\mathscr{S}$-set.

SUFFICIENCY. Let $x(t)$ be a nonzero bounded solution on $\mathscr{I}$ of equation (3.1), i.e. a solution satisfying the condition $x(0) \in \mathfrak{B}_1$. We put

$$y(t) = x(t) \int_0^t \chi(\tau)\|x(\tau)\|^{-1}d\tau_1,$$

where

$$\chi(t) = \begin{cases} 1, & 0 \leqq t \leqq t_0 + \tau, \\ 1 - (t - t_0 - \tau), & t_0 + \tau \leqq t \leqq t_0 + \tau + 1, \\ 0, & t \geqq t_0 + \tau + 1. \end{cases}$$

A simple calculation shows that $y(t)$ is a solution of equation (3.14) for

$$f(t) = \chi(t)\, x(t)/\|x(t)\|.$$

This function is bounded and continuous and such that $y(0) = 0 \in \mathfrak{B}_2$. Therefore, by virtue of Lemma 3.3,

$$\|y(t)\| \leqq K \sup_t \|f(t)\| \leqq K \qquad (t \in \mathscr{I})$$

and, in particular, when $t = t_0 + \tau$

$$\|y(t_0 + \tau)\| = \|x(t_0 + \tau)\| \int_0^{t_0+\tau} \|x(s)\|^{-1} ds \leqq K. \tag{3.23}$$

We introduce the function

$$\phi(t) = \int_0^t \|x(s)\|^{-1} ds.$$

From (3.23) we obtain the inequality

$$\phi'(t_0 + \tau)/\phi(t_0 + \tau) \geqq \frac{1}{K},$$

which after an integration with respect to τ over the interval $[1, \tau]$ reduces to the estimate

$$\phi(t_0 + \tau) \geqq \phi(t_0 + 1)e^{(\tau-1)/K} \qquad (\tau \geqq 1). \tag{3.24}$$

Further, for $s \in [t_0, t_0 + 1]$

$$\|x(s)\| \leqq \|x(t_0)\| \exp\left\{ \int_{t_0}^{t_0+1} \|A(\tau)\| d\tau \right\} \leqq e^{M_1} \|x(t_0)\|,$$

and hence

$$\phi(t_0 + 1) = \int_{t_0}^{t_0+1} \|x(s)\|^{-1} ds \geqq \|x(t_0)\|^{-1} e^{-M_1}.$$

Combining this inequality with (3.23) and (3.24), we obtain for $\tau \geqq 1$ the estimate

$$\|x(t_0 + \tau)\| \leqq \frac{K}{\phi(t_0 + \tau)} \leqq \frac{Ke^{-(\tau-1)/K}}{\phi(t_0 + 1)} \leqq N_1 e^{-\tau/K} \|x(t_0)\|,$$

where $N_1 = Ke^{1/K+M_1}$ Since for $\tau \leqq 1$

$$\|x(t_0 + \tau)\| \leqq e^{M_1} \|x(t_0)\| \leqq e^{1/K+M_1} e^{-\tau/K} \|x(t_0)\|,$$

we obtain the estimate

$$\|x(t)\| \leqq Ne^{-\nu(t-t_0)} \|x(t_0)\|, \tag{3.25}$$

where $\nu = 1/K$ and $N = \max\{1, K\}e^{1/K+M_1}$, i.e. inequality (3.6a).

Suppose now $x(t)$ is a nonzero solution of equation (3.1) with its initial value $x(0) \in \mathfrak{B}_2$. Then the function

$$y(t) = x(t) \int_t^{\infty} \chi(s) \|x(s)\|^{-1} ds$$

is a solution of equation (3.14) for $f(t) = -\chi(t)x(t)/\|x(t)\|$. This solution is bounded (it vanishes for $t \geqq t_0 + \tau$) and its initial value $y(0) \in \mathfrak{B}_2$. Therefore

$$\|y(t)\| = \|x(t)\| \int_t^\infty \chi(s)\|x(s)\|^{-1}ds \leqq K \,|||f||| = K.$$

Letting τ tend to infinity, we obtain the inequality

$$\int_t^\infty \|x(s)\|^{-1}ds \leqq K \|x(t)\|^{-1}, \tag{3.26}$$

which, setting $\psi(t) = \int_t^\infty \|x(s)\|^{-1}ds$, is equivalent to the inequality $\psi'(t) \leqq \psi(t)/K$. An integration of the latter leads to the estimate

$$\psi(t) \leqq \psi(t_0) \exp \frac{t - t_0}{K}. \tag{3.27}$$

Further, since for $\tau \geqq t$

$$\|x(\tau)\| \leqq \|x(t)\| \exp\left\{ \int_t^\tau \|A(s)\|ds \right\},$$

it follows that

$$\|x(t)\|\psi(t) = \|x(t)\| \int_t^\infty \|x(s)\|^{-1}ds \geqq \int_t^\infty \exp\left\{ - \int_t^\tau \|A(s)\|ds \right\} d\tau$$

$$\geqq \sum_{k=1}^\infty \int_{t+(k-1)}^{t+k} \exp\left\{ - \int_t^{t+k} \|A(s)\|ds \right\} d\tau \geqq \sum_{k=1}^\infty e^{-kM_1} = \frac{1}{e^{M_1}-1} = c$$

and, by virtue of (3.26) and (3.27), that

$$\|x(t)\| \geqq \frac{c}{\psi(t)} \geqq \frac{c}{\psi(t_0)} e^{\frac{1}{K}(t-t_0)} \geqq \frac{c}{K} e^{\frac{1}{K}(t-t_0)} \|x(t_0)\|.$$

We have obtained an estimate of form (3.6b). The theorem is proved.

COROLLARY 3.1. *In a finite-dimensional phase space equation* (3.1) *with an integrally bounded* $A(t)$ *is e-dichotomic on* $[0, \infty)$ *precisely when there corresponds to each function* $f(t) \in C(\mathfrak{B})$ *at least one bounded solution on* $[0, \infty)$ *of the inhomogeneous equation.*

5. *e-dichotomic equation on the real line:* $\mathcal{J} = (-\infty, \infty)$. In this case the subspace $\mathfrak{B}_1$ ($\mathfrak{B}_2$) consists of the initial values of those solutions that remain bounded as $t \to +\infty$ ($t \to -\infty$) Since these subspaces have only the zero element in common, the equation does not have nontrivial solutions that are bounded on the real line. Further, the solutions $x(t)$ for which $x(0) \notin \mathfrak{B}_1$ ($x(0) \notin \mathfrak{B}_2$) unboundedly increase as $t \to +\infty$ ($t \to -\infty$).

Let us show that a theorem analogous to Theorem 3.3 is valid. We recall that the right and left $\mathcal{S}$-sets $\mathfrak{B}_1$ and $\mathfrak{B}_2$ of equation (3.1) are the linear manifolds consisting of the initial values of those solutions of this equation that are bounded on the right and left halflines respectively.

THEOREM 3.3′. *Suppose the operator function* $A(t)$ $(-\infty < t < \infty)$ *is integrally bounded. In order for equation* (3.1) *to be e-dichotomic on the real line it is necessary and sufficient that the right and left* $\mathscr{S}$*-sets of this equation be mutually complementary subspaces and that there correspond to each function* $f(t) \in \mathbf{C}(\mathfrak{B})$ *at least one bounded solution on* $(-\infty, \infty)$ *of the inhomogeneous equation* (3.14). *This solution is the only one bounded on the real line.*

PROOF. The necessity follows from Theorem 3.2 without the assumption that $A(t)$ is integrally bounded. To prove the sufficiency we note that under the hypothesis of the theorem there corresponds to every function $f(t)$ that is bounded on either the right or left halfline a solution $x(t)$ that is bounded on this halfline. Applying Theorem 3.3 to each of these halflines, we see that equation (3.1) is e-dichotomic on each of them for one and the same pair of projections P_1 and P_2. This means that the equation is e-dichotomic on the whole real line.

The uniqueness of a bounded solution follows from the absence of nontrivial bounded solutions on the real line of a homogeneous e-dichotomic equation.

REMARK 3.7. We note that an equation that is e-dichotomic on both the right and left halflines is not necessarily e-dichotomic on the real line inasmuch as the corresponding pairs of mutually complementary projections need not be the same.

COROLLARY 3.2. *In a finite-dimensional phase space equation* (3.1) *with an integrally bounded coefficient* $A(t)$ *is e-dichotomic on the real line* $(-\infty, \infty)$ *precisely when there corresponds to each function* $f(t) \in \mathbf{C}(\mathfrak{B})$ *at least one (exactly one) bounded solution on* $(-\infty, \infty)$ *of the inhomogeneous equation.*

§ 4. Exponential splitting

1. *Definition and basic properties.* We consider the following natural generalization of the notion of an exponential dichotomy.

We will say that *equation* (3.1) *admits an exponential splitting of order n on* $\mathscr{J}$ if there exists a direct decomposition

$$\mathfrak{B} = \mathfrak{B}_1 \dotplus \mathfrak{B}_2 \dotplus \cdots \dotplus \mathfrak{B}_n \tag{4.1}$$

such that the Bohl intervals $[\kappa_B'(P_k), \kappa_B(P_k)]$ of equation (3.1) corresponding to the subspaces $\mathfrak{B}_k$ are mutually disjoint and each of the corresponding projections P_k is uniformly conjugated on $\mathscr{J}$ by the operator $U(t)$:

$$\|U(t)P_kU^{-1}(t)\| \leqq M \qquad (k = 1,2,\cdots,n;\ t \in \mathscr{J}). \tag{4.2}$$

The intervals $(\kappa_B(P_k), \kappa_B'(P_{k+1}))$ will be called *gaps in the Bohl spectrum.*[2]

By combining subspaces with neighboring Bohl intervals into direct sums in (4.1), we see that the equation also admits an exponential splitting of any smaller order, in particular, of the second.

[2] We agree to enumerate the projections in the order of increase of the exponents.

It is not difficult to see that an equation is e-dichotomic precisely when it admits an exponential splitting of second order and the point $\nu = 0$ falls into a gap between the Bohl intervals.

We note some properties of these notions.

1) The replacement in equation (3.1) of the operator $A(t)$ by an operator of the form $A(t) + \lambda I$ produces an equation all of the Bohl exponents of the solutions of which have been shifted by the amount λ. This maneuver permits one to shift the point $\nu = 0$ into any gap between the Bohl intervals and thereby obtain an e-dichotomic equation.

2) Kinematically similar equations have the same Bohl intervals. Thus a kinematic similarity transformation does not disturb the exponential splitting property. In particular, this property is preserved on $[0, \infty)$ with preservation of the Bohl exponents if the coefficient $A(t)$ is altered on a finite interval.

3) When $\mathscr{J}$ is the right (left) halfline, only the subspace $\mathfrak{B}_1$ ($\mathfrak{B}_n$) consisting of the initial values of those solutions "growing" most slowly (rapidly) is uniquely determined by the equation.

Any other subspace $\mathfrak{B}_k$ can always be "rotated" (i.e. the corresponding projection P_k can always be conjugated) within the direct sum of subspaces with subscripts not greater (less) than k under the condition that the sum remain direct.

4) From the integral boundedness of the operator $A(t)$ it follows as in Lemma 3.2 that the angular distance between the subspaces $\mathfrak{B}_k(t) = U(t)\mathfrak{B}_k$ is bounded from below in $\mathscr{J}$. We recall once again that this property of the subspaces $\mathfrak{B}_k(t)$ is equivalent to the boundedness from above of the projection functions $P_k(t) = U(t)P_kU^{-1}(t)$.

5) A stationary equation $dx/dt = Ax$ obviously admits an nth order exponential splitting of its solutions if (and only if) the spectrum $\sigma(A)$ decomposes into n spectral sets lying in mutually disjoint vertical strips $\nu_k' < \operatorname{Re}\lambda < \nu_k$ $(k = 1,\cdots,n)$.

We consider separately the case $n = 3$ when the left (right) Bohl interval lies in the interior of the left (right) halfline and the center interval lies in a neighborhood of the origin.

Here, for any $-\nu_1 > \kappa_{\mathrm{B}}(P_1)$, $-\nu_3 < \kappa_{\mathrm{B}}'(P_3)$, $\tilde\nu_3 > \kappa_{\mathrm{B}}(P_3)$ and $\nu_2 < \kappa_{\mathrm{B}}'(P_2)$ $(\nu_1 > \nu_3 > 0,\ \nu_2 > \tilde\nu_3 > 0)$, there exist positive numbers N_k $(k = 1,3)$ such that the following estimates are satisfied on $\mathscr{J}$ for any $x \in \mathfrak{B}$ and $t \geqq s$:

$$\left.\begin{aligned} \|U(t)P_1x\| &\leqq N_1e^{-\nu_1(t-s)}\|U(s)P_1x\|;\\ \|U(t)P_2x\| &\geqq N_2^{-1}e^{\nu_2(t-s)}\|U(s)P_2x\|;\\ N_3^{-1}e^{-\tilde\nu_3(t-s)}\|U(s)P_3x\| &\leqq \|U(t)P_3x\| \leqq N_3e^{\nu_3(t-s)}\|U(s)P_3x\|. \end{aligned}\right\} \tag{4.3}$$

Let us determine some conditions under which in this case the equation

$$dx/dt = A(t)x + f(t) \tag{4.4}$$

has bounded solutions on $\mathscr{J}$.

We introduce a Green function by setting

$$G(t, s) = \begin{cases} U(t)P_1U^{-1}(s) & \text{for} \quad s < t, \\ -U(t)(P_2 + P_3)U^{-1}(s) & \text{for} \quad t < s. \end{cases} \tag{4.5}$$

Then at least for bounded functions $f(t)$ with compact support the formula

$$x(t) = \int_{\mathscr{I}} G(t, s)f(s)ds \tag{4.6}$$

provides certain solutions of equation (4.4). It can be written more explicitly in the form

$$\begin{aligned} x(t) = & \int_{s \leqq t} U(t)P_1U^{-1}(s)f(s)ds \\ & - \int_{s \geqq t} U(t)P_2U^{-1}(s)f(s)ds - \int_{s \geqq t} U(t)P_3U^{-1}(s)f(s)ds. \end{aligned} \tag{4.7}$$

Estimates (4.3) imply that the first two summands of the right side of this expression are bounded for any bounded function $f(t)$. This is true for the third summand only under the condition that $f(t)$ decrease sufficiently fast at infinity.

In place of the function $G(t, s)$ we consider an "*incomplete*" *Green function* $\tilde{G}(t, s)$, which differs from $G(t, s)$ in that the projection $P_2 + P_3$ in formula (4.5) has been replaced by the projection P_2.

The function

$$x(t) = \int_{\mathscr{I}} \tilde{G}(t, s)f(s)ds \qquad (t \in \mathscr{I}) \tag{4.8}$$

turns out to be bounded for any bounded function $f(s)$. One can easily see that it satisfies the equation

$$dx/dt = A(t)x(t) + (I - P_3(t))f(t), \tag{4.9}$$

which coincides with (4.4) if $P_3(t)f(t) = 0$ or equivalently

$$P_3U^{-1}(t)f(t) = 0. \tag{4.10}$$

We note that under the satisfaction of this condition the order of growth of the solutions $U(t)P_3x_0$ is no longer important, i.e. the restrictions $\nu_3 < \nu_1$ and $\tilde{\nu}_3 < \nu_2$ are not essential.

In particular, the projection P_3 can commute with $U(t)$, i.e. we can have $P_3(t) = U(t)P_3U^{-1}(t) = P_3$. In this case condition (4.10) converts into the condition

$$P_3f(t) = 0 \tag{4.11}$$

of "orthogonality" of $f(t)$ to the constant subspace $\mathfrak{B}_3(t) \equiv \mathfrak{B}_3$ while the incomplete Green function is expressed by the formula

$$\tilde{G}(t, s) = G(t, s)(I - P_3). \tag{4.12}$$

2. *Induced exponential splitting of the solutions of an auxiliary equation.* We consider the following differential equation in a phase space $[\mathfrak{B}]$:

$$dX/dt = A(t)X - XA(t). \tag{4.13}$$

We assume that the corresponding equation in $\mathfrak{B}$

$$dx/dt = A(t)x \tag{4.14}$$

admits an exponential splitting of order n.

The Cauchy transform $\mathfrak{U}(t)$ of equation (4.13) can be expressed in terms of the Cauchy operator $U(t)$ of equation (4.14) by means of the formula (see III.1.25))

$$\mathfrak{U}(t)X = U(t)XU^{-1}(t) \qquad (X \in [\mathfrak{B}]).$$

We consider the projection transforms $\mathscr{P}_{jk}X = P_jXP_k$ and let

$$\mathscr{P}_1 = \sum_{j<k} \mathscr{P}_{jk}; \quad \mathscr{P}_2 = \sum_{j>k} \mathscr{P}_{jk}; \quad \mathscr{P}_3 = \sum_{j} \mathscr{P}_{jj}.$$

We assume that the projections are enumerated in the order of arrangement of the Bohl intervals from left to right.

From the formula

$$[\mathfrak{U}(t)\mathscr{P}_{jk}]X = U(t)P_jU^{-1}(s)\,[\mathfrak{U}(s)X]U(s)P_kU^{-1}(t)$$

it can easily be seen that the solutions of equation (4.13) of the form

$$X_k(t) = \mathfrak{U}(t)\mathscr{P}_kX \qquad (k = 1,2,3;\ X \in [\mathfrak{B}]) \tag{4.15}$$

will exponentially decrease (increase) as $t \to \infty$ for $k = 1$ $(k = 2)$, the exponents of decrease and increase being expressed in terms of differences of the Bohl exponents of the vector equation (4.14).

Let $F(t)$ be a bounded operator function satisfying the condition of type (4.10)

$$\mathscr{P}_3\mathfrak{U}^{-1}(t)F(t) = \sum_j P_jU^{-1}(t)F(t)U(t)P_j = 0$$

or equivalently

$$P_jU^{-1}(t)F(t)U(t)P_j = 0 \qquad (j = 1,2,\cdots,n). \tag{4.16}$$

By means of a formula of type (4.6) a bounded solution of the equation

$$dx/dt = A(t)X - XA(t) + F(t) \tag{4.17}$$

satisfying the condition

$$\mathscr{P}_3\mathfrak{U}^{-1}(t)X(t) = 0$$

will be obtained independently of the behavior of the solution $X_3(t)$.

We write out this formula in detail under the assumption that $A(t)P_k = P_kA(t)$ $(k = 1,\cdots,n)$, and hence that $U(t)P_k = P_kU(t)$.

The incomplete Green function in this case has the form

$$\mathscr{G}(t,s)X = \begin{cases} [\mathfrak{U}(t)\mathscr{P}_1\mathfrak{U}^{-1}(s)]X = \sum_{j<k} P_jU(t,s)XU(s,t)P_k & (t \geqq s), \\ -[\mathfrak{U}(t)\mathscr{P}_2\mathfrak{U}^{-1}(s)]X = \sum_{j>k} P_jU(t,s)XU(s,t)P_k & (t < s), \end{cases}$$

and consequently the formula

$$X(t) = \sum_{j<k} P_j \int_{s\leqq t} U(t, s)F(s)U(s, t)dsP_k \tag{4.18}$$
$$- \sum_{j>k} P_j \int_{s\geqq t} U(t, s)F(s)U(s, t)dsP_k \qquad (t \in \mathscr{I})$$

gives a solution of equation (4.17) for a bounded function $F(t)$ satisfying condition (4.16), which now takes the simpler form

$$P_jF(t)P_j = 0 \qquad (t \in \mathscr{I};\ \ j = 1,2,\cdots,n). \tag{4.19}$$

This condition is also satisfied by the solution:

$$P_jXP_j = 0 \qquad (t \in \mathscr{I};\ \ j = 1,2,\cdots,n). \tag{4.20}$$

Formula (4.18) will be employed in the next section to obtain a useful transformation of equations.

§ 5. Stability of the Bohl exponents of an exponential splitting

1. *Reduction to an e-dichotomy.* Suppose the equation

$$dx/dt = A(t)x \qquad (t \in \mathscr{I}) \tag{5.1}$$

with an integrally bounded operator function $A(t)$ admits an exponential splitting of order $n \geqq 2$.

We will show that this property is stable in the sense that it is preserved in going from equation (5.1) to the perturbed equation

$$dx/dt = [A(t) + B(t)]x \qquad (t \in \mathscr{I}), \tag{5.2}$$

when the perturbation $B(t)$ is sufficiently small, i.e. when the condition

$$\frac{1}{\tau_0}\int_t^{t+\tau_0} \|B(\tau)\|d\tau < \delta \qquad (t \in \mathscr{I}) \tag{5.3}$$

is satisfied for some $\tau_0 > 0$ and a sufficiently small $\delta > 0$.

We wish to determine the nature of the response of the Bohl exponents of an equation to a perturbation. Inasmuch as this has already been done for the upper and lower exponents (see Theorem III.4.6), we need only examine the behavior of the ends of each gap in the Bohl spectrum.

By transforming equation (5.1) with the use of the substitution $x = ye^{\lambda t}$, we can always shift the point $\nu = 0$ to the center of some gap. The equation then becomes e-dichotomic, i.e. its solutions will satisfy estimates (3.6) and (3.11), in which $\nu_1 = \nu_2$ (and it can be assumed that $N_1 = N_2$).

We see that it suffices to show that the exponents ν_1 and ν_2 of the equation are stable in this case.

2. *Bounded solutions of the perturbed equation.* We first consider the case when $\mathscr{I} = [0, \infty)$.

Suppose an e-dichotomy of the solutions of equation (4.1) exists for a direct decomposition $\mathfrak{B} = \mathfrak{B}_1 + \mathfrak{B}_2$ with corresponding projections P_1 and P_2. We must first establish the existence of a similar decomposition for the perturbed equation.

Let $x(t)$ be a bounded solution on $[0, \infty)$ of equation (5.2). It can be regarded as a bounded solution of the equation

$$dx/dt = A(t)x + f(t),$$

corresponding to the integrally bounded function $f(t) = B(t)x(t)$.

Formula (3.22) permits one to represent this solution in the form

$$x(t) = U(t)y + \int_0^\infty G(t, \tau)f(\tau)d\tau \qquad (y \in \mathfrak{B}_1)$$

and thus to obtain the integral equation

$$x(t) = U(t)y + \int_0^\infty G(t, \tau)B(\tau)x(\tau)d\tau. \tag{5.4}$$

Here $U(t)$ and $G(t, s)$ are the Cauchy operator and Green function of the unperturbed equation, and y is a vector of the subspace $\mathfrak{B}_1$. It is not difficult to verify that one also has the converse assertion: every bounded solution of the integral equation (5.4) also satisfies the differential equation (5.2).

LEMMA 5.1. *There exists a $\delta > 0$ such that the satisfaction of condition (5.3) implies the existence of a unique bounded solution on $[0, \infty)$ of equation (5.4). This solution is representable in the form*

$$x(t) = \Gamma(t)y \qquad (y \in \mathfrak{B}_1), \tag{5.5}$$

where $\Gamma(t)$ is a bounded operator function on $[0, \infty)$.

PROOF. We consider in the space $C(\mathfrak{B})$ of bounded continuous vector functions on $[0, \infty)$ the linear operator $Sx(t) = \int_0^\infty G(t, \tau)B(\tau)x(\tau)d\tau$. It maps $C(\mathfrak{B})$ into itself, as follows from Remark 3.6.

Using formulas (3.11) for $N_1 = N_2 = N$ and $\nu_1 = \nu_2 = \nu$, we estimate its norm. We have for $n = [t/\tau_0]$

$$\begin{aligned} |||Sx||| &\leqq N \int_0^\infty e^{-\nu|t-\tau|}\|B(\tau)\|d\tau\,|||x||| \\ &\leqq N|||x|||\tau_0\delta\left\{\sum_{k=1}^{n} e^{-\nu(n-k)\tau_0} + 1 + \sum_{k=n+2}^{\infty} e^{-\nu(k-n-2)\tau_0}\right\} \\ &\leqq N|||x|||\tau_0\delta\left\{1 + \frac{2}{1 - e^{-\nu\tau_0}}\right\}. \end{aligned} \tag{5.6}$$

Thus under the condition

$$\delta < \frac{1}{N\tau_0\left\{1 + \dfrac{2}{1 - e^{-\nu\tau_0}}\right\}} = \frac{\nu}{2N} \cdot \frac{1}{\dfrac{\nu\tau_0}{2} + \dfrac{\nu\tau_0}{1 - e^{-\nu\tau_0}}} \tag{5.7}$$

$$\left(= \frac{\nu}{2N}[1 + O(\nu\tau_0)] \quad \text{for} \quad \tau_0 \to 0\right)$$

we obtain the inequality $|||S||| < 1$.

From this inequality it follows that equation (5.4) has for each $y \in \mathfrak{B}_1$ exactly one solution $x(t)$ on $[0, \infty)$ belonging to $C(\mathfrak{B})$. Clearly, this solution linearly depends on y, and by virtue of (4.6)

$$|||x||| \leqq \frac{N}{1 - N\tau_0\delta\left(1 + \dfrac{2}{1 - e^{-\nu\tau_0}}\right)} \|y\|. \tag{5.8}$$

The lemma is proved.

3. *e-dichotomicity of the perturbed equation.* Using Lemma 5.1, we construct for equation (5.2) the direct decomposition.

$$\mathfrak{B} = \tilde{\mathfrak{B}}_1 + \tilde{\mathfrak{B}}_2, \tag{5.9}$$

which leads to an e-dichotomy of its solutions.

The subspace $\tilde{\mathfrak{B}}_1$ must consist of the initial values $x_0 = x(0)$ of those solutions of equation (5.2) which remain bounded on $[0, \infty)$.

From formulas (5.4) and (5.5) it follows that these vectors are given by the formula

$$x_0 = x(0) = y + \int_0^\infty G(0, \tau)B(\tau)\Gamma(\tau)y d\tau = (I - P_2RP_1)y,$$

where $y \in \mathfrak{B}_1$ and

$$R = \int_0^\infty P_2U^{-1}(\tau)B(\tau)\Gamma(\tau)P_1 d\tau$$

is a bounded operator inasmuch as (3.11), (5.3) and (5.8) imply that

$$\|R\| \leqq \frac{N\tau_0\delta}{1 - e^{-\nu\tau_0}} \cdot \frac{N\|P_1\|}{1 - N\tau_0\delta\left(1 + \dfrac{2}{1 - e^{-\nu\tau_0}}\right)}. \tag{5.10}$$

Thus the bounded operator $I - P_2RP_1$ maps the subspace $\mathfrak{B}_1$ onto $\tilde{\mathfrak{B}}_1$. This operator has the bounded inverse

$$(I - P_2RP_1)^{-1} = I + P_2RP_1,$$

and therefore the subspace $\tilde{\mathfrak{B}}_1$ is closed. The operator

$$\begin{aligned}\tilde{P}_1 &= (I - P_2RP_1)P_1(I - P_2RP_1)^{-1} \\ &= (I - P_2RP_1)P_1(I + P_2RP_1) = P_1 - P_2RP_1\end{aligned}$$

is a projection whose range coincides with $\tilde{\mathfrak{B}}_1$. The complementary projection has the form

$$\tilde{P}_2 = I - \tilde{P}_1 = P_2 + P_2RP_1 = P_2(I + RP_1),$$

showing that $\tilde{\mathfrak{B}}_2 = \mathfrak{B}_2$.

To obtain the desired result we must estimate the solutions of equation (5.2) that are initially in the subspaces $\tilde{\mathfrak{B}}_1$ and $\tilde{\mathfrak{B}}_2$.

We first consider a solution $x_1(t)$ satisfying the condition $x_1(0) \in \tilde{\mathfrak{B}}_1$, i.e. an arbitrary bounded solutions of equation (5.2). These solutions are given by formula (5.4). Since

$$x_1(s) = U(s)y + \int_0^\infty G(s, \tau)B(\tau)x_1(\tau)d\tau,$$

it follows that

$$\begin{aligned} y &= U^{-1}(s)x_1(s) - U^{-1}(s)\int_0^\infty G(s, \tau)B(\tau)x_1(\tau)d\tau \\ &= P_1U^{-1}(s)x_1(s) - P_1U^{-1}(s)\int_0^\infty G(s, \tau)B(\tau)x_1(\tau)d\tau \\ &= P_1U^{-1}(s)x_1(s) - \int_0^s P_1U^{-1}(\tau)B(\tau)x_1(\tau)d\tau. \end{aligned}$$

Substituting this expression into (5.4), we obtain for $t > s$ the equation

$$\begin{aligned} x_1(t) &= U(t)P_1U^{-1}(s)x_1(s) \\ &\quad + \int_0^\infty G(t, \tau)B(\tau)x_1(\tau)d\tau - \int_0^s U(t)P_1U^{-1}(\tau)B(\tau)x_1(\tau)d\tau \\ &= U(t)P_1U^{-1}(s)x_1(s) + \int_s^\infty G(t, \tau)B(\tau)x_1(\tau)d\tau. \end{aligned}$$

By virtue of estimate (3.6) it implies the integral inequality

$$\|x(t)\| \leqq Ne^{-\nu(t-s)}\|x(s)\| + N\int_s^\infty e^{-\nu|t-\tau|}\|B(\tau)\|\,\|x(\tau)\|d\tau. \tag{5.11}$$

But this implies according to Lemma III.2.2 that

$$\|x(t)\| \leqq N_1e^{-\mu(t-s)}\|x(s)\| \qquad (t \geqq s), \tag{5.12}$$

where

$$N_1 = \frac{N(\nu^2 - \mu^2)}{\nu^2 - \mu^2 - 2N\nu\delta + O(\tau_0\delta)},$$

and μ satisfies the single condition

$$(0 <)\, \mu < \sqrt{\nu^2 - 2N\nu\delta + O(\tau_0\delta)}. \tag{5.13}$$

We note that, by decreasing δ, we can make μ as close as we like to ν.

Suppose now $x_2(t)$ is a solution satisfying the condition $x_2(0) \in \tilde{\mathfrak{B}}_2 = \mathfrak{B}_2$. We make use of the following formulas under the restriction $t < s$:

$$x(t) = U(t)x_2(0) + \int_0^t U(t, \tau)B(\tau)x(\tau)d\tau,$$

$$x(s) = U(s)x_2(0) + \int_0^s U(s, \tau)B(\tau)x(\tau)d\tau.$$

Employing the relation $x_2(0) = \tilde{P}_2x_2(0) = P_2x_2(0)$ in a manner analogous to the way in which we obtained the above estimate of $x_1(t)$, we deduce the equation

$$x(t) = U(t)P_2U^{-1}(s)x(s) + \int_0^t U(t)P_1U^{-1}(\tau)B(\tau)x(\tau)d\tau - \int_t^s U(t)P_2U^{-1}(\tau)B(\tau)x(\tau)d\tau$$

and then the integral inequality

$$\begin{aligned}\|x(t)\| &\leqq Ne^{-\nu(s-t)}\|x(s)\| + N\int_0^t e^{-\nu(t-\tau)}\|B(\tau)\|\,\|x(\tau)\|d\tau \\ &\quad + N\int_t^s e^{-\nu(\tau-t)}\|B(\tau)\|\,\|x(\tau)\|d\tau \\ &= Ne^{-\nu(s-t)}\|x(s)\| + N\int_0^s e^{-\nu|t-\tau|}\|B(\tau)\|\,\|x(\tau)\|d\tau.\end{aligned}$$

From this inequality it follows in the same way as above that

$$\|x(t)\| \leqq N_2e^{-\mu(s-t)}\|x(s)\| \qquad (t \leqq s) \tag{5.14}$$

for μ satisfying condition (5.13) and some N_2.

We thus arrive at the following result.

THEOREM 5.1. *Suppose equation* (5.1) *with an integrally bounded coefficient* $A(t)$ *admits an exponential splitting of order* n. *For any* $\varepsilon \in (0, \varepsilon_0)$, *where* ε_0 *is one half of the minimal width of the gaps in the Bohl spectrum, there exists a number* $\delta > 0$ *depending only on equation* (5.1) *and the number* ε *and such that under condition* (5.3) *equation* (5.2) *also admits on exponential splitting of order* n.

The Bohl exponents $\kappa_B(\tilde{P}_k)$ *and* $\kappa_B'(\tilde{P}_k)$ *of equation* (5.2) *then satisfy the estimates*

$$\kappa_B'(\tilde{P}_k) \geqq \kappa_B'(P_k) - \varepsilon, \quad \kappa_B(\tilde{P}_k) \leqq \kappa_B(P_k) + \varepsilon.$$

REMARK 5.1. If the operator function $B(t)$ satisfies the condition $\|B(t)\| < \delta$ $(t \in \mathscr{I})$, we can put $\tau_0 = 0$ in estimates (5.7) and (5.13) and thereby obtain the inequalities

$$\delta < \nu/2N \quad \text{and} \quad \mu < \sqrt{\nu^2 - 2N\nu\delta}.$$

REMARK 5.2. The integral boundedness condition (5.3) need only be satisfied for t greater than some $t_0 > 0$, i.e. it can be replaced by a condition of type (III.4.16).

REMARK 5.3. An analogous result holds when equation (5.1) admits an exponential splitting on the real line. As before, it suffices to consider the case of an e-dichotomy.

We can consider equation (5.2) on each halfline (right and left) separately. By virtue of Theorem 5.1 it can be asserted that for sufficiently small $\delta > 0$ equation (5.2) is e-dichotomic on each of the halflines, the dichotomy on the right (left) halfline being connected with the subspaces $\tilde{\mathfrak{B}}_1$ and $\mathfrak{B}_2$ ($\mathfrak{B}_1$ and $\tilde{\mathfrak{B}}_2$), where $\tilde{\mathfrak{B}}_1 = (I - P_2RP_1)\mathfrak{B}_1$ and $\tilde{\mathfrak{B}}_2 = (I - P_1R'P_2)\mathfrak{B}_2$.

Also the operator

$$S = (I - P_2RP_1)P_1 + (I - P_1R'P_2)P_2 = I - (P_2RP_1 + P_1R'P_2)$$

is bounded for a sufficiently small (possibly smaller) $\delta > 0$, invertible (since the operators R and R' turn out to be small) and such that $\tilde{\mathfrak{B}}_1 = S\mathfrak{B}_1$ and $\tilde{\mathfrak{B}}_2 = S\mathfrak{B}_2$. Therefore $\mathfrak{B} = \tilde{\mathfrak{B}}_1 + \tilde{\mathfrak{B}}_2$.

It follows from Remark 3.4 that the pair $\tilde{\mathfrak{B}}_1, \tilde{\mathfrak{B}}_2$ is connected with an e-dichotomy of the solutions of (5.1) on each of the halflines and hence on the whole real line.

4. *Stably decomposing equations.* In this subsection we consider equation (5.1) under the assumption that it admits an exponential splitting on an infinite or semi-infinite interval and that the operator $A(t)$ commutes with the projections $P_1, \cdots, P_n$ inducing this splitting: $A(t)P_k = P_kA(t)$ $(k = 1, \cdots, n)$.

Such an equation decomposes into a system of independent equations in the phase spaces $\mathfrak{B}_k = P_k\mathfrak{B}$:

$$dx_k/dt = P_kA(t)x_k \qquad (k = 1,2,\cdots,n),$$

and the Bohl intervals $[\nu_k', \nu_k]$ of these equations are pairwise disjoint.

We now consider the perturbed equation (5.2) and show (Theorem 5.2) that if the integral boundedness condition (5.3) is satisfied with a sufficiently small $\delta > 0$, equation (5.2) is kinematically similar to an equation of the form

$$dx/dt = [A(t) + C(t)]x, \tag{5.15}$$

where the operator $C(t)$ is sufficiently small and commutes with the projections P_k.

From this result it will follow independently of Theorem 5.1 that the exponential splitting property is stable for an equation of the described special (quasidiagonal) form.

We note that this permits us to eliminate the requirement of integral boundedness of $A(t)$.

It will be shown in the next section that any equation of form (5.1) admitting an exponential splitting in a Hilbert space is kinematically similar to a system of independent equations of the described type.

We first note that if equations (5.2) and (5.15) are kinematically similar the, transforming operator $Q(t) = U(t)V^{-1}(t)$, where $U(t)$ and $V(t)$ are fundamental operators of these equations, must satisfy the equation

$$\begin{aligned} Q'(t) &= [A(t) + B(t)]Q(t) - Q(t)[A(t) + C(t)] \\ &= A(t)Q(t) - Q(t)A(t) + B(t)Q(t) - Q(t)C(t). \end{aligned} \tag{5.16}$$

We need only select the operator $C(t)$ so that equation (5.16) has a solution $Q(t)$ that is bounded together with the operator function $Q^{-1}(t)$.

It is necessary in this connection that $C(t)$ be of quasidiagonal form:

$$C(t) = \sum_k P_k C(t) P_k. \tag{5.17}$$

We will seek a solution in the form $Q(t) = I + S(t)$, where the operator $S(t)$ is small enough to guarantee the invertibility of the operator $Q(t)$. For $S(t)$ we have from (5.16) the equation

$$S'(t) = A(t)S(t) - S(t)A(t) + B(t)[I + S(t)] - [I + S(t)]C(t). \tag{5.18}$$

A solution $S(t)$ of equation (5.18) is a solution of the equation

$$S'(t) = A(t)S(t) - S(t)A(t) + F(t) \tag{5.19}$$

where

$$F(t) = B(t)[I + S(t)] - [I + S(t)]C(t). \tag{5.20}$$

Equation (5.19) has been considered in the preceding section. The formula

$$\begin{aligned} S(t) = &\sum_{j<k} P_j \int_{s\leqq t} U(t, s)F(s)U(s, t)ds P_k \\ &- \sum_{j>k} P_j \int_{s\geqq t} U(t, s)F(s)U(t, s)ds P_k \end{aligned} \tag{5.21}$$

represents the bounded solutions of equation (5.19) satisfying the condition

$$P_m S(t) P_m = 0 \qquad (m = 1,2,\cdots,n), \tag{5.22}$$

provided $F(t)$ satisfies the same condition: $P_m F(t) P_m = 0$ $(m = 1,\cdots,n)$.

The latter condition, as follows from (5.17), (5.20) and (5.22), reduces to the relation

$$\begin{aligned} P_m B(t)[I + S(t)]P_m &= P_m[I + S(t)]\Big(\sum_k P_k C(t) P_k\Big) P_m \\ &= P_m[I + S(t)]P_m C(t) P_m. \end{aligned}$$

It is satisfied if one puts

$$C(t) = B(t)[I + S(t)] = \sum_k P_k B(t)[I + S(t)]P_k.$$

Taking into account relation (5.22), we obtain

$$F(t) = B(t)[I + S(t)] - [I + S(t)]\Big(\sum_k P_k B(t)[I + S(t)]P_k\Big)$$
$$= \sum_{k\neq j} P_j B(t)[I + S(t)]P_k - S(t) \sum_k P_k B(t)[I + S(t)]P_k \tag{5.23}$$
$$= \sum_{k\neq j} P_j[I - S(t)]B(t)[I + S(t)]P_k.$$

The problem consequently reduces to the determination of a solution $S(t)$ ($|||S||| < 1$) of the integral equation obtained from (5.21) by substituting for F the transform $F[S]$ defined by the right side of expression (5.23).

Denoting the transform defined by the right side of (5.21) by $L(F)$, we write this equation in the form

$$S = L(F[S]). \tag{5.24}$$

LEMMA 5.2. *For any* $0 < q < 1$ *there exists a number* $\delta > 0$ *such that under condition* (5.3) *equation* (5.24) *has a unique solution subject to the estimate*

$$\|S(t)\| \leqq q \qquad (t \in \mathcal{J}). \tag{5.25}$$

This solution satisfies condition (5.22).

PROOF. We will prove the lemma for $n = 2$. Since the reduction of an equation to a quasidiagonal form can be carried out in successive steps in each of which the space is decomposed into a direct sum of two summands, the general case will then follow.

In this simplest case equations (5.21) can be written in the form

$$S(t) = P_1 \int_{s\leqq t} U(t, s)[I - S(t)]B(t)[I + S(t)]U(s, t)ds P_2 \tag{5.21$'$}$$
$$- P_2 \int_{s\geqq t} U(t, s)[I - S(t)]B(t)[I + S(t)]U(s, t)ds P_1 \overset{\text{def}}{=} L_1[S].$$

We recall that, inasmuch as equation (5.1) admits an exponential splitting, we have the estimates

$$\left.\begin{aligned}\|P_1U(t, s)\| &= \|U(t)P_1U^{-1}(s)\| \leqq N_1e^{\nu_1(t-s)} \quad (t \geqq s),\\ \|P_2U(t, s)\| &= \|U(t)P_2U^{-1}(s)\| \leqq N_2e^{-\nu_2(t-s)} \quad (t \leqq s),\end{aligned}\right\} \tag{5.26}$$

in which $\nu_1 < \nu_2$.

We consider the metric space $\boldsymbol{K}_q$ of operator functions $S(t)$ satisfying condition (5.25) with metric

$$|||S_2 - S_1||| = \sup_{\mathcal{J}} \|S_2(t) - S_1(t)\|.$$

From (5.26) and (5.3) we obtain the estimate

$$|||L_1[S]||| \leqq N_1N_2(1 + |||S|||)^2 \int_{\mathcal{J}} e^{-(\nu_2-\nu_1)|t-s|}\|B(s)\|ds$$
$$\leqq \frac{2\tau_0\delta}{1 - e^{-\tau_0(\nu_2-\nu_1)}} N_1N_2(1 + q)^2,$$

showing that the operator L_1 maps the space $\boldsymbol{K}_q$ into itself for

$$\delta < \frac{q}{(1+q)^2} \cdot \frac{1}{N_1 N_2} \cdot \frac{1-e^{-\tau_0(\nu_2-\nu_1)}}{2\tau_0} = \frac{q}{(1+q)^2} \cdot \frac{2(\nu_2-\nu_1)}{N_1 N_2} + O(\tau_0). \quad (5.27)$$

Analogously, using the fact that

$$\begin{aligned} &\|[I - S_2(t)]B(t)[I + S_2(t)] - [I - S_1(t)]B(t)[I + S_1(t)]\| \\ &\quad \leqq \|[I - S_2(t)]B(t)[S_2(t) - S_1(t)]\| + \|[S_1(t) - S_2(t)]B(t)[I + S_1(t)]\| \\ &\quad \leqq (1+q)|||S_2 - S_1||| \cdot \|B(t)\| \end{aligned}$$

for $S_1, S_2 \in \boldsymbol{K}_q$, we obtain under condition (5.27) the estimate

$$\begin{aligned} &|||L_1[S_2] - L_1[S_1]||| \\ &\qquad \leqq N_1 N_2 (1+q)|||S_2 - S_1||| \int_{\mathscr{I}} e^{-(\nu_2-\nu_1)|t-s|}\|B(s)\|ds \\ &\qquad \leqq \frac{2\tau_0 \delta N_1 N_2 (1+q)}{1 - e^{-\tau_0(\nu_2-\nu_1)}} |||S_2 - S_1||| \leqq \frac{q}{1+q} |||S_2 - S_1|||. \end{aligned}$$

Thus under condition (5.27) the operator L_1 is a contraction in $\boldsymbol{K}_q$ and hence has in this space exactly one fixed point, which is a solution of equation (5.21) having the required properties.

Lemma 5.2 and the arguments preceding it imply the following result.

THEOREM 5.2. *Suppose equation* (5.1) *admits an exponential splitting on an interval* $\mathscr{I}$ *with the projections* $P_1, \cdots, P_n$ *commuting with the operator function* $A(t)$.

Then for a fixed τ_0 *there exists a number* $\delta > 0$ *depending only on equation* (5.1) *and such that under the condition*

$$\frac{1}{\tau_0} \int_t^{t+\tau_0} \|B(s)\|ds < \delta \qquad (t \in \mathscr{I}) \qquad (5.28)$$

a kinematic similarity transformation $x(t) = Q(t)y(t)$ *can be found which reduces the perturbed equation* (5.2) *to the quasidiagonal form*

$$\frac{dy}{dt} = \left[A(t) + \sum_{k=1}^{n} P_k B(t) Q(t) P_k \right] y \qquad (t \in \mathscr{I}), \qquad (5.29)$$

with a coefficient differing sufficiently little from the original coefficient $A(t)$. *The operator* $Q(t)$ *is a solution of the nonlinear equation*

$$Q'(t) = A(t)Q(t) - Q(t)A(t) + B(t)Q(t) - Q(t) \sum_k P_k B(t) Q(t) P_k,$$

satisfying the condition $\|I - Q(t)\| \leqq q < 1$ *and can be obtained from the integral equation* (5.21), (5.23) (*for* $S = Q - I$) *by means of the method of successive approximations.*

REMARK 5.4. Under the conditions of the theorem there exists a direct decomposition $\mathfrak{B} = \mathfrak{B}_1 \dotplus \cdots \dotplus \mathfrak{B}_n$ such that both of the equations (5.1) and (5.29) decompose into systems of independent equations in the subspaces $\mathfrak{B}_k$:

$$dx_k/dt = P_kA(t)x_k \quad \text{and} \quad dx_k/dt = P_kA(t)x_k + P_kB(t)Q(t)P_kx_k.$$

Since the perturbation is sufficiently small for each of them, the shift in the Bohl interval is also small. This result extends to the kinematically similar (to (5.29)) equation (5.2). We note that it does not depend on $A(t)$ being an integrally bounded operator function but requires that $A(t)P_k = P_kA(t)$ $(k = 1,\cdots, n)$.

REMARK 5.5. When $\mathscr{J}$ is a semi-infinite interval, the condition of integral smallness of the operator function $B(t)$ need only be satisfied for sufficiently large t, i.e.

$$\overline{\lim_{t\to\infty}} \frac{1}{\tau_0} \int_t^{t+\tau_0} \|B(\tau)\| d\tau < \delta.$$

§ 6. Exponential splitting in a Hilbert phase space

1. *Reduction to a system of independent equations.* When the phase space $\mathfrak{B}$ is a Hilbert space $\mathfrak{H}$, we can strengthen some of the results obtained in recent sections by applying the methods developed in § 1 in connection with the conjugation of projections.

We begin by noting the following result, which by virtue of Theorem 2.3 is a direct consequence of the definition of an exponential splitting of the solutions of an equation.

THEOREM 6.1. *If the equation*

$$\frac{dx}{dt} = A(t)x \qquad (t \in \mathscr{J}) \tag{6.1}$$

in $\mathfrak{H}$ admits an exponential splitting of order n on $\mathscr{J}$, it is kinematically similar to a system of independent equations

$$\frac{dx_k}{dt} = A_k(t)x_k \qquad (k = 1,2,\cdots,n; \quad t \in \mathscr{J}) \tag{6.2}$$

in the fixed subspaces $\mathfrak{H}_k = P_k\mathfrak{H}$ $(k = 1,\cdots,n)$ inducing the exponential splitting of the solutions of equation (6.1).

REMARK 6.1. Theorem 6.1 in combination with Theorem 5.2 leads to the following result: *For a fixed τ_0 there exists a number $\delta > 0$ depending only on equation* (6.1) *and such that if*

$$\frac{1}{\tau_0} \int_t^{t+\tau_0} \|B(s)\| ds < \delta \qquad (t \in \mathscr{J}), \tag{6.3}$$

the equation

$$dx/dt = A(t)x + B(t)x \qquad (t \in \mathscr{I}) \tag{6.4}$$

is kinematically similar to a system of independent equations

$$dx_k/dt = A_k(t)x_k + B_k(t)x_k \qquad (k = 1,2,\cdots,n) \tag{6.5}$$

in the same subspaces $\mathfrak{H}_k$ *and in which* $\|B_k(s)\| \leqq c_k\|B(s)\|$, *where the* c_k *depend only on the operators* P_k $(k = 1,\cdots,n)$.

The latter estimate follows from the formula $B_k(t) = P_kB(t)Q(t)P_k$, where $\|Q(t) - I\| \leqq q < 1$ and the relation $\|P_k\| = 1$ holds when $P_k = P_k^*$.

To each of the equations (6.5) we can apply Theorem III.4.6 on the stability of the upper and lower Bohl exponents. This leads us to the following result, which generalizes Theorem 5.1 in the case of Hilbert space in the sense that the integral boundedness of the operator function $A(t)$ is not required.

THEOREM 6.2. *Suppose equation* (6.1) *admits an exponential splitting of order n on an infinite or semi-infinite interval* $\mathscr{I}$. *For any* $\varepsilon \in (0, \varepsilon_0)$, *where* ε_0 *is one half of the minimal width of the gaps in the Bohl spectrum corresponding to this splitting, there exists a number* $\delta > 0$ *depending only on equation* (6.1) *and the number* ε *such that under condition* (6.3) *equation* (6.4) *admits an exponential splitting of the same order.*

The Bohl exponents of equations (6.4) *then satisfy the estimates*

$$\kappa'_{\text{B}}(\tilde{P}_k) \geqq \kappa'_{\text{B}}(P_k) - \varepsilon; \quad \kappa_{\text{B}}(\tilde{P}_k) \leqq \kappa_{\text{B}}(P_k) + \varepsilon.$$

REMARK 6.2. When $\mathscr{I}$ is a semi-infinite interval, condition (6.3) need only be satisfied for sufficiently large t.

2. *Exponential splitting of the solutions of equations with a precompactly valued operator function.* The importance of the exponential splitting property for equation (6.1) has been demonstrated in the preceding sections. Until now we could assert only that this property is possessed by stationary equations $dx/dt = Ax$, with a spectrum $\sigma(A)$ decomposing into components lying in mutually disjoint vertical strips, and the equations differing sufficiently little from them.

We now obtain for equations in Hilbert space on the halfline $\mathscr{I} = [0, \infty)$ another test for the presence of this property, which generalizes the test for negativeness of the upper Bohl exponent presented in § III.6

We recall that, by definition, an operator function $A(t)$ satisfies condition $S_{\delta,L}$ $(\delta > 0,\ L > 0)$ if there exists a number $T > 0$ such that the inequality $\|A(s) - A(t)\| \leqq \delta$ is satisfied for $s, t \geqq T$ and $|s - t| \leqq L$.

A function $A(t)$ satisfying condition $S_{\delta,L}$ can be represented in the form

$$A(t) = A_1(t) + A_2(t), \tag{6.6}$$

where

$$A_1(t) = \frac{1}{L}\int_t^{t+L}[A(t) - A(\tau)]d\tau; \quad A_2(t) = \frac{1}{L}\int_t^{t+L}A(\tau)d\tau, \tag{6.7}$$

the following conditions being satisfied for sufficiently large t ($t \geqq T$):

$$\|A_1(t)\| \leqq \delta, \quad \|A_2'(t)\| \leqq \delta/L. \tag{6.8}$$

THEOREM 6.3. *Suppose $A(t)$ ($0 \leqq t < \infty$) is a continuous precompactly valued operator function and suppose the spectra of all of its ω-limit operators lie in a set Π consisting of n mutually disjoint vertical strips*

$$\Pi_k = \{\lambda | \nu_k' \leqq \operatorname{Re}\lambda \leqq \nu_k\}$$
$$(k = 1,2,\cdots,n; \nu_1' \leqq \nu_1 \leqq \nu_2' \leqq \nu_2 \leqq \cdots \leqq \nu_n' \leqq \nu_n),$$

each of which contains at least one point of the spectra of these operators.

For any $\varepsilon < \min(\nu_{k+1}' - \nu_k)/2$ there exist numbers $\delta > 0$ and $L > 0$ depending only on ε and the ω-limit operators of $A(t)$ and such that if $A(t)$ satisfies conditions $S_{\delta,L}$, equation (6.1) *admits an exponential splitting of order n with Bohl intervals satisfying the condition*

$$[\kappa'(P_k), \kappa(P_k)] \subset [\nu_k' - \varepsilon, \nu_k + \varepsilon].$$

PROOF. For some $h > 0$ the spectra of all of the ω-limit operators lie in the horizontal strip $|\operatorname{Im}\lambda| \leqq h$ (by virtue of the boundedness of this set of operators). We consider the rectangles

$$\Pi_k^{\varepsilon,h} = \{\lambda | \nu_k' - \varepsilon/2 < \operatorname{Re}\lambda < \nu_k + \varepsilon/2; \quad |\operatorname{Im}\lambda| < h\}.$$

It follows from Theorem I.2.1. that, beginning with some $t > t_0$, the spectra $\sigma(A(t))$ lie in the interior of the union of these rectangles while their resolvents are uniformly bounded on the contours Γ_k of these rectangles, the corresponding constants depending only on the quantity ε and the ω-limit operators of $A(t)$.

Since $A_2(t) = A(t) - A_1(t)$, it follows from the first estimate of (6.8) for a sufficiently small $\delta > 0$ that these same properties will also be possessed by the operator function $A_2(t)$.

By varying $A(t)$ on a finite interval (and hence changing the equation to a kinematically similar one), we can establish that all of these properties as well as estimates (6.8) will be fulfilled on the whole interval $[0, \infty)$. In view of Theorem 6.2 it suffices to prove the theorem for the equation

$$dx/dt = A_2(t)x. \tag{6.9}$$

We introduce the projections

$$P_k(t) = -\frac{1}{2\pi i}\oint_{\Gamma_k}[A_2(t) - \lambda I]^{-1}d\lambda. \tag{6.10}$$

They are continuous and differentiable together with the operator $A_2(t)$; in fact,

$$\|P_k'(t)\| \leqq c\,\|A_2'(t)\| \leqq c\delta; \quad \|P_k(t)\| \leqq c_1, \tag{6.11}$$

where

$$c = \frac{l_k}{2\pi}\cdot \sup_{\substack{\lambda\in\Gamma_k\\ t\in[0,\infty)}} \|[A_2(t)-\lambda I]^{-1}\|^2,$$

$$c_1 = \frac{l_k}{2\pi} \sup_{\substack{\lambda\in\Gamma_k\\ t\in[0,\infty)}} \|[A_2(t)-\lambda I]^{-1}\|$$

(l_k is the length of the contour Γ_k).

With respect to the projections (6.10) it is possible to construct a conjugation operator $\Omega(t)$ solving the Cauchy problem

$$\Omega'(t) = \sum_{k=1}^{n} P_k'(t)P_k(t)\Omega(t), \quad \Omega(0) = I, \tag{6.12}$$

and having the property

$$P_k(t) = \Omega(t)P_k(0)\Omega^{-1}(t). \tag{6.13}$$

Since the operator functions $P_k(t)$ $(k = 1,\cdots, n)$ are bounded, $\Omega(t)$ uniformly conjugates the projections $P_k(0)$.

From Theorem 2.3 it follows that equation (6.12) is kinematically similar to an equation $dy/dt = B(t)y$ the coefficient $B(t)$ and Cauchy operator $W(t)$ of which commute with the projections $P_k(0)$. Let $Q(t)$ be the operator effecting the similarity transformation. Then

$$\Omega(t) = Q(t)W(t) \tag{6.14}$$

and from (6.13) it follows that

$$P_k(t) = Q(t)W(t)P_k(0)W^{-1}(t)Q^{-1}(t) = Q(t)P_k(0)Q^{-1}(t). \tag{6.15}$$

We note that by the definition of kinematic similarity there exists a constant $q > 0$ such that

$$\|Q(t)\| \leqq q, \quad \|Q^{-1}(t)\| \leqq q. \tag{6.16}$$

We now carry out the transformation $x = Q(t)y$ in equation (6.9). It goes over into the form

$$\frac{dy}{dt} = [Q^{-1}(t)A_2(t)Q(t) - Q^{-1}(t)Q'(t)]y. \tag{6.17}$$

The operator $\tilde{A}_2(t) = Q^{-1}(t)A_2(t)Q(t)$, as follows from (6.15), commutes with the projections $P_k(0)$, and the spectrum of its restriction to a subspace $\mathfrak{H}_k = P_k(0)\mathfrak{H}$ coincides with the part of the spectrum $\sigma(A_2(t))$ lying in the interior of $\Pi_k^{\varepsilon,h}$.

It is not difficult to get, using (6.14), that

$$Q'(t)Q^{-1}(t) = \Omega'(t)\Omega^{-1}(t) - Q(t)W'(t)W^{-1}(t)Q^{-1}(t)$$
$$= \sum_{k=1}^{n} P_k'(t)P_k(t) - Q(t)B(t)Q^{-1}(t).$$

From Theorem 2.3 it follows that for some $c_2 > 0$

$$\|B(t)\| \leqq c_2 \left\| \sum_{k=1}^{n} P_k'(t)P_k(t) \right\|.$$

Using estimates (6.10) and (6.15), we now get that there exists a constant $c_3 > 0$ for which

$$\|Q'(t)Q^{-1}(t)\| \leqq c_3\delta.$$

We can now estimate the derivative of the operator function $\tilde{A}_2(t)$:

$$\tilde{A}_2'(t) = Q^{-1}(t)A_2'(t)Q(t) - Q^{-1}(t)Q'(t)Q^{-1}(t)A_2(t)Q(t)$$
$$- Q^{-1}(t)A_2(t)Q'(t)$$
$$= Q^{-1}(t)A_2'(t)Q(t) - Q^{-1}(t)Q'(t)Q^{-1}(t)A_2(t)Q(t)$$
$$- Q^{-1}(t)A_2(t)Q'(t)Q^{-1}(t)Q(t),$$

and hence

$$\|\tilde{A}_2'(t)\| \leqq q^2\|A_2'(t)\| + 2q^2\|A_2(t)\|\,\|Q'(t)Q^{-1}(t)\| \leqq c_4\delta,$$

where c_4 is a positive constant.

With the above estimates we can complete the proof.

The equation $dy/dt = \tilde{A}_2(t)y$ is equivalent to the system of independent equations

$$dy_k/dt = \tilde{A}_2(t)P_k(0)y_k \qquad (k = 1,2,\cdots,n)$$

in the subspaces $\mathfrak{H}_k = P_k(0)\mathfrak{H}$. We can apply Theorem III.6.6 to each of these equations and conclude that for a sufficiently small $\delta > 0$ their Bohl intervals lie within the intervals $(\nu_k' - \varepsilon/2, \nu_k + \varepsilon/2)$.

It remains to apply Theorem 5.2 to equation (6.17), which by virtue of (6.15) differs sufficiently little from (6.16).

It is not difficult to verify that all of the constants contained in the estimates depend only on ε and on estimates of $A(t)$ and its resolvent for sufficiently large t, i.e. in the final analysis on the number ε and the ω-limit operators of $A(t)$. The theorem is proved.

EXERCISES

1. Prove the existence of the conjugation operator $\Omega(t, \tau)$ described in Theorem 1.1 by assuming (in place of differentiability) that the $P_k(t)$ are continuous functions of bounded variation (Ju. L. Daleckiĭ [3]).

Hint. Replace the differential equation (1.10) by an integral equation and make use of the multiplicative Stieltjes integral.

2. Show that the bounded variation condition in Exercise 1 is essential even in a two-dimensional space (L. A. Ivanov [1]).

Hint. Consider the operator $P(t)$ of orthogonal projection on the straight line forming an angle $\theta(t)$ with the axis of abscissas. Show that the product $P(t_n)P(t_{n-1})\cdots P(t_1)$ has a limit if and only if $\theta(t)$ is a function of bounded variation.

3. Let $P(t)$ $(a \leqq t \leqq b)$ be a continuous projection function in a Hilbert space $\mathfrak{H}$. Construct a unitary operator $\Omega_q(t)$ having the property $\mathfrak{H}(t) = \Omega_q(t)\mathfrak{H}(a)$, where $\mathfrak{H}(t) = P(t)\mathfrak{H}$.

Hint. Make use of Theorem 1.2 and the following result (I. C. Gohberg and M. G. Kreĭn [3]): the norm of the difference of two projections on a pair of subspaces achieves a minimum when these projections are Hermitian.

4. Let $A(z)$ be a matrix function of order n that analytically depends on z in some simply connected domain G of the complex plane. Suppose that for each $z \in G$ the spectrum $\sigma(A(z))$ decomposes into spectral sets:

$$\sigma(z) \equiv \sigma(A(z)) = \sigma_1(z) \cup \sigma_2(z) \cup \cdots \cup \sigma_p(z),$$

each of which continuously depends on z (in the sense that a contour γ_k separating $\sigma_k(z)$ from the rest of $\sigma(z)$ for some $z \in G$ will continue to do so under a small variation of z).

Show that there exists a matrix $U(z)$ analytically depending on z in G that reduces the matrix $A(z)$ to the quasidiagonal form

$$U^{-1}(z)A(z)U(z) = \begin{pmatrix} \Lambda_1(z) & & & \\ & \Lambda_2(z) & & \\ & & \ddots & \\ & & & \Lambda_p(z) \end{pmatrix},$$

where $\Lambda_k(z)$ is a matrix function of order n_k $(k = 1, \cdots, p)$ not depending on z ($n_k = \dim \mathfrak{B}_k(z)$, where $\mathfrak{B}_k(z)$ is the invariant subspace of $A(z)$ corresponding to the spectral set $\sigma_k(z)$).

Hint. Consider the differential conjugation equation (1.10) in a domain of the complex variable z. The matrix $U(z)$ can be expressed in terms of its solution, which is analytic in z.

The result noted in this exercise is essentially contained in a paper of Ju. L. Daleckiĭ and S. G. Kreĭn [2]. It was later published by Y. Sibuya (see W. Wasow [1]).

5. Show that if the equation $\dot{x} = A(t)x$ with an integrally bounded coefficient admits an exponential splitting and $\int_0^\infty \|B(t)\| dt < \infty$, the equation

$$dx/dt = [A(t) + B(t)]x \qquad (0 \leqq t < \infty)$$

admits an exponential splitting with the same Bohl intervals.

6. Suppose the equation $\dot{x} = A(t)x$ with an almost periodic coefficient $A(t)$ is e-dichotomic on the real line and $f(t)$ is an almost periodic vector function.

a) Show that the only bounded solution of the equation $dx/dt = A(t)x + f(t)$ is almost periodic.

b) The *module* $M(f)$ *of an almost periodic function* is the totality of finite integral linear combinations of the points of its spectrum. The following result is due to J. Favard [1].

Let f and g be almost periodic functions. The relation $M(f) \subset M(g)$ is satisfied precisely when for any $\varepsilon > 0$ there exists a $\delta > 0$ such that every δ-translation number of g is an ε-translation number of f (see § II.4.5).

Show that the module of a solution $x(t)$ of the equation $\dot{x} = A(t)x + f(t)$ is contained in the module of the almost periodic function $(A(t), f(t))$ (taking values in the direct sum $[\mathfrak{B}] \dotplus \mathfrak{B}$), which coincides with the minimal module containing the union of the spectra of the almost periodic functions $f(t)$ and $A(t)$.

Hint. Consider the family of equations for the translates $x_\tau(t) = x(t + \tau)$.

7. Let A be an operator of simplest type (see Exercise I.1) and let $R(t)$ be an operator function satisfying the condition

$$\int_0^\infty \|R(t)\| dt < \infty. \tag{0.1}$$

a) Show that the equations

$$dx/dt = Ax \tag{0.2}$$

and

$$dx/dt = Ax + R(t)x \tag{0.3}$$

are kinematically similar (this means that equation (0.3) is reducible to the form (0.2).

Hint. By representing the operator $Q(t)$ effecting the kinematic similarity transformation in the form $Q(t) = I + S(t)$, obtain for $S(t)$ the equation

$$S'(t) = [A + R(t)]S(t) - S(t)A + R(t). \tag{0.4}$$

Prove the existence of a solution of this equation satisfying the condition $\|S(t)\| < 1$ by using the results of § 4.2 and Exercise 5.

b) Obtain an analogous result for the case when A is an operator of algebraic type (see Exercise I.9) and condition (0.1) is replaced by the stronger condition

$$\int_0^\infty t^{2n-2} \|R(t)\| dt < \infty,$$

where n is the maximal index of the eigenvalues of A.

c) Extend the results of a) and b) to be case of an equation $dx/dt = A(t)x + R(t)x$ with a variable operator of simplest type $A(t)$ having eigenspaces not depending on t and eigenvalues $\lambda_k(t)$ with the property that $\operatorname{Re}[\lambda_k(t) - \lambda_j(t)]$ does not change in sign for $j, k = 1,\cdots,n$.

Finite-dimensional systems of this form with an integrable $R(t)$ are customarily called *L-diagonal systems* (see I. M. Rapoport [1]).

Hint. See the hint to Exercise a). To prove the existence of a bounded solution of equation (0.4) construct a special Green function by dividing the totality of spectral projections $\mathcal{P}_{jk}X = P_jXP_k$ of the transform $TX = AX - XB$ into two groups according to the summability of the functions $\operatorname{Re}[\lambda_k(t) - \lambda_j(t)]$.

8. Consider in a Hilbert phase space the equation

$$\frac{dx}{dt} = A(\varepsilon t)x, \tag{0.5}$$

where $A(\tau)$ is a uniformly continuous precompactly valued operator function. Extend Exercise III.17 by showing that if the spectra of the ω-limit operators of this operator function lie in a system of vertical strips, equation (0.5) admits an exponential splitting for sufficiently small $\varepsilon > 0$.

9. Theorem 6.3 on exponential splittability conditions for equations with precompactly valued operator functions was proved in a Hilbert phase space only. This would not appear to be completely natural inasmuch as the analogous theorem of Chapter III was proved for an arbitrary Banach space. Investigate the possibility that Theorem 6.3 can be extended to the case of a Banach space.

10. Prove that the equation

$$dx/dt = A(t)x \qquad (-\infty < t < \infty) \tag{0.6}$$

in a Hilbert space $\mathfrak{H}$ is e-dichotomic whenever

1) $\operatorname{Re} A(t)$ is an integrally bounded operator function and

2) there exists an indefinite operator $W \in [\mathfrak{H}]$ such that $\operatorname{Re}(WA(t)) \ll 0$ uniformly in $t \in (-\infty,\infty)$.

The latter condition implies the existence of a positive α such that

$$\operatorname{Re}(WA(t)) \leqslant -\alpha I \qquad (-\infty < t < \infty).$$

For the stationary case $A(t)=\text{const}$ this assertion admits a converse assertion and both assertions together constitute Theorem II.2.1. In Chapter V this assertion is proved (Theorem V.2.3) for the case of a periodic operator function $A(t) = A(t+T)$. In this case the proof is significantly simpler.

When $\mathfrak{H}$ is a finite-dimensional real space and $A(t)$ is a bounded real operator function (W is a symmetric real matrix), the assertion is a corollary of a more general theorem of A. D. Maĭzel' [1] (see Exercise 16).

Hint. Make use of the results of the exercise presented below and appropriately extend the arguments used in establishing the assertion of Exercise 14 (it is a special case of the assertion considered here and comparatively detailed hints are given for its proof).

11. Show that

a) If condition 2) of Exercise 10 is satisfied, the evolution operator $U(t,\tau)$ of equation (0.6) is a u-dichotomic uniform W-contraction for all $(-\infty <)\ \tau < t\ (< \infty)$.

b) If conditions 1) and 2) are satisfied, there exists for any $l > 0$ a $q > 0$ such that

$$W - U^*(t,\tau)WU(t,\tau) \geqq qI \qquad \text{for } t-\tau \geqq l. \tag{0.7}$$

Hint. Differentiate the left side with respect to t (see identity (V.2.4)) and make use of the result of Exercise I.41 as well as the fact that

$$\int_\tau^{\tau+l} \frac{dt}{\|U^{-1}(t,\tau)\|} \geq \int_\tau^{\tau+l} \exp\left(-\int_\tau^t \|\operatorname{Re}A(s)\|ds\right) dt$$

(the latter expression is bounded from below by a constant $q(l) > 0$ for $l > 0$ by virtue of condition 1)).

Let $\mathscr{P}(t)$ $(-\infty < t < \infty)$ be a locally integrable operator function with values in $[\mathfrak{H}]$.

Exercises 12–14 below will be formulated for the following equation in $\mathfrak{H}$:

$$d^2y/dt^2 + \mathscr{P}(t)y = 0, \tag{0.8}$$

which by the simple substitution $z = dy/dt$ reduces to the system

$$\begin{cases} dy/dt = z, \\ dz/dt = -\mathscr{P}(t)y, \end{cases} \tag{0.9}$$

i.e. to equation (0.6) for the vector function $x = y \oplus z$ in the phase space $\mathfrak{H}^{(2)} = \mathfrak{H} \oplus \mathfrak{H}$ with

$$\mathscr{A}(t) = \begin{pmatrix} 0 & I \\ -\mathscr{P}(t) & 0 \end{pmatrix}.$$

In conjunction with equation (0.8) it is convenient to consider the following equation in $[\mathfrak{H}]$:

$$d^2Y/dt^2 + \mathscr{P}(t)Y = 0. \tag{0.10}$$

12. Show that if the values of a solution $Y(t)$ of equation (0.10) are invertible in $\mathfrak{H}$ for t in an interval $[a, b]$, the operator function $K(t) = Y'(t)Y^{-1}(t)$ $(a \leqq t \leqq b)$ satisfies the Riccati operator equation

$$dK/dt + K^2 + \mathscr{P}(t) = 0. \tag{0.11}$$

We note that a priori a solution $K(t)$ of equation (0.11) can "go to infinity" or "come from infinity" in a finite period of time. One can easily check this by considering the scalar equation (0.11) (when $\mathfrak{H}$ is one dimensional).

But if a solution $K(t)$ of this equation exists in the whole interval $[a, b]$, it defines to within an invertible right constant factor in $[\mathfrak{H}]$ a solution $Y(t)$ of equation (0.10) on $[a, b]$ with invertible values in $[\mathfrak{H}]$.

13. Show that the following assertions are true.

a) Suppose the following condition is satisfied for some $\alpha > 0$:

1) $\operatorname{Re} \mathscr{P}(t) \ll -\alpha I \; (-\infty < t < \infty)$.

Then for any $(-\infty <) \; \tau < t \; (< \infty)$ the evolution operator $U(t, \tau)$ of system (0.9) is a u-dichotomic uniform W-contraction, where

$$W = \begin{pmatrix} 0 & I \\ I & 0 \end{pmatrix}.$$

b) Suppose in addition that the following integral boundedness condition for $\mathscr{P}(t)$ is satisfied:

2) $(M =) \sup \int_t^{t+1} \|\mathscr{P}(s)\| ds < \infty$.

Then for any $l > 0$ there exists a q such that $U(t, \tau)$ will satisfy condition (0.7).

Hint. Obtain this assertion as a corollary of the assertions of Exercise 11.

In the following assertions we assume that conditions 1) and 2) are satisfied.

c) For any $K_0 \in [\mathfrak{H}]$ with $\operatorname{Re} K_0 \gg I$ and $\tau \in (-\infty, \infty)$ there exists in the interval $[\tau, \infty)$ a solution $K(t)$ of the Riccati equation (0.11) satisfying the condition $K(\tau) = K_0$; in this connection

$$\operatorname{Re} K(t) \geq \gamma I \qquad (\tau \leq t < \infty), \tag{0.12}$$

where γ depends only on $\|K_0\|$ and α.

Hint. Make use of assertions a) and b) and the assertion of Exercise I.41.

d) Let $K_j(t) \; (\tau \leq t < \infty; j = 1, 2)$ be two solutions of equation (0.11) satisfying condition (0.12). Then

$$\|K_2(t) - K_1(t)\| \leq \|K_2(\tau) - K_1(\tau)\| e^{-\gamma(t-\tau)} \qquad (\tau \leq t < \infty).$$

Hint. Use the fact that K_2 and K_1 are solutions of one and the same Riccati equation, calculate the derivative $d(K_2 - K_1)/dt$ and then show that

$$K_2(t) - K_1(t) = X_1(t)[K_2(\tau) - K_1(\tau)]X_2(t),$$

where

$$X_j(t) = \int_\tau^t \exp[-K_j(t)] dt \qquad (j = 1, 2).$$

After this it remains to apply Wintner's estimate (III.4.19).

14. Equation (0.8) is said to be *e-dichotomic* if system (0.9) is e-dichotomic. Show that equation (0.8) is e-dichotomic whenever 2) $\mathscr{P}(t)$ is an integrally bounded operator function and 3) $\operatorname{Re} \mathscr{P}(t) \ll 0$ uniformly in $t \in (-\infty, \infty)$.

It is easily seen that if conditions 2) and 3) are satisfied, conditions 1) and 2) of Exercise 10 will be satisfied for system (0.9) and that therefore the above assertion is a corollary of the assertion of Exercise 10. But we insert a hint below leading to an independent solution of Exercise 14 in order to highlight some interesting details.

Hint. Choosing au arbitrary $\tau \in (-\infty, \infty)$, consider the solution $K_\tau(t) \; (\tau \leq t < \infty)$ of the Riccati equation (0.11) satisfying the condition $K_\tau(\tau) = I$. The limit $K_+(t) = \lim_{\tau \to -\infty} K_\tau(t)$ will also be a solution of this equation (see Exercise 13) satisfying the condition $\operatorname{Re} K_+(t) \gg 0$ uniformly in $t \in (-\infty, \infty)$. Let $\mathfrak{L}_+$ denote the subspace in $\mathfrak{H}^{(2)}$ consisting of all $x = y \oplus z$ such that $z = K_+(0)$. Then $\mathfrak{L}_+$ will be the left $\mathscr{S}$-set of system (0.9).

The right $\mathscr{S}$-set can be constructed analogously by reversing the time t.

The assertion of this exercise was proved for $\dim \mathfrak{H} < \infty$ and a bounded real operator function $\mathscr{P}(t)$ in an article of D. V. Anosov and Ja. G. Sinaĭ [1]. In this case it is a direct corollary of a theorem of A. D. Maĭzel' (see Exercise 16). The arguments of Anosov and Sinaĭ carry over directly to the case of a bounded Hermitian operator function $\mathscr{P}(t) = \mathscr{P}^*(t)$. The assertions of Exercises 13c) and 13d) are generalizations of corresponding propositions in the article of these authors.

Exercises 10–14 were taken from an article of M. G. Kreĭn [12].

15. Suppose the equation

$$dx/dt = A(t)x \qquad (-\infty < t < \infty) \tag{0.13}$$

in a Hilbert phase space $\mathfrak{H}$ is e-dichotomic on the real line.

a) Show that there exists a bounded indefinite operator function $W(t)$ such that the following estimate holds for the solutions of equation (0.13):

$$\frac{d(W(t)x(t), x(t))}{dt} \leqq -\alpha\|x(t)\|^2. \tag{0.14}$$

An analogous result holds for a halfline.

b) Extend the assertion to the case of a Banach space. (J. Massera and J. Schäffer [1]).

Hint. See the hint to Exercise III.15. For an equation on the real line we have

$$W(t) = \int_t^\infty U^{*-1}(t)P_-^* U^*(\tau)H(\tau)U(\tau)P_- U^{-1}(t)d\tau$$
$$- \int_{-\infty}^t U^{*-1}(t)P_+^* U^*(\tau)H(\tau)U(\tau)P_+ U^{-1}(t)d\tau,$$

where $P_\pm$ are the projections on the right and left $\mathscr{S}$-sets of equation (0.13) and $H(t)$ is an arbitrary uniformly positive operator function.

16. An assertion analogous to the assertion of Exercise III.15 is valid for equation (0.13) in a finite-dimensional real space and with an integrally bounded operator Re $A(t)$. Namely,

The following assertions are equivalent:

a) Equation (0.13) is e-dichotomic on the real line (on $[0, \infty)$).

b) The equation $dx/dt = A(t)x + f(t)$ has at least one bounded solution on the real line (on $[0, \infty)$) for each bounded $f(t)$.

c) There exists a bounded indefinite operator $W(t)$ satisfying condition (0.14).

The above result was established for bounded $A(t)$ by A. D. Maĭzel' [1], who generalized a result of I. G. Malkin (see Exercise III.15) by employing the technique (used also by Malkin) of Perron's transformation.

A generalization of this result has been obtained by J. Massera and J. Schäffer [1]. The equivalence of conditions a) and b) in a Banach space under additional assumptions on the $\mathscr{S}$-sets constitutes the assertion of Theorems 3.3 and 3.3′. These additional assumptions are automatically satisfied in the finite-dimensional case. The fact that a) implies c) was indicated in Exercise 15. As was shown by Schäffer, the converse assertion c)⇒a) is not true in the general case of a Hilbert space.

NOTES

e-dichotomic systems of equations with variable coefficients were essentially considered by O. Perron [2], who studied nonlinear perturbations of such equations. Still earlier, an analogous problem for a nonlinear equation with a stationary linear part was considered by P. Bohl. Perron's work was a generalization to the two-dimensional discrete case of a work of J. Hadamard [1]. The condition of e-dichotomicity did not appear explicitly in it. Instead there was given the condition of the existence of bounded solutions of the inhomogeneous linear equation (3.14) for bounded functions $f(t)$.

The equivalence of this condition to the condition of e-dichotomicity was first established by A. D. Maĭzel' [1] (see the comments to Exercise 16).

The e-dichotomicity of solutions of equations in Banach space has been studied by J. Massera and J. Schäffer [1] under more general assumptions than those adopted in the present book (to simplify the presentation we have adopted the unnecessary requirement that the $\mathscr{S}$-set of the equation be complemented). Our proof of the theorem on the stability of the e-dichotomicity property differs from the one presented by Massera and Schäffer. It should be mentioned that these brief notes do not begin to reflect the very extensive investigations of e-dichotomies performed by Massera and Schäffer for equations of first order, and by P. Hartman [1] for equations of higher order.

A lucid account of a number of results in this direction in the finite-dimensional case is contained in the works of W. Coppel [1 – 3]. Coppel was the first to apply the methods connected with conjugation operators for subspaces. To him is due Theorem 1.2 on the normalization of a conjugation operator, Theorem 2.3 on the decomposition of an equation, a proof of Theorem 5.2 based on an idea of N. Ja. Ljaščenko [1] and a special case of Theorem 6.3 (when $A(t)$ is a differentiable function with a small derivative).

The proofs of all of these results of Coppel made explicit use of the finite dimensionality of the phase space, so that some modification was required in order to transfer them to Hilbert space.

The simplest conjugation operators considered in § 1.1 were encountered long ago in papers on perturbation theory.

The differential conjugation equation was first introduced by Ju. L. Daleckiĭ and S. G. Kreĭn [2]. These results were later repeated by T. Kato [1]. Ju. L. Daleckiĭ [3] was the first to consider the more general problem when the projections are not assumed to be differentiable.

Similarity transformations and the reducible systems connected with them were first considered by A. M. Ljapunov.[3)]

A number of results on such systems and the asymptotic behavior of their solutions have been obtained by N. P. Erugin [1], A. Wintner [2], H. Weyl [1], V. A. Jakubovič [2, 3, 5] and others.

Unfortunately, we cannot dwell here on the interesting investigations of D. V. Anosov [1] and D. V. Anosov and Ja. G. Sinaĭ [1], in which e-dichotomic equations were used for the study of flows and cascades on a smooth manifold. The monograph of Anosov [1] contains a brief but eloquent historical commentary (which, unfortunately, fails to mention P. Bohl).

Some additional historical comments can be found in the exercises for this chapter.

[3)] The term *kinematic similarity* was introduced by L. Markus [1].

CHAPTER V

THE EQUATION WITH A PERIODIC OPERATOR FUNCTION

In this chapter we consider an important class of linear equations which we will call periodic, i.e. equations with a periodic coefficient $A(t)$. The behavior of the solutions of such an equation is determined by the spectral properties of its monodromy operator. In § 1 we present the more or less traditional material: we introduce the monodromy operator, consider its elementary properties and indicate a condition for the validity of the well-known Floquet representation of a Cauchy operator. In § 2 we study e-dichotomicity conditions for the periodic equation. In § 3 we establish various theorems on the localization of the spectrum of a monodromy operator. The presentation in this section makes use of the notions of deviation and amplitude of an operator introduced in Exercises I.32–I.34.

We next (§ 4) consider the so-called canonical equations, whose coefficients are skew-Hermitian in an indefinite metric. The methods connected with the theory of operators in a space with an indefinite metric that were developed in §§ 7 and 8 of Chapter I underlie the results presented here (these methods are also partially used in § 2, where they permit one to establish a simple e-dichotomicity test).

The basic contents of § 4 are connected with the consideration of stable boundedness (strong stability) conditions for the solutions of canonical equations. We establish exact estimates for the central stability band of a canonical equation with a real parameter. In § 5 we present analogous results for equations of second order. Finally, in § 6 we give a useful method of calculating the monodromy operator for an equation whose coefficient depends analytically on a small parameter, by expanding the logarithm of this operator in powers of this parameter.

The creator of the general theory of λ-stability bands for the scalar Hill equation, A. M. Ljapunov, attached great importance in his works not only to the establishment of general theorems but also to the deduction from them of individual exact stability tests. We have endeavored to carry on this tradition in the main text as well as in the exercises, which contain important additions to this chapter.

§ 1. Monodromy operator. Reducibility

1. *The upper Ljapunov and upper Bohl exponents of the periodic equation.* In this chapter we consider the important equation

$$dx/dt = A(t)x, \tag{1.1}$$

in which the operator $A(t)$ is a T-periodic operator function, i.e. for some $T > 0$

$$A(t + T) = A(t) \qquad (0 \leqq t < \infty). \tag{1.2}$$

The Cauchy operator $U(t)$ of equation (1.1) solves the Cauchy problem

$$\begin{cases} \dfrac{dU(t)}{dt} = A(t)U(t), \\ U(0) = I. \end{cases} \tag{1.3}$$

It is easily seen that the same problem is also solved by the operator

$$U_1(t) = U(t+T)U^{-1}(T).$$

By virtue of the uniqueness of the solution of problem (1.3) we have $U_1(t) \equiv U(t)$, which implies $U(t+T) = U(t)U(T)$.

The operator $U(T)$ is called the *monodromy operator* of equation (1.1). We recall that the *spectral radius* of $U(T)$ is the radius r of the minimal circle containing its spectrum. This radius is given by the formula $r = \lim_{k\to\infty} \sqrt[k]{\|U^k(T)\|}$. (see Chapter I, § 2).

We now prove the following assertion.

THEOREM 1.1. *The periodic equation* (1.1) *has strict upper Ljapunov and strict upper Bohl exponents which coincide with the logarithm of the spectral radius of its monodromy operator divided by the period*:

$$\kappa_L = \kappa_B = T^{-1}\ln r.$$

PROOF. By virtue of the continuous dependence on t of the operators U and U^{-1} there exists a q such that $\|U(\tau)\| \leqq q, \|U^{-1}(\tau)\| \leqq q$ for $0 \leqq \tau \leqq T$.

Suppose $t = nT + \tau$ and $t' = mT + \tau'$ ($\tau, \tau' \in [0, T]$). Then $U(t) = U(\tau)U^n(T)$ and $U(t') = U(\tau')U^m(T)$, which implies

$$U(t, t') = U(t)U^{-1}(t') = U(\tau)U^{n-m}(T)U^{-1}(\tau'),$$

and hence

$$(1/q^2)\|U^{n-m}(T)\| \leqq \|U(t, t')\| \leqq q^2\|U^{n-m}(T)\|.$$

Taking the logarithm of this system of inequalities and dividing each term by $t - t'$ (> 0), we get

$$\begin{aligned} -\frac{\ln q^2}{t-t'} &+ \frac{\ln\|U^{n-m}(T)\|}{(n-m)T + (\tau - \tau')} \\ &\leqq \frac{\ln\|U(t,t')\|}{t-t'} \leqq \frac{\ln\|U^{n-m}(T)\|}{(n-m)T + (\tau-\tau')} + \frac{\ln q^2}{t-t'}. \end{aligned}$$

Passing to the limit for $t - t' \to \infty$, we find that the following limit exists uniformly with respect to t':

$$\begin{aligned} \lim_{t-t'\to\infty} \frac{\ln\|U(t,t')\|}{t-t'} &= \lim_{n-m\to\infty} \frac{\ln\|U^{n-m}(T)\|}{(n-m)T} \\ &= \lim_{k\to\infty} \frac{\ln \sqrt[k]{\|U^k(t)\|}}{T} = \frac{\ln r}{T}. \end{aligned}$$

In particular, when $t' = 0$ ($t' \to \infty$) the left side is the strict upper Ljapunov (Bohl) exponent of the equation.

COROLLARY 1.1. *In order for the periodic equation to have a negative upper Bohl exponent it is necessary and sufficient that the spectrum of its monodromy operator lie in the interior of the unit disk.*

With the use of the monodromy operator one can easily formulate a condition under which the solutions of the differential equation (1.1) are bounded on the real line.

LEMMA 1.1. *In order for every solution of the differential equation* (1.1) *to be bounded on the real line it is necessary and sufficient that its monodromy operator $U(T)$ be stable.*

PROOF. If a solution $x(t) = U(t)x_0$ of equation (1.1) is bounded, then

$$\begin{aligned}\sup_n \|U^n(T)x_0\| &= \sup_n \|U(nT)x_0\| = \sup_n \|x(nT)\| \\ &\leqq \sup_{-\infty<t<\infty} \|x(t)\| < \infty.\end{aligned}$$

Thus the sequence $\{U^n(T)\}_{n=-\infty}^{\infty}$ is bounded at every element of the space $\mathfrak{B}$ and hence, by the uniform boundedness principle, is bounded in $[\mathfrak{B}]$.

Conversely, from the formula $U(t + nT) = U(t)U^n(T)$ it follows that for $\|U^n(T)\| \leqq c$

$$\sup_{-\infty<t<\infty} \|U(t)x_0\| \leqq \sup_{0\leqq t\leqq T} \|U(t)\| \cdot c\|x_0\|.$$

REMARK 1.1. It is easily seen that the condition $\|U^n(T)\| \leqq c$ ($n = 1,2,\cdots$) is equivalent to the boundedness of every solution of equation (1.1) on the halfline $t \geqq 0$.

2. *Reducibility condition for the periodic equation. Floquet representation.* We assume that the monodromy operator $U(T)$ of equation (1.1) has a logarithm. This means that an operator $S = \ln U(T)$ exists for which $U(T) = e^S$.

We introduce the operator $\Gamma = T^{-1} \ln U(T)$. Then $U(T) = e^{T\Gamma}$.

We now put

$$Q(t) = U(t)e^{-t\Gamma}. \tag{1.4}$$

The operator function $Q(t)$ is T-periodic:

$$\begin{aligned}Q(t + T) &= U(t + T)e^{-(t+T)\Gamma} \\ &= U(t)U(T)e^{-T\Gamma}e^{-t\Gamma} = U(t)e^{-t\Gamma} = Q(t).\end{aligned}$$

On the segment $0 \leqq t \leqq T$ this operator function is continuous, differentiable and has a continuous inverse $Q^{-1}(t)$.

From (1.4) we obtain a *Floquet representation* of the Cauchy operator

$$U(t) = Q(t)e^{t\Gamma} \tag{1.5}$$

in the form of a product of a differentiable periodic operator function $Q(t)$ having a bounded inverse $Q^{-1}(t)$ and an exponential operator function $e^{t\Gamma}$ with a constant operator Γ.

From the foregoing we obtain a theorem on the Floquet representation.

THEOREM 1.2. *In order for the Cauchy operator of the periodic equation* (1.1) *to admit a Floquet representation* (1.5) *it is necessary and sufficient that the monodromy operator of this equation have a logarithm. This will be the case, in particular, if the spectrum of the monodromy operator does not surround the origin.*[1)]

PROOF. The sufficiency of the existence of the logarithm has been proved above. The necessity follows from the relations

$$Q(T) = Q(0) = U(0) = I \quad \text{and} \quad U(T) = Q(T)e^{T\Gamma} = e^{T\Gamma}.$$

To prove the last assertion we choose a closed Jordan contour γ surrounding the spectrum $\sigma(U(T))$ but not surrounding the origin, and a branch of $\ln \lambda$ having a single value on this contour.

A logarithm of the monodromy operator can then be constructed with the use of the formula

$$\ln U(T) = -\frac{1}{2\pi i} \oint_\gamma \ln \lambda [U(T) - \lambda I]^{-1} d\lambda. \tag{1.6}$$

The theorem is proved.

Some other tests for the existence of a Floquet representation are indicated below in § 3 (with the use of the theorem on the localization of the spectrum of the monodromy operator) and in § 4 (for canonical equations).

REMARK 1.2. Recalling a necessary and sufficient condition for the reducibility of an equation (see § 2 of Chapter IV), we obtain the following result of Ljapunov, which is a rephrasing of Theorem 1.2: *in order for the periodic equation* (1.1) *to be reducible by means of the T-periodic operator $Q(t)$ to the form $dy/dt = \Gamma y$, it is necessary and sufficient that there exist a logarithm of the monodromy operator.*

Let us consider an important special case. Suppose $\mathfrak{B}$ is a Hilbert space $\mathfrak{H}$ and suppose the following condition is satisfied for some invertible Hermitian operator

[1)] The spectrum $\sigma(U(t))$ of course cannot surround the origin if the space $\mathfrak{B}$ is finite dimensional or if $A(t)$ is an operator function with completely continuous values.

$W \in [\mathfrak{H}]$:

$$WA(t) + A^*(t)W = 0, \tag{1.7}$$

i.e. the operator $A(t)$ is W-skew-Hermitian.

By virtue of Lemma III.1.1 we at once obtain the first part of the following theorem.

THEOREM 1.3. *If condition* (1.7) *is satisfied, the Cauchy operator $U(t)$ of equation* (1.1) (*in particular, its monodromy operator $U(T)$*) *is W-unitary. The spectrum of this operator is symmetric relative to the unit circle.*

If, in addition, the spectrum $\sigma(U(T))$ does not surround the origin, then in the representation (1.5) *the operator Γ is W-skew-Hermitian while the operator $Q(t)$ is W-unitary.*

PROOF. It remains to prove that Γ is W-skew-Hermitian; the latter assertion then follows from formula (1.4) since in this case the operator $e^{-t\Gamma}$ is also W-unitary.

We make use of formula (1.6). From the symmetry of the spectrum of the W-unitary operator $U(T)$ relative to the unit circle it follows that the contour γ can be chosen so that it is symmetric relative to the unit circle.

Inasmuch as the fact that $U(T)$ is a W-unitary operator implies the equality $U(T)W^{-1} = W^{-1}[U^{-1}(T)]^*$ (see § I.8.1), we have

$$\begin{aligned} -W\Gamma &= \frac{1}{2\pi i T}\oint_\gamma \ln\lambda \cdot W[U(T) - \lambda I]^{-1}d\lambda \\ &= \frac{1}{2\pi i T}\oint_\gamma \ln\lambda [U(T)W^{-1} - \lambda W^{-1}]^{-1}d\lambda \\ &= \frac{1}{2\pi i T}\oint_\gamma \ln\lambda[(U^*(T))^{-1} - \lambda I]^{-1}d\lambda \cdot W. \end{aligned}$$

Further,

$$\begin{aligned} [(U^*(T))^{-1} - \lambda I]^{-1} &= [I - \lambda U^*(T)]^{-1}U^*(T) \\ &= [I - \lambda U^*(T)]^{-1}\left[\frac{1}{\lambda}(I - (I - \lambda U^*(T)))\right] \\ &= \frac{1}{\lambda}[(I - \lambda U^*(T))^{-1} - I] = -\frac{1}{\lambda}I - \frac{1}{\lambda^2}\left(U^*(T) - \frac{1}{\lambda}I\right)^{-1} \end{aligned}$$

and, as a result of the analyticity of the function $\lambda^{-1}\ln\lambda$ in the interior of γ,

$$W\Gamma = \frac{1}{2\pi i T}\oint_\gamma \frac{\ln\lambda}{\lambda^2}\left[U^*(T) - \frac{1}{\lambda}I\right]^{-1}d\lambda \cdot W. \tag{1.8}$$

Let γ' be the contour described by the point $\mu = 1/\lambda$ as λ runs along γ. This contour, like γ, is symmetric relative to the unit circle and obviously surrounds the spectrum of the operator $U^*(T)$ while separating it from the origin.

Carrying out the change of variable $1/\lambda = \mu$ in the integral (1.8), we obtain the required assertion:

$$W\Gamma = -\frac{1}{2\pi i T}\oint_{\gamma'} \ln \mu[U^*(T) - \mu I]^{-1}d\mu \cdot W = -\Gamma^* W,$$

since (1.6) and the symmetry of γ imply the equality

$$\Gamma^* = \frac{1}{2\pi i T}\oint_{\gamma} \ln \bar{\lambda}[U^*(T) - \bar{\lambda} I]^{-1}d\bar{\lambda}$$

$$= \frac{1}{2\pi i T}\oint_{\gamma} \ln \mu[U^*(T) - \mu I]^{-1}d\mu.$$

§ 2. Exponential dichotomy of the solutions of the periodic equation

1. *e-dichotomicity criterion.* With the use of the monodromy operator we can give a simply formulated condition for the e-dichotomicity of the periodic equation on the real line.

THEOREM 2.1. *In order for the periodic equation* (1.1) *to be e-dichotomic on the real line it is necessary and sufficient that the monodromy operator $U(t)$ be u-dichotomic* (*i.e. that the spectrum $\sigma(U(T))$ not intersect the unit circle and contain components lying in both the interior and exterior of the unit circle*).

PROOF. NECESSITY. If equation (1.1) is e-dichotomic, there corresponds to each bounded continuous vector function $f(t)$ on the real line exactly one bounded solution of the equation

$$\frac{dx}{dt} = A(t)x + f(t). \tag{2.1}$$

From the uniqueness of the solution it follows that there corresponds to a T-periodic function $f(t)$ a T-periodic solution $x(t)$, since the same equation (2.1) is also satisfied by the function $x(t + T)$ in place of $x(t)$.

For an arbitrary $y \in \mathfrak{B}$ we choose the vector function $f_y(t) = -(6/T^3)t(T - t) \cdot U^{-1}(T, t)y$ for $0 \leqq t \leqq T$ and periodically extend it on the real line.

The unique periodic solution $x^{(y)}(t)$ of equation (2.1) corresponding to it can be written in the form

$$x^{(y)}(t) = U(t)x_0^{(y)} + \int_0^t U(t, \tau)f_y(\tau)d\tau$$

$$= U(t)x_0^{(y)} - \frac{6}{T^3}\int_0^t U(t, T)\tau(T - \tau)d\tau \cdot y,$$

from which for $t = T$ we obtain $x_0^{(y)} = U(T)x_0^{(y)} - y$ or

$$[U(T) - I]x_0^{(y)} = y. \tag{2.2}$$

Thus equation (2.2) has exactly one solution for any $y \in \mathfrak{B}$. This means that $[U(T) - I]^{-1} \in [\mathfrak{B}]$ and hence that the point $\lambda = 1$ is not contained in the spectrum $\sigma(U(t))$.

We now consider the equation

$$\frac{dx}{dt} = A(t)x + f(t)e^{2\pi i\omega t/T}.$$

Setting $x(t) = z(t)e^{2\pi i\omega t/T}$, we reduce it to the form

$$dz/dt = [A(t) - 2\pi i\omega/T]z + f(t).$$

This equation has the same properties as (2.1), and hence the spectrum of its monodromy operator $U(T)e^{-2\pi i\omega}$ does not contain the point $\lambda = 1$. But this means that the spectrum $\sigma(U(T))$ does not contain the point $\lambda = e^{2\pi i\omega}$. Since ω is arbitrary, the necessity of the condition of the theorem is proved.

SUFFICIENCY. Suppose the spectrum of the monodromy operator admits a decomposition $\sigma(U(T)) = \sigma_1 \cup \sigma_2$ in which the set σ_1 (σ_2) lies in the interior (exterior) of the unit disk. We denote by P_1, P_2, $\mathfrak{B}_1$, $\mathfrak{B}_2$ the spectral projections and invariant subspaces corresponding to this decomposition.

The spectrum of the restriction $U_1 = U(T)|\mathfrak{B}_1$ coincides with the set σ_1 lying in the interior of the unit disk, and hence the operator U_1 has a spectral radius $r_1 < 1$. In exactly the same way the spectral radius r_2 of the operator U_2^{-1}, where $U_2 = U(T)|\mathfrak{B}_2$, turns out to be less than one.

There therefore exist numbers $N > 0$ $(0 < q < 1)$ for which

$$\|U_1\|_{\mathfrak{B}_1}^n \leq Nq^n, \quad \|U_2^{-n}\|_{\mathfrak{B}_2} \leq Nq^n \qquad (n = 1,2,\cdots). \tag{2.3}$$

To prove the required assertion it suffices to show that the upper Bohl exponent $\kappa_B(P_1)$ is negative while the lower Bohl exponent $\kappa_B'(P_2)$ is positive.

We make use of formulas (III.4.10).

Let $t_1 = n_1T + \tau_1$ and $t_2 = n_2T + \tau_2$ $(\tau_1, \tau_2 \in [0, T])$. Then

$$\begin{aligned} U(t_2)P_kU^{-1}(t_1) &= U(\tau_2)[U^{n_2}(T)P_kU^{-n_1}(T)]U(\tau_1) \\ &= U(\tau_2)[U^{n_2-n_1}P_k]U(\tau_1), \end{aligned}$$

and consequently

$$\|U(t_2)P_kU^{-1}(t_1)\| \leq M^2\|U^{n_2-n_1}\|_{\mathfrak{B}_k},$$

where $M = \sup_{0\leq\tau\leq T}\|U(\tau)\|$.

Therefore

$$\begin{aligned} \kappa_B(P_1) &= \overline{\lim_{t_2-t_1\to\infty}} \frac{\ln\|U(t_2)P_1U^{-1}(t_1)\|}{t_2 - t_1} \\ &\leq \overline{\lim_{t_2-t_1\to\infty}} \frac{\ln(M^2Nq^{n_2-n_1})}{(n_2 - n_1)T + (\tau_2 - \tau_1)} = \frac{\ln q}{T} < 0. \end{aligned}$$

The second assertion can be proved analogously. The theorem is proved.

It is not difficult to verify that the annulus $\kappa_B(P_1) < |\lambda| < \kappa'_B(P_2)$ is the maximal open annulus containing the unit circle and not intersecting the spectrum $\sigma(U(T))$.

2. *e-dichotomicity tests for* $\mathfrak{B} = \mathfrak{H}$. Suppose the space $\mathfrak{B}$ is a Hilbert space $\mathfrak{H}$. In this case it is possible to indicate a simple e-dichotomicity test for equation (1.1) which is connected with the behavior of the coefficient $A(t)$ when $\mathfrak{H}$ is rigged with an indefinite metric.

Let $W \in [\mathfrak{H}]$ be an invertible indefinite Hermitian operator.

We first derive an auxiliary relation. Let

$$H(t) = -\operatorname{Re} WA(t) \quad \left(= -\frac{1}{2}[WA(t) + A^*(t)W]\right).$$

Integrating the obvious relation

$$\begin{aligned}\frac{d}{dt}[U^*(t)WU(t)] &= \frac{dU^*}{dt} WU(t) + U^*(t)W\frac{dU}{dt} \\ &= U^*(t)A^*(t)WU(t) + U^*(t)WA(t)U(t) \\ &= -2U^*(t)H(t)U(t),\end{aligned}$$

from 0 to T, we obtain the formula

$$W - U^*(T)WU(T) = 2\int_0^T U^*(t)H(t)U(t)dt. \tag{2.4}$$

From it one easily deduces the following result.

THEOREM 2.2. *Suppose for some indefinite* $W = W^*$ $(\in [\mathfrak{H}])$

$$\int_0^T \operatorname{Re}(WA(t))dt \ll 0.$$

Then there exists an $\varepsilon > 0$ *such that for any* λ $(-\varepsilon < \lambda < \varepsilon, \lambda \neq 0)$ *the equation*

$$\frac{dx}{dt} = \lambda A(t)x \qquad (A(t+T) = A(t)) \tag{2.5}$$

will be e-dichotomic and, in addition, the phase space $\mathfrak{H}$ *will decompose into a direct sum*

$$\mathfrak{H} = \mathfrak{H}_+(\lambda) \dotplus \mathfrak{H}_-(\lambda)$$

of uniformly W-definite subspaces that are invariant under the monodromy operator $U(T; \lambda)$, *the spectra of the restrictions* $U(t; \lambda)|\mathfrak{H}_\pm$ *lying respectively in the interior and exterior of the unit disk.*

PROOF. If we set $H(t) = -\operatorname{Re}(WA(t))$, we will get according to (2.4) that

$$W - U^*(T; \lambda)WU(T;\lambda) = 2\lambda\mathscr{H}(\lambda), \tag{2.6}$$

where

$$\mathscr{H}(\lambda) = \int_0^T U^*(t; \lambda)H(t)U(t; \lambda)dt. \tag{2.7}$$

Since (see (III.1.11))

$$U(t; \lambda) = I + \lambda \int_0^t A(s)ds + \cdots = I + O(\lambda)$$

uniformly in $t \in [0, T]$, the Hermitian operator $\mathscr{H}(\lambda)$ will differ arbitrarily little from the operator $\int_0^T H(t)dt \ll 0$ for real λ that is sufficiently small in absolute value.

There therefore exists an $\varepsilon > 0$ such that the operator $\mathscr{H}(\lambda)$ will be $\ll 0$ for $-\varepsilon < \lambda < \varepsilon$. But then by virtue of (2.6) the operator $U(t; \lambda)$ will be a uniform W-contraction (W-expansion) for $0 < \lambda < \varepsilon$ $(-\varepsilon < \lambda < 0)$. Since, moreover, the spectrum of the operator $U(T; \lambda)$ $(|\lambda| < \varepsilon)$ can be made to lie in any neighborhood of the point 1 and a fortiori not cover the unit circle by taking ε sufficiently small, the assertion of the theorem is a corollary of Theorem I.7.1′.

REMARK 2.1. We have simultaneously obtained the following sharpening of the second assertion of the theorem: *for* $0 < \lambda < \varepsilon$ $(-\varepsilon < \lambda < 0)$ *the subspace* $\mathfrak{H}_+$, *for which the spectrum* $\sigma(U(T; \lambda)|\mathfrak{H}_+)$ *lies in the interior of the unit disk, is uniformly W-positive (W-negative) while, conversely, the subspace* $\mathfrak{H}_-$ *is uniformly W-negative (W-positive).*

THEOREM 2.3. *Suppose the following inequality is satisfied for some indefinite* $W = W^*$:

$$\operatorname{Re}(WA(t)) \ll 0 \qquad \textit{uniformly in } t \in (-\infty, \infty). \tag{2.8}$$

Then the periodic equation (1.1) *is e-dichotomic and, in addition, the phase space* $\mathfrak{H}$ *decomposes into a direct sum* $\mathfrak{H} = \mathfrak{H}_+ \dotplus \mathfrak{H}_-$ *of subspaces invariant under* $U(T)$, *of which* $\mathfrak{H}_+$ *is uniformly W-positive while* $\mathfrak{H}_-$ *is uniformly W-negative, the spectrum* $\sigma(U(T)|\mathfrak{H}_+)$ $(\sigma(U(T)|\mathfrak{H}_-))$ *lying in the interior (exterior) of the unit disk.*

PROOF. Condition (2.8) implies the existence of a positive α such that $H(t) = \operatorname{Re}(WA(t)) \leqslant -\alpha I$. It follows according to (2.7) that for $\mathscr{H}(\lambda)$ we will have

$$(\mathscr{H}(\lambda)x, x) = \int_0^T (H(t)U(t; \lambda)x, U(t; \lambda)x)dt$$
$$\leqq -\alpha \int_0^T \|U(t; \lambda)x\|^2 dt \leqq -\alpha \int_0^T \frac{dt}{\|U^{-1}(t; \lambda)\|} \|x\|^2,$$

so that $\mathscr{H}(\lambda) \ll 0$ $(-\infty < \lambda < \infty)$.

Therefore, by virtue of (2.6) the operator $U(T; \lambda)$ is a uniform W-contraction for $\lambda > 0$; in addition, according to the preceding theorem the operator $U(T; \lambda)$

is a u-dichotomic W-contraction for sufficiently small $\lambda > 0$. But then it is a u-dichotomic W-contraction for all $\lambda > 0$ (see Corollary I.7.2).

Thus the assertions of Theorem 2.2 and Remark 2.1 hold for the operator $U(T;\lambda)$ for all $\lambda > 0$ and, in particular, for $\lambda = 1$. The theorem is proved.

3. *The periodic equation of second order.* We consider in $\mathfrak{H}$ the equation

$$d^2y/dt^2 + \mathscr{P}(t)y = 0 \tag{2.9}$$

with a periodic operator function $\mathscr{P}(t)$ $(= \mathscr{P}(t + T))$.

Setting

$$z = \frac{dy}{dt}, \quad x = \begin{pmatrix} y \\ z \end{pmatrix} \quad (= y \oplus z),$$

we reduce equation (2.9) to the following equation of first order in the phase space $\mathfrak{H}^{(2)} = \mathfrak{H} \oplus \mathfrak{H}$:

$$dx/dt = \mathscr{A}(t)x, \tag{2.10}$$

where

$$\mathscr{A}(t) = \begin{pmatrix} 0 & I \\ -\mathscr{P}(t) & 0 \end{pmatrix} \qquad (I = I_{\mathfrak{H}}). \tag{2.10$'$}$$

The monodromy operator $U(T)$ of equation (2.10) will also be called the *monodromy operator of equation* (2.9). It is easily seen that it has the form

$$U(T) = \begin{pmatrix} \Phi(T) & \Psi(T) \\ \Phi'(T) & \Psi'(T) \end{pmatrix}, \tag{2.11}$$

where $\Phi(t)$ and $\Psi(t)$ are the solutions of the equation in $[\mathfrak{H}]$

$$d^2Y/dt^2 + \mathscr{P}(t)Y = 0$$

distinguished by the initial conditions

$$\Phi(0) = I, \quad \Phi'(0) = 0;$$
$$\Psi(0) = 0, \quad \Psi'(0) = I.$$

We introduce an indefinite operator which belongs to $[\mathfrak{H}^{(2)}]$ and whose spectrum consists of the points ± 1:

$$W = \begin{pmatrix} 0 & -I \\ -I & 0 \end{pmatrix}. \tag{2.12}$$

We have

$$W\mathscr{A} + \mathscr{A}^*W = \begin{pmatrix} \mathscr{P} + \mathscr{P}^* & 0 \\ 0 & -2I \end{pmatrix}.$$

This relation together with Theorem 2.3 permits us to prove the following proposition.

THEOREM 2.4. *Suppose*

$$\mathrm{Re}\mathscr{P}(t) \ll 0 \qquad \textit{uniformly in } t \in (-\infty, \infty). \tag{2.13}$$

Then the phase space $\mathfrak{H}^{(2)}$ *decomposes into a direct sum of two subspaces* $\mathfrak{L}_\pm$ *that are invariant under the monodromy operator* $U(T)$ *of equation* (2.9) *and such that the spectrum of* $U(T)|\mathfrak{L}_+$ ($U(T)|\mathfrak{L}_-$) *lies in the interior (exterior) of the unit disk.*

Each of the subspaces $\mathfrak{L}_\pm$ *can be given by the equation* $z = \pm K_\pm y$,[2] *where the operators* $K_\pm$ ($\in [\mathfrak{H}]$) *are invertible uniformly dissipative operators.*

The operator K_+ (K_-) will be called the *plus* (*minus*) *equation operator* of the periodic equation (2.9).

PROOF. The fulfilment of condition (2.13) implies the fulfilment of the condition of Theorem 2.3 for the corresponding equation (2.10). Therefore the first assertion of Theorem 2.4 is a corollary of Theorem 2.3.

Moreover, according to Theorem 2.3 it can be asserted that the subspace $\mathfrak{L}_+$ ($\mathfrak{L}_-$) in the decomposition $\mathfrak{H}^{(2)} = \mathfrak{L}_+ \dot{+} \mathfrak{L}_-$ is uniformly W-positive (W-negative).

Thus in regard to $\mathfrak{L}_+$, for example, we can assert the existence of a constant m (> 0) such that $(Wx, x) \geqq m(x, x)$ for $x \in \mathfrak{L}_+$. Since W has the form (2.12), this property means that

$$-(y, z) - (z, y) \geqq m(|y|^2 + |z|^2) \qquad (x = y \oplus z \in \mathfrak{L}_+)$$

or, equivalently

$$\begin{aligned} \|y - z\|^2 - \|y + z\|^2 &\geqq 2m(\|y\|^2 + \|z\|^2) \\ &= m(\|y - z\|^2 + \|y + z\|^2) \qquad (x = y \oplus z \in \mathfrak{L}_+). \end{aligned} \tag{2.14}$$

Let $S_\pm$ denote the operators mapping $\mathfrak{L}_+$ ($\subset \mathfrak{H}^{(2)}$) into $\mathfrak{H}$ according to the rule $S_\pm x = y \mp z$ ($x \in \mathfrak{L}_+$). If $S_+ x = y - z = 0$ for some $x \in \mathfrak{L}_+$, it will also be true by virtue of (2.14) that $y + z = 0$, i.e. $x = y = z = 0$. Thus the operator S_+ is a one-to-one mapping of $\mathfrak{L}_+$ onto $\mathfrak{M} = S_+\mathfrak{L}_+$, i.e. to any $u \in \mathfrak{M}$ there corresponds a unique pair $y, z \in \mathfrak{H}$ such that $u = y - z$, $y \oplus z \in \mathfrak{L}_+$.

We put $\mathscr{T}_+ u = y + z$ for any $u \in \mathfrak{M}$. Then inequality (2.14) can be written in the form

$$\|u\|^2 - \|\mathscr{T}_+ u\|^2 \geqq m(\|u\|^2 + \|\mathscr{T}_+ u\|^2),$$

from which it follows that $m < 1$ and

$$\|\mathscr{T}_+ u\| \leqq \sqrt{\frac{1 - m}{1 + m}}\, \|u\| \qquad (u \in \mathfrak{M}).$$

Thus the linear operator $\mathscr{T}_+$ acting from $\mathfrak{M}$ into $\mathfrak{H}$ is a uniform contraction.

Since it follows from $y - z = u$ and $y + z = \mathscr{T}_+ u$ that

[2] We say that a subspace $\mathfrak{L} \subset \mathfrak{H}_1 \oplus \mathfrak{H}_2$ has the equation $x_2 = Kx_1$, where $K \in [\mathfrak{H}_1, \mathfrak{H}_2]$, if $\mathfrak{L} = \{x \oplus Kx \mid x \in \mathfrak{H}_1\}$. The operator K is called the *equation operator for* $\mathfrak{L}$ (more precisely, the *equation operator for* $\mathfrak{L}$ *with respect to* $(\mathfrak{H}_1, \mathfrak{H}_2)$).

$$y = \frac{1}{2}(u + \mathscr{T}_+u), \quad z = \frac{1}{2}(\mathscr{T}_+u - u),$$

we have

$$\mathfrak{L}_+ = \{(u + \mathscr{T}_+u) \oplus (\mathscr{T}_+u - u) | u \in \mathfrak{M}\}. \tag{2.15}$$

Let us show that $\mathfrak{M} = \mathfrak{H}$. Suppose $\mathfrak{M}$ is a proper part of $\mathfrak{H}$. Then $\mathscr{T}_+$ has an extension $\mathscr{T} \in [\mathfrak{H}]$ with norm $\|\mathscr{T}\| = \|\mathscr{T}_+\|$ ($\leq [(1-m)/(1+m)]^{1/2}$). Such an extension can be obtained, for example, by defining $\mathscr{T}$ on the closure $\bar{\mathfrak{M}}$ as the closure of $\mathscr{T}_+$ by continuity and setting $\mathscr{T}$ equal to zero on $\mathfrak{H} \ominus \mathfrak{M}$.

After this, by setting

$$\mathfrak{L} = \{(u + \mathscr{T}u) \oplus (\mathscr{T}u - u) | u \in \mathfrak{H}\},$$

we obtain a subspace $\mathfrak{L}$ which contains $\mathfrak{L}_+$ as a proper part (the operators $I \pm \mathscr{T}$ are continuous one-to-one mappings of $\mathfrak{H}$ onto itself) and, in addition, is uniformly W-positive (also inequality (2.14) will be satisfied with the same m for all $x \in \mathfrak{L}$).

But if $\mathfrak{L}$ is an extension of $\mathfrak{L}_+$, it follows from $\mathfrak{L}_+ \dot{+} \mathfrak{L}_- = \mathfrak{H}^{(2)}$ that the intersection of $\mathfrak{L}$ with $\mathfrak{L}_-$ contains nonzero elements. Since $\mathfrak{L}_-$ is a W-negative subspace, we have arrived at a contradiction.

Thus $\mathfrak{M} = \mathfrak{H}$, and it therefore follows from (2.15) that

$$\mathfrak{L}_+ = \{y \oplus K_+y | y \in \mathfrak{H}\}, \tag{2.15'}$$

where $K_+ = (\mathscr{T}_+ - I)(I + \mathscr{T}_+)^{-1}$. Since $\|\mathscr{T}_+\| < 1$, we get $\operatorname{Re} K_+ \ll 0$ (see § I.7). Thus the assertion of the theorem concerning the subspace $\mathfrak{L}_+$ is proved.

The assertion concerning the subspace $\mathfrak{L}_-$ can be proved analogously. The theorem is proved.

4. *Equation operators as fixed points.* We consider a vector $x = y \oplus K_+y \in \mathfrak{L}_+$. It is mapped by the monodromy operator (2.11) into the vector

$$U(T)x = (\Phi(T)y + \Psi(T)K_+y) \oplus (\Phi'(T)y + \Psi'(T)K_+y).$$

Since this vector also lies in $\mathfrak{L}_+$,

$$(\Phi'(T) + \Psi'(T)K_+)y = K_+(\Phi(T) + \Psi(T)K_+)y. \tag{2.16}$$

In view of its arbitrariness the vector y in this equation can be dropped.

We put

$$\phi(Z) = (\Phi'(T) + \Psi'(T)Z)(\Phi(T) + \Psi(T)Z)^{-1}$$

for those Z for which the right side is meaningful. It can be shown (see Exercise I.41) that under condition (2.13) of Theorem 2.4 the linear fractional function $\phi(Z)$ is meaningful for any Z in the interior

$$\Pi_-^0(\mathfrak{H}) = \{Z \mid Z \in [\mathfrak{H}], \operatorname{Re} Z \ll 0\}$$

of the left operator halfplane and maps it into itself.

Relation (2.16) shows that K_+ ($\in \Pi_-^0$) is a fixed point of the mapping ϕ: $\phi(K_+) = K_+$. It can be shown that K_+ is the only fixed point of ϕ in Π_-^0. Moreover (see Exercises IV.13 and I.41), for any

$Z_0 \in \Pi_l^0$

$$K_+ = \lim_{n\to\infty} Z_n, \qquad \text{where } Z_n = \phi(Z_{n-1}) \qquad (n = 1,2,\cdots). \tag{2.17}$$

We consider the linear fractional transformation

$$\phi^{-1}(Z) = -(\Psi'(T) - Z\Psi(T))^{-1}(\Phi'(T) - Z\Phi(T)),$$

which is the inverse of ϕ: $\phi(\phi^{-1}(Z)) = \phi^{-1}(\phi(Z)) = Z$. It maps the interior

$$\Pi_r^0(\mathfrak{H}) = \{Z \mid Z \in [\mathfrak{H}], \operatorname{Re} Z \gg 0\}$$

of the right operator halfplane into itself and its only fixed point in Π_r^0 is $-K_-$, for which there exists a formula analogous to (2.17).

5. *Splitting of the solutions by means of the initial conditions.* Let $r_\pm$ denote the spectral radii of the restrictions of the operators $U(T)$ and $U^{-1}(T)$ to $\mathfrak{L}_+$ and $\mathfrak{L}_-$ respectively. Then the quantities $\kappa_B(P_+) = \ln r_+$ (< 0) and $\kappa_B'(P_-) = -\ln r_-$ (> 0), where $P_\pm$ are the spectral projections of $U(T)$ corresponding to the subspaces $\mathfrak{L}_\pm$, will be Bohl exponents of equation (2.10) in the phase space $\mathfrak{H}^{(2)}$. Thus for any $\varepsilon > 0$ there exists an $N_\varepsilon > 0$ such that

$$\|x(t)\| \leqq N_\varepsilon e^{(\kappa_B(P_+)+\varepsilon)(t-s)}\|x(s)\| \qquad (\|x(s)\| \leqq N_\varepsilon e^{-(\kappa'_B(P_-)-\varepsilon)(t-s)}\|x(t)\|)$$

whenever $x(t)$ is a solution of equation (2.10) with $x(0) \in \mathfrak{L}_+$ $(x(0) \in \mathfrak{L}_-)$ and $-\infty < s < t < \infty$.

The second assertion of Theorem 2.4 permits us to reformulate these relations in terms of the solutions of the original equation (2.9) as follows: *suppose $y(t)$ is a solution of equation* (2.9) *and* $-\infty < s < t < \infty$; *then if* $y'(0) = K_+ y(0)$,

$$(\|y(t)\|^2 + \|y'(t)\|^2)^{1/2} \leqq N_\varepsilon e^{(\kappa_B(P_+)+\varepsilon)(t-s)}(\|y(s)\|^2 + \|y'(s)\|^2)^{1/2}, \tag{2.18}$$

while if $y'(0) = -K_- y(0)$,

$$(\|y(s)\|^2 + \|y'(s)\|^2)^{1/2} \leqq N_\varepsilon e^{-(\kappa'_B(P_-)-\varepsilon)(t-s)}(\|y(t)\|^2 + \|y'(t)\|^2)^{1/2} \tag{2.19}$$

Let $\mathfrak{L}_+(t) = U(t)\mathfrak{L}_+$. Since $U(t+T) = U(t)U(T)$ while $U(T)\mathfrak{L}_+ = \mathfrak{L}_+$, we have $\mathfrak{L}_+(t+T) = \mathfrak{L}_+(t)$. By means of arguments similar to the ones that led us to relation (2.15′) it is not difficult to see that the subspace $\mathfrak{L}_+(t)$ has the equation $z = K_+(t)y$, where $K_+(t)$ is a continuous operator function with values in Π_l ($\operatorname{Re} K_+(t) \ll 0$) obtained by substituting K_+ for $K_+(0)$ in the formula

$$K_+(t) = (\Phi'(t) + \Psi'(t)K_+(0))(\Phi(t) + \Psi(t)K_+(0))^{-1}.$$

Clearly, $K_+(t+T) = K_+(t)$.

Since $y'(0) = K_+(0)y(0)$ in (2.18), it readily follows that the assertion *for any* $\varepsilon > 0$ *and a corresponding* $N_\varepsilon > 0$ *estimate* (2.18) *holds for all solutions* $y(t)$ *of equation* (2.9) *with* $y'(0) = K_+ y(0)$ is equivalent to the assertion *for any* $\varepsilon > 0$ *and a corresponding* $M_\varepsilon > 0$ *the following estimate holds for these solutions*:

$$\|y(t)\| \leqq M_\varepsilon e^{(\kappa_B(P_+)+\varepsilon)(t-s)}\|y(s)\| \qquad (-\infty < s < t < \infty).$$

It is now not difficult to conclude that *the least upper bound of Bohl exponents of all solutions* $y(t)$ *of equation* (2.9) *with* $y'(0) = K_+ y(0)$ *coincides with* $\ln r_+$.

An analogous assertion can be formulated in regard to the solutions $y(t)$ of equation (2.9) with $y'(0) = -K_- y(0)$.

6. *The equation of second order with a Hermitian coefficient* $\mathscr{P}(t)$. When $\mathscr{P}(t) = \mathscr{P}^*(t)$ $(-\infty < t < \infty)$, Theorem 2.4 can be sharpened.

For suppose $\mathscr{P}(t) = \mathscr{P}^*(t)$ $(-\infty < t < \infty)$. Then for $\mathscr{A}(t)$ of the form (2.10′) we have

$$\mathscr{J}\mathscr{A}(t) = i\begin{pmatrix} \mathscr{P}(t) & 0 \\ 0 & I \end{pmatrix} = -\mathscr{A}^*(t)\mathscr{J},$$

where

$$\mathscr{J} = i\begin{pmatrix} 0 & -I \\ I & 0 \end{pmatrix} \qquad (= \mathscr{J}^*).$$

It accordingly follows from Theorem 1.3 that $U(T)$ is a $\mathscr{J}$-unitary operator:

$$U^*\mathscr{J}U = \mathscr{J}; \quad U = \mathscr{J}^{-1}U^{*-1}\mathscr{J},$$

and hence that its spectrum $\sigma(U(T))$ is situated symmetrically with respect to the unit circle.

This implies the latter assertion of the following proposition.

THEOREM 2.5. *Suppose in equation* (2.9)

$$\mathscr{P}(t) = \mathscr{P}^*(t) \ll 0 \qquad \textit{uniformly in } t \in (-\infty, \infty). \tag{2.20}$$

Then the equation operators $K_\pm$ *of equation* (2.9) *are uniformly negative Hermitian operators, i.e.* $K_\pm = K_\pm^* \ll 0$.

The spectra of the restrictions of the monodromy operator $U(T)$ *to* $\mathfrak{L}_+$ *and* $\mathfrak{L}_-$ *are specularly situated with respect to the unit circle.*

PROOF. Since the fulfilment of condition (2.20) implies the fulfilment of the original condition of Theorem 2.4, the existence of uniformly dissipative operators $K_\pm$ such that $z = \pm K_\pm y$ are the equations for the invariant subspaces $\mathfrak{L}_\pm$ $(\subset \mathfrak{H}^{(2)})$ of $U(T)$ is guaranteed by Theorem 2.4.

Since the spectrum $\sigma^+ = \sigma(U(T)|\mathfrak{L}_+)$ $(\sigma^- = \sigma(U(T)|\mathfrak{L}_-))$ lies entirely in the interior (exterior) of the unit disk and $U(T)$ is a $\mathscr{J}$-unitary operator, it follows from this fact that $\mathfrak{L}_+$ and $\mathfrak{L}_-$ are $\mathscr{J}$-neutral subspaces (see Exercise I.29).

Let $x_1 = y_1 \oplus K_+ y_1$ and $x_2 = y_2 \oplus K_+ y_2$ be any two vectors of $\mathfrak{L}_+$. The relation $(\mathscr{J}x_1, x_2) = 0$ implies

$$(K_+ y_1, y_2) - (y_1, K_+ y_2) = 0.$$

In view of the arbitrariness of $y_1, y_2 \in \mathfrak{H}$ it follows that $K_+ = K_+^*$. The Hermitianness of K_- can be proved analogously.

We give another proof of the Hermitianness of the $K_\pm$ which does not make use of the result of Exercise I.29.

Let P_+ and P_- denote the complementary spectral projections of $U(T)$ corresponding to the decomposition $\sigma(U(T)) = \sigma^+ \cup \sigma^-$ of its spectrum, so that $P_\pm \mathfrak{H}^{(2)} = \mathfrak{L}_\pm$. Clearly, $P_\pm^*$ will be the spectral projections of $U^*(T)$ corresponding to the decomposition $\sigma(U^*(T)) = \bar{\sigma}^+ \cup \bar{\sigma}^-$ in which the spectral sets $\bar{\sigma}^\pm$ are the specular images with respect to the real axis of the sets $\sigma^\pm$. The corresponding invariant subspaces $\mathfrak{L}_\pm^*$ of $U^*(T)$ can be found by means of the formulas $\mathfrak{L}_\pm^* = P_\pm^* \mathfrak{H}^{(2)}$. It is easily seen that these formulas have the following geometric meaning:

$$\mathfrak{L}_+^* = \mathfrak{H}^{(2)} \ominus \mathfrak{L}_-; \qquad \mathfrak{L}_-^* = \mathfrak{H}^{(2)} \ominus \mathfrak{L}_+. \tag{2.21}$$

Since, on the other hand, $U(T) = \mathcal{J}(U^*(T))^{-1}\mathcal{J}^{-1}$, we have $\mathcal{J}\mathfrak{L}_-^* = \mathfrak{L}_+$ and $\mathcal{J}\mathfrak{L}_+^* = \mathfrak{L}_-$.

In fact, the subspaces $\mathcal{J}\mathfrak{L}_\pm^*$ are invariant under the operator $\mathcal{J}(U^*(T))^{-1}\mathcal{J}$, the restrictions of which to these subspaces have the spectra $(\bar{\sigma}^\pm)^{-1} = \sigma^\mp$.

According to (2.21) a vector $y \oplus z$ $(\in \mathfrak{H}^{(2)})$ belongs to $\mathfrak{L}_-^*$ if and only if it is orthogonal to all of the vectors $y_1 \oplus K_+ y_1$ $(y_1 \in \mathfrak{H})$, i.e.

$$(y, y_1) + (z, K_+ y_1) = 0 \qquad (y_1 \in \mathfrak{H}).$$

From this result we obtain the equation $y = -K_+^* z$ $(y \in \mathfrak{H})$ for the subspace $\mathfrak{L}_-^*$.[3] But then, as is easily seen, $\mathfrak{L}_+$ $(= \mathcal{J}\mathfrak{L}_-^*)$ has the equation $z = K_+^* y$, whereas we already know that it has the equation $z = K_+ y$. Therefore $K_+^* = K_+$.

Analogously, from the relations $\mathfrak{L}_- = \mathcal{J}\mathfrak{L}_+^*$ and $\mathfrak{L}_+^* = \mathfrak{H} \ominus \mathfrak{L}_-$ we can get that $K_-^* = K_-$.

The theorem is proved.

7. *The equation of second order with a parameter*. Since for $\mathcal{A}(t) = \begin{pmatrix} 0 & I \\ -\mathcal{P}(t) & 0 \end{pmatrix}$ and $W = \begin{pmatrix} 0 & -I \\ -I & 0 \end{pmatrix}$ we have

$$\int_0^T \operatorname{Re}(W\mathcal{A}(t))\,dt = \begin{pmatrix} \int_0^T \operatorname{Re}\mathcal{P}(t)\,dt & 0 \\ 0 & -I \end{pmatrix},$$

the following proposition can be deduced from Theorem 2.2 by means of the same arguments that were used to obtain Theorems 2.4 and 2.5 (from Theorem 2.3).

THEOREM 2.6. *If*

$$\int_0^T \operatorname{Re}\mathcal{P}(t)\,dt \gg 0, \tag{2.22}$$

there exists an $\varepsilon > 0$ *such that the assertions of Theorem* 2.4 *and also, if* $P(t)$ *is a*

[3] I.e. $\mathfrak{L}_-^* = \{-K_+^* z \oplus z \mid z \in \mathfrak{H}\}$.

Hermitian operator function, the assertions of Theorem 2.5 will hold for the periodic equation

$$d^2y/dt^2 + \lambda\mathscr{P}(t)\,y = 0 \qquad (\mathscr{P}(t+T) = \mathscr{P}(t)) \tag{2.23}$$

when $-\varepsilon < \lambda < 0$.

Thus the monodromy operator $U(T;\lambda)$ of the equation will be *unstable* when λ is sufficiently small in absolute value and *negative*. It will be shown below (§ 6; see also Exercise 8) for a Hermitian operator function satisfying condition (2.22) that the monodromy operator $U(T;\lambda)$ will be *strongly stable* when λ is sufficiently small and *positive*.

§ 3. Localization theorems on the spectrum of the monodromy operator ($\mathfrak{B} = \mathfrak{H}$)

1. *Annular localization.* According to Wintner's estimate (III.4.19), if $\mathfrak{B} = \mathfrak{H}$, the Cauchy operator of equation (1.1) satisfies the inequalities

$$\|U(T)\| \leqq \exp\Big(\int_0^T \lambda_M(A_{\mathfrak{R}}(t))dt\Big),$$

$$\|U^{-1}(T)\| \leqq \exp\Big(-\int_0^T \lambda_m(A_{\mathfrak{R}}(t))dt\Big).$$

This immediately implies the following assertion.

THEOREM 3.1. *The spectrum of the monodromy operator* $U(T)$ *lies in the annulus* $K_A = \{\zeta \mid r_A \leqq |\zeta| \leqq R_A\}$, *where*

$$r_A = \exp\int_0^T \lambda_m(A_{\mathfrak{R}}(t))dt, \quad R_A = \exp\int_0^T \lambda_M(A_{\mathfrak{R}}(t))dt.$$

Despite the apparent roughness of this localization theorem it has some interesting applications (see Exercise 5 and the commentary on it).

2. *Angular localization.* If the operator function $A(t)$ is sufficiently "small in the large" on the segment $(0, T)$ then $U(t)$ must be sufficiently close to the identity operator and hence the spectrum of the monodromy operator must lie in the intersection of the above annulus K_A and some sector $|\arg\rho| \leqq \alpha$ $(\alpha < \pi)$.

We will cite below two theorems which permit one to distinguish a sector (or a rotated sector) of this type. To deduce these theorems we must make use of the properties of the following functionals, which were introduced in the exercises of Chapter I:

$$\operatorname{dev} U \stackrel{\text{def}}{=} \sup\{\measuredangle(Ux, x) \mid x \neq 0, x \in \mathfrak{H}\},$$

$$\operatorname{am} U \stackrel{\text{def}}{=} \min\{\operatorname{dev}(\zeta U) \mid |\zeta| = 1\}.$$

We recall that the spectrum of U lies in the angle $|\arg \rho| \leqq \operatorname{dev} U$. Let

$$\lambda_r(A_{\Re}) = |\min(0, -\lambda_m(A_{\Re}), \lambda_M(A_{\Re}))|.$$

THEOREM 3.2. *The following estimate holds*:

$$\operatorname{dev}(U(T)) \leqq \int_0^T \sqrt{\|A(t)\|^2 - [\lambda_r(A_{\Re}(t))]^2}\, dt \quad (= \mathscr{T}_A). \tag{3.1}$$

Thus if $\mathscr{T}_A < \pi$, the spectrum of the monodromy operator of equation (1.1) *lies in the sector*

$$|\arg \rho| \leqq \mathscr{T}_A. \tag{3.2}$$

In particular, it lies in the sector

$$|\arg \rho| \leqq \int_0^T \|A(t)\| dt. \tag{3.3}$$

In proving this and the following theorem (and the intermediate lemmas) we will assume that the operator function $A(t) = A(t + T)$ is continuous. Once the theorems have been proved under this assumption, they can be obtained in the general case by applying them to the approximating equations

$$\frac{dx}{dt} = A_h(t)x \qquad \left(A_h(t) = \frac{1}{2h}\int_{-h}^{h} A(t+s)\, ds\right)$$

and passing to the limit for $h \downarrow 0$ in the obtained estimates.

PROOF. Any solution $x(t) = U(t)x_0$ of equation (1.1) clearly satisfies the inequality

$$\|dx/dt\| \leqq \|A(t)\|\,\|x(t)\|.$$

In addition,[4)]

$$\left|\frac{d \ln \|x(t)\|}{dt}\right| \geqq \lambda_r(A_{\Re}(t)), \tag{3.4}$$

since by virtue of Wintner's estimates

$$\int_t^{t+h} \lambda_m(A_{\Re}(s))ds \leqq \ln \frac{\|x(t+h)\|}{\|x(t)\|} \leqq \int_t^{t+h} \lambda_M(A_{\Re}(s))ds.$$

If we put $e(t) = \|x(t)\|^{-1}x(t)$ or $x(t) = \|x(t)\|e(t)$, we will have

$$\frac{dx}{dt} = \frac{d}{dt}\|x\| \cdot e + \|x\| \frac{de}{dt}.$$

Since $(e, e) = 1$, it follows that

---

4) Whenever $x(t)$ is a continuously differentiable function, $\|x(t)\|$ has a left and right derivative (see Exercise I. 18). It can be assumed for the sake of definiteness that the right derivative of $\ln \|x(t)\|$ has been taken in (3.4).

$$(de/dt, e) + (e, de/dt) = 0,$$

and consequently

$$\left\| \frac{dx}{dt} \right\|^2 = \left(\frac{d\|x\|}{dt} \right)^2 + \|x\|^2 \left(\frac{de}{dt} \right)^2,$$

$$\left\| \frac{de}{dt} \right\| = \sqrt{\left\| \frac{dx}{dt} \right\|^2 \cdot \frac{1}{\|x\|^2} - \left(\frac{d}{dt} \ln \|x(t)\| \right)^2}$$

$$\leqq \sqrt{\|A\|^2 - [\lambda_r(A_{\mathfrak{R}}(t))]^2}.$$

Thus (see Exercise I.31)

$$\measuredangle\,(e(T), e(0)) \leqq \int_0^T \left\| \frac{de}{dt} \right\| dt \leqq \mathscr{T}_A.$$

Since $\measuredangle\,(e(T), e(0)) = \measuredangle\,(U(T)x_0, x_0)$ for an arbitrary nonzero vector $x_0 \in \mathfrak{H}$, the above estimate implies (3.1) and hence the assertions of the theorem.

3. *Angular localization with regard for amplitude.* As we know (see § III.2), equation (1.1) is kinematically similar to the equation

$$dy/dt = Q^{-1}(t)A_{\mathfrak{R}}(t)Q(t)y, \tag{3.5}$$

where $Q(t)$ is the unitary operator function solving the Cauchy problem

$$dQ/dt = iA_{\mathfrak{J}}(t)Q, \qquad Q(0) = I.$$

Under the substitution $y = Qx$ equation (1.1) goes over into equation (3.5), for which $W(t) = Q^{-1}(t)U(t)$ is a fundamental operator function. The equality $U(T) = Q(T)W(T)$ implies that

$$\operatorname{am} U(T) \leqq \operatorname{am} Q(T) + \operatorname{am} W(T) \leqq \operatorname{am} Q(T) + \operatorname{dev} W(T).$$

We can obtain simple estimates for $\operatorname{am} Q(T)$ and $\operatorname{dev} W(T)$.

The spectral length of a Hermitian operator H, i.e. the length of the shortest segment containing its spectrum $\sigma(H)$, will be denoted by $l(H)$:

$$l(H) \overset{\text{def}}{=\!=} \lambda_M(H) - \lambda_m(H).$$

LEMMA 3.1. *Suppose $H(t)$ $(0 \leqq t \leqq T)$ is a Hermitian operator function and*

$$W(T) = \int_0^{T\,\curvearrowleft} e^{H(t)\,dt}. \tag{3.6}$$

Then

$$\operatorname{dev} W(t) \leqq \frac{1}{2} \int_0^T l(H(t))dt. \tag{3.7}$$

PROOF. It can be assumed without loss of generality that $H(t)$ is a continuous operator function. For sufficiently large natural n the product

$$\prod_1^n \exp(H(t_k)\Delta t_k) \qquad (t_k = kT/n, \Delta t_k = T/n; k = 1,2,\cdots,n)$$

will be arbitrarily close to $W(T)$. The deviation of this product (see Exercise I.32) will not be larger than

$$\sum_{k=1}^n \operatorname{dev} \exp(H(t_k)\,\Delta t_k). \tag{3.8}$$

By virtue of the result of L. V. Kantorovič (see Exercise I.33), for arbitrary $G = G^*$

$$\cos(\operatorname{dev}(\exp G)) = \frac{1}{\cosh(l(G)/2)},$$

and consequently $\sin(\operatorname{dev}(\exp G)) = \tanh(l(G)/2)$, so that

$$\operatorname{dev}(\exp G) \leqq \tan(\operatorname{dev}(\exp G)) = \sinh(l(G)/2) = l(G)/2 + O(l^3(G)).$$

Therefore the sum (3.8) does not exceed

$$\frac{1}{2}\sum_1^n l(H(t_k))\,\Delta t_k + O\left(\frac{1}{n^2}\right).$$

But this implies (3.7).

LEMMA 3.2. *Suppose $H(t)$ $(0 \leqq t \leqq T)$ is a Hermitian operator function and*

$$Q(T) = \int_0^{\curvearrowleft T} e^{iH(t)dt}. \tag{3.9}$$

Then

$$\operatorname{dev}(\zeta_H^{-1}Q(T)) \leqq \frac{1}{2}\int_0^T l(H(t))dt, \tag{3.10}$$

where

$$\zeta_H = \exp(i\phi(T)),\ \phi(t) = \frac{1}{2}\int_0^t [\lambda_M(H) + \lambda_m(H)]dt \qquad (0 \leqq t \leqq T).$$

PROOF. Let Ω $(= \Omega(H))$ denote the right side of (3.10), which is a trivial estimate unless $\Omega < \pi$. As in the preceding proof, it can be assumed that $H(t)$ is a continuous operator function, so that for sufficiently large n

$$Q_n = \prod_1^n e^{iH(t_k)\Delta t_k} = U_nU_{n-1}\cdots U_1 \quad \left(t_k = \frac{kT}{n},\ \Delta t_k = \frac{T}{n};\ k = 1,2,\cdots,n\right) \tag{3.11}$$

will differ arbitrarily little from $Q(T)$.

The spectrum of $U_k = \exp(iH(t_k)\,\Delta t_k)$ lies on the arc

$$\gamma_k = \{e^{i\lambda} | \lambda_m(H(t_k))\Delta t_k \leqq \lambda \leqq \lambda_M(H(t_k))\Delta t_k\}.$$

The sum of the lengths of these arcs is equal to $\sum_{k=1}^{n} l(H(t_k))\Delta t_k$ $(= 2\omega_n)$, which for sufficiently large n will be arbitrarily close to Ω and hence less than π when $\Omega < \pi$.

Therefore (see Exercise I.34) the spectrum of the product (3.11) will lie on the product of the arcs γ_k $(k = 1,\cdots,n)$, i.e. on the arc of length $2\omega_n$ with center

$$\zeta_n = \exp i\left\{\frac{1}{2}\sum_{i=1}^{n} [\lambda_M(H(t_k)) + \lambda_m(H(t_k))]\Delta t_k\right\}.$$

Passing to the limit for $n \to \infty$, we conclude that the spectrum of $Q(t)$ will lie on the arc of length $2\Omega(H)$ with center ζ_H. But this is the fact that inequality (3.10) expresses.

THEOREM 3.3. *Suppose the following condition is satisfied for equation* (1.1):

$$(\omega \stackrel{\text{def}}{=}) \frac{1}{2} \int_0^T [l(A_{\Re}) + l(A_{\Im})]dt < \pi. \tag{3.12}$$

Then the spectrum of the monodromy operator of equation (1.1) *lies in the sector of the complex ρ-plane*

$$|\arg \rho - \alpha| \leqq \omega,$$

where $\alpha \in [-\pi, \pi]$ is determined from the congruence

$$\alpha \equiv \frac{1}{2} \int_0^T [\lambda_M(A_{\Im}) + \lambda_m(A_{\Im})]dt \pmod{2\pi}.$$

PROOF. We return to the relation $U(T) = Q(T)W(T)$, where $Q(t)$ has the form (3.9) with $H(t) = A_{\Im}(t)$ while $W(t)$ has the form (3.6) with $H(t) = Q^{-1}(t)A_{\Re}(t)Q(t)$. In view of the fact that Q is unitary we have according to Lemma 3.1 that

$$\text{dev } W(T) \leqq \frac{1}{2} \int_0^T l(Q^{-1}A_{\Re}Q)\, dt = \frac{1}{2} \int_0^T l(A_{\Re})dt.$$

According to Lemma 3.2

$$\text{dev}(e^{-i\alpha}U(T)) \leqq \frac{1}{2} \int_0^T l(A_{\Im})dt,$$

and since $\text{dev}(e^{-i\alpha}U(T)) \leqq \text{dev}(e^{-i\alpha}Q(T)) + \text{dev } W(T)$, it follows that

$$\text{dev}(e^{-i\alpha}U(T)) \leqq \omega.$$

The latter estimate contains the assertion of the theorem.

REMARK 3.1. If either of the estimates

$$\int_0^T \sqrt{\|A(t)\|^2 - [\lambda_r(A_{\Re}(t))]^2}\, dt < \pi \tag{3.13}$$

or

$$\frac{1}{2}\int_0^T [l(A_{\mathfrak{R}}(t)) + l(A_{\mathfrak{J}}(t))]dt < \pi \tag{3.14}$$

is satisfied, the spectrum of the monodromy operator of equation (1.1) will lie in a certain sector, so that the Floquet representation will hold for it, and this equation will consequently be reducible.

This fact will be the case if, in particular,

$$\int_0^T \|A(t)\|dt < \pi. \tag{3.15}$$

§ 4. Canonical differential equations

1. *Stable canonical equations.* In this section we consider in a Hilbert space $\mathfrak{H}$ a differential equation of the form

$$dx/dt = iA(t)x, \tag{4.1}$$

where $A(t)$ is a periodic operator function that is also W-Hermitian for some indefinite invertible operator W.

By suitably renorming the space it is always possible to replace the operator W with an operator $\mathcal{J}$ such that (see Exercise I.36) $\mathcal{J} = \mathcal{J}^* = \mathcal{J}^{-1}$.

The fact that A is a $\mathcal{J}$-Hermitian operator means that $\mathcal{H} = \mathcal{J}A$ is a Hermitian operator. Inasmuch as then $A = \mathcal{J}\mathcal{H}$, we conclude that it is possible without loss of generality to consider in place of equation (4.1) the equation

$$dx/dt = i\mathcal{J}\mathcal{H}(t)x, \tag{4.2}$$

where $\mathcal{H}(t)$ is a periodic Hermitian operator function. An equation of this form is said to be *canonical*, and the operator $\mathcal{H}$ is called the *Hermitian*.

The similar (to (4.2)) equation

$$dx/dt = J_r\mathcal{H}(t)x \tag{4.2'}$$

is frequently considered in a real space $\mathfrak{H}$. Here the role of the operator $i\mathcal{J}$ is played by a skew-Hermitian unitary operator J_r $(= -J_r^* = -J_r^{-1})$.

The Hermitian operator function $\mathcal{H}(t)$ in this case is called the *Hamiltonian*, while the canonical equation (4.2′) is called *Hamilton's equation*.

It can easily be shown that to the operator J_r there always corresponds an orthogonal decomposition of $\mathfrak{H}$:[5]

$$\mathfrak{H} = \mathfrak{H}_1 \oplus \mathfrak{H}_2$$

into subspaces of equal dimension ($\dim \mathfrak{H}_1 = \dim \mathfrak{H}_2$) and with respect to which

[5] To this end it is necessary to carry out in the complex hull $\tilde{\mathfrak{H}}$ of $\mathfrak{H}$ the same arguments as in Exercise III. 6.

the operator J_Γ has the matrix representation

$$J_\Gamma = \begin{pmatrix} 0 & I_{12} \\ -I_{21} & 0 \end{pmatrix},$$

where I_{21} is a unitary operator mapping $\mathfrak{H}_2$ onto $\mathfrak{H}_1$ and $I_{12} = I_{21}^{-1}$.

Thus Hamilton's equation can always be considered in a doubled Hilbert space, assuming that the operator J_Γ has the indicated matrix representation.

From Theorem 1.3 it follows that the evolution operator $U(t, \tau)$ of equation (4.2), and hence its monodromy operator $U(T) = U(T, 0)$, is $\mathcal{J}$-unitary, i.e.

$$U^*(T)\mathcal{J}U(t) = \mathcal{J}, \tag{4.3}$$

or, equivalently,

$$U^{-1}(T) = \mathcal{J}U^*(T)\mathcal{J}. \tag{4.4}$$

In particular, the spectrum of $U(T)$ is symmetric with respect to the unit circle. From formula (4.4) we obtain the equality

$$\|U^{-n}(T)\| = \|\mathcal{J}U^{n*}(T)\mathcal{J}\| = \|U^n(T)\|$$

(since $\|\mathcal{J}A\mathcal{J}\| = \|\mathcal{J}A\mathcal{J}^{-1}\| = \|A\|$ for any bounded operator A by virtue of the fact that $\mathcal{J}$ is a unitary operator), which shows that the stability of a $\mathcal{J}$-unitary operator is equivalent to the uniform boundedness of its positive powers:

$$\sup_{n\geq 0} \|U^n(t)\| < \infty.$$

From Lemma 1.1 and Remark 1.1 it follows that for canonical equations the condition of boundedness of the solutions on the real line is equivalent to the condition of boundedness of them on $[0, \infty)$.

In accordance with the general definition of Chapter III we will say that equation (4.2) is *stable* if each of its solutions is bounded on the real line (on $[0, \infty)$).

From Lemma 1.1 we conclude that the stability of equation (4.2) is equivalent to the stability of its monodromy operator. With the use of Theorem I.8.1 we at once obtain the following result.

THEOREM 4.1. *In order for equation* (4.2) *to be stable it is necessary and sufficient that the space* $\mathfrak{H}$ *decompose into a* $\mathcal{J}$*-orthogonal sum* $\mathfrak{H} = \mathfrak{H}_1 + \mathfrak{H}_2$ *of subspaces that are invariant under the monodromy operator, the subspace* $\mathfrak{H}_1$ ($\mathfrak{H}_2$) *being uniformly* $\mathcal{J}$*-positive* ($\mathcal{J}$*-negative*).

COROLLARY 4.1. *A stable canonical equation is reducible; more precisely, there exists a periodic differential operator function* $Q(t)$ *having a bounded inverse* $Q^{-1}(t)$ *and such that the substitution* $x = Q(t)y$ *transforms the given equation into a canonical equation* $dy/dt = i\mathcal{J}Gy$ *with a constant Hermitian* G.

By virtue of Theorem 1.2 it suffices to show that the monodromy operator has a logarithm.

Let U_1 and U_2 denote the restrictions of $U(T)$ to $\mathfrak{H}_1$ and $\mathfrak{H}_2$ respectively. With respect to the scalar product $(x,y)_1 = (\mathscr{J}x, y)$, $x, y \in \mathfrak{H}_1$ (the scalar product $(x, y)_2 = -(\mathscr{J}x, y)$, $x, y \in \mathfrak{H}_2$) the space $\mathfrak{H}_1$ ($\mathfrak{H}_2$) is a Hilbert space while the operator U_1 (U_2) is an ordinary unitary operator. Therefore we have the representations $U_k = \exp(iC_k)$ $(k = 1, 2)$, where C_k is a selfadjoint operator acting in $\mathfrak{H}_k$. The operators C_k can be chosen so that they are positive and their spectra lie on the segment $[0, 2\pi]$.

We consider in $\mathfrak{H}$ the direct sum C of C_1 and C_2: $Cx = C_1x_1 + C_2x_2$ if $x = x_1 + x_2$ $(x_1 \in \mathfrak{H}_1, x_2 \in \mathfrak{H}_2)$.

The operator C is a $\mathscr{J}$-Hermitian operator. Setting $G = T^{-1}\mathscr{J}C$, we get

$$U(T) = \exp(iT\mathscr{J}G). \tag{4.5}$$

Later on (see § 6) we indicate a method for calculating the operators $Q(t)$ and G.

REMARK 4.1. The spectrum of the monodromy operator of a stable equation lies on the unit circle. In fact, this property is possessed by every stable operator inasmuch as by Theorem I.6.1. it is similar to a unitary operator.

2. *Strongly stable canonical equations.* We recall that a $\mathscr{J}$-unitary operator U is said to be *strongly stable* if all of the $\mathscr{J}$-unitary operators in a sufficiently small neighborhood of it are stable.

We say that equation (4.2) is *strongly stable* if there exists a $\delta > 0$ such that every equation

$$\frac{dx}{dt} = i\mathscr{J}\mathscr{H}_1(t)x$$

whose Hermitian satisfies the inequality

$$\int_0^T \|\mathscr{H}(t) - \mathscr{H}_1(t)\|dt < \delta \tag{4.6}$$

is stable.

LEMMA 4.1. *In order for equation* (4.2) *to be strongly stable it is necessary and sufficient that its monodromy operator* $U(T)$ *be a strongly stable operator.*

PROOF. To prove the sufficiency we note that to the Hermitian $\mathscr{H}(t)$ there correspond two constants $c_1 > 0$ and $c_2 > 0$ such that

$$\|U_1(T) - U(T)\| \leqq c_1\left[\exp\left\{c_2\int_0^T \|\mathscr{H}(t) - \mathscr{H}_1(t)\|dt\right\} - 1\right]$$

(see Lemma III.2.3). This shows that the mapping $\mathscr{H}(t) \to U(T)$ that takes each Hermitian into its monodromy operator is continuous. Thus under a small variation of the Hermitian we obtain a monodromy operator $U_1(T)$ lying in a small neighborhood of $U(T)$. From the strong stability of the latter we obtain the stability of $U_1(T)$.

To prove the necessity we must show that any $\mathcal{J}$-unitaty operator $U_1(T)$ lying in a small neighborhood of $U(T)$ is the monodromy operator of an equation with a Hermitian $\mathcal{H}_1(t)$ differing little from $\mathcal{H}(t)$. In other words, we must show that the mapping described above is open. We omit a proof of this fact (see Exercise 4).

Lemma 4.1 and Theorem I.8.3 imply the following fact.

THEOREM 4.2. *In order for a canonical equation to be strongly stable it is necessary and sufficient that its monodromy operator $U(T)$ be normally $\mathcal{J}$-decomposable, i.e. that under a decomposition*[6] $\mathfrak{H} = \mathfrak{H}_1 + \mathfrak{H}_2$ *of* $\mathfrak{H}$ *into uniformly* $\mathcal{J}$*-definite invariant (under* $U(T)$*) subspaces the spectra* $\sigma(U(T)|\mathcal{H}_1)$ *and* $\sigma(U(T)|\mathcal{H}_2)$ *do not intersect.*

3. *Central stability band of the canonical equation with a parameter.* We now consider the canonical equation with a real parameter λ:

$$dx/dt = i\lambda \mathcal{J}\mathcal{H}(t)x. \tag{4.7}$$

The *stability* (*strong stability*) *points* of this equation are those values of λ for which it is stable (strongly stable), i.e. for which its monodromy operator $U(T; \lambda)$ is stable (strongly stable). The set of strong stability points of equation (4.7) is obviously open and therefore, if it is not empty, decomposes into a system of open intervals—the stability intervals of equation (4.7).

We note that the following representation is valid:

$$U(T; \lambda) = I + i\lambda \mathcal{J} \int_0^T \mathcal{H}(s)ds + O(\lambda^2) \quad (\lambda \to 0). \tag{4.8}$$

For if we integrate each side of the equality $dU/dt = i\lambda\mathcal{J}\mathcal{H}(t)\,U(t; \lambda)$, we get

$$U(t; \lambda) = I + i\lambda \int_0^t \mathcal{J}\mathcal{H}(s)U(s; \lambda)ds.$$

By iterating this equality and setting $t = T$, we obtain (4.8) (cf. (III.1.11)).

THEOREM 4.3. *If the operator*

$$\mathcal{H}_{av} = \frac{1}{T}\int_0^T \mathcal{H}(s)ds$$

is uniformly positive, there exists a maximal interval $(\Lambda_{-1}, \Lambda_1)$ $(\ni 0)$ *(the central stability band of the equation) all of whose nonzero points* λ *are strong stability points of equation* (4.7).

[6] See Theorem 4.1.

PROOF. It suffices to show that λ is a strong stability point if it is different from zero and sufficiently small in absolute value.

From (4.8) we obtain the equality

$$U(T; \lambda) = I + i\lambda[T\mathcal{J}\mathcal{H}_{\mathrm{av}} + O(\lambda)].$$

From the properties of normally W-decomposable operators cited in § I.8 it follows that under the conditions of the theorem the operator $T\mathcal{J}\mathcal{H}_{\mathrm{av}}$ is normally $\mathcal{J}$-decomposable. Since the property of normal $\mathcal{J}$-decomposability is stable (Theorem I.8.2), it is not affected by adding a small operator $O(\lambda)$ to $T\mathcal{J}\mathcal{H}_{\mathrm{av}}$. It is also obviously preserved under multiplication by a scalar $i\lambda \neq 0$ and the subsequent addition of the identity operator.

Thus the $\mathcal{J}$-unitary operator $U(T; \lambda)$ is normally $\mathcal{J}$-decomposable. By Theorem I.8.3 it is strongly stable. The theorem is proved.

The purpose of the following arguments is an estimation of the central stability band of a canonical equation whose Hermitian satisfies the conditions

$$\mathcal{H}(t) \geqslant 0 \quad (0 \leqq t \leqq T); \quad \mathcal{H}_{\mathrm{av}} = \frac{1}{T}\int_0^T \mathcal{H}(s)ds \gg 0.$$

Let us agree to call such an operator function $\mathcal{H}(t)$, a *Hermitian of positive type*, and the canonical equation corresponding to it, a *canonical equation of positive type*.

LEMMA 4.2. *Let* $[a, b]$ *be a closed interval containing zero at each point* λ *of which there exists the operator*

$$[U(T; \lambda) + I]^{-1} \qquad (\in [\mathfrak{H}]). \tag{4.9}$$

If the Hermitian $\mathcal{H}(t)$ *in equation* (4.7) *is of positive type, every nonzero point of* $[a, b]$ *is a strong stability point of this equation.*

PROOF. We consider the operator

$$V(\lambda) = i[I - U(T; \lambda)][I + U(T; \lambda)]^{-1}.$$

We shall prove that $V(\lambda)$ can be represented in the form

$$V(\lambda) = \mathcal{J}N(\lambda) \qquad (\lambda \in [a, b]), \tag{4.10}$$

where $N(\lambda)$ is a uniformly positive operator. It will then follow from the properties of normally decomposable operators (see § I.8.2) that $V(\lambda)$ is normally $\mathcal{J}$-decomposable. Since $U(T; \lambda)$ can be expressed in terms of $V(\lambda)$ in the form of a linear fractional transformation, it too is normally $\mathcal{J}$-decomposable. In addition, this operator is $\mathcal{J}$-unitary. By virtue of Theorem I.8.3 this operator will be strongly stable.

To obtain formula (4.10) we consider the derivative

$$\begin{aligned}\dot{V}(\lambda) &= \frac{dV}{d\lambda} = i\frac{d}{d\lambda}\{2[I + U(T;\lambda)]^{-1} - I\} \\ &= -2i[I + U(T;\lambda)]^{-1}\frac{dU(T;\lambda)}{d\lambda}[I + U(T;\lambda)]^{-1}.\end{aligned} \tag{4.11}$$

Let us calculate $\dot{U}(T;\lambda) = dU(T;\lambda)/d\lambda$.
Differentiating the equation

$$dU(t;\lambda)/dt = i\lambda\mathcal{J}\mathcal{H}(t)U(t;\lambda)$$

and the equality $U(0;\lambda) = I$ with respect to λ, we get

$$\frac{d\dot{U}(t;\lambda)}{dt} = i\lambda\mathcal{J}\mathcal{H}(t)\dot{U}(t;\lambda) + i\mathcal{J}\mathcal{H}(t)U(t;\lambda), \quad \dot{U}(0;\lambda) = 0.$$

We have thus obtained an operator equation of the form of equation (III.1.1), where $f(t) = i\mathcal{J}\mathcal{H}(t)U(t;\lambda)$. Its solution is given by formula (III.1.19):

$$\begin{aligned}\dot{U}(t;\lambda) &= \int_0^t U(t;\lambda)U^{-1}(s;\lambda)\mathcal{J}\mathcal{H}(s)U(s;\lambda)ds \\ &= iU(t;\lambda)\mathcal{J}\int_0^t U^*(s;\lambda)\mathcal{H}(s)U(s;\lambda)ds.\end{aligned}$$

Introducing the notation

$$R(\lambda) = \int_0^T U^*(s;\lambda)\mathcal{H}(s)U(s;\lambda)ds, \tag{4.12}$$

we get

$$dU(T;\lambda)/d\lambda = iU(T;\lambda)\mathcal{J}R(\lambda).$$

We substitute this expression in (4.11) and make use of the relation

$$(I + U)^{-1}U\mathcal{J} = (U^{-1} + I)\mathcal{J} = (\mathcal{J}U^*\mathcal{J} + I)^{-1}\mathcal{J} = \mathcal{J}(U^* + I)^{-1}.$$

Then

$$\begin{aligned}\dot{V}(\lambda) &= -2i[I + U(T;\lambda)]^{-1}\cdot iU(T;\lambda)\mathcal{J}R(\lambda)[I + U(T;\lambda)]^{-1} \\ &= 2\mathcal{J}[U^*(T;\lambda) + I]^{-1}R(\lambda)[T;\lambda) + I]^{-1}.\end{aligned}$$

Since $V(0) = 0$ it follows that

$$V(\lambda) = 2\mathcal{J}\int_0^\lambda [U^*(T;\mu) + I]^{-1}R(\mu)[U(T;\mu) + I]^{-1}d\mu,$$

and we obtain relation (4.10) with the operator

$$N(\lambda) = 2\int_0^\lambda [U^*(T;\mu) + I]^{-1}R(\mu)[U(T;\mu) + I]^{-1}d\mu. \tag{4.13}$$

Finally, the operator $R(\mu)$ in the integrand is positive and, for sufficiently small

$\mu > 0$, uniformly positive. For it follows from (4.12) that $R(\mu) > 0$; and if μ is sufficiently small, the operator $U(\tau; \mu)$ is close to I, so that $R(\mu)$ is close to $\int_0^T \mathscr{H}(s)\,ds$, and therefore $R(\mu) \gg 0$. But this means that $N(\lambda) \gg 0$ $(0 < \lambda < b)$.

We note that the operator $N(\lambda)$ receives a positive increase from an increase in the value of λ. An analogous examination can be made of the case $a < \lambda < 0$. The lemma is proved.

We now establish a connection between the existence of the operator (4.9) and the solvability of a special boundary problem.

We consider the boundary problem

$$dx/dt = i\lambda \mathscr{J}\mathscr{H}(t)x, \quad x(0) + x(T) = f. \tag{4.14}$$

A number λ is called a *regular point of problem* (4.14) if for any $f \in \mathfrak{H}$ there exists a unique solution $x(t)$ $(0 \leqq t \leqq T)$ depending continuously on f in the sense of the norm $\sup_{0 \leqq t \leqq T} \|x(t)\|$.

The complement of the set of regular points of problem (4.14) is called its *spectrum*.

LEMMA 4.3. *In order for a point λ to be a regular point of problem* (4.14) *it is necessary and sufficient that the operator* $[U(T; \lambda) + I]^{-1}$ $(\in [\mathfrak{H}])$ *exist.*

PROOF. We write the solution of equation (4.14) in the form $x(t) = U(t; \lambda)x_0$. Substituting this expression in the boundary condition, we obtain the relation $[U(T; \lambda) + I]x_0 = f$, which proves the lemma.

REMARK 4.2. Since the operator $U(T; \lambda)$ depends continuously on λ, the set of points λ for which the operator $[U(T; \lambda) + I]^{-1}$ is bounded is open. Consequently, the spectrum of problem (4.14) is closed.

Let λ_{-1} and λ_1 denote respectively the maximal negative and minimal positive points of the spectrum of problem (4.14).[7] The operator $[U(T; \lambda) + I]^{-1}$ is bounded in the interval $(\lambda_{-1}, \lambda_1)$. Using Lemma 4.2, we obtain the final result.

THEOREM 4.4. *Suppose $\mathscr{H}(t)$ is a Hermitian of positive type. Every nonzero point of the interval $(\lambda_{-1}, \lambda_1)$ of regularity of problem* (4.14) *is a strong stability point of the canonical equation* (4.7), *i.e.*

$$\Lambda_{-1} \leqq \lambda_{-1} < 0 < \lambda_1 \leqq \Lambda_1.$$

4. *Characteristic multipliers of the finite-dimensional canonical equation.* We consider the canonical equation under the assumption that the space $\mathfrak{H}$ is finite dimensional. The eigenvalues ρ of the monodromy operator $U(T; \lambda)$ are called the *characteristic multipliers* of equation (4.7).

When the equation is stable, the characteristic multipliers ρ lie on the unit circle.

---

[7] We will not dwell on a proof of the existence of the spectrum of the problem on each of the halflines. In the finite-dimensional case this fact follows from formula (0.21) of Exercise 12; if, in addition, the equation is a Hamiltonian equation, it can be shown that $\Lambda_{\pm 1} = \lambda_{\pm 1}$.

Of interest is their behavior under a variation of λ.

Let us discuss the behavior of the characteristic multipliers when λ is in the interval $(\lambda_{-1}, \lambda_1)$. We consider the operator

$$V(\lambda) = i[I - U(T; \lambda)][I + U(T; \lambda)]^{-1}.$$

In the proof of Lemma 4.2 it was shown that $V(\lambda) = \mathscr{J}N(\lambda)$, where $N(\lambda)$ is a positive operator having a positive increase under an increase in the value of λ.

The equation for the eigenvalues of $V(\lambda)$ can be written in the form $N(\lambda)x = \mu\mathscr{J}x$.

As is well known, the eigenvalues μ can be found with the use of the so-called minimax relations

$$\frac{1}{\mu} = \min \max \frac{(\mathscr{J}x, x)}{(N(\lambda)x, x)},$$

where the maximum is taken over all vectors satisfying additional linear relations and the minimum, over all such relations.

Since the denominator increases under an increase in the value of λ, we obtain the following result: *the eigenvalues of $V(\lambda)$ corresponding to $\mathscr{J}$-positive ($\mathscr{J}$-negative) eigenvectors increase (decrease) under an increase in the value of λ.*

It is easily seen that the characteristic multipliers ρ are connected with the eigenvalues μ of $V(\lambda)$ by the formula

$$\mu = i\frac{1-\rho}{1+\rho} \quad \text{or} \quad \rho = \frac{i-\mu}{i+\mu}.$$

This leads to the following rule: *under an increase in the value of λ ($\lambda \in (\lambda_{-1}, \lambda_1)$) the characteristic multipliers of the first kind (corresponding to $\mathscr{J}$-positive eigenvectors of the monodromy operator) undergo a counterclockwise movement while the characteristic multipliers of the second kind (corresponding to $\mathscr{J}$-negative eigenvectors of the monodromy operator) undergo a clockwise movement.*[8)]

§ 5. Second order equations (infinite-dimensional analog of Hill's equation)

1. *Central stability band.* We consider in a Hilbert space $\mathfrak{H}$ the differential equation of second order with a real parameter μ

$$d^2y/dt^2 + \mu\mathscr{P}(t)y = 0, \tag{5.1}$$

where $\mathscr{P}(t)$ is a periodic Hermitian valued operator function: $\mathscr{P}(t + T) = \mathscr{P}(t) = \mathscr{P}^*(t)$.

As in the case of a first order equation, we say that this equation is *stable* for a given μ if all of its solutions are bounded on the real line, and *strongly stable* if it

[8)] Additional rules for the movement of characteristic multipliers in more complicated situations can be found in the papers of M. G. Kreĭn [6] and M. G. Kreĭn and G. Ja. Ljubarskiĭ [1].

remains stable under a small variation of the operator function $\mathscr{P}(t)$.[9] In accordance with this we will refer to the values of μ as *stability* or *strong stability points*.

If $\mu \neq 0$, we can use the substitution $\lambda = \sqrt{|\mu|}$, $y_1 = \lambda y$, $y_2 = dy/dt$, $x = \binom{y_1}{y_2}$ to reduce equation (5.1) to a first order equation in the space $\mathfrak{H}^{(2)} = \mathfrak{H} \oplus \mathfrak{H}$:

$$\frac{dx}{dt} = \lambda \begin{pmatrix} 0 & I \\ \mp \mathscr{P}(t) & 0 \end{pmatrix} x,$$

which can also be written in the Hamilton canonical form if $\mathscr{P}(t)$ is real:

$$\frac{dx}{dt} = i\lambda \begin{pmatrix} 0 & -iI \\ iI & 0 \end{pmatrix} \begin{pmatrix} \pm \mathscr{P}(t) & 0 \\ 0 & I \end{pmatrix} x = i\lambda \mathscr{J}\mathscr{H}x. \tag{5.2}$$

Here the upper sign is taken when $\mu = \lambda^2 > 0$ while the lower is taken when $\mu = -\lambda^2 < 0$.

It is obvious that the solutions of equation (5.1) are bounded if the solutions of equation (5.2) are bounded, and hence, if λ is a strong stability point for equation (5.2), $\mu = \pm\lambda^2$ will be a strong stability point for equation (5.1).

With the use of this fact we can make a number of deductions concerning second order equations.

Theorem 4.3 immediately implies the following result.

THEOREM 5.1. *If $\mathscr{P}(t)$ is an operator function with a uniformly positive mean value $\mathscr{P}_{av} \gg 0$, there exists a number $\mu_1 > 0$ such that all of the points of the open interval $(0, \mu_1)$ are strong stability points of equation* (5.1).[10]

We will further assume that $\mu > 0$ and $\mathscr{P}(t)$ is an *operator function of positive type*.[11] In this case we will also call equation (5.1) an *equation of positive type*. Its Hamiltonian

$$\mathscr{H}(t) = \begin{pmatrix} \mathscr{P}(t) & 0 \\ 0 & I \end{pmatrix}$$

will also be of positive type.

In order to characterize the number μ_1 we must in accordance with Lemmas 4.2 and 4.3 consider the boundary problem

$$\frac{dx}{dt} = i\lambda \mathscr{J}\mathscr{H}(t)x, \quad x(0) + x(T) = f,$$

where

$$f = \begin{pmatrix} f_1 \\ f_2 \end{pmatrix} \quad (f_1, f_2 \in \mathfrak{H}), \qquad \mathscr{J} = \begin{pmatrix} 0 & -iI \\ iI & 0 \end{pmatrix},$$

9) In the metric $d(\mathscr{P}_1, \mathscr{P}_2) = \int_0^T \|\mathscr{P}_1(t) - \mathscr{P}_2(t)\| dt$.

10) A generalization and sharpening of this theorem as well as a mechanical interpretation of it are given in Exercise 8.

11) I.e. $\mathscr{P}(t) \geqslant 0$ and $\mathscr{P}_{av} \gg 0$.

and find its regular points λ.

Recalling that $x = \binom{y_1}{y_2}$ and $i\mathcal{J}\mathcal{H} = \begin{pmatrix} 0 & I \\ -\mathcal{P} & 0 \end{pmatrix}$, we reduce this problem to the form

$$\frac{dy_1}{dt} = \lambda y_2; \quad \frac{dy_2}{dt} = -\lambda\mathcal{P}(t)y_1,$$

$$y_1(0) + y_1(T) = f_1; \quad y_2(0) + y_2(T) = f_2,$$

or, finally, by setting $y = y_1/\lambda$ and $y' = y_2$, to the form

$$\begin{gathered} \frac{d^2y}{dt^2} + \mu\mathcal{P}(t)y = 0, \\ y(0) + y(T) = g_1, \quad y'(0) + y'(T) = g_2, \end{gathered} \tag{5.3}$$

where in the present case $g_1 = f_1/\lambda$, $g_2 = f_2$, $\lambda = \sqrt{\mu}$.

In accordance with the definition adopted earlier it is natural to regard a given value of μ as a *regular point of problem* (5.3) if there exists a unique solution $y(t)$ for any $g_1, g_2 \in \mathfrak{H}$ that together with its derivative $y'(t)$ depends continuously on g_1 and g_2.

On the basis of the above arguments we immediately deduce the following result from Theorem 4.4.

THEOREM 5.2. *Suppose $\mathcal{P}(t)$ is an operator function of positive type. If μ_1 is the minimal point of the spectrum of problem* (5.3), *every point of the open interval* $(0, \mu_1)$ *is a strong stability point of the equation*

$$d^2x/dt^2 + \mu\mathcal{P}(t)x = 0.$$

2. *Generalization of Ljapunov's test.* In this subsection we derive an effective estimate for the central stability band of equation (5.1). To this end we consider in more detail the boundary problem (5.3). By means of the substitution

$$y(t) = z(t) + \phi(t),$$

where $\phi(t) = (f_1/2\lambda - (T/4)f_2) + \frac{1}{2}tf_2$, we reduce it to the problem of finding a function $z(t)$ satisfying the equation

$$\frac{d^2z}{dt^2} + \mu\mathcal{P}(t)z = -\mu\mathcal{P}(t)\phi(t) \tag{5.4}$$

and the homogeneous boundary conditions

$$z(0) + z(T) = 0, \quad z'(0) + z'(T) = 0. \tag{5.5}$$

We now take advantage of the fact that the solution of the differential equation

$$-(d^2z/dt^2) = h(t) \tag{5.6}$$

with conditions (5.5) can be written in the form

$$z(t) = \int_0^T g(t-s)h(s)ds,$$

where $g(t) = T/4 - |t|/2$ is the Green function of problem (5.6), (5.5) (see, for example, R. Courant and D. Hilbert [1]).

It is easily verified that the Green function $g(t)$ has the series representation

$$g(t) = \sum_{k=-\infty}^{\infty} c_k e^{i(2k+1)t}, \tag{5.7}$$

where $c_k = (T/\pi^2)(2k+1)^2$. We note that

$$\sum_{k=-\infty}^{\infty} c_k = g(0) = T/4.$$

Setting $h(t) = \mu\mathscr{P}(t)(z+\phi)$, we conclude that the boundary problem (5.4), (5.5) is equivalent to the integral equation

$$z(t) - \mu\int_0^T g(t-s)\mathscr{P}(s)z(s)ds = \mu\int_0^T g(t-s)\mathscr{P}(s)\phi(s)ds. \tag{5.8}$$

We consider the Hilbert space $\bar{\mathfrak{H}}$ of functions $z(t)$ $(0 \leqq t \leqq T)$ with values in $\mathfrak{H}$ and scalar square

$$\langle\!\langle z\rangle\!\rangle^2 = \langle z, z\rangle = \int_0^T (\mathscr{P}(s)z(s), z(s))ds = \int_0^T \|\mathscr{P}^{1/2}(s)z(s)\|^2 ds.$$

In addition, we consider the Banach space $\mathfrak{B}$ of continuous functions $z(t)$ with norm $|\!|\!|z|\!|\!| = \max_{0\leqq t\leqq T}\|z(t)\|$.

For $z \in \mathfrak{B}$ we have

$$\langle\!\langle z\rangle\!\rangle^2 \leqq \int_0^T \|\mathscr{P}(s)\|\,\|z(s)\|^2 ds \leqq |\!|\!|z|\!|\!|^2 \int_0^T \|\mathscr{P}(s)\|ds. \tag{5.9}$$

We consider the operator

$$Kz = \int_0^T g(t-s)\mathscr{P}(s)z(s)ds$$

and verify that it acts from $\bar{\mathfrak{H}}$ into $\mathfrak{B}$.

In fact, for any $\psi \in \mathfrak{H}$

$$\begin{aligned}
|(Kz,\psi)|^2 &= \left|\int_0^T g(t-s)(\mathscr{P}(s)z(s),\psi)ds\right|^2 \\
&\leqq \frac{T^2}{16}\left\{\int_0^T \|\mathscr{P}^{1/2}(s)z(s)\|\cdot\|\mathscr{P}^{1/2}(s)\psi\|ds\right\}^2 \\
&\leqq \frac{T^2}{16}\int_0^T \|\mathscr{P}^{1/2}(s)z(s)\|^2 ds\cdot\int_0^T \|\mathscr{P}^{1/2}(s)\psi\|^2 ds \\
&\leqq \frac{T^2}{16}\|\psi\|^2\int_0^T \|\mathscr{P}(s)\|ds\cdot\langle\!\langle z\rangle\!\rangle^2,
\end{aligned}$$

which implies

$$|||Kz||| = \sup_{0 \leq t \leq T} \|Kz\| \leqq \frac{T}{4} \sqrt{\int_0^T \|\mathscr{P}(s)\| ds} \quad \langle\langle z \rangle\rangle. \tag{5.10}$$

It is now obvious how to prove the continuity of $(Kz)(t)$.

Let C denote the norm of the operator K acting from $\bar{\mathfrak{H}}$ into $\mathfrak{B}$: $C = \sup_{z \in \mathfrak{H}} |||Kz|||/\langle\langle z \rangle\rangle$, and let $C_1 = \|K\|$ denote the norm of this operator in the space $\bar{\mathfrak{H}}$.

Equation (5.8) can be written in the form

$$z - \mu Kz = \mu K\phi. \tag{5.11}$$

If the condition

$$(0 <)\mu < 1/C_1 \tag{5.12}$$

is satisfied, there exists a unique solution of equation (5.11). This solution is represented by the series $z = \sum_1^\infty \mu^s K^s \phi$, which converges in the space B by virtue of the majorization

$$|||z||| \leqq C \sum_{s=1}^{\infty} \mu^s C_1^{s-1} \langle\langle \phi \rangle\rangle \leqq \frac{\mu C}{1 - \mu C_1} \langle\langle \phi \rangle\rangle. \tag{5.13}$$

Recalling the expression for $\phi(t)$, we find that

$$\begin{aligned} \langle\langle \phi \rangle\rangle^2 &= \int_0^T \left\{ \frac{1}{2\lambda} (\mathscr{P}(s) f_1, f_1) - \frac{T}{4} (\mathscr{P}(s) f_2, f_2) + \frac{t}{2} (\mathscr{P}(s) f_2, f_2) \right\} ds \\ &\leqq \frac{T}{2\lambda} (\mathscr{P}_{\mathrm{av}} f_1, f_1) + \frac{T^2}{4} (\mathscr{P}_{\mathrm{av}} f_2, f_2) \quad (\lambda = \sqrt{\mu}). \end{aligned} \tag{5.14}$$

Estimates (5.13) and (5.14) show that the solution of problem (5.4), (5.5), and hence also the solution of problem (5.3), depends continuously on f_1 and f_2 in the norm $|||y||| = \sup \|y(t)\|$.

Differentiating relation (5.8) with respect to t, we obtain for the derivative $z'(t)$ the expression

$$z'(t) = -\mu \int_0^T \operatorname{sign}(t - s) \mathscr{P}(s) z(s) ds - \mu \int_0^T \operatorname{sign}(t - s) \mathscr{P}(s) \phi(s) ds$$

and the estimate

$$|||z_2' - z_1'||| \leqq \mu \int_0^T \|\mathscr{P}(s)\| ds \, |||z_2 - z_1|||,$$

from which it follows that the derivative $y'(t)$ also depends continuously on f_1 and f_2. But this means that the points $\mu \in (0, 1/C_1)$ are regular points of problem (5.3).

On the basis of Theorem 5.2 we can formulate the following result.

LEMMA 5.1. *The points* μ *of the interval* $(0, 1/\|K\|)$ *are strong stability points of equation* (5.1).

From the above arguments (see formulas (5.9) and (5.10)) it follows that

$$\|K\| \leqq C\left[\int_0^T \|\mathscr{P}(s)\|ds\right]^{1/2} \leqq \frac{T}{4}\int_0^T \|\mathscr{P}(s)\|ds.$$

On the basis of Lemma 5.1 we conclude that the interval

$$0 < \mu < \frac{1}{\frac{T}{4}\int_0^T \|\mathscr{P}(s)\|ds}$$

consists of strong stability points.

But this estimate is too rough, and we now show that it can be replaced by the more exact estimate

$$0 < \mu < \frac{1}{\frac{T}{4}\left\|\int_0^T \mathscr{P}(s)ds\right\|}.$$

THEOREM 5.3. *Suppose* $\mathscr{P}(t)$ *is an operator function of positive type. Then every point* $\mu \in (0, 4/T^2\|\mathscr{P}_{\text{av}}\|)$ *is a strong stability point of equation* (5.1).

PROOF. By virtue of Lemma 5.1 it suffices to show that $\|K\| \leqq T^2\|\mathscr{P}_{\text{av}}\|/4$.

Using the expansion (5.7), we write the operator K in the form

$$K = \sum_{j=-\infty}^{\infty} c_j K_j, \tag{5.15}$$

where

$$K_j y = e^{i(2j+1)t}\int_0^T e^{-i(2j+1)s}\mathscr{P}(s)y(s)ds.$$

The operators K_j have the following properties.

a) The range $\mathfrak{R}(K_j)$ of an operator K_j consists of elements of the form $fe^{i(2j+1)t}$, where f is a constant element of $\mathfrak{H}$.

b) On its range $\mathfrak{R}(K_j)$ an operator K_j has the form $K_j z = T\mathscr{P}_{\text{av}}z$.

For if $z \in \mathfrak{R}(K_j)$ then $z = fe^{i(2j+1)t}$ and

$$K_j z = e^{i(2j+1)t}\int_0^T e^{-i(2j+1)s}\mathscr{P}(s)fe^{i(2j+1)s}ds$$

$$= e^{i(2j+1)t}\int_0^T \mathscr{P}(s)dsf = \int_0^T \mathscr{P}(s)dsz = T\mathscr{P}_{\text{av}}z.$$

c) The K_j are Hermitian operators in $\tilde{\mathfrak{H}}$.

For

$$\langle K_j z, z\rangle = \int_0^T \Big(\mathscr{P}(t)e^{i(2j+1)t}\int_0^T e^{-i(2j+1)s}\mathscr{P}(s)z(s)ds,\ z(t)\Big)dt$$

$$= \Big(\int_0^T e^{-i(2j+1)s}\mathscr{P}(s)z(s)ds,\ \int_0^T e^{-i(2j+1)t}\mathscr{P}(t)z(t)dt\Big)$$

$$= \Big\|\int_0^T e^{-i(2j+1)s}\mathscr{P}(s)z(s)ds\Big\|^2.$$

In order to calculate the norm of an operator K_j we write an arbitrary element $x \in \tilde{\mathfrak{H}}$ in the form $x = x_1 + x_2$, where $x_1 \in \mathfrak{R}(K_j)$ and $x_2 \perp \mathfrak{R}(K_j)$.

Since then $K_j x_2 = 0$ and $x_1 = f_1 e^{i(2j+1)t}$, where $f_1 \in \mathfrak{H}$, we have

$$\langle K_j x, x\rangle = \langle K_j x_1, x_1\rangle = T\langle \mathscr{P}_{av} x_1, x_1\rangle$$

$$= T\langle \mathscr{P}_{av} e^{i(2j+1)t} f_1, e^{i(2j+1)t} f_1\rangle = T\langle \mathscr{P}_{av} f_1, f_1\rangle$$

$$= T\int_0^T (\mathscr{P}(s)\mathscr{P}_{av} f_1, f_1)ds = T^2\|\mathscr{P}_{av} f_1\|^2$$

and

$$\langle x, x\rangle \geqq \langle x_1, x_1\rangle = \langle f_1, f_1\rangle = \int_0^T (\mathscr{P}(s)f_1, f_1)ds = T(\mathscr{P}_{av} f_1, f_1).$$

Therefore

$$\|K_j\| = \sup_x \frac{\langle K_j x, x\rangle}{\langle x, x\rangle} \leqq \sup_{x_1} \frac{\langle K_j x_1, x_1\rangle}{\langle x_1, x_1\rangle} = T\sup_{f_1} \frac{\|\mathscr{P}_{av} f_1\|^2}{(\mathscr{P}_{av} f_1, f_1)} = T\|\mathscr{P}_{av}\|.$$

From formula (5.15) we obtain the estimate

$$\|K\| \leqq \sum_j |c_j|\|K_j\| \leqq T\|\mathscr{P}_{av}\|\cdot \sum_j |c_j| = \frac{T^2}{4}\|\mathscr{P}_{av}\|,$$

which proves the theorem.

§ 6. Expansion of the logarithm of the monodromy operator in powers of a small parameter

1. *The equation with a periodic operator function analytically depending on a parameter.* We consider in a Banach phase space $\mathfrak{B}$ the differential equation

$$dx/dt = A(t;\lambda)x \tag{6.1}$$

with a periodic operator function

$$A(t+T;\lambda) = A(t;\lambda) \qquad (-\infty < t < \infty;\ |\lambda| < R)$$

which in some disk $|\lambda| < R$ is a holomorphic vector function with values in the space $L_T[\mathfrak{B}]$ of all periodic locally integrable operator functions $A(t) = A(t+T)$ with norm

$$|||A|||_L = \int_0^T \|A(t)\|dt.$$

Under the indicated assumptions the operator function $A(t; \lambda)$ will have an expansion

$$A(t; \lambda) = \sum_{k=0}^{\infty} A_k(t)\lambda^k$$

with coefficients $A_k \in L_T[\mathfrak{B}]$ $(k = 0, 1, \cdots)$ which converges absolutely in the metric of $L_T[\mathfrak{B}]$ and uniformly in λ in each disk $|\lambda| \leqq R - \varepsilon$ $(\varepsilon > 0)$.

From the expansion (III.1.11) and the corresponding estimates it readily follows that the Cauchy operator $U(t; \lambda)$ $(0 \leqq t \leqq T)$ of equation (6.1) will be a holomorphic vector function in the disk $|\lambda| < R$ with values in the space $C_T[\mathfrak{B}]$ of continuous functions $F(t)$ $(0 \leqq t \leqq T)$ with values in $[\mathfrak{B}]$ and norm

$$|||F|||_C = \max_{0 \leqq t \leqq T} \|F(t)\|.$$

Therefore

$$U(t; \lambda) = \sum_{k=0}^{\infty} U_k(t)\lambda^k, \tag{6.2}$$

where $U_k \in C_T[\mathfrak{B}]$ $(k = 0,1,\cdots)$, the series in the right side of (6.2) converging absolutely and uniformly in $t \in [0, T]$ and λ in a disk $|\lambda| \leqq R - \varepsilon$ $(\varepsilon > 0)$.

We consider the monodromy operator $U_0(T) = U(T; 0)$ of equation (6.1) for $\lambda = 0$. It will be assumed below that its spectrum does not surround the origin and hence that there exists a simple curve l that goes from the origin to infinity without intersecting the spectrum $\sigma(U_0(T))$ of $U_0(T)$.

We will further understand that the function $\ln \mu$ is a well-defined branch of the logarithm that is single valued in the plane cut by the curve l. Let γ be a simple closed contour lying in this cut plane and surrounding the spectrum $\sigma(U_0(T))$. The spectrum of $U(T; \lambda)$ will also lie inside this contour for sufficiently small $|\lambda|$. For such λ it is therefore possible to construct the operator function

$$\Gamma(\lambda) = \frac{1}{T} \ln U(T; \lambda) = -\frac{1}{2\pi i T} \oint_\gamma \ln \mu (U(T; \lambda) - \mu I)^{-1} d\mu, \tag{6.3}$$

which will obviously be a holomorphic function of λ in the small neighborhood of the origin in question.

Let R_1 denote the radius of the maximal disk with center at the origin in which $\Gamma(\lambda)$ will be a holomorphic function. We will then have

$$\Gamma(\lambda) = \Gamma_0 + \lambda\Gamma_1 + \lambda^2\Gamma_2 + \cdots, \tag{6.4}$$

where $\Gamma_k \in [\mathfrak{B}]$ ($k = 0, 1, \cdots$), and the expansion converges in $[\mathfrak{B}]$ absolutely and uniformly in each disk $|\lambda| \leqq R_1 - \varepsilon$ ($\varepsilon > 0$).

In general, $R_1 < R$.

If the space $\mathfrak{B}$ is a Hilbert space $\mathfrak{H}$, the localization theorems of § 3 can be used to estimate the radius R_1 from below (see Remark 3.1).

It is not difficult to see that the estimate

$$R_1 \geqq R_0 = \sup \left\{ \rho \;\middle|\; \sup_{|\lambda| \leqq \rho} \int_0^T \sqrt{\|A(t;\lambda)\|^2 - [\lambda_r(A_R(t;\lambda))]^2}\, dt < \pi \right\}$$

is valid if the set indicated by the braces is not empty.

In fact, for the points λ of a disk of radius $r < R_0$ the spectrum $\sigma(U(T;\lambda))$ lies outside an angle containing the negative halfline.

A rougher estimate is

$$R_1 \geqq \sup \left\{ \rho \;\middle|\; \sup_{|\lambda| \leqq \rho} \int_0^T \|A(t;\lambda)\| dt < \pi \right\}.$$

We note that if $\mathfrak{B}$ is a Hilbert space $\mathfrak{H}$ and for some Hermitian operator $W \in [\mathfrak{H}]$ the operator $A(t;\lambda)$ is W-skew-Hermitian for real λ it will follow that $\Gamma(\lambda)$ is also a W-skew-Hermitian operator (see Theorem 1.3) for real λ ($|\lambda| < R_1$). But then the operators $\Gamma_k = (k!)^{-1}\Gamma^{(k)}(0)$ ($k = 1, 2, \cdots$) will also have this property.

We consider the operator function

$$Q(t;\lambda) = U(t;\lambda)e^{-t\Gamma(\lambda)}. \tag{6.5}$$

This operator function is periodic with period T (see § 1). From (6.2) and (6.4) it follows that it is analytic for $|\lambda| < R_1$ and the series

$$Q(t;\lambda) = \sum_{k=0}^{\infty} Q_k(t)\lambda^k \tag{6.6}$$

converges uniformly for $t \in (-\infty, \infty)$ and $|\lambda| < R_1 - \varepsilon$ ($\varepsilon > 0$).

We note that

$$Q_k(t + T) = Q_k(t) \qquad (-\infty < t < \infty). \tag{6.7}$$

2. *Simplest case*. We first consider in more detail an equation of the form

$$dx/dt = \lambda A(t)x. \tag{6.8}$$

Here $A_0(t) = 0$, $A_1(t) = A(t)$ and $A_k(t) = 0$ ($k = 2, 3, \cdots$).

In the case of a Hilbert phase space estimate (3.3) implies the following estimate for the radius of convergence of the series (6.4):

$$R_1 \geqq \pi \Big/ \int_0^T \|A(t)\| dt.$$

The monodromy operator corresponding to $\lambda = 0$ is converted into the identity operator: $U_0(T) \equiv I$, and hence

$$\Gamma_0 = \ln I = 0. \tag{6.9}$$

Differentiating (6.5), we obtain the equation

$$dQ(t;\lambda)/dt = \lambda A(t)Q(t;\lambda) - Q(t;\lambda)\Gamma(\lambda). \tag{6.10}$$

We substitute expansions (6.4) and (6.6) in it and equate the coefficients of like terms in the left and right sides.

Taking into account (6.9), we obtain first of all the relation $dQ_0(t)/dt = 0$, which implies the identity $Q_0(t) \equiv I$ since $Q_0(0) = I$. We write the remaining relations in the form

$$\frac{dQ_1(t)}{dt} = A(t) - \Gamma_1; \tag{6.11}$$

$$\frac{dQ_k(t)}{dt} = A(t)Q_{k-1}(t) - \sum_{s=1}^{k-1} Q_s(t)\Gamma_{k-s} - \Gamma_k \qquad (k > 1). \tag{6.12}$$

The integrals of these relations should satisfy the conditions

$$Q_k(T) = Q_k(0) = 0 \qquad (k = 1,2,\cdots), \tag{6.13}$$

which follow from the periodicity of $Q(t)$ and the identity $Q(0;\lambda) \equiv I$.

Relations (6.11)–(6.13) permit us to successively determine all of the desired coefficients Γ_k and Q_k. From (6.11) and (6.13) it follows that

$$Q_1(t) = \int_0^t [A(t) - \Gamma_1]dt, \tag{6.14}$$

and hence, since $Q_1(T) = 0$,

$$\Gamma_1 = \frac{1}{T}\int_0^T A(t)dt. \tag{6.15}$$

In the same way, if all of the coefficients up to order $k-1$ have been found, we have

$$Q_k(t) = \int_0^t \left[A(t)Q_{k-1}(t) - \sum_{s=1}^{k-1} Q_s(t)\,\Gamma_{k-s} - \Gamma_k\right] dt \tag{6.16}$$

and

$$\Gamma_k = \frac{1}{T}\int_0^T \left[A(t)Q_{k-1}(t) - \sum_{s=1}^{k-1} Q_s(t)\Gamma_{k-s}\right] dt. \tag{6.17}$$

3. *Case when $A(t)$ is holomorphic in the upper halfplane*. We consider an interesting special case. Suppose the Fourier series of the periodic operator function $A(t)$ does not contain terms with negative indices:

$$A(t) \sim \sum_{m=0}^{\infty} A_m e^{m(2\pi i/T)t}. \tag{6.18}$$

From (6.15) it follows that $\Gamma_1 = A_0$, while from (6.14) and (6.18) we get

$$Q_1(t) = \int_0^t \sum_{m=1}^{\infty} A_m e^{m(2\pi i/T)t}\, dt = \sum_{m=1}^{\infty} \frac{T}{2\pi i} \frac{A_m}{m} (e^{m(2\pi i/T)t} - 1).$$

We could carry out the calculations indicated by formulas (6.16) and (6.17), but there is another approach of interest to us.

An interesting modification of the method under consideration consists in replacing the conditions $Q_k(0) = 0$ by the conditions

$$\int_0^T Q_k(t)dt = 0 \qquad (k = 1,2,\cdots). \tag{6.19}$$

Relations (6.16) and (6.17) then take the form

$$\tilde{Q}_k(t) = \int_0^t \Big[A(t)\tilde{Q}_{k-1}(t) - \sum_{s=1}^{k-1} \tilde{Q}_s(t)\tilde{\Gamma}_{k-s} - \tilde{\Gamma}_k \Big] dt + \tilde{Q}_k(0), \tag{6.16a}$$

where $\tilde{Q}_k(0)$ is determined from conditions (6.19), and

$$\tilde{\Gamma}_k = \frac{1}{T} \int_0^T \Big[A(t)\tilde{Q}_{k-1}(t) - \sum_{s=1}^{k-1} \tilde{Q}_s(t)\tilde{\Gamma}_{k-s} \Big] dt. \tag{6.17a}$$

In particular, we again have $\tilde{\Gamma}_1 = A_0$, but now

$$\tilde{Q}_1(t) = \sum_{m=1}^{\infty} \frac{T}{2\pi i} \frac{A_m}{m} e^{m(2\pi i/T)t}.$$

This function differs from $Q_1(t)$ in that its Fourier series contains only terms with positive indices. From (6.18) and (6.16a) it follows that this property will also be possessed by all subsequent functions $\tilde{Q}_k(t)$. But then the integrand in (6.17a) has a Fourier series without a free term, and therefore $\tilde{\Gamma}_k = 0$ $(k = 2, 3,\cdots)$.

Thus under this method of calculating

$$\tilde{\Gamma}(\lambda) = \lambda\Gamma_1 = \lambda A_0.$$

But, in contrast to the case when conditions (6.13) are adopted, we can no longer assert that the series

$$\tilde{Q}(t;\lambda) = \sum_{k=0}^{\infty} \tilde{Q}_k(t)\lambda^k \tag{6.20}$$

converges in any neighborhood whatsoever of the point $\lambda = 0$.

If this series nevertheless converges in some neighborhood of the origin, it gives a solution of the equation

$$d\tilde{Q}(t;\lambda)/dt = \lambda A(t)\tilde{Q}(t;\lambda) - \lambda\tilde{Q}(t;\lambda)A_0$$

(cf. equation (6.10)). Hence

$$\tilde{Q}(t;\lambda) = U(t;\lambda)\tilde{Q}(0;\lambda)e^{-\lambda A_0 t}.$$

The latter formula leads to the representation

$$U(t;\lambda) = \tilde{Q}(t;\lambda)e^{\lambda A_* t}\tilde{Q}^{-1}(0;\lambda). \tag{6.21}$$

These arguments can be interpreted as follows. We put $e^{2\pi it/T} = z$ in (6.18). Then equation (6.8) can be written for $\lambda = 2\pi i/T$ in the form

$$z\frac{dx}{dz} = \left(\sum_{k=0}^{\infty} A_k z^k\right)x, \tag{6.22}$$

and formula (6.21), when it makes sense, leads to the representation of a fundamental operator of this equation in the form $U(z) = V(z)z^{A_*}C$, where $V(z)$ is a holomorphic operator function in a neighborhood of $z = 0$.

Similar results are rigorously obtained in the next chapter, where equation (6.22) will be studied in detail under certain additional assumptions in the class of analytic operator functions.

4. *Determination of the coefficients of the expansions in the general case.* We will assume that the Cauchy operator $U(t; 0)$ of the equation resulting from (6.1) for $\lambda = 0$ is known, which implies that the operators $\Gamma_0 = T^{-1}\ln U(T; 0)$ and $Q_0(t)$ are known. As long as there exists a ray emanating from the origin that does not intersect the spectrum $\sigma(U(T; 0))$ it can be assumed that the spectrum of Γ_0 has the property

$$|\mathrm{Im}(\lambda - \mu)| < 2\pi/T \qquad (\lambda, \mu \in \sigma(\Gamma_0)). \tag{6.23}$$

We will indicate below a method of successively calculating the other coefficients of the expansions (6.4) and (6.6).

It will be convenient for us to consider in place of $Q(t;\lambda)$ the function

$$S(t;\lambda) = Q_0^{-1}(t)Q(t;\lambda).$$

From (6.6) we have the expansion

$$S(t;\lambda) = I + \sum_{k=1}^{\infty} \lambda^k S_k(t), \tag{6.24}$$

where

$$S_k(t) = Q_0^{-1}(t)Q_k(t). \tag{6.25}$$

Differentiating (6.5), we obtain the equation

$$dQ(t;\lambda)/dt = A(t;\lambda)Q(t;\lambda) - Q(t;\lambda)\Gamma(\lambda); \tag{6.26}$$

and in particular, for $\lambda = 0$,

$$dQ_0(t)/dt = A_0(t)Q_0(t) - Q_0(t)\Gamma_0. \tag{6.27}$$

Since

$$Q(t;\lambda) = Q_0(t)S(t;\lambda), \tag{6.28}$$

it follows from (6.26) that $S(t;\lambda)$ satisfies the equation

$$dS(t;\lambda)/dt = [Q_0^{-1}(t)A(t;\lambda)Q_0(t) - Q_0^{-1}(t)Q_0'(t)]S(t;\lambda) - S(t;\lambda)\Gamma(\lambda),$$

which after an application of (6.27) takes the form

$$dS(t;\lambda)/dt = \Gamma_0 S(t;\lambda) - S(t;\lambda)\Gamma_0 + \left[\sum_{k=1}^{\infty}\lambda^k Q_0^{-1}(t)A_k(t)Q_0(t)\right]S(t;\lambda) - S(t;\lambda)\sum_{k=1}^{\infty}\lambda^k\Gamma_k. \tag{6.29}$$

We substitute the expansion (6.24) in (6.29) and equate the coefficients of like powers of λ. This gives us the system of equations

$$dS_k(t)/dt = \Gamma_0 S_k(t) - S_k(t)\Gamma_0 + F_k(t) - \Gamma_k \qquad (k = 1,2,\cdots), \tag{6.30}$$

where

$$F_k(t) = Q_0^{-1}(t)A_k(t)Q_0(t) + \sum_{j=1}^{k-1} Q_0^{-1}(t)A_j(t)Q_0(t)S_{k-j}(t) - \sum_{j=1}^{k-1} S_j(t)\Gamma_{k-j}. \tag{6.31}$$

Inasmuch as the operator function $F_k(t)$ depends only on those $S_j(t)$ with indices $1 \leqq j \leqq k-1$, the equations (6.30) are recursion equations and can be solved successively. Let us describe how this is done.

From the condition $S(0;\lambda) = Q_0^{-1}(0)Q(0,\lambda) \equiv I$ it follows that equation (6.30) should be considered under the condition

$$S_k(0) = 0 \qquad (k = 1,2,\cdots). \tag{6.32}$$

The solution of equation (6.30) under condition (6.32) is given by formula (II.1.11) and has the form

$$S_k(t) = \int_0^t e^{\Gamma_0(t-\tau)}[F_k(\tau) - \Gamma_k]e^{-\Gamma_0(t-\tau)}d\tau \qquad (k = 1,2,\cdots). \tag{6.33}$$

We note further that (6.7) and (6.25) imply that the operator functions $S_k(t)$ $(k = 1,2,\cdots)$ are periodic, and hence by virtue of (6.32)

$$S_k(T) = 0 \qquad (k = 1,2,\cdots). \tag{6.34}$$

Substituting expression (6.33) for $S_k(t)$ in (6.34), we obtain the relation

$$\int_0^T e^{\Gamma_0(T-\tau)}\Gamma_k e^{-\Gamma_0(T-\tau)}d\tau = \int_0^T e^{\Gamma_0(T-\tau)}F_k(\tau)e^{-\Gamma_0(T-\tau)}d\tau, \tag{6.35}$$

which can be regarded as an equation for the operator Γ_k.

Suppose now that all of the operators $S_j(\tau)$ and Γ_j $(j = 1,\cdots, k-1)$ have been determined. Then owing to condition (6.23) we can find Γ_k from (6.35) with the use of formula (I.3.17):

$$
\begin{aligned}
\Gamma_k = & -\frac{1}{4\pi^2}\oint_{\gamma_0}\oint_{\gamma_0}\frac{\lambda-\mu}{e^{(\lambda-\mu)T}-1}(\Gamma_0-\lambda I)^{-1} \\
& \times\left(\int_0^T e^{-\Gamma_0\tau}F_k(\tau)e^{\Gamma_0\tau}d\tau\right)(\Gamma_0-\mu I)^{-1}d\lambda\,d\mu \\
= & -\frac{1}{4\pi^2}\int_0^T\oint_{\gamma_0}\oint_{\gamma_0}\frac{(\lambda-\mu)e^{-(\lambda-\mu)\tau}}{e^{(\lambda-\mu)T}-1}(\Gamma_0-\lambda I)^{-1} \\
& \times F_k(\tau)(\Gamma_0-\mu I)^{-1}d\lambda\,d\mu\,d\tau,
\end{aligned}
\tag{6.36}
$$

where γ_0 is a contour surrounding $\sigma(\Gamma_0)$ sufficiently closely.

After this, formula (6.33) permits us to find $S_k(t)$. Thus we can actually successively calculate the coefficients of the expansions (6.4) and (6.24) with sufficiently large indices and thereby find to within any degree of accuracy approximate expressions for the operators $Q(t;\lambda)$ and $\ln U(T;\lambda)$.

5. *Use of Fourier series.* There exists another method of solving system (6.30) which is based on an application of Fourier series.

Suppose that to the operator functions $F_k(t)$ and $S_k(t)$ there correspond the Fourier series

$$F_k(t)\sim\sum_{p=-\infty}^{\infty}C_pe^{2p\pi it/T},\qquad \text{where } C_p=\frac{1}{T}\int_0^T F_k(t)e^{-2p\pi it/T}\,dt,$$

and

$$S_k(t)\sim\sum_{p=-\infty}^{\infty}D_pe^{2p\pi it/T},\tag{6.37}$$

where $D_p=(1/T)\int_0^T S_k(t)e^{-2p\pi it/T}\,dt$.

Multiplying equation (6.30) by $e^{-2p\pi it/T}$ and integrating with respect to t from 0 to T, we find that the operator D_p $(p\neq 0)$ satisfies the equation

$$(2\pi pi/T)D_p-\Gamma_0D_p+D_p\Gamma_0=C_p\qquad (p\neq 0).\tag{6.38}$$

From condition (6.23) it follows that $2\pi pi/T-\lambda+\mu\neq 0$ for $\lambda,\ \mu\in\sigma(\Gamma_0)$, and therefore equation (6.38) has the unique solution

$$D_p=-\frac{1}{4\pi^2}\oint_{\gamma_0}\oint_{\gamma_0}\frac{(\Gamma_0-\lambda I)^{-1}C_p(\Gamma_0-\mu I)^{-1}}{2\pi pi/T-\lambda+\mu}\,d\lambda\,d\mu.\tag{6.39}$$

The series (6.37) converges for each t since $S_k(t)$ is an absolutely continuous function (see (6.30)).

If $\mathfrak{B}=\mathfrak{H}$ and the function $\|A(t)x\|$ $(0\leqq t\leqq T)$ belongs to $L_2(0,T)$ (and hence $F_k(t)x\in L_2(0,T)$) for each $x\in\mathfrak{H}$, the series (6.37) converges absolutely and uniformly in t for each $x\in\mathfrak{H}$. For when $N>0$ is sufficiently large the inequality

$$\|C_px\|(2\pi pi/T-\lambda+\mu)^{-1}\leqq N\,\|C_px\|\,|p|^{-1}\tag{6.40}$$

holds for any $\lambda,\ \mu\in\sigma(\Gamma_0)$ and integral p.

As a result, the series

$$\sum_p C_p x(2\pi pi/T - \lambda + \mu)^{-1} e^{2\pi pit/T}, \tag{6.41}$$

which is majorized by the series

$$N\Big(\sum_p \|C_p x\| |p|^{-1}\Big) \leqq N\Big(\sum_p p^{-2}\Big)^{1/2}\Big(\sum_p \|C_p x\|^2\Big)^{1/2}$$

$$< \text{const}\Big(\int_0^T \|F_k(t)x\|^2 dt\Big)^{1/2} < \infty.$$

converges uniformly in t, λ, μ for any $x \in \mathfrak{H}$.

It remains to note that the series (6.37) is obtained from (6.41) by multiplying by bounded operator functions and integrating.

We must still determine the coefficient D_0. It is obtained from the relation $S_k(0) = \sum_p D_p = 0$, which reduces to the equality $D_0 = -\sum_{p\neq 0} D_p$.

Finally, integrating (6.30) from 0 to T under the condition $\int_0^T (dS_k(t)/dt)\, dt = 0$, we find the operator Γ_k:

$$\Gamma_k = \frac{1}{T}\int_0^T [\Gamma_0 S_k(t) - S_k(t)\Gamma_0 + F_k(t)]\, dt = \Gamma_0 D_0 - D_0\Gamma_0 + C_0.$$

We note that the series (6.41) can be summed. It can be verified that we arrive in this way at formula (6.33).

EXERCISES

1. Consider the T-periodic equation

$$dx/dt = A(t)x \tag{0.1}$$

in a *real* Hilbert space $\mathfrak{H}$.

a) Show that the Floquet representation holds for equation (0.1) if the spectrum of its monodromy operator does not contain the point $\lambda = -1$ and does not surround the origin.

Hint. Make use of the symmetry of the spectrum of the real operator $U(T)$ relative to the real axis and construct a symmetric contour separating this spectrum from the origin. Use this contour to calculate $\ln U(T)$.

b) Show (see J. Massera and J. Schäffer [1]) that the Floquet representation holds for (0.1) if

$$\int_0^T \|A(\tau)\| d\tau < \pi.$$

Hint. Make use of the result of the preceding exercise. We note that the corresponding fact for a complex space is indicated in Remark 3.1.

c) Suppose the real Hilbert space $\mathfrak{H}$ is finite dimensional. If $-1 \in \sigma(U(T))$, there exists a real Floquet representation of order 2, i.e. a representation $U(t) = Q(t)\exp(\Gamma t)$ with real $Q(t)$ and Γ, where $Q(t)$ is a periodic function with period $2T$: $Q(t + 2T) = Q(t)$.

Hint. Consider the operator $U^2(t)$ and show that it has a real logarithm (see E. A. Coddington and N. Levinson [1]).

2. Show that the Cauchy operator $U(t)$ of a T-periodic equation (0.1) in a phase space $\mathfrak{H}$ (complex or real) has a Floquet representation if at least one of the following two conditions is satisfied.

a) For some operator $C \in [\mathfrak{H}]$ of simplest type with spectrum contained in the sequence

$$\Big\{\frac{2j\pi i}{T}\Big|\, j = 0, \pm 1, \pm 2, \cdots\Big\}, \qquad \int_0^T \|e^{-Ct}A(t)e^{Ct} - C\| dt < \pi.$$

If $C^* = -C$, the latter condition is equivalent to the condition

$$\int_0^T \|A(t) - C\| dt < \pi.$$

Hint. Perform the substitution $U = V \exp(Ct)$ in the equation $\dot{U} = AU$ and make use of Remark 3.1 and the result of Exercise 1b).

REMARK. It is helpful to think through the meaning of the assertion of the exercise for the case when to an orthogonal decomposition $\mathfrak{H} = \mathfrak{H}_1 \oplus \mathfrak{H}_2 \oplus \cdots \oplus \mathfrak{H}_n$ there correspond the decompositions $A(t) = A_1(t) \oplus A_2(t) \oplus \cdots \oplus A_n(t)$ and $C = C_1 \oplus \cdots \oplus C_n$, where $C_k = 2\pi i T^{-1} j_k I_k$, the j_k $(k = 1, \cdots, n)$ being integers.

b) There exists an invertible operator C and a periodic scalar function $\phi/(t)$ with zero mean value for which

$$\int_0^T \|C^{-1}A(t)C - \phi(t)I\| dt < \pi.$$

Hint. Perform the substitution $U(t) = Ce^{\int_0^t \phi(\tau)d\tau} V(t)C^{-1}$ (J. Massera and J. Schäffer [1], page 353).

3. Consider in a real or complex Hilbert space $\mathfrak{H}$ with an orthonormal basis e_n $(n = 0, \pm 1, \pm 2, \cdots)$ the orthogonal or unitary operator V defined by the relation $Ve_n = e_{n+1}$ $(n = 0, \pm 1, \pm 2, \cdots)$. Let R be the operator defined by the relation

$$Re_n = \begin{cases} 0, & n < 0, \\ \varepsilon e_n, & n \geqq 0, \ \varepsilon > 0. \end{cases}$$

The operator

$$Ve^R \tag{0.2}$$

differs only by a scalar factor from the operator studied in § 1.7.3. It is known (J. Schäffer [1] and P. Halmos and G. Lumer [1]) that an operator having the property noted there (its spectrum contains an annulus consisting of points of regular type with finite deficiency) does not have a square root and hence a logarithm.

Construct a periodic equation for which (0.2) is the monodromy operator and hence for which a Floquet representation does not hold (J. Massera and J. Schäffer [1], page 354).

Hint. Consider the periodic equation with coefficient $A(t) = A(t + 1)$ defined by the formula

$$A(t) = -W - 6t(1 - t)e^{tW}Re^{-tW} \qquad (0 \leqq t \leqq 1),$$

where W is a skew-Hermitian operator satisfying the condition $e^W = V$.

4. Let $U(T)$ be the monodromy operator of a strongly stable equation $\dot{x} = i\mathcal{J}\mathcal{H}(t)x$ and let U_1 be a $\mathcal{J}$-unitary operator satisfying the condition $\|U_1 - U(T)\| < \delta$.

Show that for sufficiently small $\delta > 0$ there exists a Hermitian $\mathcal{H}_1(t)$ such $U_1 = U_1(T)$ is the monodromy operator of the equation

$$dx/dt = i\mathcal{J}\mathcal{H}_1(t)x.$$

In this connection the condition

$$\int_0^T \|\mathcal{H}(t) - \mathcal{H}_1(t)\| dt < \varepsilon$$

can be satisfied for any $\varepsilon > 0$ by choosing δ sufficiently small.

Hint. Make use of the Floquet representation and the considerations of Corollary 4.1.

5. Show that the spectrum of the monodromy operator of the equation

$$d^2y/dt^2 + \mathcal{P}(t)y = 0 \tag{0.3}$$

in $\mathfrak{H}$ with periodic coefficient $\mathscr{P}(t)$ $(= \mathscr{P}(t+T))$ lies in the annulus $1/r \leqq \rho \leqq r$, where

$$r = \exp\left(T\int_0^T \|\mathscr{P}(t)\|dt\right)^{1/2}. \qquad (0.4)$$

The estimate is exact, for example, when $\mathscr{P}(t) = -pI$ $(p > 0)$.[12]

Hint. By setting $dy/dt = lz$ $(l > 0)$ and $x = y \oplus z$, transform equation (0.3) into equation (0.1) in the phase space $\mathfrak{H} \oplus \mathfrak{H}$, where

$$\mathscr{A}(t) = \begin{pmatrix} 0 & iI \\ l^{-1}\mathscr{P}(t) & 0 \end{pmatrix}.$$

The norm $\|U(T)\|$ of the monodromy operator of equation (0.1) can be taken as r. It is estimated by means of Wintner's inequality (III.4.19), after which the estimate is minimized with respect to l.

Suppose $\mathscr{P}(t) \ll 0$ $(-\infty < t < \infty)$. It is not known in this case whether the assertion remains in force if (0.4) is replaced by the equality

$$r = \exp\left(T\left\|\int_0^T \mathscr{P}(t)dt\right\|\right)^{1/2}$$

(i.e. if the operations of integration and taking of the norm are interchanged in (0.4)).

A conjecture of Ljapunov's. Consider the scalar Hill equation

$$y'' + \lambda p(t)y = 0 \qquad (p(t+T) = p(t)). \qquad (0.5)$$

Let $\phi(t; \lambda)$ and $\psi(t; \lambda)$ be the solutions of it that satisfy the conditions

$$\phi(0; \lambda) = 1, \phi'(0; \lambda) = 0; \qquad \psi(0; \lambda) = 0, \psi'(0; \lambda) = 1.$$

The characteristic multipliers $\rho_1(\lambda)$ and $\rho_2(\lambda)$ of equation (0.5) will be the eigenvalues of the monodromy matrix

$$\begin{pmatrix} \phi(T; \lambda) & \phi'(T; \lambda) \\ \psi(T; \lambda) & \psi'(T; \lambda) \end{pmatrix}.$$

Since its determinant is equal to 1, they will be the roots of the quadratic equation

$$\rho^2 - 2A(\lambda)\rho + 1 = 0, \qquad (0'.6)$$

where $2A(\lambda) = \psi(T; \lambda) + \psi'(T; \lambda)$.

Since the free term of this equation is equal to 1, we have $\rho_1\rho_2 = 1$. Let the characteristic multipliers be renumbered so that $|\rho_1(\lambda)| \geqq |\rho_2(\lambda)|$ and consequently $|\rho_1(\lambda)| \geqq 1 \geqq 1/|\rho_1(\lambda)| = |\rho_2(\lambda)|$. According to estimate (0.4)

$$|\rho_1(\lambda)| \leqq \exp\left(|\lambda|T\int_0^T |p(s)|ds\right)^{1/2},$$

which implies, since $2A(\lambda) = \rho_1(\lambda) + 1/\rho_1(\lambda)$ and hence $2|A(\lambda)| \leqq |\rho_1(\lambda)| + 1/|\rho_1(\lambda)|$, that

$$|A(\lambda)| \leqq \cosh\left(|\lambda|T\int_0^T |p(s)|ds\right)^{1/2}. \qquad (0.7)$$

Let

$$A(\lambda) = \sum_{k=1}^{\infty} a_k(-\lambda)^k.$$

It is easily seen that

[12] In this and only this case for the scalar equation (see L. Ja Miročnik [1]).

$$a_0 = 1, \ a_1 = \frac{T}{2}\int_0^T p(s)ds. \tag{0.8}$$

A. M. Ljapunov conjectured[13] back in 1902 ([1], page 419) that

$$|a_n| \leq \frac{(T\int_0^T |p(s)|ds)^n}{(2n)!} \quad (n = 0, 1, \cdots). \tag{0.9}$$

It follows from (0.8) that this estimate is valid for $n = 0, 1$. Ljapunov showed by means of an ingenious calculation that it is also valid for $n = 2, 3$.

The assumed estimates (0.9) clearly imply estimate (0.7), but not conversely.[14]

A very simple direct proof of estimate (0.7) has recently been proposed by L. Ja. Miročnik [1].

6. Show that all of the points λ of the interval

$$0 < |\lambda| < \frac{\pi}{\int_0^T \|\mathscr{H}(s)\|ds} \tag{0.10}$$

are strong stability points for the periodic canonical equation of positive type

$$dx/dt = i\lambda \mathscr{J}\mathscr{H}(t)x \qquad (\mathscr{H}(t+T) = \mathscr{H}(t)). \tag{0.11}$$

Moreover, if condition (0.10) is satisfied, the spectrum of the strongly stable monodromy operator $U(T; \lambda)$ lies on the arc

$$\gamma = \left\{e^{i\theta} | 0 < \theta \leq \lambda \int_0^T \|\mathscr{H}(s)\|ds\right\}$$

of the unit circle, and to each of the two parts of this spectrum lying above and below the real axis there corresponds a uniformly $\mathscr{J}$-definite invariant subspace of $U(T; \lambda)$.

Hint. Make use of Remark 3.1, Theorem 4.4 and the arguments in its proof.

The assertion was obtained for the case of a Hamiltonian equation in the two-dimensional phase space R^2 by V. A. Jakubovič [7], and in a $2m$-dimensional space by M. G. Neĭgauz and V. B. Lidskiĭ [1], who used the rule of M. G. Kreĭn for the "traffic flow" of characteristic multipliers of the first and second kind.

7. Show that the equation

$$d^2y/dt^2 + \lambda\mathscr{P}(t)y = 0 \qquad (\mathscr{P}(t) = \mathscr{P}(t+T)) \tag{0.12}$$

can be transformed into the canonical form

$$dx/dt = \mathscr{J}_\Gamma(\mathscr{H}_0 + \lambda\mathscr{H}_1(t))x, \tag{0.12'}$$

where

$$\mathscr{J}_\Gamma = \begin{pmatrix} 0 & I \\ -I & 0 \end{pmatrix}, \quad \mathscr{H}_0 = \begin{pmatrix} \mathscr{P}_{av} & 0 \\ 0 & 0 \end{pmatrix}, \quad \mathscr{H}_1(t) = \begin{pmatrix} Q^2(t) & -Q(t) \\ -Q(t) & I \end{pmatrix},$$

in which

$$Q(t) = \int^t [\mathscr{P}(t) - \mathscr{P}_{av}]\, dt, \qquad Q_{av} = 0.$$

Hint. Put $z = Q(t)y \oplus (1/\lambda)\, dy/dt \ (\lambda > 0)$ and $x = \begin{pmatrix} y \\ z \end{pmatrix}$.

We note that if

---

[13] Ljapunov considered the case when $p(t) \geq 0$. In this case it follows from formulas obtained by him for the a_n that every $a_n > 0$. From these same formulas it directly follows that estimate (0.9) holds for complex $p(t)$ as long as it holds for nonnegative $p(t)$.

[14] *Note added for the translation.* The conjecture of A. M. Ljapunov was successfully proved by M. G. Kreĭn [13] at the end of 1971.

$$\mathscr{P}(t) \sim \sum_{k=-\infty}^{\infty} e^{2\pi ikt/T} C_k \qquad (C_0 = \mathscr{P}_{av}),$$

i.e.

$$C_k = \frac{1}{T}\int_0^T e^{-2\pi ikt/T}\mathscr{P}(t)dt \qquad (\in \mathfrak{H}; k = 0, \pm 1, \cdots),$$

then

$$Q(t) \sim \frac{T}{2\pi i}\sum_{k\neq 0}\frac{1}{k} e^{2\pi ikt/T} C_k.$$

In the following exercises it is assumed everywhere that $\mathscr{P}(t)$ $(= \mathscr{P}(t+T))$ and $Q(t)$ $(= Q(t+T))$ are connected by the above relations.

8. Generalize Theorem 5.1 by showing that if in the case $\mathscr{P}(t) = \mathscr{P}^*(t) = \mathscr{P}(t+T)$ at least one of the conditions

$$\mathscr{P}_{av} \gg 0 \tag{0.13}$$

or

$$((Q^2)_{av} =) \ \frac{1}{T}\int_0^T Q^2(t)dt \gg 0, \ \mathscr{P}_{av} \geqslant 0 \tag{0.14}$$

is satisfied, the points λ will be strong stability points for equation (0.12) whenever $0 < \lambda < \lambda_1$, where λ_1 is the least positive number of the spectrum of the skewperiodic boundary problem (see (5.3)).

Mechanical commentary. Let us represent $\mathscr{P}(t)$ in the form

$$\mathscr{P}(t) = \mathscr{P}_{av} + R(t).$$

Then $R_{av} = 0$. Setting $\lambda = 1/\omega^2$ and performing the substitution $t \to \omega t$, we reduce equation (0.12) to the form

$$d^2y/dt^2 + [\mathscr{P}_{av} + R(\omega t)]\, y = 0.$$

If $\mathscr{P}_{av} \gg 0$, the unperturbed equation $\ddot{y} + \mathscr{P}_{av}\, y = 0$ describes the motion of a stable conservative system. The assertion of the exercise means that, independently of the form and size of the amplitude of a parametric excitation with zero mean value, the system remains stable for an excitation of sufficiently large frequency.

Moreover, if the system is even semistable: $\mathscr{P}_{av} \geqslant 0$, the above assertion remains valid under the usually satisfied condition (0.14).

Hint. Pass from equation (0.12) to equation (0.12′). If at least one of the conditions (0.13) or (0.14) is satisfied, the operator function $\mathscr{H}_0 + \lambda\mathscr{H}_1$ is of positive type for any $\lambda > 0$. Verify that this circumstance permits one to extend the part of Theorem 4.4 concerning the right half $0 < \lambda < \lambda_1$ of the interval of regularity to the canonical equation (0.12′).

The assertion of the theorem was published for the scalar Hill equation by A. M. Ljapunov in "Soobščenijah Har'kovskogo Matematičeskogo Obščestva" back in 1896 (see [1]), and for the n-dimensional case by M. G. Kreĭn [7].

We note that if $\mathscr{P}(t) = \mathscr{P}^*(t)$ then $C_k = C^*_{-k}$ $(k = 1,2,\cdots)$ and

$$\int_0^T Q^2(t)dt = \frac{T^2}{4\pi^2}\sum_{k=1}^{\infty}\frac{C_kC_k^* + C_k^*C_k}{k^2}.$$

Condition (0.14) will be satisfied if at least one of the coefficients C_k is invertible in $[\mathfrak{H}]$.

When $\dim \mathfrak{H} < \infty$ condition (0.14) means that for any $x \in \mathfrak{H}$ the vector function $\mathscr{P}(t)x$ is different from zero on a set of positive measure.

We recall that according to Theorem 2.4, if the condition $\mathscr{P}_{av} \gg 0$ is satisfied, equation (0.12) will be e-dichotomic and a fortiori unstable for *negative* λ that are sufficiently small in absolute value; and, as we now note for the second time (see Theorem 5.1), this equation will be strongly stable for sufficiently small *positive* λ.

9. Suppose $\mathscr{P}_{av} = 0$ and $\int_0^T Q^2(t)dt \gg 0$. Show that equation (0.12) will be strongly stable whenever

$$T\int_0^T \|Q^2(t)\|dt < \frac{\pi^2}{4}. \tag{0.15}$$

Hint. Transform equation (0.12) by means of the substitution

$$lz = Q(t)y + dy/dt \qquad (l > 0)$$

into a canonical equation of positive type, make use of the test of Exercise 6 and "optimize" it with respect to l.

The test (0.15) for dim $\mathfrak{H} < \infty$ was indicated in an article of M. G. Kreĭn [7]. It is exact since it is exact in the scalar case (the latter was proved by T. M. Karaseva [1]).

The assertion of the exercise can be generalized to the case $\mathscr{P}_{av} \geqslant 0$, but in this case we do not obtain exact tests in the indicated way. For the scalar Hill equation exact tests for the strong stability of the equation in terms of the two parameters $a = \mathscr{P}_{av}$ and $b = T^2(Q^2)_{av}$ were found by T. M. Karaseva [1]. Surprisingly, the periods of hyperelliptic integrals play a role in the investigations of the boundaries of the "stability domains" in the (a, b)-plane.

10. *A paradoxical result in the theory of parametric excitation.* Suppose a periodic operator function $\mathscr{P}(t)$ satisfies the conditions of the preceding exercise and $R(t) = R(t + T)$ is a bounded Hermitian operator function.

Consider the equation

$$d^2y/dt^2 + [R(\omega t) + \varepsilon\omega^2\mathscr{P}(\omega t)]y = 0. \tag{0.16}$$

Show that there exists an $\varepsilon_0 > 0$ such that for $\varepsilon \in (0, \varepsilon_0)$ equation (0.16) is stable if ω is sufficiently great: $\omega > \Omega(\varepsilon)$, where $\Omega(\varepsilon)$ is a number depending on ε.

Hint. By the substitution $\tau = \omega t$ reduce the equation to the form

$$\frac{d^2y}{d\tau^2} + \left[\varepsilon\mathscr{P}(\tau) + \frac{1}{\omega^2}R(\tau)\right]y = 0. \tag{0.17}$$

The equation $\ddot{y} + \varepsilon\mathscr{P}(\tau)y = 0$ is strongly stable for sufficiently small $\varepsilon > 0$ by virtue of the result of the preceding exercise, while equation (0.17) for large ω is obtained from it by means of a small perturbation.

The assertion recorded above shows in particular that, for a nonpositive $R =$ const, an unstable system described by the equation

$$d^2y/dt^2 + Ry = 0,$$

can be converted into a stable one by means of a high frequency parametric excitation $\varepsilon\omega^2\mathscr{P}(\omega t)$ of zero mean value (and consequently of variable sign) (cf. the deduction of Exercise 8 as well as Exercises III.17, III.18 and IV.8).

In particular, this assertion applies to the linearized equation of motion of a mathematical pendulum of length l with point of suspension vertically vibrating according to the law $s = \varepsilon l \sin(\omega t + \alpha)$. For this equation has the form

$$\phi'' + [\pm g/l - \varepsilon\omega^2 \sin(\omega t + \alpha)]\phi = 0,$$

where ϕ is the angular displacement of the pendulum from its initial vertical position while the + or − sign is taken before g/l, depending on whether the pendulum is under or over its point of suspension. Thus the "linear theory" permits one to assert that an $\varepsilon_0 > 0$ can always be found such that for $\varepsilon < \varepsilon_0$ the pendulum will perform stable oscillations independently of whether it was originally under or over its point of suspension, provided the vibration frequency ω of its point of suspension is sufficiently great ($\omega > \Omega(\varepsilon)$).

This paradoxical deduction of the linear theory is well known. P. L. Kapica [1, 2] confirmed it experimentally and proposed a nonlinear theory for the phenomenon. An outline of a general nonlinear theory of motion in a high frequency field is given by L. D. Landau and E. M. Lifšic in [1] (but their arguments have remained unintelligible to the authors of this book).

11. Consider the equation

$$d^2y/dt^2 + [\mathscr{P}_0 + \varepsilon\mathscr{P}_1(\omega t)]\,y = 0, \tag{0.18}$$

where $\mathscr{P}_0 \gg 0$ and $\mathscr{P}_1$ is a 2π periodic Hermitian operator function.

Show (M. G. Kreĭn [6]) that this equation is strongly stable for sufficiently small $\varepsilon > 0$ if

$$\omega \neq (\lambda + \mu)/N \qquad (N = 1,2,\cdots), \tag{0.19}$$

when λ^2 and μ^2 independently range over the spectrum $\sigma(\mathscr{P}_0)$. In particular, the equation is strongly stable for sufficiently large ω ($\omega > 2\sqrt{\lambda_M(\mathscr{P}_0)}$).

In the finite-dimensional case, when the spectrum $\sigma(\mathscr{P}_0)$ consists of the squares of numbers $P_1,\cdots,P_m$ (>0), a violation of the strong stability of equation (0.18) can occur only for the following isolated values of the frequency:

$$\omega = (p_j + p_k)/N \qquad (j, k = 1,2,\cdots,m;\ N = 1,2,\cdots).$$

Hint. Reduce the equation $\ddot{y} + \mathscr{P}_0 y = 0$ to an equation of first order in the doubled space $\mathfrak{H}^{(2)}$. If it is considered as a periodic equation with period $T = 2\pi/\omega$, the spectrum of its monodromy operator will consist of the numbers $e^{\pm i\lambda T}$ ($\lambda^2 \in \sigma(\mathscr{P}_0)$, $\lambda > 0$) and consequently of the two sets $\{e^{i\lambda T} \mid \lambda^2 \in \sigma(\mathscr{P}_0)\}$ and $\{e^{-i\lambda T} \mid \lambda^2 \in \sigma(\mathscr{P}_0)\}$, which are the symmetric images of each other with respect to the real axis.

To these sets there correspond $\mathscr{J}_r$-definite invariant subspaces of the monodromy operator in question. One should then make use of the results of § I.8, according to which this operator turns out to be strongly stable if the above two sets do not intersect. For a further development of this theme see the paper of V. A. Jakubovič [6] and the survey of M. G. Kreĭn and V. A. Jakubovič [1].

In the following problems the space $\mathfrak{H}$ is assumed to be finite dimensional ($\mathfrak{H} = C^n$ or R^n). The assertions of these exercises can be extended to the infinite-dimensional case only under special conditions requiring that the values of the operator coefficients be completely continuous operators of this or that class.

12. Consider the skewperiodic boundary problem

$$\begin{cases} dx/dt = i\lambda\mathscr{J}\mathscr{H}(t)x, \\ x(0) + x(T) = f \end{cases} \tag{0.20}$$

in the phase space C^n with a Hermitian $\mathscr{H}(t)$ of positive type.

It is elementary to prove that all of the eigenvalues of this boundary problem are real (the problem is selfadjoint).

Show that the following two equalities hold for the complete sequence $\{\lambda_j\}$ of eigenvalues of the problem, taken with regard for their multiplicities (M. G. Kreĭn [6, 7]):

$$\sum_j \frac{1}{\lambda_j^2} = \frac{T^2}{4}\,\mathrm{tr}\,(\mathscr{J}\mathscr{H}_{av})^2; \tag{0.21}$$

$$\lim_{r\to 0} \sum_{|\lambda_j|<r} \frac{1}{\lambda_j} = 0. \tag{0.22}$$

Hint. To obtain the first equality consider the equivalent (to the boundary problem) integral equation

$$x(t) = \frac{\lambda}{2}\int_0^T \operatorname{sign}(t-s)\mathcal{J}\mathcal{H}(s)x(s)ds$$

and calculate the trace of the second iterated kernel, which turns out to be continuous.

The proof of the second equality is complicated; two different proofs of it are given in the articles [6, 7] of M. G. Kreĭn, and a third is given in the book [2] of I. C. Gohberg and M. G. Kreĭn.

All three proofs make use of subtle facts of the theory of entire functions of exponential type.

In the cited book of Gohberg and Kreĭn there is a generalization of equalities (0.21) and (0.22) to the case of an infinite-dimensional phase space $\mathfrak{H}$ when $\mathcal{H}(t)$ and $\mathcal{H}_{av}$ are nuclear operators.

We note that equality (0.21) implies that *the boundary problem* (0.20) *always has at least one positive and at least one negative eigenvalue.*

13. Show (M. G. Kreĭn [7]) that a T-periodic canonical equation of positive type $\dot{x} = i\lambda\mathcal{J}\mathcal{H}(t)x$ is strongly stable whenever

$$T^2 \operatorname{tr}(\mathcal{J}\mathcal{H}_{av})^2 < 4. \tag{0.23}$$

If the Hermitian $\mathcal{H}(t)$ is even: $\mathcal{H}(t) = \mathcal{H}(-t)$, the number 4 in this test should be replaced by the number 8.

Hint. Make use of formula (0.21) and Theorem 4.4 on the central stability band.

To obtain the second assertion one should make use of the fact that when the Hermitian is even: $\mathcal{H}(t) = \mathcal{H}(-t)$, the spectrum of the boundary problem (0.20) is symmetrically arranged with respect to the origin and hence (0.21) implies an upper bound for $2/\lambda_1^2$.

The use of the first or second test is facilitated by bearing in mind that if $i\mathcal{J} = \mathcal{J}_r$ and $\mathcal{H}_{av} = \|A_{jk}\|_1^2$, then

$$-\operatorname{tr}(\mathcal{J}_r\mathcal{H}_{av})^2 = 2\operatorname{tr}(A_{11}A_{12} - \operatorname{Re} A_{12}A_{21}) \qquad (A_{21} = A_{12}^*).$$

REMARK. The test (0.23) is not exact. This will follow from the assertion of Exercise 15.

On the other hand, the second test for even Hermitians turns out to be exact (see Exercise 17).

14. If $\mathcal{H}_{av} > 0$, the operator $T\mathcal{J}\mathcal{H}_{av}$ will have exactly n_+ positive eigenvalues: $\omega_1^+ \geqq \cdots \geqq \omega_{n_+}^+$ (>0), and n_- negative ones: $-\omega_1^- \leqq \cdots \leqq -\omega_{n_-}^-$, where n_+ and n_- are respectively the multiplicities of the eigenvalues $+1$ and -1 of the operator $\mathcal{J} = \mathcal{J}^* = \mathcal{J}^{-1}$.

The following theorem holds (I. C. Gohberg and M. G. Kreĭn [2]).

I. *Suppose that for a T-periodic canonical equation of positive type $x = i\mathcal{J}\mathcal{H}(t)x$ the following condition is satisfied for some κ_0 $(0 < \kappa_0 < 1)$:*

$$\sum_{j=1}^{n_+} \frac{\omega_j^+}{j + \kappa_0 - 1} + \sum_{j=1}^{n_-} \frac{\omega_j^-}{j - \kappa_0} < 2\pi. \tag{0.24}$$

Then the equation is strongly stable. Furthermore, all of its characteristic multipliers of the first kind lie on the halfopen arc of the unit circle $\gamma_+ = \{\exp(i\theta) \mid 0 < \theta \leqq 2\pi\kappa_1\}$ while those of the second kind lie on the halfopen arc $\gamma_- = \{\exp(i\theta) \mid -2\pi(1-\kappa_2) \leqq \theta < 0\}$, where $(0 <)\ \kappa_1 < \kappa_2$ (< 1) are the unique pair of roots in the interval (0, 1) *of the equation*

$$\sum_{j=1}^{n_+} \frac{\omega_j^+}{j + \kappa - 1} + \sum_{j=1}^{n_-} \frac{\omega_j^-}{j - \kappa} = 2\pi. \tag{0.25}$$

The proof of this theorem is based on the rules of motion of characteristic multipliers and a number of propositions in the theory of nonselfadjoint operators, a formulation of which would take us too far astray.

It is elementary to verify that under condition (0.24) equation (0.25) has a total of two simple roots in the interval (0, 1).

The following proposition is obtained as a simple corollary of this theorem.

II. *A periodic Hamilton equation* $\dot{x} = J_\Gamma \mathscr{H}(t)x$ *of positive type in a* $2m$*-dimensional phase space is strongly stable whenever*

$$\sum_{j=1}^{m} \frac{\omega_j^+}{2j-1} < \frac{\pi}{2}.$$

In this case its characteristic multipliers of the first kind lie on the arc $\gamma_+ = \{\exp(i\theta) | 0 < \theta \leq 2\pi\kappa_1\}$ *while those of the second kind lie on the mirror image of this arc with respect to the real axis, where* κ_1 *is the unique root in the interval* $(0, 1/2)$ *of the equation*

$$\sum_{j=1}^{m} \omega_j^+ \left\{ \frac{1}{j+\kappa-1} + \frac{1}{j-\kappa} \right\} = 2\pi.$$

Hint. Use the fact that when $\mathscr{J} = iJ_\Gamma$ and $\dim \mathfrak{H} = 2m$ one has $n_+ = n_- = m$ and $\omega_j^+ = \omega_j^-$ $(j = 1, 2, \cdots, m)$. Verify that if κ_0 is a root of equation (0.25), the second root is $1 - \kappa_0$.

15. Show that the first assertion of Exercise 13 is a corollary of Theorem I and, in addition, that the test (0.23) can be replaced by the stronger test

$$T^2 \operatorname{tr} (\mathscr{J}\mathscr{H}_{av})^2 < \pi^2 \left(\sum_{j=-n_-+1}^{n_+} \frac{1}{(2j-1)^2} \right)^{-1}$$

(the right side of this inequality tends to $\pi^2/4$ as $n_\pm \to \infty$).

Hint. Estimate the left side of (0.24) for $\kappa_0 = 1/2$ by means of Cauchy's inequality and make use of the fact that

$$T^2 \operatorname{tr} (\mathscr{J}\mathscr{H}_{av})^2 = \sum (\omega_j^+)^2 + \sum (\omega_j^-)^2.$$

16. Suppose that in a T-periodic m-dimensional Hill equation $\ddot{y} + \mathscr{P}(t)y = 0$ the coefficient $\mathscr{P}(t) = \mathscr{P}^*(t)$ is an odd function: $\mathscr{P}(-t) = -\mathscr{P}(t)$, and $Q = \int \mathscr{P}\, dt$, $Q_{av} = 0$. Then (M. G. Krein [7]) this equation will be strongly stable whenever

$$(T^2 (\operatorname{tr} Q^2)_{av} =)\, T \int_0^T \operatorname{tr} (Q^2) dt < 4. \tag{0.26}$$

For the scalar case ($m = 1$) this proposition was obtained (in a completely different way, of course) by A. M. Ljapunov [1] back in 1899 (see the reference in Exercise 8).

Hint. Transform the Hill equation into a canonical equation with an even Hermitian $\mathfrak{H}(t)$ and make use of the second assertion of Exercise 13.

17. Show that the test (0.26) for strong stability of the Hill equation is exact (M. G. Krein [7]).

Hint. Consider the scalar Hill equation $\ddot{y} + \lambda p_a(t)y = 0$ with a periodic generalized function $p_a(t)$ defined in the interval $(-T/2, T/2)$ by the equality

$$p_a(t) = \delta(t+a) - \delta(t-a) \qquad (-T/2 < t < T/2),$$

where $a \in (0, T/2)$.

Verify that the skewperiodic boundary problem

$$\begin{cases} d^2y/dt^2 + \lambda p_a(t)y = 0, \\ y(0) + y(T) = y'(0) + y'(T) = 0 \end{cases}$$

has a total of two eigenvalues $\lambda_+(a)$ (> 0) and $\lambda_-(a) = -\lambda_+(a)$, and

$$\lambda_+(a) T \int_0^T q^2 dt = 4,$$

where $\dot{q} = p_a$, $q_{av} = 0$.

The equation $\ddot{y} + \lambda p_a(t)y = 0$ with real λ will be strongly stable (e-dichotomic) for $0 < |\lambda| < \lambda_+(a)$ $(|\lambda| > \lambda_+(a))$.

REMARK. If the equation $\ddot{y} + \lambda p_a(t)y = 0$ is transformed into a Hamiltonian equation with an even Hamiltonian (see Exercise 7), the strong stability test indicated in Exercise 13, when applied to the latter equation, becomes a necessary and sufficient condition for $\lambda \neq 0$ to be a strong stability point.

18. Suppose that in a T-periodic m-dimensional Hill equation $\ddot{y} + \mathscr{P}(t)y = 0$ the coefficient $\mathscr{P}(t) = \mathscr{P}^*(t)$ has a mean value $\mathscr{P}_{\text{av}} = 0$ while $Q(t)$ is defined as in the preceding exercise. Let q_j^2 $(j = 1,\cdots, m; q_1 \geqq \cdots \geqq q_m)$ denote the successive eigenvalues of the operator $T^2(Q^2)_{\text{av}}$.

Show that the Hill equation is strongly stable whenever

$$\sum_{j=1}^{m} \frac{q_j}{2j-1} < \frac{\pi}{2},$$

and a fortiori whenever (cf. (0.26))

$$T^2(\operatorname{tr} Q^2)_{\text{av}} < \frac{\pi^2}{4}\left(\sum_{j=1}^{m} \frac{1}{(2j-1)^2}\right)^{-1}.$$

Hint. Transform the Hill equation into a canonical one and make use of Theorem I (see Exercise 14).

19. Show that when $\dim \mathfrak{H} < \infty$ Theorem 2.4 can be sharpened by replacing the condition $\operatorname{Re} \mathscr{P}(t) \ll 0$ in the theorem by the condition $\operatorname{Re} \mathscr{P}(t) \leqslant 0$, $\operatorname{Re} \mathscr{P}_{\text{av}} \ll 0$.

Hint. Using the notation in the proof of the theorem, show that the strict inequality

$$(Wx_0, x_0) - (WU(T)x_0, U(T)x_0) > 0$$

is satisfied for $x_0 \in \mathfrak{H}$, $x_0 \neq 0$.

It will then follow that $W - U^*(T)WU(T) \gg 0$ (the relations $\mathscr{H} > 0$ and $\mathscr{H} \gg 0$ are equivalent in a finite-dimensional $\mathfrak{H}$).

20. Let $\partial_T(\mathfrak{H})$ denote the Banach space of T-periodic Hermitians $\mathscr{H}(t)$ $(= \mathscr{H}^*(t) = \mathscr{H}(t + T))$ with values in $[\mathfrak{H}]$ and norm

$$|||\mathscr{H}||| = \int_0^T \|\mathscr{H}(t)\|dt.$$

Let $\mathscr{J}$ $(\in [\mathfrak{H}])$ denote a *signature operator*: $\mathscr{J} = \mathscr{J}^* = \mathscr{J}^{-1}$, and let $\mathfrak{U}(\mathscr{J})$ denote the group of all $\mathscr{J}$-unitary operators $U \in [\mathfrak{H}]$.

The operator $\mathscr{J}$ induces a mapping $\mathscr{T}\colon \partial_T(\mathfrak{H}) \to \mathfrak{U}(\mathscr{J})$ which takes a Hermitian $\mathscr{H}(t)$ into the monodromy operator of the corresponding equation $\dot{x} = i\mathscr{J}\mathscr{H}(t)x$. Let $S_T(\mathfrak{H}; \mathscr{J})$ denote the set of all $\mathscr{H} \in \partial_T(\mathfrak{H})$ whose images under $\mathscr{T}$ are strongly stable $\mathscr{J}$-unitary operators $U \in \mathfrak{U}(\mathscr{J})$. According to Lemma 4.1, $S_T(\mathfrak{H}; \mathscr{J})$ is an open set in $\partial_T(\mathfrak{H})$ and therefore divides into domains (connected components) called *stability domains.*

When $\mathfrak{H}$ is finite dimensional, these domains form a countable set. For real $\mathfrak{H} = R^{2m}$ and $\mathscr{J} = J_r$, a complete description of the stability domains for $m = 1$ has been given by V. A. Jakubovič [7], and for any natural m, by I. M. Gel'fand and V. B. Lidskiĭ in their important paper [1]. The results of the latter were supplemented and then generalized by V. A. Jakubovič [8, 9] to the general case of a complex finite-dimensional $\mathfrak{H}$ and any signature operator $\mathscr{J}$ (in connection with these papers we note the paper of V. B. Lidskiĭ and P. A. Frolov [1], in which related questions are solved for the T-periodic equation

$$\frac{dx}{dt} = \left(\mathscr{H}(t) - \frac{1}{2}\frac{dQ}{dt}\right)x \qquad (Q^*(t) = -Q(t))$$

in a finite-dimensional $\mathfrak{H}$ and which contains some corrections to the paper [9] of V. A. Jakubovič). See also the papers of W. A. Coppel and A. Howe [1] and S. Diliberto [1.]

Among the stability domains in $S_T(\mathfrak{H}; \mathscr{J})$ there is always (independently of the dimension $\dim \mathfrak{H} \leqq \infty$) the domain $Q_T^+(\mathfrak{H}; \mathscr{J})$ which contains all of the Hermitians $\mathscr{H}(t)$ of positive type for which $\lambda_1(\mathscr{H}) > 1$ (see Theorem 4.4 on the central stability band). The domain $Q_{\ddagger}(\mathfrak{H}; \mathscr{I})$ has the obvious property that, if it contains a Hermitian of positive type $\mathscr{H}(t)$, it also contains the halfopen "segment" $q\mathscr{H}$ $(0 < q \leqq 1)$.

In addition, if $\mathscr{H}$ belongs to $Q_{\ddagger}$, then so does every Hermitian $\mathscr{H}_1$ of positive type such that $\mathscr{H}_1(t) \leqq \mathscr{H}(t)$ (this assertion is offered as an exercise). All of the tests for strong stability of a canonical equation of positive type indicated in the preceding exercises are tests of whether or not $\lambda_1(\mathscr{H}) > 1$, i.e. tests of whether or not a Hermitian $\mathscr{H}$ belongs to $Q_{\ddagger}(\mathfrak{H}; \mathscr{I})$. These tests are called *Z-tests.*

There exist techniques permitting one to obtain from the Z-tests tests of whether or not a Hermitian $\mathscr{H}$ belongs to the other domains of $S_T(\mathfrak{H}; \mathscr{I})$. In regard to these techniques and other methods of obtaining tests different from the Z-tests for the strong stability of a canonical equation see the paper of M. G. Kreĭn [7] and the survey paper of M. G. Kreĭn and V. A. Jakubovič [1] (see also the survey article of V. M. Staržinskiĭ [1]).

NOTES

Equations with periodic coefficients were first investigated in a finite-dimensional space by G. Floquet [1], who essentially introduced the monodromy matrix and obtained formula (1.5). A. M. Ljapunov and H. Poincaré discovered that in the case of a Hamiltonian equation the monodromy matrix is a J_r-unitary matrix, and it is from this fact that we obtain the symmetry properties of the spectrum of the monodromy matrix.

The e-dichotomicity criterion (Theorem 2.1) is taken from the book of J. Massera and J. Schäffer [1]. The subsequent results of § 2, which make use of the properties of operators in Hilbert space with an indefinite metric, are due to M. G. Kreĭn. For the elementary aspects of Theorem 2.2 see his paper [5]. (See also the commentary to Exercise IV.14.)

The localization theorems of § 3 are due to M. G. Kreĭn [11]. Theorem 3.2 was obtained by a sharpening and development of the arguments of J. Massera and J. Schäffer [1] showing that under condition (3.15) the spectrum $\sigma(U(T))$ does not intersect the negative half of the real axis. Massera and Schäffer managed to find an analogous test for Banach spaces (see their book [1]).

The theory of stability bands for the scalar Hill equation with a parameter was constructed by A. M. Ljapunov for both the cases of a coefficient of constant sign and one of variable sign. He was also the first to obtain existence theorems and various estimates for the central stability band.

M. G. Kreĭn [3, 6] constructed the theory of characteristic multipliers of the first and second kind, which permitted him to generalize the results of A. M. Ljapunov and obtain a number of new results for finite-dimensional canonical equations. At the present time there is apparently no stability test for the scalar Hill equation which does not have analogs (often several of them) in the multidimensional, and frequently also in the infinite-dimensional, case. Moreover, the methods developed for systems of equations have in many cases provided tests that are also new for the scalar case (see M. G. Kreĭn and V. A. Jakubovič [1]).

The important paper of I. M. Gel'fand and V. B. Lidskiĭ [1] is devoted to the classification and calculation of the homotopy invariants of stable Hamiltonian systems (see also the Notes to Chapter I and Exercise 20 above).

The papers [1, 8, 9] of V. A. Jakubovič are devoted to the further development of the theory of periodic canonical equations in this direction.

The first investigations of canonical systems in infinite-dimensional Hilbert space were undertaken by V. I. Derguzov [1, 2], who investigated specific classes of canonical equations with an unbounded selfadjoint Hamiltonian $\mathscr{H}(t)$. It is to canonical systems of this type that concrete problems in the study of parametric resonance of elastic systems with an infinite number of degrees of freedom reduce.

The papers [1–5] of V. N. Fomin are devoted to the further investigation of canonical systems of this type.

The papers of V. I. Derguzov moved M. G. Kreĭn to investigate the possibility of adapting the methods at his disposal to the infinite-dimensional case. The results obtained were partially published in [**9**, **10**]. These results of M. G. Kreĭn are presented with revisions and additions in § § 4 and 5 and the corresponding exercises.

The method in § 6.2 of calculating the logarithm of the monodromy operator for the elementary equation $\dot{x} = \lambda A(t)x$ was pointed out by N. P. Erugin [**1**]. A similar method for equations with almost periodic coefficients of the form $\dot{x} = [A_0 + \lambda A(t)]x$ was applied still earlier by I. Z. Štokalo [**1**] to obtain asymptotic expansions of the solutions. A simpler presentation of these results has been given by N. P. Erugin [**2**]. The general case considered here for equations with a periodic coefficient $A(t; \lambda)$ was considered by M. G. Kreĭn in a seminar lecture in 1948, but these results were never published. It was in connection with this work that formulas (I.3.7) were first obtained.

In 1959 there appeared an article by V. A. Jakubovič [**2**] that is also devoted to the application of the small parameter method to canonical systems. His results were generalized to the infinite-dimensional case by I. I. Kovtun in [**1**], where, in particular, he proposed a method for calculating coefficients that is based on the use of formula (I.3.17).

For the further development of asymptotic methods in application to canonical equations and the relevant literature see the survey of M. G. Kreĭn and V. A. Jakubovič [**1**].

CHAPTER VI

LINEAR DIFFERENTIAL EQUATIONS IN THE COMPLEX PLANE

In this chapter we consider a linear differential equation the operator coefficient $A(z)$ of which is an analytic function whose only singularity is a pole of first order at the origin. In § 1 we consider the simplest case, when the spectrum of the principal part A_0 of $A(z)$ does not contain pairs of points with integral differences. In § 2 we weaken this condition by assuming only that integral differences correspond to isolated points of the spectrum. In § 3 we show how to calculate in the finite-dimensional case the coefficients of a series representing a solution by using the Jordan form of the matrix A_0. In § 4 we consider a method of solution which under certain conditions permits one to also obtain a solution of the equation in the case when there are integral differences of points of the continuous spectrum.

§ 1. The equation with a regular singularity. Simplest case

1. *The equation with a regular coefficient.* In this chapter we consider some questions concerning the Cauchy problem

$$dx/dz = A(z)x, \quad x(z_0) = x_0 \tag{1.1}$$

in a complex domain of the z-plane.

The case when the operator function $A(z)$ is regular in a simply connected domain G containing the point z_0 can be investigated by means of considerations similar to those at the beginning of Chapter III.

The solution of problem (1.1) can be represented in the form

$$x(z) = U(z, z_0)x_0, \tag{1.2}$$

where the operator $U(z, z_0)$ satisfies the condition

$$U(z_0, z_0) = I \tag{1.3}$$

and the equation

$$dU/dz = A(z)U \tag{1.4}$$

in the phase space [$\mathfrak{B}$].

This operator has the series expansion

$$U(z, z_0) = \sum_{k=0}^{\infty} W_k(z), \tag{1.5}$$

where

$$W_0(z) = I, \quad W_k(z) = \int_{z_0}^{z} A(z)W_{k-1}(z)dz,$$

the integrals being independent of the path of integration in the domain G.

The series (1.5) converges uniformly in every closed part $F \subset G$ whose points can be reached from z_0 along paths of bounded length.

The solution (1.2) of problem (1.1) is unique.

The general solution of equation (1.4) is $V(z) = U(z, z_0)C$, where C $(= V(z_0))$ is an arbitrary bounded operator.

2. *Simplest equation with a singularity.* Of greater interest are equations with a nonregular coefficient $A(z)$.

We confine ourselves to a study of the behavior of a solution of equation (1.1) in a neighborhood of a simple pole z_0 of the operator function $A(z)$. It can be assumed without loss of generality that $z_0 = 0$ and hence that $A(z)$ has the form

$$A(z) = (1/z)\,A_0 + A_1 + A_2 z + \cdots, \tag{1.6}$$

the series (1.6) converging in a domain $0 < |z| < \rho$. It can easily be shown by the usual methods that if a power series with operator coefficients converges in the interior of a disk, it converges there absolutely, i.e. $\sum \|A_k\|\,|z|^k < \infty$. Under the above assumptions it is customary in the analytic theory of differential equations to call equation (1.1) an *equation with a regular singularity.*

We first consider the simplest equation

$$\frac{dU}{dz} = \frac{A_0}{z}U. \tag{1.7}$$

It is not difficult to verify by a direct substitution that the operator function

$$U(z) = e^{A_0 \ln z}C = z^{A_0}C, \tag{1.8}$$

where C is an arbitrary operator, is a solution of (1.7) for $z \neq 0$.

This solution is not single valued; it is multiplied by the operator $e^{2\pi i A_0}$ when one makes a complete circuit about the point $z_0 = 0$ in the positive direction.

Any solution of equation (1.7) that is single valued in a simply connected domain not containing the point $z_0 = 0$ in its interior can be represented in the form (1.8) under a suitable choice of C and a single valued branch of the logarithm.

For this reason the function z^{A_0} and, more generally, any solution of equation (1.4) having the analogous property will be called a *fundamental solution.*

3. *Case when the spectrum of the principal part does not contain pairs of points with integral differences.* We pass to the consideration of the equation with operator (1.6)

$$z\frac{dU}{dz} = \left(\sum_{k=0}^{\infty} A_k z^k\right)U, \tag{1.9}$$

assuming that the series in the right side converges in a disk $0 \leqq |z| < \rho$.

We first suppose that the operator A_0 satisfies the following condition: *there does not exist a pair of points in the spectrum of A_0 whose difference is equal to a natural number*:

$$\lambda - \mu \neq n \qquad (n = 1,2,3,\cdots;\ \lambda, \mu \in \sigma(A_0)). \tag{1.10}$$

By analogy with formula (1.8) we will seek a solution of equation (1.9) in the form

$$U(z) = \left(\sum_{k=0}^{\infty} U_k z^k\right) z^{A_0}. \tag{1.11}$$

Formally substituting the expansion (1.11) into equation (1.9) and equating the coefficients of like powers of z, we obtain the system of recursion relations

$$\begin{gathered} A_0U_0 - U_0A_0 = 0, \\ mU_m - (A_0U_m - U_mA_0) = \sum_{k=1}^{m} A_kU_{m-k} \qquad (m = 1,2,\cdots). \end{gathered} \tag{1.12}$$

We put $U_0 = I$. To successively determine the operators U_m $(m > 0)$ we apply formula (I.3.10). We write (1.12) in the form

$$U_mA_0 - (A_0 - mI)U_m = F_m \qquad (m = 1,2,\cdots), \tag{1.13}$$

where $F_m = \sum_{k=1}^{m} A_kU_{m-k}$.

From condition (1.10) it follows that the spectra of the operators A_0 and $A_0 - mI$ $(m = 1, 2,\cdots)$ do not intersect. Therefore the solution of equation (1.13) has the form (see (I.3.10))

$$\begin{aligned} U_m &= \frac{1}{4\pi^2} \oint_{\Gamma_0} \oint_{\Gamma_0^{(m)}} \frac{(A_0 - mI - \lambda I)^{-1}F_m(A_0 - \mu I)^{-1}}{\lambda - \mu}\, d\lambda\, d\mu \\ &= \frac{1}{4\pi^2} \oint_{\Gamma_0} \oint_{\Gamma_0} \frac{(A_0 - \lambda I)^{-1} F_m (A_0 - \mu I)^{-1}}{\lambda + m - \mu}\, d\lambda\, d\mu, \end{aligned} \tag{1.14}$$

where Γ_0 and $\Gamma_0^{(m)}$ are smooth contours surrounding the spectra of the operators A_0 and $A_0 - mI$ respectively, the contour $\Gamma_0^{(m)}$ being a parallel displacement of the contour Γ_0.

In order to prove the convergence of the series in (1.11) we estimate the norms of the operators U_m. Formula (1.14) readily implies that the following estimate is valid for sufficiently large m:

$$\|U_m\| \leqq (c/m)\|F_m\|.$$

We note in addition that, under our assumptions, for any $\rho_1 < \rho$

$$\lim_{k\to\infty} \|A_k\|\rho_1^k = 0,$$

and hence

$$\lim_{m\to\infty} \frac{1}{m} \sum_{k=1}^{m} \|A_k\|\rho_1^k = 0.$$

Since

$$\|U_m\|\rho_1^m \leqq \frac{c}{m} \|F_m\|\rho_1^m \leqq \frac{c}{m} \sum_{k=1}^{m} \|A_k\|\rho_1^k\|U_{m-k}\|\rho_1^{m-k}$$

$$\leqq \max_{1\leqq s\leqq m-1} \{\|U_s\|\rho_1^s\} c \cdot \frac{1}{m} \sum_{k=1}^{m} \|A_k\|\rho_1^k,$$

we have, beginning with some $m = m_0$,

$$\|U_m\|\rho_1^m \leqq \max_{1\leqq s\leqq m-1} \{\|U_s\|\rho_1^s\}$$

and consequently the sequence $\|U_m\|\rho_1^m$ is bounded for any $\rho_1 < \rho$. Using this fact for some ρ_2 ($\rho_1 < \rho_2 < \rho$), we see that in each disk $|z| \leqq \rho_1$ $(< \rho)$ the series in (1.11) is majorized by a convergent geometric series and is therefore absolutely convergent.

Thus the following theorem is valid.

THEOREM 1.1. *If the operator A_0 satisfies condition* (1.10) *and the series $\sum_0^\infty A_k z^k$ converges in a disk $|z| < \rho$, equation* (1.9) *has the solution*

$$U(z) = \Big(\sum_{k=0}^{\infty} U_k z^k\Big) z^{A_0} = B(z) z^{A_0} \qquad (U_0 = B(0) = I), \tag{1.11}$$

the series $\sum_0^\infty U_k z^k$ converging in the same disk.

REMARK 1.1. Let G_ρ be an open disk of radius ρ cut along some radius. By choosing a branch of z^{A_0} that is single valued in G_ρ, we can obtain from (1.11) a single valued solution of equation (1.9).

We assume that the boundary of G_ρ contains the origin. Then any solution that is single valued in this domain can be represented in the form $V(z) = U(z)C$, where C is a constant operator, i.e. $U(z)$ is a fundamental solution.

In fact, since $B(0) = I$, for sufficiently small z_0 the operator $B(z_0)$ has a bounded inverse, and therefore the operator

$$U^{-1}(z_0) = z_0^{-A_0} B^{-1}(z_0)$$

exists.

In the simply connected domain G_ρ the operator $U(z, z_0) = U(z)U^{-1}(z_0)$ satisfies the condition $U(z_0, z_0) = I$ and the differential equation (1.9), which does not have any singularities in this domain. Therefore

$$V(z) = U(z, z_0)V(z_0) = U(z)U^{-1}(z_0)V(z_0) = U(z)C.$$

§ 2. Case of integral differences of eigenvalues

1. *Solutions of minimal type of the inhomogeneous equation with a constant coefficient.* In this section we consider equation (1.9) without the assumption (1.10) on the spectrum of A_0.

We first collect some auxiliary facts.

We consider the equation

$$dx/dz = Ax + f(z), \tag{2.1}$$

under the assumption that A is an invertible operator.

Suppose

$$f(z) = \sum_{k=0}^{n} f_k z^k \qquad (f_k \in \mathfrak{B}; k = 0,1,\cdots,n) \tag{2.2}$$

is a polynomial of degree n with vector coefficients.

In this case there exists a solution

$$x(z) = \sum_{k=0}^{n} x_k z^k \qquad (x_k \in \mathfrak{B}; k = 0,1,\cdots,n) \tag{2.3}$$

of the same form. The coefficients x_k are most readily found by the method of undetermined coefficients. Substituting expressions (2.2) and (2.3) into (2.1) and equating coefficients of like powers of z, we obtain the system of recursion relations

$$Ax_n + f_n = 0, \quad Ax_{k-1} = kx_k - f_{k-1} \qquad (k = n, n-1,\cdots,1),$$

which implies the equalities

$$\left.\begin{aligned} x_n &= -A^{-1}f_n, \\ x_{n-1} &= -A^{-1}f_{n-1} - nA^{-2}f_n, \\ &\cdots\cdots\cdots\cdots \\ x_j &= -\frac{1}{j!}A^{j-1}\sum_{s=j}^{n} s!A^{-s}f_s. \end{aligned}\right\} \tag{2.4}$$

We recall that an entire function $f(z)$ is called a *function of mininal type* (more precisely, *of minimal exponential type*) if for any $\varepsilon > 0$ there exists a constant N_ε such that the estimate $\|f(z)\| \leq N_\varepsilon e^{\varepsilon|z|}$ holds throughout the complex plane.

Suppose

$$f(z) = \sum_{k=0}^{\infty} f_k z^k. \tag{2.5}$$

It is known (see Exercise I.22) that this function is of minimal type if and only if

$$\lim_{k\to\infty} (k!\|f_k\|)^{1/k} = 0. \tag{2.6}$$

The above result concerning polynomial solutions admits the following generalization.

LEMMA 2.1. *Equation* (2.1) *has a unique solution in the class of solutions of minimal type if the operator A has a bounded inverse and the vector function $f(z)$ is of minimal type.*

PROOF. We put in analogy with (2.4)

$$x_j = -\frac{1}{j!}A^{j-1}\sum_{s=j}^{\infty} s!A^{-s}f_s \qquad (j = 0,1,\cdots). \tag{2.7}$$

For any $\varepsilon > 0$ there exists by virtue of (2.6) an index j_0 such that $s!\|f_s\| \leqq \varepsilon^s$ when $s > j_0$. Therefore

$$\begin{aligned} j!\|x_j\| &\leqq \sum_{s=j}^{\infty} s!\|f_s\| \cdot \|A^{-1}\|^{s-(j-1)} \\ &\leqq \sum_{s=j}^{\infty} \varepsilon^s\|A^{-1}\|^{s-(j-1)} = \frac{\varepsilon^j\|A^{-1}\|}{1-\varepsilon\|A^{-1}\|} \end{aligned}$$

when $j > j_0$, and hence

$$\lim_{j\to\infty}\{j!\|x_j\|\}^{1/j} = 0.$$

Thus the function $x(z) = \sum_0^\infty x_j z^j$ is of minimal type.

It follows from (2.7) that

$$kx_k = Ax_{k-1} + f_{k-1} \qquad (k = 1,2,\cdots). \tag{2.8}$$

Multiplying (2.8) by z^{k-1} and summing, we get that $x(z)$ is a solution of equation (2.1).

For a proof of uniqueness it suffices to show that when A is invertible the homogeneous equation $dx/dz = Ax$ cannot have nonzero solutions of minimal type. Suppose $x(z) = \sum_k x_k z^k$ is a nonzero solution. Then $x_k = A^k x_0/k!$ and

$$\|x_0\| = \|A^{-k}A^k x_0\| \leqq \|A^{-1}\|^k \cdot \|x_k\| \cdot k! \leqq \varepsilon^k\|A^{-1}\|^k$$

for arbitrarily large k and sufficiently small ε, which implies $x_0 = 0$.

LEMMA 2.2. *Equation* (2.1) *has a solution of minimal type if the spectrum of A has an isolated point at $\lambda = 0$ and $f(z)$ is of minimal type. Two such solutions differ by a function whose values belong to the invariant subspace $\mathfrak{B}_0$ of A corresponding to the point $\lambda = 0$.*

PROOF. Let P_0 be the projection corresponding to the point $\lambda = 0$ of $\sigma(A)$. Equation (2.1) decomposes into the pair of equations

$$dy_0/dz = A_0 y_0 + g_0(z), \tag{2.9}$$

$$dy_1/dz = A_1 y_1 + g_1(z), \tag{2.10}$$

where

$$y_0 = P_0x, y_1 = (I - P_0)x, g_0 = P_0f, g_1 = (I - P_0)f, A_0 = AP_0, A_1 = A(I - P_0).$$

In the subspace $\mathfrak{B}_1 = (I - P_0)\mathfrak{B}$ equation (2.10) satisfies the conditions of Lemma 2.1 and consequently has a solution $y_1(z)$ of minimal type.

We consider equation (2.9) in the subspace $\mathfrak{B}_0 = P_0\mathfrak{B}$. In this equation the operator A_0 has a spectrum consisting of only the point $\lambda = 0$, while the function $g_0(z)$ is of minimal type.

Let us show that all of the solutions of this equation are also of minimal type. To this end we estimate the values of the derivatives of a solution of equation (2.9) at the origin. It is easily seen from (2.9) that

$$y_0^{(k+1)}(0) = A_0^{k+1}y_0(0) + \sum_{s=0}^{k} A_0^s g_0^{(k-s)}(0).$$

We note that under the above assumptions there exists for any $\varepsilon > 0$ a constant N_ε such that

$$\|A_0^s\| \leqq N_\varepsilon \varepsilon^s; \qquad \|g_0^{(j)}(0)\| \leqq N_\varepsilon \varepsilon^j.$$

In fact the second estimate follows from (2.5) since $g_0(z)$ is of minimal type. The first estimate can be deduced from the fact that the operator function $(I - zA_0)^{-1}$ is an entire function and the series $\sum_0^\infty z^k A_0^k$ representing it converges for any z.

Further,

$$\begin{aligned}\|y_0^{(k+1)}(0)\| &\leqq \|A_0^{k+1}\| \cdot \|y_0(0)\| + \sum_{s=0}^{k} \|A_0^s\| \cdot \|g_0^{(k-s)}(0)\| \\ &\leqq N_\varepsilon \|y_0(0)\| \cdot \varepsilon^{k+1} + N_\varepsilon^2(k+1)\varepsilon^k.\end{aligned}$$

From this inequality one easily deduces the estimate

$$\overline{\lim_{k\to\infty}} \sqrt[k]{\|y_0^{(k+1)}(0)\|} \leqq \varepsilon,$$

which by virtue of the arbitrariness of ε implies the equality

$$\lim_{k\to\infty} \{\|y_0^{(k+1)}(0)\|\}^{1/(k+1)} = 0,$$

proving the lemma.

REMARK 2.1. It is not difficult to see that equation (2.1) has a polynomial solution $x(z)$ if $f(z)$ is a polynomial and the dimension of the invariant subspace of A_0 corresponding to the point $\lambda = 0$ is finite.

2. *Case when the integral differences correspond to isolated points of the spectrum of* A_0. We consider the equation

$$z\frac{dU}{dz} = \left(\sum_{k=0}^{\infty} A_k z^k\right) U, \tag{2.11}$$

without assuming that condition (1.10) is satisfied.

We note that an integer n for which there exist $\lambda, \mu \in \sigma(A_0)$ such that

$$\lambda - \mu = n \tag{2.12}$$

is a point of the spectrum of the transform

$$\mathfrak{A}X = A_0X - XA_0 \tag{2.13}$$

acting in the Banach space $[\mathfrak{B}]$ ($\rho(\mathfrak{A})$ does not contain differences of the points of $\sigma(A_0)$; see § 3 below).

We relax condition (1.10) and replace it by the following condition (α): each of the integers $n_1 < n_2 < \cdots < n_p$ ($p < \infty$) that are representable in the form $n_k = \lambda - \mu$ for $\lambda, \mu \in \sigma(A_0)$ is an isolated point of the spectrum of $\mathfrak{A}$.

A solution of equation (2.11) will be sought in the form

$$U(z) = \left(\sum_{k=0}^{\infty} U_k(\ln z)z^k\right)z^{A_0} \qquad \left(U_k(\zeta) = \sum_{s=0}^{\infty} U_{ks}\zeta^s\right). \tag{2.14}$$

This expression is analogous to (1.11), although the coefficients here are no longer constants but functions of $\zeta = \ln z$.

Substituting the expansion (2.14) into equation (2.11) and equating coefficients of like powers of z, we obtain the system of relations

$$U_k(\ln z)(A_0 + kI) + U_k'(\ln z) = \sum_{s=0}^{k} A_{k-s}U_s(\ln z) \qquad (k = 0, 1, \cdots), \tag{2.15}$$

which is more conveniently written in the form

$$U_k'(\zeta) = A_0U_k(\zeta) - U_k(\zeta)(A_0 + kI) + F_k(\zeta),$$

where $\zeta = \ln z$, $F_k(\zeta) = \sum_{s=0}^{k-1} A_{k-s}U_s(\zeta)$.

The relation

$$U_0'(\zeta) = A_0U_0(\zeta) - U_0(\zeta)A_0$$

corresponding to $k = 0$ is satisfied if one takes $U_0(\zeta) = I$.

For $0 < k < n_1$ we can take $U_k(\zeta) = U_k$, i.e. we can assume that $U_k(\zeta)$ does not depend on ζ. Equation (2.15) then takes the form

$$- A_0U_k + U_k(A_0 + kI) = \sum_{s=0}^{k-1} A_{k-s}U_s$$

and can be investigated in exactly the same way as under condition (1.10).

We next consider equation (2.15) for $k = n_1$. Going over to the space $[\mathfrak{B}]$, it can be written in the form (setting $X(\zeta) = U_{n_1}(\zeta)$)

$$X'(\zeta) = (\mathfrak{A} - n_1I)X(\zeta) + F_{n_1},$$

where $\mathfrak{A}$ is the transform (2.13) and $F_{n_1} \in [\mathfrak{B}]$.

This equation has a solution $X(\zeta)$ which is a function of minimal type, i.e. an entire function satisfying the condition

$$\|X(\zeta)\| \leqq C_\varepsilon e^{\varepsilon|\zeta|} \tag{2.16}$$

for every $\varepsilon > 0$. For the sake of definiteness (although not out of necessity) we will assume that it satisfies the condition

$$P_{n_1}X(0) = 0, \tag{2.17}$$

where P_{n_1} is the spectral projection on the invariant subspace of $\mathfrak{A}$ corresponding to the point n_1 of its spectrum. Under this condition a solution in the class of functions of minimal type is unique.

For larger k we obtain an equation of the form

$$X'(\zeta) = (\mathfrak{A} - kI)X(\zeta) + F_k(\zeta), \tag{2.18}$$

where the function $F_k(\zeta)$ is of minimal type. The operator $\mathfrak{A} - kI$ here has at the origin either a regular point (this is always true for $k > n_p$) or an isolated point of its spectrum. In either case equation (2.18) has solutions of minimal type. We can make the process of choosing such solutions completely determinate if we require that conditions of type (2.17) be satisfied.

We have thus indicated a method for successively constructing the coefficients of the series (2.14).

We will consider this series in a disk G_0 of radius $R < \rho$ cut, for example, along the radius $(-R, 0)$. We choose the branch of the logarithm that varies for $z \in G_0$ in the halfstrip

$$D_0 = \{\zeta | \operatorname{Re}\zeta < \ln R, \quad -\pi < \ln\zeta < \pi\}.$$

From condition (2.16) it follows that the function $U_n(\ln z)z^n = U_n(\zeta)e^{n\zeta}$ satisfies the condition $\lim_{z\to 0}\|U_n(\ln z)z^n\| = 0$. In fact,

$$\begin{aligned}\|U_n(\ln z)z^n\| &\leqq C_\varepsilon e^{n\ln|z| + \varepsilon\sqrt{\ln^2|z| + \pi^2}}\\ &\leqq C'_\varepsilon e^{(n-2\varepsilon)\ln|z|} = C'_\varepsilon |z|^{n-2\varepsilon}.\end{aligned}$$

To prove the convergence of the series (2.14) we consider a domain G of the form $\{z | 0 < \sigma \leqq |z| \leqq R - \pi < \arg z < \pi\}$. Each of the functions $U_n(\ln z)$ is bounded in this domain.

We consider the series (2.14) on the circle $|z| = R$ and suppose $R < R_1 < \rho$. By virtue of the estimate

$$\|U_n(\ln z)z^n\| \leqq \|U_n(\ln z)\| \cdot R^n = (R/R_1)^n \|U_n(\ln z)\| R_1^n$$

the convergence of (2.14) will follow if we prove the uniform boundedness of the $\|U_n(\ln z)\| R_1^n$.

We consider the functions $U_k(\zeta)R_1^k$ $(k = 1,\cdots,m)$. Since each of them is of minimal type there exists a constant C_ε such that

$$\|U_{ks}s!\|R_1^k \leqq C_\varepsilon\varepsilon^s \qquad (k = 1,2,\cdots,m). \tag{2.19}$$

Hence for $F_{m+1,s} = \sum_{r=0}^{m} A_{m+1-r}U_{rs}$ we have

$$\begin{aligned}\|F_{m+1,s}s!\|R_1^{m+1} &\leqq \sum_{r=0}^{m} \|A_{m+1-r}\|R_1^{m+1-r} \cdot \|U_{rs}s!\|R_1^r \\ &\leqq C_\varepsilon\varepsilon^s \sum_{r=1}^{m+1} \|A_r\| \cdot R_1^r.\end{aligned}$$

For sufficiently large m the operator $(\mathfrak{A} - mI)^{-1}$ is invertible and satisfies the estimate $\|(\mathfrak{A} - mI)^{-1}\| \leqq b/m$.

The solution of minimal growth of the equation

$$U'_{m+1} = (\mathfrak{A} - (m + 1)I)U_{m+1} + F_{m+1}$$

is given by a formula of form (2.7):

$$U_{m+1,s}s! = -(\mathfrak{A} - (m + 1)I)^{s-1} \sum_{r=s}^{\infty} r!(\mathfrak{A} - (m + 1)I)^{-r}F_{m+1,r},$$

which implies the estimate

$$\begin{aligned}&\|U_{m+1,s}s!\|R_1^{m+1} \\ &\leqq C_\varepsilon \sum_{r=1}^{m+1} \|A_r\|R_1^r \cdot \sum_{r=s}^{\infty} \varepsilon^r \|(\mathfrak{A} - (m + 1)I)^{-1}\|^{r-(s-1)} \\ &= C_\varepsilon \sum_{r=1}^{m+1} \|A_r\|R_1^r \frac{\varepsilon^s\|(\mathfrak{A} - (m + 1)I)^{-1}\|}{1 - \varepsilon\|(\mathfrak{A} - (m + 1)I)^{-1}\|} \\ &\leqq \frac{b}{m + 1} \sum_{r=1}^{m+1} \|A_r\|R_1^r \cdot C_\varepsilon\varepsilon^s \frac{1}{1 - \varepsilon\|(\mathfrak{A} - (m + 1)I)^{-1}\|}.\end{aligned}$$

Since $\|A_r\|R_1^r \to 0$ as $r \to \infty$, for sufficiently large m

$$\|U_{m+1,s}s!\|R_1^{m+1} \leqq C_\varepsilon\varepsilon^s$$

and thus estimate (2.19) holds for all $k = 1, 2,\cdots$. But then

$$\begin{aligned}\|U_k(\zeta)R_1^k\| &\leqq \sum_{s=0}^{\infty} \|U_{ks}s!\| \frac{|\zeta|^s}{s!} \leqq C_\varepsilon \sum_{s=0}^{\infty} \frac{\varepsilon^s|\zeta|^s}{s!} \\ &= C_\varepsilon e^{\varepsilon|\zeta|} \leqq C_\varepsilon e^{\varepsilon\sqrt{R^2+\pi^2}}.\end{aligned}$$

We state our result in the form of a theorem.

THEOREM 2.1. *Suppose the following condition* (α) *is satisfied: there exist integers that are representable in the form* $n = \lambda - \mu$, *where* λ, $\mu \in \sigma(A_0)$, *each of them being an isolated point of the spectrum of the transform*

$$\mathfrak{A}X = A_0X - XA_0. \tag{2.13}$$

Suppose, further, that the series $\sum_0^\infty A_k z^k$ *converges in a disk* $|z| < \rho$. *Then the equation*

$$z\frac{dU}{dz} = \left(\sum_{k=0}^{\infty} A_k z^k\right) U \tag{2.11}$$

has for $0 < |z| < \rho$ *a solution of the form*

$$U(z) = \left(I + \sum_{k=1}^{\infty} U_k(\ln z) z^k\right) z^{A_0} = B(z) z^{A_0}, \tag{2.14}$$

where the U_k *are entire functions of* $\zeta = \ln z$ *satisfying for every* $\varepsilon > 0$ *the condition*

$$\|U_k(\ln z)\| = \|U_k(\zeta)\| \leqq C_\varepsilon e^{\varepsilon|\zeta|}. \tag{2.20}$$

If the integral points of the spectrum of the transform (2.13) *are poles of the resolvent* $R_\lambda(\mathfrak{A})$, *the functions* $U_k(\zeta)$ *are polynomials.*

The validity of the last assertion of the theorem readily follows from the arguments presented in the proof of the theorem if one takes into account Remark 2.1.

REMARK 2.2. Condition (α) means that there exist isolated points λ_{kj} and λ'_{kj} of $\sigma(A_0)$ whose difference is an integer:

$$n_k = \lambda_{kj} - \lambda'_{kj} \qquad (\lambda_{kj}, \lambda'_{kj} \in \sigma(A_0);\ j = 1,2,\cdots,r_k;\ k = 1,2,\cdots,p),$$

and that there are no other points in $\sigma(A_0)$ with integral differences.

REMARK 2.3. Condition (2.20) implies that $\lim_{z\to 0} B(z) = I$. It can be deduced from this relation in the same way as in Remark 1.1 that any solution $V(z)$ of equation (2.11) that is single valued in a disk $|z| < \rho$ cut along some radius can be represented in the form $V(z) = U(z)C$, where $U(z)$ is a solution obtained from (2.14) by choosing a single valued branch of the logarithm, i.e. $U(z)$ is a fundamental solution of the equation.

§ 3. **Finite-dimensional case** (dim $\mathfrak{B} < \infty$)

In this section we discuss the calculations needed to determine a solution of equation (2.11) in the finite-dimensional case.

We represent the operator A_0 in the form of a Jordan decomposition (see Exercise I.3):

$$A_0 = \sum_{j=1}^{n} (\lambda_j P_j + Q_j). \tag{3.1}$$

Here the λ_j ($j = 1,\cdots,n$) are the eigenvalues of A_0, the P_j are the corresponding spectral projections and the Q_j are the corresponding nilpotent operators satisfying the relations

$$P_j Q_j = Q_j, \quad Q_j^{m_j} = 0 \qquad (j = 1,\cdots,n). \tag{3.2}$$

Everywhere below we will make use of not the finite dimensionality of $\mathfrak{B}$ but relations (3.1) and (3.2), and thus our arguments remain valid in the infinite-dimensional case whenever A_0 is an operator of algebraic type (Exercise I.9).

We first consider the case when condition (1.10) is satisfied, i.e. when $\lambda_j - \lambda_k \neq n$ $(\neq 0)$ $(\lambda_j, \lambda_k \in \sigma(A_0))$ for any integer n.

In this case a solution is representable in the form of the series (1.11). To successively calculate the coefficients we must solve the equations

$$U_m A_0 - A_0 U_m + m U_m = F_m. \tag{1.13}$$

The solution of this equation can be represented in the form

$$U_m = -(\mathfrak{A} - mI)^{-1} F_m,$$

where $\mathfrak{A}$ is the transform (2.13) acting in the space $[\mathfrak{B}]$.

Let $\phi_{1,m}(\lambda, \mu) = (\lambda - \mu - m)^{-1}$. Then it is not difficult to conclude (see § I.3) that

$$(\mathfrak{A} - m\,I)^{-1} = \phi_{1,m}((A_0)_l, (A_0)_r).$$

We make use of formula (I.0.5), viz.

$$\phi(A_l, B_r)X = \sum_{j_1,j_2} \sum_{r_1,r_2} \frac{1}{r_1!\, r_2!} P'_{j_1} Q'^{r_1}_{j_1} X P''_{j_2} Q''^{r_2}_{j_2} \frac{\partial^{r_1+r_2}\phi(\lambda_{j_1}, \mu_{j_2})}{\partial \lambda^{r_1}_{j_1}\, \partial \mu^{r_2}_{j_2}}. \tag{3.3}$$

Applying this formula to the case when $A = B = A_0$ and $\phi = \phi_{1,m}$, we obtain the equality

$$U_m = -\sum_{j_1,j_2=1}^{n} \sum_{r_1=0}^{n_{j_1}-1} \sum_{r_2=0}^{n_{j_2}-1} (-1)^{r_1} \frac{(r_1+r_2)!}{r_1!\, r_2!} \frac{P_{j_1} Q^{r_1}_{j_1} F_m P_{j_2} Q^{r_2}_{j_2}}{(\lambda_{j_1} - \lambda_{j_2} - m)^{r_1+r_2+1}}. \tag{3.4}$$

In the special case when all of the elementary divisors are simple ($n_j = 1$ for $j = 1,\cdots,n$), i.e. when $A_0 = \sum_1^n \lambda_j P_j$, this formula acquires the simpler form

$$U_m = -\sum_{j_1,j_2=1}^{n} \frac{P_{j_1} F_m P_{j_2}}{\lambda_{j_1} - \lambda_{j_2} - m}. \tag{3.5}$$

We proceed now to the more complicated case when condition (1.10) is violated. In the present case, when the whole spectrum $\sigma(A_0)$ consists of a finite number of points, the condition (α) used in Theorem 2.1 is automatically satisfied.

A solution of equation (2.11) is now representable in the form of the series (2.14). To calculate the coefficient functions $U_k(\zeta)$, which, as we already know, are polynomials in the present case, we must find the solutions of minimal type of the differential equations

$$U'_m(\zeta) = (\mathfrak{A} - mI)U_m(\zeta) + F_m(\zeta) \tag{2.18}$$

in $[\mathfrak{B}]$.

We have seen that when $m < n_1$, where n_1 is the least integer that is representable in the form of a difference of the points λ_j, we can assume that $F_m(\zeta) = F_m =$

const and $U_m(\zeta) = \text{const}$. In this case $U_m(\zeta) = -(\mathfrak{A} - mI)^{-1}F_m$ and all of the calculations are carried out according to formula (3.4) given above.

In the remaining calculations we can no longer assume that U_m is a constant operator. Our procedure for solving equation (2.18) depends on whether or not the operator $\mathfrak{A} - mI$ is invertible.

In the first case, i.e. when $m \neq \lambda_j - \lambda_k$ $(j, k = 1,\cdots,n; j \neq k)$, we apply formulas (2.4), permitting us to construct from the polynomial

$$F_m(\zeta) = \sum_{k=0}^{d_m} F_{mk}\zeta^k$$

a solution

$$U_m(\zeta) = \sum_{k=0}^{d_m} U_{mk}\zeta^k,$$

in the form of a polynomial of the same degree. By virtue of (2.4) the coefficients of this polynomial have the form

$$U_{mk} = -\frac{1}{k!}\sum_{s=k}^{d_m} s!(\mathfrak{A} - mI)^{-(s-k+1)} F_{ms}. \tag{3.6}$$

We now make use of formula (3.3) for the function

$$\phi_{q,m}(\lambda, \mu) = (\lambda - \mu - m)^{-q}.$$

We get

$$(\mathfrak{A} - mI)^{-q} F_{ms} = \phi_{q,m}((A_0)_l, (A_0)_r)F_{ms}$$
$$= \sum_{j_1,j_2=1}^{n}\sum_{r_1=0}^{n_{j_1}-1}\sum_{r_2=0}^{n_{j_2}-1}(-1)^{r_1}\frac{(q + r_1 + r_2 - 1)!}{(q-1)!r_1!r_2!}\frac{P_{j_1}Q_{j_1}^{r_1}F_{ms}P_{j_2}Q_{j_2}^{r_2}}{(\lambda_{j_1} - \lambda_{j_2} - m)^{q+r_1+r_2}}.$$

Substituting this expression into (3.6), we find, finally, that

$$U_{mk} = -\frac{1}{k!}\sum_{s=k}^{d_m}\sum_{j_1,j_2=1}^{n}\sum_{r_1=0}^{n_{j_1}-1}\sum_{r_2=0}^{n_{j_2}-1}(-1)^{r_1}\frac{s!(s-k+r_1+r_2)!}{(s-k)!r_1!r_2!}$$
$$\times\frac{P_{j_1}Q_{j_1}^{r_1}F_{ms}P_{j_2}Q_{j_2}^{r_2}}{(\lambda_{j_1} - \lambda_{j_2} - m)^{r_1+r_2+s-k+1}}. \tag{3.7}$$

We now consider the second case, when $m = n_k$ $(k = 1,\cdots,p)$, i.e. when

$$m = \lambda_{kj} - \lambda'_{kj} \qquad (\lambda_{kj}, \lambda'_{kj} \in \sigma(A_0); j = 1,\cdots,r_k).$$

Here a solution of equation (2.18) should be sought by the method indicated in the proof of Lemma 2.2. It is convenient to do this in the following way. We write out (2.18) in more detail:

$$U'_m(\zeta) = A_0U_m(\zeta) - U_m(\zeta)A_0 - mU_m(\zeta) + F_m(\zeta). \tag{3.8}$$

Multiplying this equality from the left and from the right by projections P_{i_1} and

P_{i_2} and making use of formula (3.1), we obtain for the operator $X_{i_1i_2} + P_{i_1}U_mP_{i_2}$ the equation

$$X'_{i_1i_2} = Q_{i_1}X_{i_1i_2} - X_{i_1i_2}Q_{i_2} + (\lambda_{i_1} - \lambda_{i_2} - m)X_{i_1i_2} + P_{i_1}F_mP_{i_2}. \tag{3.9}$$

For those cases when

$$\lambda_{i_1} - \lambda_{i_2} - m \neq 0, \tag{3.10}$$

a solution is found by means of the same formula (3.7). The sum of these solutions is a polynomial U'_m of the same degree as F_m with coefficients

$$\begin{aligned} U'_{mk} = &- \frac{1}{k!}\sum_{s=k}^{d_m}\sum_{j_1,j_2=1}^{n}\sum_{r_1=0}^{n_{j_1}-1}\sum_{r_2=0}^{n_{j_2}-1}(-1)^{r_1}\frac{s!(s-k+r_1+r_2)!}{(s-k)!r_1!r_2!} \\ &\times \frac{P_{j_1}Q_{j_1}^{r_1}F_{ms}P_{j_2}Q_{j_2}^{r_2}}{(\lambda_{j_1}-\lambda_{j_2}-m)^{r_1+r_2+s-k+1}} \qquad (\lambda_{j_1}-\lambda_{j_2}\neq m). \end{aligned} \tag{3.11}$$

If on the other hand (3.10) is not satisfied, equation (3.9) takes the form

$$X'_{i_1i_2} = Q_{i_1}X_{i_1i_2} - X_{i_1i_2}Q_{i_2} + P_{i_1}F_mP_{i_2}.$$

According to formula (II.1.11) the solution of this equation satisfying the condition $X_{i_1i_2}(0) = 0$ has the form

$$X_{i_1i_2}(\zeta) = \int_0^{\zeta} e^{Q_{i_1}(\zeta-\tau)}P_{i_1}F_m(\tau)P_{i_2}e^{-Q_{i_2}(\zeta-\tau)}d\tau. \tag{3.12}$$

We note that expression (3.12) is a polynomial since

$$e^{Q_jt} = \sum_{s=0}^{p_j}\frac{Q_j^s}{s!}t^s.$$

The degree of this polynomial, as can be easily verified, is equal to $d_m + p_{i_1} + p_{i_2} + 1$.

Summing expressions (3.12) over all i_1 and i_2 for which $\lambda_{i_1} - \lambda_{i_2} = m$, we obtain an operator U_m which when added to U'_m gives us the desired solution of equation (3.8).

We see in addition that this solution is a polynomial if F_m is a polynomial. Since the function $F_{n_1}(\zeta) = F''_{n_1}$ is a polynomial of zero degree, the functions $U_m(\zeta)$ are also polynomials for all m. We note that the degrees of these polynomials are the same for $m > n_p$.

The coefficients of the polynomials U_m are found by means of simple algebraic operations after reducing the matrix of the operator A_0 to a Jordan normal form.

We note that in the simplest case, when $Q_j = 0$ $(j = 1,\cdots,n)$, formula (3.7) takes the form

$$U_{mk} = -\frac{1}{k!}\sum_{s=k}^{d_m}\sum_{j_1,j_2=1}^{n} s!\frac{P_{j_1}F_{ms}P_{j_2}}{(\lambda_{j_1}-\lambda_{j_2}-m)^{s-k+1}},$$

while formula (3.12) takes the form

$$X_{i_1 i_2} = \int_0^{\zeta} P_{i_1} F_m(\tau) P_{i_2} \, d\tau.$$

§ 4. Case of integral differences of points of the continuous spectrum of A_0

1. *Construction of the solution.* In this section the restrictions on the spectrum of A_0 under which we consider equation (2.11) will be weakened still further. We will be able to consider equation (2.11) in certain cases when the continuous spectrum of A_0 contains points with integral differences.

We will say that a decomposition

$$\sigma(A_0) = \sigma_1 \cup \sigma_2 \cup \cdots \cup \sigma_n \tag{4.1}$$

of the spectrum $\sigma(A_0)$ into spectral sets is *normal* if the difference $\lambda - \mu$ is not equal to an integer when λ and μ belong to one and the same spectral set σ_j $(j = 1,\cdots,n)$.

By an *integral shift* of a spectral set σ_j we mean a transformation taking it into a set of the form $\sigma_j - k = \{\mu = \lambda - k \mid \lambda \in \sigma_j\}$, where k is a nonzero integer. The shift by one unit to the left will be called the *elementary shift.*

An integral shift can result in two or more of the components of the decomposition (4.1) intersecting. By taking the union of the intersecting components, we obtain a new decomposition with fewer components. Such a shift will be called a *coalescing shift.*

We say that the decomposition (4.1) is *stably normal* if every succession of coalescing integral shifts of its spectral sets again leads to a normal decomposition.

Clearly, in order for the decomposition (4.1) to be stably normal it is necessary (but not sufficient) that the set of differences

$$\Lambda_{jk} = \{\lambda = \lambda_j - \lambda_k \mid \lambda_j \in \sigma_j, \lambda_k \in \sigma_k\}$$

for each fixed pair of indices j, k $(j \neq k)$ contain not more than one integer. Thus there exists at most one integral shift of a spectral set σ_k that coalesces it with another spectral set σ_j $(j, k = 1,\cdots, n;\ j \neq k)$.

It follows from the definition that if $\sigma(A_0)$ admits a stably normal decomposition, we can transform it by means of a finite number of integral shifts of its components into a set not containing pairs of points with integral differences (such a set will be said to be *normal*).

We now prove an auxiliary proposition that plays an essential role in future constructions.

LEMMA 4.1. *Let $\sigma(A_0) = \sigma_1 \cup \sigma_2$ be a (not necessarily normal) decomposition of $\sigma(A_0)$ into spectral sets and let P_k $(k = 1, 2)$ be the corresponding spectral projections. The substitution*

$$U = (P_2 z + P_1) Y \tag{4.2}$$

takes the equation

$$z \frac{dU}{dz} = \left(\sum_{k=0}^{\infty} A_k z^k \right) U \qquad (|z| < \rho) \tag{4.3}$$

into an equation

$$z \frac{dY}{dz} = \left(\sum_{k=0}^{\infty} B_k z^k \right) Y \qquad (|z| < \rho) \tag{4.4}$$

of the same form, the series in the right side of which converges in the same disk as the series in the right side of equation (4.3).

The operator B_0 *satisfies the relation*

$$B_0 = A_0 - P_2 + P_2 A_1 P_1 \tag{4.5}$$

and leaves the subspace $\mathfrak{B}_2 = P_2 \mathfrak{B}$ *invariant. Its spectrum is the set* $\sigma(B_0) = \sigma_1 \cup (\sigma_2 - 1)$, *which is obtained from* $\sigma(A_0)$ *by means of an elementary shift of the spectral set* σ_2.

PROOF. Substituting (4.2) into (4.3), we obtain the equation

$$z P_2 Y + z(P_2 z + P_1) \frac{dY}{dz} = \sum_{k=0}^{\infty} A_k z^k (P_2 z + P_1) Y$$

or, since

$$(P_2 z + P_1)^{-1} = P_2 / z + P_1 \qquad (z \neq 0), \tag{4.6}$$

the equation

$$z \frac{dY}{dz} = \left(\frac{P_2}{z} + P_1 \right) \left[\sum_{k=0}^{\infty} A_k z^k (P_2 z + P_1) - z P_2 \right] Y.$$

Removing the parentheses and brackets, we obtain an equation of form (4.4) since the only term containing a negative power of z is equal to zero:

$$P_2 A_0 P_1 / z = A_0 P_2 P_1 / z = 0.$$

Relation (4.5) is verified by a direct calculation.

From (4.5) it follows that $B_0 P_2 = A_0 P_2 - P_2 = P_2 B_0 P_2$ and hence $B_0 x \in \mathfrak{B}_2$ for $x \in \mathfrak{B}_2$ $(x = P_2 x)$.

The last assertion of the lemma readily follows from the matrix representation of B_0,

$$B_0 = \begin{pmatrix} A_0 & 0 \\ P_2 A_1 & A_0 - I \end{pmatrix},$$

corresponding to the direct decomposition

$$\mathfrak{B} = \mathfrak{B}_1 + \mathfrak{B}_2\,(= P_1\mathfrak{B} + P_2\mathfrak{B}).$$

REMARK 4.1. It is easy to see that the following formulas are valid:

$$z^{P_2} = e^{P_2 \ln z} = P_2 z + P_1; \quad (-z)^{P_2} = -P_2 z + P_1, \tag{4.7a}$$

$$z^{-P_2} = P_2/z + P_1; \quad (-z)^{-P_2} = -P_2/z + P_1. \tag{4.7b}$$

REMARK 4.2. Suppose there exists a decomposition $\sigma_1 = \sigma_1' \cup \sigma_1''$ of σ_1 into spectral sets of A_0 such that $\sigma_2 - 1 \cap \sigma_1'' = \varnothing$.

Let $\tilde{P}_1''$ denote the spectral projection of B_0 corresponding to the spectral set σ_1. Then

$$\tilde{P}_1'' P_2 = 0. \tag{4.8}$$

For suppose $\tilde{P}_2'$ is the spectral projection of B_0 corresponding to the set $\sigma_2 - 1 \cup \sigma_1'$. Then $\tilde{P}_1'' \tilde{P}_2' = 0$.

On the other hand, we have $P_2 x \in \mathfrak{B}_2$ for any $x \in \mathfrak{B}$ and, since both $\tilde{P}_2'$ and B_0 leave $\mathfrak{B}_2$ invariant, $\tilde{P}_2' P_2 x = P_2 x$.

Hence $\tilde{P}_1'' P_2 x = \tilde{P}_1'' \tilde{P}_2' \tilde{P}_2 x = 0$ for any $x \in \mathfrak{B}$.

Lemma 4.1 directly implies the following result.

THEOREM 4.1. *If the spectrum $\sigma(A_0)$ admits a stably normal decomposition, every fundamental solution of equation* (4.3) *is representable in the form*

$$U(z) = \left(\sum_{k=0}^{\infty} U_k z^k\right) z^B C, \tag{4.9}$$

where the U_k $(k = 1,2,\cdots)$ and B are operators in $[\mathfrak{B}]$ determined by the equation, $C \in [\mathfrak{B}]$ is an invertible operator and the series converges in the disk $|z| < \rho$.

PROOF. There exists a finite sequence of elementary shifts of the components of the decomposition (4.1) that takes $\sigma(A_0)$ into a normal set Λ. Let $B_0^{(0)} = A_0$, $B_0^{(1)},\cdots,B_0^{(q)}$ be the sequence of operators (4.5) corresponding to these shifts, so that

$$B_0^{(k+1)} = B_0^{(k)} - P_2^{(k)} + P_2^{(k)} B_1^{(k)} P_1^{(k)}, \tag{4.10}$$

where $P_2^{(k)}$ is the spectral projection of $B_0^{(k)}$ corresponding to the spectral set $\sigma_2(B_0^{(k)})$ being shifted under the recurrent transformation (4.2) and $P_1^{(k)} = I - P_2^{(k)}$.

Carrying out the series of transformations (4.2):

$$Y^{(k)} = (P_2^{(k)} z + P_1^{(k)}) Y^{(k+1)} \qquad (Y^{(0)} = U; k = 0,\cdots, q-1), \tag{4.11}$$

we arrive at the equation

$$z\frac{dY^{(q)}}{dz} = \left(\sum_{j=0}^{\infty} B_j^{(q)} z^j\right) Y^{(q)}. \tag{4.12}$$

Since the spectrum of $B_0^{(q)}$ coincides by construction with the set Λ and hence does not contain pairs of points with integral differences, a fundamental solution of this equation has the form

$$Y^{(q)}(z) = \left(\sum_{j=0}^{\infty} Y_j^{(q)} z^j\right) z^{B_0^{(q)}} C \qquad (Y_0^{(q)} = I) \quad (|z| < \rho).$$

By successively carrying out the transformations (4.11), we obtain the representation

$$U(z) = \prod_{k=q}^{0} (P_2^{(k)} z + P_1^{(k)}) \sum_{j=0}^{\infty} Y_j^{(q)} z^j \cdot z^{B_0^{(q)}} C. \tag{4.13}$$

Removing the parentheses in (4.13), we get the required representation (4.9).

The theorem is proved.

2. *More exact analysis of the solution.* The representation (4.9) is not sufficiently convenient for our investigation, and we therefore subject it to an additional transformation. In this connection we require the following auxiliary proposition.

LEMMA 4.2. *Suppose that under the conditions of Lemma* 4.1 *the set* $\sigma_1 = \sigma_1' \cup \sigma_1''$, *where* σ_1' *and* σ_1'' *are spectral sets of* A_0 *such that*

$$\sigma_2 - 1 \cap \sigma_1' \neq \emptyset, \quad \sigma_2 - 1 \cap \sigma_1'' = \emptyset.$$

There exists an invertible operator $C \in [\mathfrak{B}]$ *such that the following represention is valid:*

$$z^{B_0} C z^{-A_0} = P_2/z + P_1 + P_2 \mathcal{F}(\ln z) P_1, \tag{4.14}$$

where $\mathcal{F}(\zeta)$ ($\zeta = \ln z$) *is an entire function of exponential type with values in* $[\mathfrak{B}]$. *The type of this function coincides with the spectral radius of the restriction of the transform* $\mathfrak{A} - I$ *to its invariant subspace*

$$\mathfrak{B}_{21}'' = \{P_2 X P_1'' \mid X \in [\mathfrak{B}]\}.$$

PROOF. The operator $S(\zeta) = z^{B_0} C = e^{B_0 \zeta} C$ satisfies the equation

$$dS/d\zeta = B_0 S = (A_0 - P_2 + P_2 A_1 P_1) S \tag{4.15}$$

and the condition $S(0) = C$. From the formulas in § II.1 we see that S satisfies the integral equation

$$S = e^{(A_0 - P_2)\zeta} C + \int_0^{\zeta} e^{(A_0 - P_2)(\zeta - t)} P_2 A_1 P_1 S \, dt. \tag{4.16}$$

Multiplying this equality from the left by P_1 and taking into account the equality $P_1 P_2 = 0$ and (4.7), we get

$$P_1 S = e^{(A_0 - P_2)\zeta} P_1 C = e^{A_0 \zeta} (P_2/z + P_1) P_1 C = e^{A_0 \zeta} P_1 C.$$

Substituting this expression into (4.16), we obtain the representation

$$S = e^{(A_0-P_2)\zeta}C + \int_0^{\zeta} e^{(A_0-P_2)(\zeta-t)}P_2A_1P_1e^{A_0t}\,dtC. \tag{4.17}$$

Multiplying this equality from the right by $z^{-A_0} = e^{-A_0\zeta}$, we find that

$$z^{B_0}Cz^{-A_0} = e^{(A_0-P_2)\zeta}Ce^{-A_0\zeta} + \int_0^{\zeta} e^{(A_0-P_2)(\zeta-t)}P_2A_1P_1e^{A_0t}\,dtCe^{-A_0\zeta}. \tag{4.18}$$

Let P_1' and P_1'' denote the spectral projections of A_0 corresponding to the spectral sets σ_1' and σ_1''. We will seek the operator C in the form $C = I + P_2X_1P_1''$.

In this case $P_1''C = P_1$ and equality (4.18) takes the form

$$\begin{aligned} z^{B_0}Cz^{-A_0} &= e^{-P_2\zeta} + e^{(A_0-P_2)\zeta}P_2XP_1''e^{-A_0\zeta} \\ &\quad + \int_0^{\zeta} e^{(A_0-P_2)(\zeta-t)}P_2A_1P_1e^{-A_0(\zeta-t)}\,dt \\ &= \frac{P_2}{z} + P_1'' + \int_0^{\zeta} e^{(A_0-P_2)(\zeta-t)}P_2A_1P_1'e^{-A_0(\zeta-t)}\,dt \\ &\quad + e^{(A_0-P_2)\zeta}P_2XP_1''e^{-A_0\zeta} + \int_0^{\zeta} e^{(A_0-P_2)(\zeta-t)}P_2A_1P_1''e^{-A_0(\zeta-t)}\,dt. \end{aligned} \tag{4.19}$$

We now show that the operator X can be chosen in such a way that the sum R of the last two summands in (4.19) becomes a constant.

To this end we consider the Banach space $[\mathfrak{B}]$ and the transform

$$\mathfrak{A}X = (A_0 - P_2)X - XA_0$$

acting in it.

From (II.1.9) we have the representation

$$e^{\zeta\mathfrak{A}}X = e^{(A_0-P_2)\zeta}\,Xe^{-A_0\zeta},$$

which permits us to write R in the form

$$R = e^{\zeta\mathfrak{A}}P_2XP_1'' + \int_0^{\zeta} e^{(\zeta-t)\mathfrak{A}}P_2A_1P_1''\,dt.$$

Let $[\mathfrak{B}]_{21}''$ denote the subspace of $[\mathfrak{B}]$ consisting of operators of the form P_2XP_1'' $(X \in [\mathfrak{B}])$. It is easily seen that this subspace is invariant under the transform $\mathfrak{A}$. From (I.3.10) it follows by virtue of the condition $(\sigma_2 - 1) \cap \sigma_1'' = \varnothing$ that its restriction $\mathfrak{A}_{12}$ (which coincides with the restriction of $\mathfrak{A} - I$) to this subspace is invertible.

Therefore

$$\begin{aligned} \int_0^{\zeta} e^{(\zeta-t)\mathfrak{A}}P_2A_1P_1\,dt &= \int_0^{\zeta} e^{(\zeta-t)\mathfrak{A}_{12}}P_2A_1P_1''\,dt \\ &= -\,e^{(\zeta-t)\mathfrak{A}_{12}}\mathfrak{A}_{12}^{-1}P_2A_1P_1''\Big|_0^{\zeta} = e^{\zeta\mathfrak{A}_{12}}\mathfrak{A}_{12}^{-1}P_2A_1P_1'' - \mathfrak{A}_{12}^{-1}P_2A_1P_1''. \end{aligned}$$

Setting $X = -\mathfrak{A}_{12}^{-1}P_2A_1P_1''$, we find that $R = -\mathfrak{A}_{12}^{-1}P_2A_1P_1''$.
As a result, $z^{B_0}Cz^{-A_0} = P_2/z + P_1 + \mathscr{F}(\zeta)$,

$$\mathscr{F}(\zeta) = -\mathfrak{A}_{12}^{-1}P_2A_1P_1'' + \int_0^\zeta e^{(\zeta-t)\mathfrak{A}}P_2A_1P_1'\,dt. \tag{4.20}$$

It is easily seen that the exponential type of the entire function $\mathscr{F}(\zeta)$ coincides with the exponential type of the entire function

$$\int_0^\zeta e^{(\zeta-t)\mathfrak{A}_{12}}\,dt = \sum_{n=0}^{\infty}\frac{\mathfrak{A}_{12}^n}{(n+1)!}\zeta^{n+1},$$

which can be calculated by means of the formula (see Exercise I.22)

$$\nu = \overline{\lim_{n\to\infty}}\,\|\mathfrak{A}_{12}^n/(n+1)\|^{1/n} = \lim_{n\to\infty}\|\mathfrak{A}_{12}^n\|^{1/n}.$$

REMARK 4.3. If $\sigma_1' = \varnothing$, i.e. $\sigma_1 = \sigma_1''$, and hence

$$\sigma_2\cap\sigma_1 = \varnothing, \tag{4.21}$$

it follows from (4.20) that

$$\mathscr{F}(\zeta) = -\mathfrak{A}_{12}^{-1}P_2AP_1 = \text{const.}$$

Thus in this case

$$z^{B_0}Cz^{-A_0} = P_2/z + P_1 + P_2KP_1, \tag{4.22}$$

where $K \in [\mathfrak{B}]$ is a constant. We note that condition (4.21) is equivalent to the fact that the elementary shift in question is not a coalescing one.

REMARK 4.4. If each of the sets σ_2 and σ_1' consists of a single point, the spectrum of the restriction of the transform $\mathfrak{A}$ to the invariant subspace $[\mathfrak{B}]_{21}' = \{P_2XP_1' \mid X \in [\mathfrak{B}]\}$ consists of only zero. Therefore $\mathscr{F}(\zeta)$ is in this case a function of minimal type.

If in addition the subspaces $P_2\mathfrak{B}$ $(\subset \mathfrak{B})$ and $P_1'\mathfrak{B}$ $(\subset \mathfrak{B})$ are finite dimensional, the subspace $[\mathfrak{B}]_{21}'$ $(\subset [\mathfrak{B}])$ is also finite dimensional. An operator in a finite-dimensional space with simply zero for a spectrum is nilpotent. This implies that the entire function $\mathscr{F}(s)$ degenerates into an operator polynomial.

REMARK 4.5. Suppose the set σ_2 also decomposes into spectral sets of A_0: $\sigma_2 = \sigma_2' \cup \sigma_2''$, such that $\sigma_2'' - 1 \cap \sigma_1 = \varnothing$, and suppose P_2' and P_2'' are the corresponding spectral projections of A_0. Let $\mathfrak{A}'$ denote the transform

$$\mathfrak{A}'X = (A_{01} - P_2')X - XA_{02}, \tag{4.23}$$

where A_{01} and A_{02} are the restrictions of A_0 to the invariant subspaces $P_1'\mathfrak{B}$ and $P_2'\mathfrak{B}$.

By arguing in the same way as in the proof of the lemma, we can establish the existence of an operator C such that the type of the function $\mathscr{F}(\zeta)$ coincides with the spectral radius of the transform $\mathfrak{A}'$.

We are now ready to prove the following theorem.

THEOREM 4.2. *If the spectrum $\sigma(A_0)$ admits a stably normal decomposition* (4.1), *a fundamental solution of equation* (4.3) *is representable in the form*

$$U(z) = \left[I + F_0(z) + \sum_{k=1}^{r} F_k(z)\mathcal{F}_k(\ln z) \right] z^{A_0},$$

where $F_k(z) = \sum_1^\infty F_{kj}\, z_j$ $(k = 0,1,\cdots,r;\ |z| < \rho)$ is an analytic operator function in the disk $|z| < \rho$ and the $\mathcal{F}_k(\zeta)$ are entire functions of exponential type. The number r coincides with the number of coalescing shifts while the type of each of the functions $\mathcal{F}_k(\zeta)$ can be determined from the spectral sets that coalesce under these shifts with the use of Remark 4.5.

PROOF. We will make use of the notation introduced in the proof of Theorem 4.1. We show below that there exists a sequence of transformations (4.2) of the original equation (4.3) such that the following condition is satisfied at each step:

$$P_1^{(k+1)} P_2^{(k)} = 0 \qquad (k = 0,1,\cdots,q-1). \tag{4.24}$$

We recall that $P_2^{(k)}$ is the spectral projection of $B_0^{(k)}$ corresponding to the spectral set $\sigma_2^{(k)}$ being shifted by an elementary transformation under the recurrent substitution

$$Y^{(k)}(z) = (P_2^{(k)} z + P_1^{(k)}) Y^{(k+1)}(z) \qquad (Y^{(0)}(z) = U(z);\ k = 0,1,\cdots,q-1), \tag{4.11}$$

$P_1^{(k)} = I - P_2^{(k)}$ is the complementary projection and hence $P_1^{(k+1)}$ is the spectral projection of $B_0^{(k+1)}$ corresponding to the spectral set $\sigma_1^{(k+1)}$ that remains fixed under the $(k + 1)$th transformation. Since the operators $B_0^{(k)}$ and $B_0^{(k+1)}$ are connected by relation (4.10) it follows from Remark 4.2 that (4.24) will be satisfied if $\sigma_1^{(k+1)} \subseteq \sigma_1^{(k)}$ $(k = 0,1,\cdots,q-1)$ or, equivalently,

$$\sigma_2^{(k+1)} \supseteq \sigma_2^{(k)} \qquad (k = 0,1,\cdots,q-1). \tag{4.25}$$

We put aside for the present a proof of the existence of a sequence of elementary shifts satisfying condition (4.25) and use relations (4.24) to transform expression (4.13). We write this expression in the form

$$U(z) = \prod_{k=q-1}^{0} (P_2^{(k)} z + P_1^{(k)}) \left[\sum_{j=0}^{\infty} Y_j^{(q)} z^j \right] \left[\prod_{k=0}^{q-1} (z^{B_0^{(k+1)}} C_k z^{-B_0^{(k)}}) \right] z^{A_0},$$

where the C_k are operator constants chosen in the manner indicated in Lemma 4.2.

We successively remove the parentheses in this expression. Let

$$U_s(z) = \prod_{k=q-1}^{s} (P_2^{(k)} z + P_1^{(k)}) \left[\sum_{i=0}^{\infty} Y_i^{(q)} z^i \right] \left[\prod_{k=s}^{q-1} (z^{B_0^{(k+1)}} C_k z^{-B_0^{(k)}}) \right].$$

Then

$$\begin{aligned} U_{s-1}(z) &= (P_2^{(s-1)} z + P_1^{(s-1)})\, U_s(z) z^{B_0^{(s)}} C_{s-1} z^{-B_0^{(s-1)}} \\ &= (P_2^{(s-1)} z + P_1^{(s-1)}) U_s(z) \left(\frac{P_2^{(s-1)}}{z} + P_1^{(s-1)} + P_2^{(s-1)} \mathcal{F}_s(\ln z) P_1^{(s-1)} \right), \end{aligned} \tag{4.26}$$

where $\mathscr{F}_s(\zeta)$ is the entire function described in Lemma 4.2.

For $s = q - 1$ we have

$$U_{q-1}(z) = \sum_{j=0}^{\infty} Y_j^{(q)} z^j = I + \sum_{j=1}^{\infty} Y_j^{(q)} z^j = I + F_0^{(q)}(z),$$

where $F_0^{(q)}(z)$ is an operator function that is analytic in the disk $|z| < \rho$ ($F_0^{(q)}(0) = 0$).

Let us assume that we have already established the representation

$$U_s(z) = I + F_0^{(s+1)}(z) + \sum_{k=1}^{r_s} F_k^{(s+1)}(z)\mathscr{F}_k^{(s+1)}(\ln z), \tag{4.27}$$

in which the entire functions $\mathscr{F}_k^{(s+1)}$ satisfy the relations

$$\mathscr{F}_k^{(s+1)}(\ln z) = \mathscr{F}_k^{(s+1)}(\ln z)P_1^{(s)}. \tag{4.28}$$

We will show that an analytic representation (4.27) with the same (one more than the same) number of terms is valid for $U_{s-1}(z)$ if the shift corresponding to the transformation is not (is) a coalescing shift. The theorem will then follow since $U(z) = U_0(z)z^{A_0}$.

Substituting (4.27) into (4.26) and using (4.28) and (4.24), we get

$$\begin{aligned} U_{s-1}(z) &= (P_2^{(s-1)}z + P_1^{(s-1)})\Big[I + F_0^{(s+1)}(z) + \sum_{k=1}^{r_s} F_k^{(s+1)}(z)\mathscr{F}_k^{(s+1)}(\ln z)\Big] \\ &\quad \times [P_2^{(s-1)}/z + P_1^{(s-1)} + P_2^{(s-1)}\mathscr{F}_s(\ln z)P_1^{(s-1)}] \\ &= I + P_1^{(s-1)}F_{01}^{(s+1)}P_2^{(s-1)} \\ &\quad + (P_2^{(s-1)}z + P_1^{(s-1)})\Big[F_0^{(s+1)}(z) + \sum_{k=1}^{r_s} F_k^{(s+1)}(z)\mathscr{F}_k^{(s+1)}(\ln z)\Big]P_1^{(s-1)} \\ &\quad + (P_2^{(s-1)}z + P_1^{(s-1)})\sum_{k=2}^{\infty} F_{0k}^{(s+1)}z^{k-1}P_2^{(s-1)} \\ &\quad + [(P_2^{(s-1)}z + P_1^{(s-1)})F_0^{(s+1)}(z) + P_2^{(s-1)}z]P_2^{(s-1)}\mathscr{F}_s(\ln z)P_1^{(s-1)}. \end{aligned}$$

We put

$$F_k^{(s)}(z) = (P_2^{(s-1)}z + P_1^{(s-1)})F_k^{(s+1)}(z) \quad\text{and}\quad \mathscr{F}_k^{(s)}(\zeta) = \mathscr{F}_k^{(s+1)}(\zeta)P_1^{(s-1)}$$
$$(k = 1,2,\cdots,r_s)$$

and, for the case of a coalescing shift,

$$F_{r_{s+1}}^{(s)} = [(P_2^{(s-1)}z + P_1^{(s-1)})F_0^{(s+1)}(z) + P_2^{(s-1)}z]P_2^{(s-1)} \qquad (r_{s+1} = r_s + 1);$$

$$F_0^{(s)}(z) = (P_2^{(s-1)}z + P_1^{(s-1)})\Big(F_0^{(s+1)}(z)P_1^{(s-1)} + \sum_{k=2}^{\infty} F_{0k}^{(s+1)}z^{k-1}P_2^{(s-1)}\Big);$$

$$\mathscr{F}_{r_{s+1}}^{(s)}(\zeta) = \mathscr{F}_s(\zeta)P_1^{(s-1)}.$$

In the case of a noncoalescing shift the operator $F_s(\zeta)$ is a constant and the corresponding summands should be attached to $F_0^{(s)}(z)$; here $r_{s+1} = r_s$.

As a result, we obtain the formulas

$$U_{s-1}(z) = I + P_1^{(s-1)}F_{01}^{(s-1)}P_2^{(s-1)} + F_0^{(s)}(z) + \sum_{k=1}^{r_{s+1}} F_k^{(s)}(z)\mathscr{F}_k^{(s)}(\ln z),$$

condition (4.28) being satisfied by construction:

$$\mathscr{F}_k^{(s)}(\zeta) = \mathscr{F}_k^{(s)}(\zeta)P_1^{(s-1)}.$$

If we show that

$$P_1^{(s-1)}F_{01}^{(s-1)}P_2^{(s-1)} = 0, \tag{4.29}$$

we will have proved formula (4.27) by induction.

In order to prove (4.29) we note that the function

$$Y_{s-1}(z) = U_{s-1}(z)z^{B_0^{(s-1)}} \tag{4.30}$$

satisfies the equation

$$z\frac{dY_{s-1}}{dz} = \sum_{k=0}^{\infty} B_k^{(s-1)}z^k Y_{s-1}. \tag{4.31}$$

Substituting (4.30) into (4.31) and separating out the terms not depending on z, we see (cf. (1.12)) that $U_{s-1,0} = I + P_1^{(s-1)}F_{01}^{(s-1)}P_2^{(s-1)}$ commutes with the operator $B_0^{(s-1)}$ and hence with its spectral projection $P_1^{(s-1)}$.

But then

$$\begin{aligned} P_1^{(s-1)}F_{01}^{(s-1)}P_2^{(s-1)} &= P_1^{(s-1)}(U_{s-1,0} - I) = (U_{s-1,0} - I)P_1^{(s-1)} \\ &= P_1^{(s-1)}F_{01}^{(s-1)}P_2^{(s-1)}P_1^{(s-1)} = 0. \end{aligned}$$

Thus the proof of the theorem will be complete if we show that it is possible to construct a system of shifts satisfying relation (4.25). This can be done by regrouping the spectral sets of the decomposition (4.1) into a new decomposition in the following manner.

We will say that two spectral sets σ_{j_1} and σ_{j_2} are *comparable* ($\sigma_{j_1} \sim \sigma_{j_2}$) if there exists an integer $k_{j_1j_2} \in \Lambda_{j_1j_2}$. We partition the totality of spectral sets into minimal classes containing together with each spectral set all of the sets that are comparable with it. Two sets σ_i and σ_j belong to one and the same class if there exists a chain

$$\sigma_i = \sigma_{j_1} \sim \sigma_{j_2} \sim \cdots \sim \sigma_{j_r} = \sigma_j.$$

In this case we put $k_{ij} = k_{ij_2} + k_{j_2j_3} + \cdots + k_{j_{r-1}j}$. From the stable normality of the decomposition it follows that the number k_{ij} is uniquely defined. Arbitrarily choosing a set σ_j in a given class, we assign to each set σ_s of this class the index

$$\alpha_s = k_{js} - \min k_{jp}.$$

Finally, we let $\tilde{\sigma}_k$ denote the union of the sets of index k from all of the classes (if they exist). As a result, we obtain the decomposition

$$\sigma(B_0) = \tilde{\sigma}_0 \cup \tilde{\sigma}_{j_1} \cup \cdots \cup \tilde{\sigma}_{j_r}.$$

The following sequence of elementary shifts has the required properties: we first shift the set $\tilde{\sigma}_{j_r}$; after $j_p - j_{p-1}$ elementary shifts we adjoin the set $\sigma_{j_{r-1}}$ to it and shift the union $\tilde{\sigma}_{j_{r-1}} \cup \tilde{\sigma}_{j_r}$, and so on. The theorem is proved.

COROLLARY 4.1. *If the spectrum $\sigma(A_0)$ admits a stably normal decomposition* (4.1), *a fundamental solution of equation* (4.3) *is representable in the form*

$$U(z) = \left(I + \sum_{k=1}^{\infty} U_k(\ln z) z^k \right) z^{A_0},$$

where the $U_k(\zeta)$ are entire functions of exponential type.

Let n_j ($j = 1,\cdots,p$) denote the natural numbers belonging to the spectrum of the transform $\mathfrak{A}$ and let σ_j denote the connected spectral sets containing the points n_j.

Let $\nu \geqq 0$ denote the least number having the property that for any $\varepsilon > 0$ the disk $|\lambda - n_j| < \nu + \varepsilon$ covers the set σ_j.

All of the functions $U_k(\zeta)$ ($k = 1,2,\cdots,\infty$) are of exponential type at most ν.

In particular, if $n_1,\cdots, n_p$ are isolated points of the spectrum of $\mathfrak{A}$, all of the functions $U_k(\zeta)$ are of minimal type (cf. Theorem 2.1).

EXERCISES

1. Suppose $A(z)$ has a pole of order k at $z = 0$. Obtain the following estimates (Hille [1]) for the behavior of the solution of problem (1.1) in a neighborhood of the origin:

a) for $k = 1$

$$\|x_0\| B^{-1} |\zeta|^M \leqq \|x(\zeta)\| \leqq \|x_0\| B |\zeta|^{-M},$$

where $M = \sup_{|\zeta|<\rho} |\zeta| \, \|A(\zeta)\|$ and B is a finite constant when $\arg \zeta$ is bounded;

b) for $k > 1$

$$\|x_0\| B^{-1} e^{-|\zeta|^{1-k}/(k-1)} \leqq \|x(\zeta)\| \leqq \|x_0\| B e^{|\zeta|^{1-k}/(k-1)}.$$

2. Suppose that condition (α) of § 2.2 is satisfied and, in addition, that the invariant subspaces corresponding to the integers that are points of the spectrum of the transform $\mathfrak{A}X = A_0X - XA_0$ are finite dimensional. We denote the sum of their dimensions by N.

Show that a solution of equation (2.11) of form (2.14) can be obtained by putting

$$U(z) = \lim_{\eta \to 0} \frac{\partial^N}{\partial \eta^N} w(\zeta, \eta), \tag{0.1}$$

where $w(\zeta,\eta) = \sum_{m=0}^{\infty} a_m(\eta) \zeta^{A_0 + (m+\eta)I}$ and the functions $a_m(\eta)$ are regular in a neighborhood of $\eta = 0$ (Hille [1]).

Hint. In the equations for the coefficients resulting from the substitution of the series (0.1) into the equation, put $a_m(\eta) = \eta^{N-k_m} b_m(\eta)$, where k_m is the sum of the dimensions of the invariant subspaces corresponding to the integral points of $\sigma(A_0)$ not exceeding m.

3. Show that with the use of the substitution $x(\zeta)=\phi(\zeta)w(\zeta)$, where $\phi(\zeta)$ is regular in a neighborhood of the origin, it is possible to reduce equation (1.1) to the form $\zeta w'(\zeta) = P(\zeta)w(\zeta)$, where $P(\zeta)$ is a polynomial containing only those terms whose exponents are points of the spectrum of $\mathfrak{A}$.

NOTES

The theory of equations with a regular singularity was constructed in the finite-dimensional case (for a system of equations of first order and for a single equation of higher order) by Fuchs [1] and Frobenius [1]. This case has been extensively studied; a presentation of it can be found, for example, in the books of Gantmaher [1] and Coddington and Levinson [1].

The more complicated variant when the operator A_0 has eigenvalues with integral differences is usually considered with the use of the Jordan normal form of the matrix A_0. Another method, which was applied by Frobenius in the case of an equation of higher order to obtain a definitive expression for the solution, requires differentiating up to an order equal to the sum of the multiplicities of the corresponding eigenvalues.

E. Hille [1] was the first to consider equation (1.1) in the infinite-dimensional case. His approach is followed in § 1. In the case when the spectrum $\sigma(A_0)$ has numbers with integral differences, Hille modifies the Frobenius method. He is able to do this only when the corresponding points of the spectrum are isolated and have finite multiplicities.

In § 2 we apply another method (Ju. L. Daleckiĭ and I. K. Korobkova [1]), which is based on the use of double logarithm-power series and does not require finite multiplicities of the eigenvalues.

The results presented in § 4 are due to P. A. Švarcman [1]. They were obtained with the use of an extension of the method applied in the finite-dimensional case by Coddington and Levinson [1]. Their presentation has been slightly altered. The authors wish to express their deep gratitude to P. A. Švarcman for his friendly criticism and help in the presentation of this chapter.

CHAPTER VII
NONLINEAR EQUATIONS

In this chapter we study nonlinear equations that, as a rule, differ little in some well defined sense from linear equations. We present (in § 1) only the simplest theorems for the existence and uniqueness of solutions of local and global type since the more refined results are not used in this book (some of the more refined theorems are indicated in the Exercises). Also in § 1 we introduce basic stability concepts and prove a general theorem on the stability of the negativeness property of the Bohl exponent of a nonlinear equation under nonlinear perturbations. This result is then twice repeated in special cases: in § 2 (§ 3) for nonlinear perturbations of a stationary (nonstationary) linear equation, in order to illustrate other methods and to obtain more exact estimates. The result of § 2 generalizes a classical stability theorem; here we also cite different generalizations of a theorem on the instability of the zero solution of an equation with a stationary principal linear part. The proofs make essential use of the technique developed in Chapter II of renorming a Banach space (with definite and indefinite norms). In § 3 we consider an equation with a nonstationary principal part and we elucidate the important role of the negativeness of the Bohl exponent of a linear approximation of this equation. In § 4 we proceed to a consideration of the nonlinear equation

$$dx/dt = A(t)x + F(t, x)$$

with an exponentially dichotomic principal linear part and a sufficiently small nonlinear addition satisfying a Lipschitz condition with a small constant. For the sake of simplicity we discuss only the case when the principal part is stationary ($A(t) \equiv A$); here the Green operator function introduced in Chapter II plays an essential role. Since an analogous function also exists for a nonstationary e-dichotomic equation, the majority of results, as noted in the appropriate places, easily carry over to the general case. The main result consists in a proof of the existence of a solution $x_0(t)$ lying in a small ball for all $t \in (-\infty, \infty)$. We also investigate conditions for the periodicity and almost periodicity of this solution. In addition, we prove the existence in a small ball of two initial manifolds that are canonically homeomorphic to neighborhoods of invariant subspaces $\mathfrak{B}_+$ and $\mathfrak{B}_-$ of A and have the property that solutions initially in them tend to $x_0(t)$ as $t \to -\infty$ and $t \to \infty$ respectively. We note that under the condition $F(t, 0) = 0$ the results of § 2 imply that when $x_0(t) \equiv 0$ these solutions must leave a small ball in the direction opposite to that indicated above (as $t \to \infty$ and $t \to -\infty$ respectively).

The more complicated case when the spectrum of A contains points lying on the imaginary axis is considered in §§5 and 6 (an instability theorem for this case is given in § 2). Under certain conditions the equation can be decomposed so that one part of it, roughly speaking, determines integral manifolds of a special form of the equation, while the other part (corresponding to the points of the spectrum lying on the imaginary axis) determines the behavior of the trajectories on these manifolds. In § 5 we prove that such a manifold exists in an arbitrarily thin hypercylinder for all $t \in (-\infty,\infty)$, and in § 6 we determine conditions for its stability. Most of the results of

§§5 and 6 can be extended to equations with a variable exponentially dichotomic principal linear part in the case when the space is a Hilbert space. In § 7 we present two variants of the so-called averaging principle for nonlinear equations. One of them concerns the behavior of solutions over a finite time interval while the other concerns their behavior over an infinite time interval.

§ 1. Existence of solutions. Basic stability concepts

1. *Existence theorems.*[1)] In this chapter we consider in a Banach space $\mathfrak{B}$ nonlinear differential equations of the form

$$dx/dt = f(t, x), \tag{1.1}$$

where $f(t, x)$ is a given function with values in $\mathfrak{B}$ of a real variable t and a variable $x \in \mathfrak{B}$. As a rule, we will assume that $f(t, x)$ is continuous in t.

We will be interested in various questions connected with the behavior of the solutions of equation (1.1) as $t \to \infty$.

There first of all arises the question of the existence of solutions of this equation. This question is thoroughly examined in the literature. We give only the simplest result, which generalizes the classical theorem of Picard.

The reader can find various tests for the existence of solutions, their uniqueness and extendability to an infinite interval in the work of M. A. Krasnosel'skiĭ and S. G. Kreĭn [2], for example (see Exercises 1 – 3).

In the sequel, when studying the behavior of solutions at infinity, we will usually assume their existence a priori without binding ourselves to additional conditions.

THEOREM 1.1 (LOCAL). *Suppose there exists a neighborhood of a point* (t_0, x_0) *in which the function* $f(t, x)$ *is continuous in* t *and satisfies the Lipschitz condition*

$$\|f(t, x_2) - f(t, x_1)\| \leqq M \|x_2 - x_1\| \tag{1.2}$$

for some finite positive constant M.

Then there exists a neighborhood of t_0 *in which equation* (1.1) *has a unique solution* $x = \phi(t)$ *satisfying the condition*

$$\phi(t_0) = x_0. \tag{1.3}$$

The proof is based on the contraction principle.

It immediately follows from the hypothesis of the theorem that there exist positive constants ε and η such that when $|t - t_0| \leqq \varepsilon$ and $\|x - x_0\| \leqq \eta$ the function $f(t, x)$ is continuous, satisfies (1.2) and is bounded:

$$\|f(t, x)\| \leqq M_1 < \infty. \tag{1.4}$$

Let $\delta = \min(\varepsilon, \eta/M_1)$ and let $C_\delta(\mathfrak{B})$ denote the Banach space of continuous functions $x(t)$ that are defined for $|t - t_0| \leqq \delta$, take their values in $\mathfrak{B}$ and have the norm

[1)] The authors wish to acknowledge that they are indebted to the translator for an improvement in the presentation of this subsection

$$|||x||| = \sup_{|t-t_0|\leqq\delta} \|x(t)\|.$$

We consider in this space the closed ball

$$B_\eta(x_0) = \{x \in C_\delta \mid |||x - x_0||| \leqq \eta\}.$$

The boundedness condition (1.4) implies that the operator

$$(Tx)(t) \equiv x_0 + \int_{t_0}^{t} f(\tau, x(\tau))\,d\tau \tag{1.5}$$

maps B_η into itself, since $\|(Tx)(t) - x_0\| \leqq \delta M_1 \leqq \eta$ for $x(t) \in B_\eta$.

Further, when $x_1, x_2 \in B_\eta$ the Lipschitz condition (1.2) implies the estimates

$$\|(Tx_2)(t) - (Tx_1)(t)\| \leqq \int_{t_0}^{t} \|f(\tau, x_2(\tau)) - f(\tau, x_1(\tau))\|\,d\tau$$
$$\leqq M \int_{t_0}^{t} \|x_2(\tau) - x_1(\tau)\|\,d\tau \leqq M(t - t_0)|||x_2 - x_1|||$$

which in turn imply the estimates

$$\|(T^2x_2)(t) - (T^2x_1)(t)\| \leqq M \int_{t_0}^{t} \|(Tx_2)(\tau) - (Tx_1)(\tau)\|\,d\tau$$
$$\leqq M^2 |||x_2 - x_1||| \int_{t_0}^{t} (\tau - t_0)\,d\tau = \frac{[M(t - t_0)]^2}{2!} |||x_2 - x_1|||$$

from which we induce that

$$\|(T^nx_2)(t) - (T^nx_1)(t)\| \leqq \frac{[M(t - t_0)]^n}{n!} |||x_2 - x_1|||,$$

and hence that

$$|||T^nx_2 - T^nx_1||| \leqq \frac{(\delta M)^n}{n!} |||x_2 - x_1|||. \tag{1.6}$$

It follows that the operator T^n is a contraction in B_η for sufficiently large n. Thus, by the contraction principle there exists a unique solution $x(t) \in B_\eta$ of the integral equation

$$x(t) = x_0 + \int_{t_0}^{t} f(\tau, x(\tau))\,d\tau, \tag{1.7}$$

which is easily seen to be equivalent to the Cauchy problem (1.1), (1.3) for $x(t) \in B_\eta$. The theorem is proved.

REMARK 1.1. Theorem 1.1 asserts the existence of solutions only in a certain neighborhood of the point t_0. But having constructed a solution in the interval $[t_0 - \delta, t_0 + \delta]$, we can attempt to extend it further. It is obvious that we will be able to continue such a procedure indefinitely if, for example, conditions (1.2) and (1.4) are satisfied for all t and $x \in \mathfrak{B}$ with one and the same constants M and M_1. In particular, if conditions (1.2) and (1.4) are satisfied for $t \in [\alpha, \infty)$, $\|x - x_0\| \leqq \eta$ and the solution $x(t)$ of equation (1.1) is known to lie in some ball $\|x(t) - x_0\| \leqq \eta_0 < \eta$, it can be extended indefinitely as $t \to \infty$.

If we impose requirements of a global character on $f(t, x)$, we can achieve unlimited extendability of the solutions without a priori assumptions on their behavior.

THEOREM 1.2 (GLOBAL). *Suppose there exists a domain* $[a, b] \times \mathfrak{B}$ *on which the function* $f(t, x)$ *is continuous in* t *and satisfies the Lipschitz condition* (1.2). *Then for any* $(t_0, x_0) \in [a, b] \times \mathfrak{B}$ *the Cauchy problem* (1.1), (1.3) *has a unique solution* $x = \phi(t)$ *defined on* $[a, b]$.

The proof is analogous to the proof of Theorem 1.1 and therefore need not be given. It suffices to note (i) that the hypothesis of the theorem implies the boundedness of $f(t, x)$ on $[a, b] \times S$, where S is an arbitrary compact subset of $\mathfrak{B}$, and (ii) that the role of B_η is played by the Banach space $C(\mathfrak{B})$ of continuous functions $x(t)$ that are defined on $[a, b]$, take their values in $\mathfrak{B}$ and have the norm

$$|||x||| = \sup_{t \in [a,b]} \|x(t)\|.$$

2. *Stability. Variational equation.* We extend to nonlinear equations some of the definitions introduces in §§3 and 4 of Chapter III.

We consider in a Banach space $\mathfrak{B}$ the differential equation

$$dx/dt = f(t, x) \qquad (t_0 \leqq t < \infty), \tag{1.1}$$

under the assumption that the function $f(t, x)$ with values in $\mathfrak{B}$ is defined for $t \in [t_0, \infty)$ and x in some subset of $\mathfrak{B}$ (possibly coinciding with the whole space).

A solution $x = \phi(t)$ of equation (1.1) on $[t_0, \infty)$ is said to be *stable (according to Ljapunov)* if for any $\varepsilon > 0$ and any $t_1 \geqq t_0$ there exists a $\delta > 0$ such that every other solution $x = \psi(t)$ defined in a neighborhood of t_1 and satisfying the inequality $\|\phi(t_1) - \psi(t_1)\| < \delta$ exists for all $t \geqq t_1$ and satisfies the inequality $\|\phi(t) - \psi(t)\| < \varepsilon$ $(t \geqq t_1)$.

A solution $\phi(t)$ is said to be *uniformly stable* if the constant δ can be chosen independently of $t_1 \geqq t_0$.

Finally, a solution $\phi(t)$ is said to be *asymptotically stable* if it is stable and for any $t_1 \geqq t_0$ there exists a $\delta_0 > 0$ such that the inequality $\|\phi(t_1) - \psi(t_1)\| < \delta_0$ implies

$$\lim_{t \to +\infty} \|\phi(t) - \psi(t)\| = 0.$$

The definition of a stable solution is consistent with the definition given in §3 of Chapter III in the sense that if the linear equation $dx/dt = A(t)x$ $(t_0 \leqq t < \infty)$ is (right) stable, every solution of it is stable. In fact, for any pair of solutions $x_1(t)$, $x_2(t)$ of this equation we have

$$\|x_2(t) - x_1(t)\| \leqq N_{t_1}\|x_2(t_1) - x_1(t)\| \qquad (t \geqq t_1),$$

where $N_{t_1} = \sup_t \|U(t, t_1)\| < \infty$, and it suffices to take $\delta = \varepsilon/N_{t_1}$.

An investigation of the stability of a solution $x = \phi(t)$ of equation (1.1) can always be reduced to an investigation of the stability of the zero solution of an auxiliary equation.

In fact, setting $y(t) = x(t) - \phi(t)$, we obtain for $y(t)$ the equation

$$dy/dt = \Psi(t, y), \tag{1.8}$$

where $\Psi(t, y) = f(t, y + \phi(t)) - f(t, \phi(t))$. We note that $\Psi(t, 0) = 0$.

Here the solution $x = \phi(t)$ of equation (1.1) goes over into the zero solution $y \equiv 0$ of equation (1.8) while any other solution $x = \psi(t)$ goes over into the difference $y = \phi(t) - \psi(t)$.

Equation (1.8) can be reduced to a more convenient form for purposes of study if the function $f(t, x)$ is continuously differentiable with respect to x in a neighborhood of the point $x - \phi(t)$. One can then write

$$\Psi(t, y) = f(t, y + \phi(t)) - f(t, \phi(t)) = A(t)y + F(t, y),$$

where $A(t) = f'_x(t, \phi(t))$ and $\|F(t, y)\| \leqq c_t(t)\|y\|$, with $\lim_{y\to 0} c_t(y) = 0$. It sometimes turns out that this convergence is uniform in t, i.e.

$$\|F(t, y)\| \leqq c(y)\|y\|, \qquad \lim_{y\to 0} c(y) = 0. \tag{1.9}$$

Such a situation holds, for example, if the function $f(t, x)$ has a bounded second derivative or, more generally, if

$$\|f'_x(t, \phi(t) + h) - f'_x(t, \phi(t))\| \leqq c\|h\|.$$

This follows from the obvious representation (see (I.9.7))

$$\begin{aligned} f(t, y + \phi(t)) - f(t, \phi(t)) - f'_x(t, \phi(t))y \\ = \int_0^1 [f'_x(t, \phi(t) + \tau y) - f'_x(t, \phi(t))]d\tau \cdot y. \end{aligned}$$

Thus equation (1.8) can be written in the form

$$dy/dt = A(t)y + F(t,y). \tag{1.10}$$

Such an equation, when condition (1.9) is satisfied, is sometimes said to be *quasilinear*.

The linear equation

$$dy/dt = A(t)y \qquad (A(t) = f'_x(t, \phi(t)) \tag{1.11}$$

which plays an especially important role in the investigation of equation (1.10), is called the *variational equation* of equation (1.1) corresponding to the solution $x = \phi(t)$.

We note that if condition (1.9) is satisfied then for any $q > 0$ there exists a $\rho > 0$ such that

$$\|F(t, y)\| \leqq q\|y\| \tag{1.12}$$

in the ball $B_\rho = \{x \in \mathfrak{B} | \, \|x\| \leqq \rho\}$.

We will sometimes proceed directly from equation (1.10) under the assumption that condition (1.12) is satisfied with a sufficiently small fixed $q > 0$ (this condition is weaker than (1.9)).

We consider some important special cases that are frequently encountered in the applications.

An equation

$$dx/dt = f(x), \tag{1.13}$$

whose right side does not depend explicitly on t, is said to be *autonomous*. Clearly, if $x = \phi(t)$ is a solution of equation (1.13), any shifted function $x = \phi(t - t_0)$ is also a solution of it.

Let x_0 be a critical point of $f(x)$: $f(x_0) = 0$.

In this case equation (1.13) has the stationary solution $x(t) \equiv x_0$, while the variational equation (1.11) turns out to be a stationary linear equation $dx/dt = Ax$ with the operator

$$A = f'(x_0). \tag{1.14}$$

We obtain another important special case when the function $f(t, x)$ is periodic in t for all values of x: $f(t + T, x) = f(t, x)$.

Suppose $x = \phi(t)$ is a periodic solution of such an equation (with the same period): $\phi(t + T) = \phi(t)$.

In this case the operator function $A(t) = f'_x(t, \phi(t))$ is obviously also T-periodic.

In particular, if an autonomous equation has a periodic solution $y = \phi(t)$, its derivative $dy/dt = \phi'(t)$ satisfies the variational equation. In fact, differentiating the identity $d\phi(t)/dt = f(\phi(t))$, we obtain the relation

$$\frac{d\phi'(t)}{dt} = f'(\phi(t))\phi'(t) = A(t)\phi'(t).$$

3. *General theorem on the stability of the Bohl exponent.* Inasmuch as it was shown above that it suffices to carry out an investigation of stability for the zero solution, we will frequently consider the equation

$$dx/dt = f(t, x) \qquad (t \geqq 0) \tag{1.1}$$

under the additional condition

$$f(t, 0) = 0. \tag{1.15}$$

We will say that an equation (1.1) satisfying condition (1.15) has *property* $\mathscr{B}(\nu, N, \rho)$ $(-\infty < \nu < +\infty, N > 0, \rho > 0)$ if every solution $x(t)$ of it for which $\|x(t_0)\| \leqq \rho$ at some moment t_0 satisfies the estimate

$$\|x(t)\| \leqq Ne^{-\nu(t-\tau)}\|x(\tau)\| \tag{1.16}$$

for all $t > \tau \geqq t_0$ for which the solution $x(t)$ is defined.

We generalize a definition of § III.4.2 by calling the greatest lower bound of numbers $\lambda = -\nu$ for which there exist numbers ρ_ν and N_ν such that equation (1.1) has property $\mathscr{B}(\nu, N_\nu, \rho_\nu)$ the (*upper*) *Bohl exponent at zero* of this equation and denote it by κ_B.

REMARK 1.2. If the conditions of the local existence theorem are satisfied in a neighborhood of zero and the Bohl exponent at zero of the equation is negative, then the solutions whose initial values lie in a sufficiently small neighborhood of zero can be extended indefinitely (see Remark 1.1). Since, furthermore, these solutions tend to zero as $t \to \infty$, the zero solution in the case in question is uniformly and asymptotically stable. Because of this fact we will speak below without further ado of the negativeness of the Bohl exponent implying stability.

By virtue of the above definition, for each $\nu > \kappa_B$ the solutions of equation (1.1) that are contained in the ball $B_{\rho_\nu} = \{x \mid \|x\| < \rho_\nu\}$ at some moment of time t_0 satisfy the condition

$$\|x(t)\| \leqq N_\nu e^{\nu(t-\tau)}\|x(\tau)\| \qquad (t > \tau \geqq t_0). \tag{1.17}$$

It is not difficult to see that for a linear equation the Bohl exponent at zero coincides with the upper Bohl exponent of the equation.

We now see that, just as in the linear case, the Bohl exponent is stable with respect to small perturbations of the equation. This fact is implied by the following result.

THEOREM 1.3. *Suppose that a function $f(t, x)$ for $t \in [0, \infty)$ and $\|x\| \leqq r$ is continuous in t and satisfies condition* (1.15) *and the Lipschitz condition*

$$\|f(t, x_2) - f(t, x_1)\| \leqq L\|x_2 - x_1\|.$$

Suppose further that equation (1.1) *has property $\mathscr{B}(\nu, N, \rho)$ for some $\nu > 0$ and $\rho < r$. Then for any $N_1 > N$ and $\nu_1 < \nu$ there exists a number $q > 0$ such that if a function $g(t, x)$ satisfies the inequality*

$$\|g(t, x)\| \leqq q\|x\| \qquad (t \in [0, \infty); \|x\| \leqq r), \tag{1.18}$$

the equation

$$dx/dt = f(t, x) + g(t, x) \tag{1.19}$$

has property $\mathscr{B}(\nu_1, N_1, \rho_1)$ for some $\rho_1 > 0$.

PROOF. It suffices to show that for some $h > 0$ every solution of equation (1.19) that is contained in a ball B_δ at some moment t_0 satisfies the estimates

$$\|x(\tau + s)\| \leqq N_1 e^{-\nu_1 s}\|x(\tau)\| \qquad (0 \leqq s \leqq h),$$
$$\|x(\tau + h)\| \leqq e^{-\nu_1 h}\|x(\tau)\| \tag{1.20}$$

for $\tau \geqq t_0$ (provided this solution is defined at the corresponding points).

For in this case the inequality

$$\|x(\tau + nh)\| \leqq e^{-\nu_1 nh}\|x(\tau)\|$$

is valid for any $\tau \geqq t_0$ and any integer n and, finally,

$$\|x(t)\| = \|x(\tau + nh + s)\| \leqq N_1 e^{-\nu_1 s}\|x(\tau + nh)\|$$
$$\leqq N_1 e^{-\nu_1(s+nh)}\|x(\tau)\| = N_1 e^{-\nu_1(t-\tau)}\|x(\tau)\|$$

for any $t > \tau \geqq t_0$ ($t = \tau + nh + s$).

We consider a solution $y(t)$ of equation (1.1) satisfying the condition $\|y(t_0)\| \leqq \delta_0$, where

$$\delta_0 = \min\{\rho/N, \rho\}. \tag{1.21}$$

Since it is subject to inequality (1.16), in particular,

$$\|y(t)\| \leqq N\|y(t_0)\| < \varepsilon.$$

From what was said in Remark 1.2 it follows that the solution $y(t)$ is defined for all $t \in [t_0, \infty)$. We now choose an $h > 0$ so that

$$\eta = 1 - Ne^{-(\nu-\nu_1)h} > 0, \tag{1.22}$$

and consider on the interval $[t_0, t_0 + h]$ a solution $x(t)$ of equation (1.19) satisfying the condition $x(t_0) = x_0 \in B_\delta$ (for those values of t at which this solution is defined).

From the relation

$$x(t) = x(t_0) + \int_{t_0}^{t} f(s, x(s))ds + \int_{t_0}^{t} g(s, x(s))ds \tag{1.23}$$

we obtain by virtue of (1.18) the estimate

$$\|x(t)\| \leqq \|x_0\| + (L + q)\int_{t_0}^{t}\|x(s)\|ds,$$

which implies by virtue of Corollary III.2.1 that

$$\|x(t)\| \leqq \|x_0\|e^{(L+q)h}. \tag{1.24}$$

Suppose now $y(t)$ is a solution of equation (1.1) satisfying the same condition $y(t_0) = x_0$. From (1.23) and the equation

$$y(t) = x_0 + \int_{t_0}^{t} (s, y(s))ds$$

we deduce, using (1.24), the estimate

$$\|x(t) - y(t)\| \leqq L \int_{t_0}^{t} \|x(s) - y(s)\| ds + qh\|x_0\| e^{(L+q)h},$$

which implies by virtue of Corollary III.2.1 again that

$$\|x(t) - y(t)\| \leqq qh\|x_0\| e^{(2L+q)h} \leqq qhe^{(2L+q+\nu_1)h} e^{-\nu_1(t-t_0)} \|x_0\|.$$

Thus for those values of $t \in [t_0, t_0 + h)$ at which the solution $x(t)$ is defined we have the estimate

$$\begin{aligned}\|x(t)\| &\leqq \|y(t)\| + \|x(t) - y(t)\| \\ &\leqq [N + qhe^{(2L+q+\nu_1)h}]\, e^{-\nu_1(t-t_0)} \|x_0\|.\end{aligned}$$

We now subject the number q (> 0) to the condition

$$qhe^{(2L+q+\nu_1)h} = \min(N_1 - N, \eta). \tag{1.25}$$

Then

$$\|x(t)\| \leqq N_1 e^{-\nu_1(t-t_0)} \|x(t_0)\| \qquad (t_0 \leqq t < t_0 + h). \tag{1.26}$$

On the other hand, taking into account (1.22) and (1.16), we get

$$\begin{aligned}\|x(t_0 + h)\| &\leqq \|y(t_0 + h)\| + \|x(t_0 + h) - y(t_0 + h)\| \\ &\leqq [1 - \eta + qhe^{(2L+q+\nu_1)h}] e^{-\nu_1 h} \|x(t_0)\|,\end{aligned}$$

which implies by virtue of (1.25) that

$$\|x(t_0 + h)\| \leqq e^{-\nu_1 h} \|x(t_0)\|. \tag{1.27}$$

Since all of the arguments can be repeated with the point t_0 replaced by the points $t_0 + h$, $t_0 + 2h, \cdots$, we can assert that estimate (1.26) is satisfied for all $t \geqq t_0$ at which the solution $x(t)$ is defined.

We now choose $\delta = \min(\delta_0/N_1, \delta_0)$. It then follows from (1.26) for $\|x(t_0)\| < \delta$ that $\|x(\tau)\| \leqq N_1\delta \leqq \delta_0$ $(\tau \geqq t_0)$ and hence that all of the above estimates can be constructed with t_0 replaced by τ. In this connection (1.26) and (1.27) go over into estimates (1.20).

Thus the theorem is proved, it following from (1.21) that

$$\delta = \min\{\rho, \rho/N, \rho/NN_1\},$$

and the number q being determined from (1.25) and (1.22).

COROLLARY 1.1. *If the Bohl exponent at zero of equation* (1.1) *is negative, then the Bohl exponent of the perturbed equation* (1.19) *for sufficiently small* $q > 0$ *is also negative.*

REMARK 1.3. Suppose the function $g(t, x)$ satisfies the condition

$$\lim_{x \to 0} \frac{\|g(t, x)\|}{\|x\|} = 0$$

uniformly in t. In this case we will briefly write $g(t, x) = o(\|x\|)$, bearing in mind that the convergence is uniform in t.

Then for any $q > 0$, no matter how small, we can find a neighborhood of zero in which condition (1.18) is satisfied. Consequently, for a sufficiently small $\rho_1 > 0$ equation (1.19) will have property $\mathscr{B}(\nu_1, N_1, \rho_1)$ with ν_1 arbitrarily close to ν. Hence *in this case the Bohl exponents at zero of equations* (1.1) *and* (1.19) *coincide.*

§ 2. Stability and instability of the solutions of the nonlinear equation with a stationary principal part

1. *Stability.* Suppose the stationary linear equation

$$dx/dt = Ax \tag{2.1}$$

has a negative Bohl exponent κ_B and the function $F(t, x)$ is subject to the estimate

$$\|F(t, x)\| \leqq q\|x\| \qquad (t \geqq 0;\ \|x\| \leqq \rho). \tag{2.2}$$

Then from Theorem 1.3 it follows that the equation

$$dx/dt = Ax + F(t, x) \tag{2.3}$$

has a negative Bohl exponent at zero whenever $q > 0$ is sufficiently small.

We will independently obtain this same result in order to illustrate a method (a natural generalization of the so-called second method of Ljapunov) which will be used later to prove an instability theorem. The idea underlying this method was presented in Chapter II.

We recall some of the geometric observations made there. If the spectrum $\sigma(A)$ lies in the interior of the left halfplane, the vector field Ax (of tangent vectors to the integral curves of equation (2.1)) at each point $x \in \mathfrak{B}$ after an appropriate renorming of the space $\mathfrak{B}$ turns out to be directed toward the interior of the sphere centered at the origin which passes through the point x. But if $\sigma(A)$ contains a component in the right halfplane there exist points $x \in \mathfrak{B}$ at which the vector Ax is directed toward the exterior of this sphere.

It is obvious that a small perturbation of the equation cannot produce a large change in the character of the integral curves. We can therefore anticipate that under certain conditions the question of stability of the zero solution of equation (2.3) under a "not too large" perturbation $F(t, x)$ can be resolved by the spectrum of the linear approximation of this equation.

These general observations receive a rigorous formulation below.

THEOREM 2.1. *Suppose the spectrum of the operator A lies in the interior of the left halfplane, so that $\|e^{At}\| \leqq N_0 e^{-\nu_0 t}$, and suppose condition* (2.2) *is satisfied for $q < \nu_0/N_0$.*

Then equation (2.3) *has a negative Bohl exponent at zero.*

PROOF. Consider a solution $x(t)$ of equation (2.3) satisfying a condition $\|x(t_0)\| < \rho_0 \leqq \rho$ (the number ρ_0 will be chosen below) and let $t_0 \leqq t < T$ be an interval in which $\|x(t)\| < \rho$.

We introduce in the space $\mathfrak{B}$ a new norm

$$\|x\|_A = \int_0^\infty \| e^{At}x\|dt.$$

As was shown in Theorem I.4.2, $m\|x\| \leqq \|x\|_A \leqq M\|x\|$, where $m = 1/\|A\|$ and $M = N_0/\nu_0$.

We put $y(t) = e^{-At}x(t)$. Then

$$dt/dy = e^{-At}(dx/dt - Ax) = e^{-At}F(t, x)$$

and

$$\|x(t)\|_A = \int_0^\infty \|e^{A(\tau+t)}y(t)\|d\tau = \int_t^\infty \|e^{A\tau}y(t)\|d\tau.$$

Let us calculate the divided difference

$$\frac{\|x(t+h)\|_A - \|x(t)\|_A}{h}$$
$$= -\frac{1}{h}\int_t^{t+h} \|e^{A\tau}y(t)\|d\tau + \int_{t+h}^\infty \frac{\|e^{A\tau}y(t+h)\| - \|e^{A\tau}y(t)\|}{h}d\tau.$$

Denoting the first summand of the right side by $\mathscr{L}_1(h)$ and the second by $\mathscr{L}_2(h)$, we estimate them, assuming that $t_0 \leqq t < T$.

We first note that

$$\lim_{h\to 0}\mathscr{L}_1(h) = -\lim_{h\to 0}\frac{1}{h}\int_t^{t+h}\|e^{A\tau}y(t)\|d\tau = -\|e^{At}y(t)\| = -\|x(t)\|. \tag{2.4}$$

In addition, setting $\tau = s + t + h$ in the integral below, we get

$$\mathscr{L}_2(h) \leqq \int_{t+h}^\infty \frac{\|e^{A\tau}[y(t+h) - y(t)]\|}{h}d\tau$$
$$= \int_0^\infty \left\| e^{As}e^{A(t+h)}\frac{y(t+h) - y(t)}{h}\right\| ds = \left\| e^{A(t+h)}\frac{y(t+h) - y(t)}{h}\right\|_A$$
$$\leqq M\left\| e^{A(t+h)}\frac{y(t+h) - y(t)}{h}\right\| \leqq M\|e^{Ah}\|\left\| e^{At}\frac{y(t+h) - y(t)}{h}\right\|.$$

The last estimate implies the inequality

$$\overline{\lim_{h\to 0}}\,\mathscr{L}_2(h) \leqq M\|e^{At}y'(t)\| = M\|F(t, x)\| \leqq qM\|x\|. \tag{2.5}$$

Thus for the right first derivative of the norm $\|x(t)\|_A$:

$$\frac{d^+\|x(t)\|_A}{dt} = \lim_{h\to+0} \frac{\|x(t+h)\|_A - \|x(t)\|_A}{h},$$

which exists, as was shown in Chapter I (see Exercises I.17 and I.18), we obtain from (2.4) and (2.5) the estimate

$$d^+\|x(t)\|_A/dt \leqq -\nu\|x(t)\|_A, \tag{2.6}$$

where $\nu = (1 - qM)/m > 0$ since $qM < (\nu_0/N_0)(N_0/\nu_0) = 1$.

From inequality (2.6) we obtain the estimate

$$\|x(t)\|_A \leqq e^{-\nu(t-\tau)}\|x(\tau)\|_A \qquad (t_0 \leqq \tau \leqq t < T),$$

and, finally, after returning to the original norm, we get

$$\|x(t)\| \leqq Ne^{-\nu(t-\tau)}\|x(\tau)\| \qquad (t_0 \leqq \tau \leqq t < T), \tag{2.7}$$

where $N = M/m$.

We put $\rho_0 = \min(\rho, \rho/N)$. Then T can be taken arbitrarily large.

For suppose $\|x(T)\| = \rho$. Since we can pass to the limit in (2.7) for $t \to T$, we obtain the contradictory inequality

$$\rho = \|x(T)\| \leqq Ne^{-\nu(T-t_0)}\|x(t_0)\| < \rho.$$

The theorem is proved.

2. *Conditions for the absence of stability.* We now consider equation (2.3) under the assumption that the spectrum $\sigma(A)$ contains points lying in the interior of the right halfplane. This will lead us to a generalization of the classical Ljapunov instability theorem.

We begin by assuming that we have a decomposition

$$\sigma(A) = \sigma_+(A) \cup \sigma_-(A), \tag{2.8}$$

and we let P_+ and P_- denote the corresponding spectral projections. Thus we assume that the spectrum of A does not intersect the imaginary axis and that $\sigma_+(A)$ is not empty.

As in the preceding subsection, we assume that

$$\|F(t, x)\| \leqq q\|x\| \qquad (\|x\| \leqq \rho;\ t \geqq t_0). \tag{2.2}$$

The subsequent arguments are analogous to those in the proof of Theorem 2.1.

We introduce in $\mathfrak{B}$ the norm

$$\|x\|_A = \|P_+x\|_A + \|P_-x\|_A = \int_0^\infty \|e^{-A\tau}P_+x\|d\tau + \int_0^\infty \|e^{A\tau}P_-x\|d\tau. \tag{2.9}$$

It was shown in § I.4.5b) that this norm is equivalent to the original one: $m\|x\| \leqq \|x\|_A \leqq M\|x\|$.

In conjunction with (2.9) we consider the indefinite norm

$$\langle\!\langle x\rangle\!\rangle_A = \|P_+x\|_A - \|P_-x\|_A.$$

Let $x(t)$ be a solution of equation (2.3) and let $[t_0, T]$ be an interval on which it satisfies the condition $\|x(t)\| < \rho$ (such an interval exists by virtue of the continuity of a solution if $\|x(t_0)\| < \rho$).

Setting $y(t) = e^{-At}x(t)$, we get as before $y'(t) = e^{-At}F(t, x)$. In this connection

$$\|P_+x(t)\|_A = \int_0^\infty \|e^{-A(\tau-t)}P_+y(t)\|d\tau = \int_{-t}^\infty \|e^{-A\tau}P_+y(t)\|d\tau.$$

Let us estimate the right derivative of the function $\|P_+x(t)\|_A$. We write

$$\frac{\|P_+x(t+h)\|_A - \|P_+x(t)\|_A}{h} = \mathcal{L}_1(h) + \mathcal{L}_2(h),$$

where

$$\mathcal{L}_1(h) = \frac{1}{h}\int_{-t-h}^{-t} \|e^{-A\tau}P_+y(t+h)\|d\tau$$

and

$$\mathcal{L}_2(h) = \int_{-t}^\infty \frac{\|e^{-A\tau}P_+y(t+h)\| - \|e^{-A\tau}P_+y(t)\|}{h}d\tau.$$

Here

$$\lim_{h\to 0} |\mathcal{L}_1(h)| = \|e^{At}P_+y(t)\| = \|P_+x(t)\| \tag{2.10}$$

and

$$\overline{\lim_{h\to 0}}\, |\mathcal{L}_2(h)| \leqq \|e^{At}P_+y'(t)\|_A = \|P_+F(t, x(t))\|_A, \tag{2.11}$$

since

$$\begin{aligned}|\mathcal{L}_2(h)| &\leqq \int_{-t}^\infty \left\| e^{-A\tau}P_+\frac{y(t+h)-y(t)}{h}\right\| d\tau \\ &= \int_0^\infty \left\| e^{-As}P_+e^{At}\frac{y(t+h)-y(t)}{h}\right\| ds = \left\| e^{At}P_+\frac{y(t+h)-y(t)}{h}\right\|_A\end{aligned}$$

From (2.10) and (2.11) we obtain the estimate

$$\begin{aligned}\frac{d^+\|P_+x\|_A}{dt} &= \lim_{h\to 0}\frac{\|P_+x(t+h)\|_A - \|P_+x(t)\|_A}{dt} \\ &\geqq \lim_{h\to 0}|\mathcal{L}_1(h)| - \overline{\lim_{h\to 0}}\,|\mathcal{L}_2(h)| \geqq \|P_+x\| - \|P_+F(t, x)\|_A \\ &\geqq \frac{1}{M}\|P_+x(t)\|_A - \|P_+F(t, x(t))\|_A,\end{aligned}$$

which reduces to the inequality

$$\|P_+x(t)\|_A - \|P_+x(\tau)\|_A \geqq \int_\tau^t \left\{\frac{1}{M}\|P_+x(s)\|_A - \|P_+F(s, x(s))\|_A\right\} ds. \tag{2.12}$$

Analogously, we deduce the inequality

$$\frac{d^+\|P_-x(t)\|_A}{dt} = \lim_{h\to 0} \frac{\|P_-x(t+h)\|_A - \|P_-x(t)\|_A}{h} \leqq -\|P_-x\| + \|P_-F(t, x(t))\|_A \leqq -(1/M)\|P_-x(t)\|_A + \|P_-F(t, x(t))\|_A$$

and the estimate

$$\|P_-x(t)\|_A - \|P_-x(\tau)\|_A \leqq \int_\tau^t \left\{-\frac{1}{M}\|P_-x(s)\|_A + \|P_-F(s, x(s))\|_A\right\} ds. \tag{2.13}$$

Subtracting (2.13) from (2.12), we obtain the estimate

$$\langle x(t)\rangle_A \geqq \langle x(\tau)\rangle_A + \beta(t, \tau), \tag{2.14}$$

where

$$\beta(t, \tau) = \int_\tau^t \left\{\frac{1}{M}\|x(s)\|_A - \|F(s, x(s))\|_A\right\} ds.$$

We note that it follows from (2.2) and the equivalence of the two norms being considered by us that for sufficiently small q $(q < q_0 = m/M)$

$$\beta(t, \tau) \geqq \alpha \int_\tau^t \|x(s)\|_A ds, \tag{2.15}$$

where α is a positive constant.

Finally, from (2.14) and (2.15) we obtain the estimate

$$\langle x(t)\rangle_A - \langle\!\langle x(\tau)\rangle\!\rangle_A \geqq \alpha \int_\tau^t \|x(s)\|_A ds \qquad (q \leqq q_0;\ t_0 \leqq \tau \leqq t < T), \tag{2.16}$$

which will be used in subsequent arguments.

Inequality (2.16) means that the indefinite norm $\langle\!\langle x(t)\rangle\!\rangle_A$ of a solution $x(t)$ increases with increasing t when the condition $\|x(t)\| < \rho$ is satisfied.

THEOREM 2.2. *Suppose the spectrum $\sigma(A)$ does not intersect the imaginary axis and the spectral set $\sigma_+(A)$ is not empty.*

There exists a number q_0 depending only on the operator A such that if the function $F(t, x)$ satisfies condition (2.2) for $q \leqq q_0$, the zero solution of the differential equation (2.3) is unstable for $t \to +\infty$.

Moreover, every nonzero solution $x(t)$ of this equation for which

$$\langle\langle x(t_0)\rangle\rangle_A \geqq 0 \tag{2.17}$$

leaves any ball $\|x\| \leqq \rho_1$ of radius $\rho_1 < \rho$ at some $t > t_0$.

Proof. From (2.17) and (2.16) it follows that the strict inequality $\gamma = \langle\langle x(t_0')\rangle\rangle_A > 0$ holds for $t_0' > t_0$.

From the inequalities (2.16) and $\|x(t)\|_A \geqq \langle\langle x(t)\rangle\rangle_A$ we obtain the inequality

$$\|x(t)\|_A \geqq \alpha \int_{t_0'}^{t} \|x(s)\|_A \, ds + \gamma,$$

which on the basis of Corollary III.2.1 reduces to the estimate

$$\|x(t)\| \geqq \frac{1}{M} \|x(t)\|_A \geqq \frac{\gamma}{M} e^{\alpha(t-t_0')} \tag{2.18}$$

since the function $u(t) = \gamma e^{\alpha(t-t_0')}$ satisfies the equation

$$u(t) = \alpha \int_{t_0'}^{t} u(s)ds + \gamma.$$

For sufficiently large t the required inequality

$$\|x(t)\| \geqq \rho_1 \tag{2.19}$$

follows from (2.19) provided $\rho_1 < \rho$.

We note, finally, that if $P_-x(t_0) = 0$, condition (2.17) is automatically satisfied provided $x(t_0) \neq 0$.

Thus we can find a solution with an arbitrarily small initial value $\|x(t_0)\|$ satisfying condition (2.19) for sufficiently large t. This implies the instability of the zero solution of equation (2.3). The theorem is proved.

REMARK 2.1. If the condition $\|F(t, x)\| \leqq q\|x\|$ is satisfied for $-\infty < t \leqq t_0$ and the spectral set $\sigma_-(A)$ is not empty, a solution $x(t)$ satisfying the condition

$$\langle\langle x(t_0)\rangle\rangle_A \leqq 0 \tag{2.20}$$

leaves a ball $\|x\| \leqq \rho_1 < \rho$ as $t \to -\infty$.

In the sequel we must consider the case when the function $F(t, x)$ is defined on the real line and condition (2.17) is satisfied for all t. In this case every nonzero solution of equation (2.3) leaves any ball $\|x\| \leqq \rho_1 < \rho$.

3. *Another type of instability condition.* The condition of Thorem 2.2 concerning $F(t, x)$ is satisfied in every case if

$$\|F(t, x)\| \leqq q\|x\|^{1+p} \qquad (p > 0;\ \|x\| \leqq \rho;\ t \geqq t_0). \tag{2.21}$$

In fact, estimate (2.21) in a sufficiently small ball implies estimate (2.2) with a sufficiently small q.

It turns out that we can deduce from (2.21) a result that in a certain sense is more general, the assumption that $\sigma(A)$ does not intersect the imaginary axis having been dropped.

THEOREM 2.3. *If the spectrum $\sigma(A)$ of A contains points lying in the interior of the right halfplane and condition* (2.21) *is satisfied, the zero solution of equation* (2.3) *is unstable.*

PROOF. Let $\lambda = \alpha + \beta i$ be a point of $\sigma(A)$ having a maximal real part: $\alpha \geqq \operatorname{Re} \lambda$ $(\lambda \in \sigma(A))$. Then for any $\eta > 0$

$$\|e^{A(t-\tau)}\| \leqq N_\eta e^{(\alpha+\eta)(t-\tau)} \qquad (\eta > 0;\, t \geqq \tau).$$

Moreover, as follows from Lemma I.4.1, for each $T > 0$ and a sufficiently small $\delta > 0$ there exists a vector $\xi = \xi_{\delta,T}$ $(\|\xi\| = 1)$ such that

$$e^{\alpha(t-t_0)}(1-\delta) \leqq \|e^{A(t-t_0)}\xi\| \leqq e^{\alpha(t-t_0)}(1+\delta) \quad (0 \leqq t - t_0 \leqq T).$$

Let $x(t)$ $(t \geqq t_0)$ be the solution of equation (2.3) satisfying the condition $x(t_0) = x_0$. We write it in the form $x(t) = x_1(t) + x_2(t)$, where

$$x_1(t) = e^{A(t-t_0)}x_0; \quad x_2(t) = \int_{t_0}^{t} e^{A(\tau-t_0)}F(\tau, x(\tau))d\tau.$$

Let $x_0 = \varepsilon\xi_{\delta,T}$ and let R be a number satisfying the condition $1 + \delta < R < \min(2, 1 + \delta + \delta_1)$, where the number $\delta_1 > 0$ will be chosen later on.

By virtue of the continuity of $x(t)$ there exists an interval $(0 \leqq t - t_0 \leqq T_1)$ on which

$$\|x(t)\| \leqq \varepsilon R e^{\alpha(t-t_0)}. \tag{2.22}$$

We estimate both parts of the solution on this interval, assuming that

$$\eta < \alpha p, \quad \varepsilon R e^{\alpha T_1} \leqq \rho. \tag{2.23}$$

Under these conditions we have

$$\begin{aligned}
\|x_2(t)\| &\leqq \int_{t_0}^{t} \|e^{A(t-\tau)}\| \, \|F(\tau, x(\tau))\| d\tau \\
&\leqq N_\eta q \int_{t_0}^{t} e^{(\alpha+\eta)(t-\tau)} \|x(\tau)\|^{1+p} d\tau \\
&\leqq (\varepsilon R)^{1+p} N_\eta q \int_{t_0}^{t} e^{(\alpha+\eta)(t-\tau)+\alpha(p+1)(\tau-t_0)}\, d\tau \\
&\leqq \frac{N_\eta q}{\alpha p - q}(\varepsilon R e^{\alpha(t-t_0)})^{1+p}
\end{aligned}$$

and

$$\|x_1(t)\| \leqq (1 + \delta)\varepsilon e^{\alpha(t-t_0)}.$$

Therefore, provided $T_1 \leqq T$,

$$\begin{aligned}\|x(t)\| &\leqq \|x_1(t)\| + \|x_2(t)\|\\ &\leqq \left[1 + \delta + \frac{N_\eta q}{\alpha p - \eta}\, \varepsilon^p R^{1+p} e^{\alpha p(t-t_0)}\right] \varepsilon e^{(t-t_0)} \quad (0 < t - t_0 \leqq T_1).\end{aligned}$$

If it turns out in this connection that

$$1 + \delta + \frac{N_\eta q}{\alpha p - \eta}\, \varepsilon^p R^{1+p} e^{\alpha p T_1} < R,$$

then the continuity of $x(t)$ implies that condition (2.22) is also satisfied on a larger interval than $[t_0, t_0 + T_1]$.

Therefore inequality (2.22) is always satisfied on the interval $[t_0, t_0 + T_1]$, where T_1 satisfies the equation

$$1 + \delta + \frac{N_\eta q}{\alpha p - \eta}\, \varepsilon^p R^{1+p} e^{\alpha p T_1} = R,$$

which can be rewritten in the form

$$\varepsilon e^{\alpha T_1} = \frac{1}{R}\left[\frac{(R - 1 - \delta)(\alpha p - \eta)}{N_\eta q R}\right]^{1/p}. \tag{2.24}$$

For sufficiently small ε this equation always has a unique solution $T_1(\varepsilon) > 0$, which when δ, R and η are given depends only on ε and not on T. Thus T can be chosen arbitrarily and we put $T = T_1(\varepsilon)$.

We now note that (2.24) implies

$$\varepsilon R e^{\alpha T_1} = \left(\frac{\alpha p - \eta}{N_\eta q R}\right)^{1/p} (R - 1 - \delta)^{1/p} \leqq \delta_1^{1/p}\left(\frac{\alpha p - \eta}{N_\eta q}\right)^{1/p}.$$

Setting $\delta_1 \leqq N_\eta q \rho^p/(\alpha p - \eta)$, we ensure the fulfilment of inequality (2.23) and thereby the validity of all of the arguments constructed above.

Thus inequality (2.22) is satisfied on the interval $0 \leqq t - t_0 \leqq T_1$, where $T_1(\varepsilon)$ is a solution of equation (2.24).

But then

$$\begin{aligned}\|x(t_0 + T_1)\| &\geqq \|x_1(t_0 + T_1)\| - \|x_2(t_0 + T_1)\|\\ &\geqq (1 - \delta)\varepsilon e^{\alpha T_1} - \frac{N_\eta q}{\alpha p - \eta}\, \varepsilon^p R^{1+p} e^{\alpha p T_1} \varepsilon e^{\alpha T_1}\\ &= \left[1 - \delta - \frac{N_\eta q}{\alpha p - \eta}\, R^{1+p} \varepsilon^p e^{\alpha p T_1}\right] \varepsilon e^{\alpha T_1}\\ &= (2 - R)\left[\frac{(R - 1 - \delta)(\alpha p - \eta)}{N_\eta q R^{1+p}}\right]^{1/p} = c > 0.\end{aligned}$$

This estimate implies the instability of $x(t)$, since $\varepsilon = \|x_0\|$ can be taken arbitrarily small and yet there still exists a value $T_1(\varepsilon)$ for which $\|x(t_0 + T_1)\| \geqq c$.

4. *Stability of a stationary solution of the autonomous equation.* We consider the autonomous nonlinear equation

$$\frac{dx}{dt} = f(x) \qquad (t_0 \leqq t < \infty). \tag{2.25}$$

If x_0 is a critical point: $f(x_0) = 0$, in a neighborhood of which the function $f(x)$ is continuously differentiable, the difference $y(t) \equiv x(t) - x_0$ will satisfy the quasilinear equation

$$dy/dt = Ay + F(y). \tag{2.26}$$

On the basis of what has been said above we can conclude that the behavior of a small solution of this equation is determined by the spectrum of the operator $A = f'(x_0)$.

If the spectrum $\sigma(A)$ lies in the interior of the left halfplane, the zero solution of equation (2.26), and hence also the stationary solution $x \equiv x_0$ of equation (2.25), is uniformly and asymptotically stable.

But if the points of $\sigma(A)$ also lie in the interior of the right halfplane, we can assert that this solution is unstable when either (i) $\sigma(A)$ does not intersect the imaginary axis or (ii) a condition of type (2.21) is satisfied (it is satisfied, for example, if the function $f(x)$ is twice differentiable).

§ 3. The nonlinear equation with a nonstationary principal part

1. *Another variation of the theorem on the stability of the Bohl exponent.* In this section we consider the question of stability of the solutions of the nonlinear equation

$$\frac{dx}{dt} = A(t)x + F(t, x), \tag{3.1}$$

without assuming that its principal part is stationary.

As indicated in §1, it suffices to study conditions for the stability of the zero solution of equation (3.1). Thus we can assume that

$$F(t, 0) = 0. \tag{3.2}$$

From Corollary 1.1 to Theorem 1.3 it follows that a small addition $F(t, x)$ does not violate the property of negativeness of the Bohl exponent of the equation. We now give an independent proof of this fact, following the same idea as in the proof of Theorem III.4.6, in order to obtain more exact values of the constants entering into the estimates. Also we replace condition (1.18) by the analogous integral condition

$$\|F(t, x)\| \leqq \eta(t)\|x\| \qquad (t \geqq t_0;\ \|x\| \leqq \rho), \tag{3.3}$$

where

$$\frac{1}{\tau_0}\int_t^{t+\tau_0} \eta(\tau)d\tau \leqq q. \tag{3.4}$$

An analogous generalization could have been made in Theorem 1.3 (see Exercise 5).

THEOREM 3.1. *Suppose the equation*

$$dx/dt = A(t)x \qquad (t_0 \leqq t < \infty) \tag{3.5}$$

has property $\mathscr{B}(\nu, N)$ $(\nu > 0)$ *while the function* $F(t, x)$ *satisfies condition* (3.3), (3.4) *for a fixed* τ_0.

Then when $q < \nu/N$ *equation* (3.1) *has property* $\mathscr{B}(\nu_1, N_1, \rho_0)$ *for some* $\rho_0 > 0$, $N_1 = Ne^{N\tau_0}$ *and* $\nu_1 = \nu - Nq > 0$.

PROOF. Suppose $\rho_0 < \min(\rho/N_1, \rho)$, $x_\tau \in B_{\rho_0} = \{x \mid \|x\| \leqq \rho_0\}$ and $x(t)$ is the solution of equation (3.1) satisfying the condition $x(\tau) = x_\tau$. We can always find a $T > 0$ such that $\|x(t)\| < \rho$ $(\tau \leqq t < \tau + T)$.

In the indicated interval $x(t)$ can be written in the form

$$x(t) = U(t, \tau)x_\tau + \int_\tau^t U(t, s)F(s, x(s))ds \qquad (\tau \leqq t < \tau + T). \tag{3.6}$$

The evolution operator $U(t, \tau)$ of equation (3.5) is by assumption subject to the estimate

$$\|U(t, \tau)\| \leqq Ne^{-\nu(t-\tau)} \qquad (t \geqq \tau).$$

Using this inequality and inequality (3.3), we obtain from (3.7) the integral inequality

$$\|x(t)\| \leqq N\|x_\tau\|e^{-\nu(t-\tau)} + \int_\tau^t Ne^{-\nu(t-s)}\eta(s)\|x(s)\|ds$$

$$(\tau \leqq t < \tau + T),$$

which implies (see Corollary III.2.2) that on the interval $0 \leqq t - \tau < T$

$$\|x(t)\| \leqq N_1\|x_\tau\|e^{-(\nu-Nq)(t-\tau)} \qquad (N_1 = Ne^{N\tau_0}). \tag{3.7}$$

From estimate (3.7) and the condition $\|x_\tau\| \leqq \rho_0$ we obtain the inequality

$$\|x(t)\| \leqq \rho e^{-(\nu-Nq)(t-\tau)} \qquad (0 \leqq t - \tau < T),$$

from which it is not difficult to deduce the fact that the quantity T can be chosen arbitrarily large.

Indeed, there would otherwise exist a number $T > 0$ such that $\|x(\tau + T)\| = \rho$, which contradicts inequality (3.7).

Thus inequality (3.7) is satisfied for any t and τ $(t \geqq \tau)$, which proves the theorem.

COROLLARY. *For sufficiently small $q > 0$ the zero solution of equation* (3.1) *is uniformly and asymptotically stable if equation* (3.5) *has a negative Bohl exponent.*

(Here it is assumed that the perturbed equation satisfies the conditions of the local existence theorem (see Remark 1.2).)

2. *Stability of the zero solution of the quasilinear equation.* If equation (3.1) is quasilinear, i.e. $\|F(t, x)\| = o(\|x\|)$, condition (3.4) is satisfied for arbitrarily small q in a sufficiently small neighborhood of $x = 0$.

Thus the quasilinear equation (3.1) has property $\mathscr{B}(\nu_1, N_1, \rho_1)$ with ν_1 arbitrarily close to ν; in particular, its zero solution is uniformly stable.

It turns out that this result has a converse.

THEOREM 3.2. *In order for the zero solution of equation* (3.1) *to be uniformly stable for any perturbation $F(t, x)$ satisfying the quasilinearity condition it is necessary and sufficient that the corresponding unperturbed equation have a negative Bohl exponent.*

PROOF. The sufficiency of the assertion follows directly from the above arguments.

We prove the necessity by considering only one perturbed equation (3.1) with $F(t, x) = x\|x\|^p$ for some fixed $p > 1$.

Let $B_\rho = \{x \mid \|x\| < \rho\}$ be a ball having the property that every solution $x(t)$ of the equation under consideration with $x_0 = x(t_0) \in B_\rho$ is indefinitely extendable for $t > t_0$. Each such solution can be represented in the form $x(t) = U(t, t_0)y(t)$, where $U(t, t_0)$ is the evolution operator of equation (3.5).

From (3.1) we obtain the following equation for $y(t)$:

$$y'(t) = \phi(t)y(t) \qquad (\phi(t) = \|U(t, t_0)y(t)\|^p;\ t \geqq t_0). \tag{3.8}$$

It follows from (3.8) that $y(t) = \psi(t)x_0$, where

$$\psi(t) = \exp\left(\int_{t_0}^{t} \phi(s)ds\right) \qquad (> 1 \quad \text{for} \quad t > t_0), \tag{3.9}$$

and therefore by virtue of (3.8)

$$\psi'(t) = \psi^{p+1}(t)\|U(t, t_0)x_0\|^p. \tag{3.10}$$

From this result, taking into account that $\psi(t_0) = 1$, we get

$$\int_{t_0}^{t} \|U(\tau, t_0)x_0\|^p\, d\tau = \int_{t_0}^{t} \frac{\psi'(\tau)}{\psi^{p+1}(\tau)}\, d\tau \leqq \frac{1}{p} \quad (x_0 \in \mathscr{B}_\rho). \tag{3.11}$$

But this estimate implies by virtue of Theorem III.6.2 and Remark III.6.3 that the Bohl exponent κ_B of equation (3.5) is negative under the additional assumption of integral boundedness of the coefficient $A(t)$.

In the general case, according to the same Theorem III.6.3, it remains for us to show that $\kappa_B < \infty$. To this end we make use of the fact that uniform stability must hold for the perturbed equation under consideration. We choose an $\varepsilon > 0$ and find a δ such that $\|x(t_0)\| < \delta$ implies $\|x(t)\| < \varepsilon$ independently of the choice of t_0.

Then for $\|x_0\| < \min(\delta, \rho)$ we will have

$$\|x(t)\| = \psi(t)\|U(t, t_0)x_0\| < \varepsilon \qquad (t \geqq t_0),$$

which implies

$$\|U(t, t_0)x_0\| \leqq \varepsilon/\psi(t) \leqq \varepsilon \qquad (t \geqq t_0),$$

and hence $\kappa_B \leqq 0$.

3. *Stability of a periodic solution.* An important class of equations of type (3.1) is the set of equations of this type with periodic coefficients. We have already noted in § 1.3 that if $\phi(t)$ is a T-periodic solution of the equation

$$dx/dt = f(t, x) \tag{3.12}$$

with a T-periodic continuously differentiable function $f(t, x)$, the difference $y(t) = x(t) - \phi(t)$ satisfies the quasilinear equation

$$dy/dt = A(t)y + F(t, y), \tag{3.13}$$

where $A(t) = f'_x(t, \phi(t))$.

If the variational equation $dy/dt = A(y)t$ in this connection has a negative Bohl exponent (i.e. the spectrum of its monodromy operator lies in the interior of the unit disk), equation (3.12) will also have a negative Bohl exponent at zero.

Thus in this case the periodic solution $\phi(t)$ is uniformly and asymptotically stable.

A different situation arises when equation (3.12) is autonomous: $f(t, x) \equiv f(x)$. In this case, as we saw in § 1, the variational equation corresponding to a periodic solution $\phi(t)$ must have a periodic solution, and hence the spectrum of its monodromy operator intersects the unit circle. We return again to this case later.

§ 4. The nonlinear equation with a principal part the spectrum of which does not intersect the imaginary axis

1. *Solutions bounded on the real line.* In this section we will consider the behavior of the solutions of the nonlinear equation

$$dx/dt = A(t)x + F(t, x), \tag{4.1}$$

under the assumption that the spectrum of the operator $A(t)$ in its principal part

$$dx/dt = A(t)x \tag{4.2}$$

does not intersect the imaginary axis. We will not as a rule assume that $F(t, 0) = 0$ and we will see that the properties of equation (4.1) resemble the properties of the inhomogeneous linear equation

$$dx/dt = A(t)x + f(t) \tag{4.3}$$

We assume here as everywhere else in this chapter that the function $F(t, x)$ is continuous in t.

In order to simplify the presentation we will consider below the more restricted equation

$$dx/dt = Ax + F(t, x) \tag{4.4}$$

with a stationary principal part such that

$$\sigma(A) = \sigma_+(A) \cup \sigma_-(A), \tag{4.5}$$

where either of the sets $\sigma_\pm(A)$ may be empty. But it is not difficult to note that only the possession by equation (4.3) of a principal Green function (see Chapter IV), which is subject to the usual estimates, is of significance in the proofs constructed below.

The passage to the more general case will be effected in special remarks.

The following terminology will be used in the sequel. A function $F(t, x)$ that is defined and continuous in t on the product space $(-\infty, \infty) \times B_\rho$, where $B_\rho = \{x \mid \|x\| \leqq \rho\}$, is of *class* (M, q, ρ) if on the indicated set it is subject to the condition

$$\|F(t, x)\| \leqq M \tag{4.6}$$

and the Lipschitz condition

$$\|F(t, x_2) - F(t, x_1)\| \leqq q\|x_2 - x_1\|. \tag{4.7}$$

A function satisfying the above conditions for $(t, x) \in (-\infty, \infty) \times \mathfrak{B}$ is of *class* (M, q). Finally, the symbols $(M, q, \rho)_\pm$ and $(M, q)_\pm$ will be used to denote the function classes resulting from the above classes when the real line $(-\infty, \infty)$ is replaced by a right halfline (a, ∞) or a left halfline $(-\infty, a)$ respectively.

We first consider equation (4.4) on the real line.

THEOREM 4.1. *Suppose the operator A satisfies condition* (4.5). *For any $\rho > 0$ there exist a constant $M > 0$ depending only on A and ρ and a constant $q > 0$ depending only on A such that if $F(t, x) \in (M, q, \rho)$, equation* (4.4) *has one and only one solution $x(t)$ that remains for all t in the ball B_ρ:*

$$\sup_{-\infty<t<\infty} \|x(t)\| \leqq \rho.$$

PROOF. Suppose $x(t)$ is a solution of equation (4.4) that remains in B_ρ. Then the function $F(t, x(t))$ is bounded on the real line and by virtue of Theorem II.4.1 a solution of equation (4.4) can be represented in the form

$$x(t) = \int_{-\infty}^{\infty} G_A(t - \tau)F(\tau, x(\tau))d\tau, \tag{4.8}$$

where G_A is the principal Green function of the operator A.

This function is subject to an estimate of the form

$$\|G_A(t)\| \leqq Ne^{-\nu|t|},$$

where N and ν are positive constants.

Thus the considered solution of equation (4.4) satisfies the integral equation (4.8). Conversely, on the basis of the same theorem every solution of the integral equation (4.8) that remains in B_ρ satisfies equation (4.4).

Consequently, for the considered solutions $x(t)$ that are defined on the real line and remain in B_ρ, equations (4.8) and (4.4) can be regarded as equivalent.

For a proof of the existence and uniqueness of the solutions of equation (4.8) we consider in the space $\boldsymbol{C}(\mathfrak{B})$ the ball $\boldsymbol{B}_\rho$ consisting of the functions $x(t)$ $(-\infty < t < \infty)$ satisfying the condition

$$|||x||| = \sup_{-\infty<t<\infty} \|x(t)\| \leqq \rho.$$

For sufficiently small $M > 0$ the transformation

$$y(t) = (Tx)(t) = \int_{-\infty}^{\infty} G_A(t-s)F(s, x(s))ds$$

acts in $\boldsymbol{B}_\rho$ and is a contraction if q is sufficiently small.

In fact, $|||F(\cdot, x(\cdot))||| \leqq M$ for $x(t) \in \boldsymbol{B}_\rho$, and therefore

$$|||y||| \leqq M \int_{-\infty}^{\infty} \|G_A(t)\| dt \leqq M \frac{2N}{\nu} \leqq \rho,$$

if $M \leqq \rho\nu/2N$. Further,

$$\begin{aligned} |||Tx_2 - Tx_1||| &\leqq q \sup_t \int_{-\infty}^{\infty} \|G_A(t-s)\| \cdot \|x_2(s) - x_1(s)\| ds \\ &\leqq q \frac{2N}{\nu} |||x_2 - x_1|||. \end{aligned}$$

If $q < \nu/2N$, the transformation T is a contraction. On the basis of the contraction principle we conclude that there exists one and only one solution of equation (4.8), and hence of equation (4.4), in $\boldsymbol{B}_\rho$. The theorem is proved.

REMARK 4.1. The frequently encountered equation

$$dx/dt = Ax + g(t) + F(x) \tag{4.9}$$

with a bounded continuous function $g(t)$ satisfying the condition $|||g||| < \rho\nu/2N$ and an autonomous function $F(x)$ satisfying conditions (4.6) and (4.7) is a special case of (4.4).

REMARK 4.2. If the condition $F(t, 0) = 0$ is satisfied and consequently the estimate $\|F(t, x)\| \leqq q\|x\|$ holds, the only bounded solution on the real line of equation (4.4) is the trivial solution $x(t) \equiv 0$. Any other solution leaves every ball B_{ρ_1} $(\rho_1 < \rho)$ either as $t \to +\infty$ or as $t \to -\infty$.

REMARK 4.3. Theorem 4.1 can be extended verbatim to equation (4.1) if equation (4.2) is e-dichotomic.

2. *Almost periodic and periodic solutions.* Theorem 4.1 can be sharpened if one assumes that the function $F(t, x)$ is not only bounded but almost periodic in t for each $x \in B_\rho$.

LEMMA 4.1. *Suppose $F(t, x)$ is an almost periodic function in t for each $x \in B_\rho$ and satisfies condition* (4.7). *Then the function $F(t, x(t))$ is almost periodic if $x(t)$ is an almost periodic function all of whose values belong to B_ρ.*

PROOF. We consider a sequence of translates $F(t + \tau_n, x)$, Let D be a compact set in B_ρ. We will prove that $\{\tau_n\}$ contains a subsequence $\{\tau_{n_k}\}$ such that the sequence of functions $F(t + \tau_{n_k}, x)$ converges uniformly on D. To this end, using the diagonal method, we choose $\{\tau_{n_k}\}$ so that the sequence $\{F(t + \tau_{n_k}, x)\}$ converges at each x of a countable subset D_1 that is dense in D. Suppose $x \in D$ and $x_1 \in D_1$. From the inequality

$$\begin{aligned}
&\|F(t + \tau_{n_k}, x) - F(t + \tau_{n_j}, x)\| \\
&\quad \leqq \|F(t + \tau_{n_k}, x) - F(t + \tau_{n_k}, x_1)\| \\
&\quad\quad + \|F(t + \tau_{n_k}, x_1) - F(t + \tau_{n_j}, x_1)\| \\
&\quad\quad + \|F(t + \tau_{n_j}, x_1) - F(t + \tau_{n_j}, x)\| \\
&\quad \leqq 2q\|x - x_1\| + \|F(t + \tau_{n_k}, x_1) - F(t + \tau_{n_j}, x_1)\|
\end{aligned}$$

it follows that for any $\varepsilon > 0$ there exists a natural number $N(x_1, \varepsilon)$ such that

$$\|F(t + \tau_{n_k}, x) - F(t + \tau_{n_j}, x)\| < \varepsilon \quad (n_k, n_j \geqq N) \tag{4.10}$$

for all $x \in U(x_1, \varepsilon) = \{x \,|\, \|x - x_1\| < \varepsilon/4q\}$. The open balls $U(x_1, \varepsilon)$ $(x_1 \in D_1)$ cover D. Choosing from among them a finite covering and taking as N the greatest of the $N(x_1, \varepsilon)$ corresponding to the members of this covering, we obtain inequality (4.10) for all $x \in D$. This proves that the subsequence $\{F(t + \tau_{n_k}, x)\}$ converges uniformly on D.

We now choose from the sequence $\{\tau_n\}$ a subsequence $\{\tau_{n_k}\}$ for which the sequence $\{x(t + \tau_{n_k})\}$ converges and for which the sequence $\{F(t + \tau_{n_k}, x)\}$ converges uniformly on the closure D of the set of values of $x(t)$ (this set is compact by virtue of the almost periodicity of $x(t)$). We then obtain the estimate

$$\begin{aligned}
&\|F(t + \tau_{n_k}, x(t + \tau_{n_k})) - F(t + \tau_{n_j}, x(t + \tau_{n_j}))\| \\
&\quad \leqq \|F(t + \tau_{n_k}, x(t + \tau_{n_k})) - F(t + \tau_{n_j}, x(t + \tau_{n_k}))\| \\
&\quad\quad + \|F(t + \tau_{n_j}, x(t + \tau_{n_k})) - F(t + \tau_{n_j}, x(t + \tau_{n_j}))\| \\
&\quad \leqq \max_D \|F(t + \tau_{n_k}, x) - F(t + \tau_{n_j}, x)\| + q\|x(t + \tau_{n_k}) - x(t + \tau_{n_j})\|,
\end{aligned} \tag{4.11}$$

which implies the convergence of the sequence $\{F(t + \tau_{n_k}, x(t + \tau_{n_k}))\}$.

We have proved the precompactness of the family $F(t + \tau, x(t + \tau))$ and hence the almost periodicity of the function $F(t, x(t))$.

THEOREM 4.2. *Suppose the spectrum $\sigma(A)$ does not intersect the imaginary axis while the function $F(t, x)$ is almost periodic for each $x \in B_\rho$ and of class (M, q, ρ) with the constants M and q restricted as in the proof of Theorem* 4.1.

Then the unique bounded solution of equation (4.4) *whose existence was proved in Theorem* 4.1 *is almost periodic.*

PROOF. The set $\tilde{B}_\rho$ of almost periodic functions lying in the ball B_ρ is closed in this ball. From the proof of Theorem II.4.4 and Lemma 4.1 it follows that the contraction T (see the proof of Theorem 4.1) leaves $\tilde{B}_\rho$ invariant. Therefore the unique fixed point of T in B_ρ lies in $\tilde{B}_\rho$.

We briefly discuss the question of periodic solutions of equation (4.4) for the case when the function $F(t,x)$ is continuous and periodic in t:

$$F(t + 2\pi, x) = F(t, x) \qquad (x \in B_\rho).$$

If the conditions of Theorem 4.1 are satisfied, there exists exactly one bounded solution $x(t)$ of this equation. It is not difficult to verify that it is a periodic function. In fact, the function $x(t + 2\pi)$ satisfies the same equation, and, by virtue of the uniqueness, $x(t) \equiv x(t + 2\pi)$.

All of what has been said applies in particular to equation (4.9) when the function $g(t)$ is almost periodic or periodic.

REMARK 4.4. The results of this and the preceding subsection can be extended without difficulty to equations of the form $dx/dt = A(t)x + F(t, x)$ with a periodic operator function $A(t)$ if the unperturbed equation is reducible.

3. *Solutions bounded on a halfline.* Let $x(t)$ be a solution of equation (4.4) that remains in the ball $B_\rho = \{x| \, \|x\| \leqq \rho\}$ for $t > t_0$.

From the results of § II.4.3. it follows that $x(t)$ satisfies the integral equation

$$x(t) = e^{A(t-t_0)}y_0 + \int_{t_0}^{\infty} G_A(t - s)F(s, x(s))ds, \tag{4.12}$$

where $y_0 = P_-x(t_0)$. The converse is also true: a solution of the integral equation (4.12) satisfies the differential equation (4.4) for $t > t_0$.

The investigation of equation (4.12) is carried out by the same methods as the above.

THEOREM 4.3. *Suppose that the spectrum $\sigma(A)$ does not intersect the imaginary axis and hence that the estimate $\|G_A(t)\| \leqq Ne^{-\nu|t|}$ $(\nu > 0, N \geqq 1)$ is satisfied.*

For any $\rho > \rho_1 > 0$ there exist positive numbers M, q and l depending only on N, ν, ρ and ρ_1 such that if $F(t, x) \in (M, q, \rho)_+$, there corresponds to each $y_0 \in \mathfrak{B}_- \cap$

$B_{\rho/2N}$ one and only one solution of equation (4.4) *satisfying the conditions $P_-x(t_0) = y_0$ and* $\sup_{t_0+l\leq t}\|x(t)\| \leq \rho_1$.

The following estimate is valid for any two solutions corresponding to different values of y_0:

$$\|x_2(t) - x_1(t)\| \leq c_\mu e^{-\mu(t-t_0)}\|y_{20} - y_{10}\|, \tag{4.13}$$

where c_μ is a positive constant not depending on t_0 while $\mu \in (0, \nu)$ and can be made arbitrarily close to ν by decreasing the number q.[2)]

PROOF. The first part of the theorem is proved with the use of the contraction principle by following the proof of Theorem 4.1. The additional condition $\|y_0\| \leq \rho/2N$ is imposed to ensure that the initial portion of the trajectory $x(t)$ does not leave the ball B_ρ. When $t - t_0 > l$ the first term in (4.12) becomes sufficiently small and the trajectory of the solution finds itself in B_{ρ_1}.

Let $x_1(t)$ and $x_2(t)$ be a pair of solutions. From (4.12) we deduce the inequality

$$\|x_2(t) - x_1(t)\| \leq \|y_{20} - y_{10}\|Ne^{-\nu(t-t_0)} + \int_{t_0}^{\infty} Ne^{-\nu|t-s|}q\|x_2(s) - x_1(s)\|ds,$$

which implies by virtue of Corollary III.2.3 that

$$\|x_2(t) - x_1(t)\| \leq \phi(t)\|y_{20} - y_{10}\|,$$

where

$$\phi(t) = \frac{2N\nu}{\nu + \sqrt{\nu^2 - 2qN\nu}}\, e^{-(t-t_0)\sqrt{\nu^2-2qN\nu}},$$

and this implies the proof of the second part of the theorem.

REMARK 4.5. To each vector $y_0 \in \mathfrak{B}_-$ with a sufficiently small norm there corresponds a unique bounded trajectory $x(t)$ ($t \geq t_0$) such that $y_0 = P_-x(t_0)$. Thus the initial points $x(t_0)$ of the trajectories that are bounded for $t \to +\infty$ form in $B_{\rho/2N}$ a manifold $\mathfrak{M}^+$ of dimension equal to the dimension of $\mathfrak{B}_-$. In fact, to each point $y_0 \in \mathfrak{B}_- \cap B_{\rho/2N}$ there corresponds one and only one point $x(t_0)$ such that $y_0 = P_-x(t_0)$ and from which there emanates a trajectory that is bounded for $t \to \infty$.

The manifold $\mathfrak{M}^+$ is conveniently imagined as a "surface" in the Cartesian product of the subspaces $P_+\mathfrak{B}$ and $P_-\mathfrak{B}$. The "equation" of this surface, i.e. the relation between $y_0 = P_-x(t_0)$ and $P_+x(t_0)$, is implicit in (4.12).

If $x(t_0) \in B_{\rho/2N}\backslash\mathfrak{M}^+$, the solution $x(t)$ leaves any ball B_{ρ_1} ($\rho_1 < \rho$) for sufficiently large t.

When the spectral set $\sigma_+(A)$ is empty, the space $\mathfrak{B}$ coincides with $\mathfrak{B}_-$ and therefore all of the solutions emanating from $B_{\rho/2N}$ terminate in B_{ρ_1} for $t \to \infty$. By virtue of (4.13) each of them is uniformly and asymptotically stable.

[2)] Hence μ can be made arbitrarily close to the distance from $\sigma(A)$ to the imaginary axis.

On the other hand, if $\sigma_-(A)$ is empty, the manifold $\mathfrak{M}^+$ becomes a single point x, for which $P_-x = 0$. In this case there is only one trajectory that is bounded for $t > t_0$. It is obtained when $y_0 = 0$.

Analogous results hold for a left halfline, the role of $\mathfrak{B}_-$ being played by $\mathfrak{B}_+$. The initial values $x(t_0)$ of the solutions that remain in B_ρ as $t \to -\infty$ fill out a manifold $\mathfrak{M}^-$ situated in $B_{\rho/2N}$. Each point x of this manifold is uniquely determined by its projection $y = P_+x$, and therefore the "dimension" of $\mathfrak{M}^-$ coincides with the dimension of $\mathfrak{B}_+$.

We again consider equation (4.4) on the real line. From Theorems 4.1 and 4.3 and the observations made in Remark 4.5 we obtain the following result.

THEOREM 4.4. *Suppose the spectrum $\sigma(A)$ does not intersect the imaginary axis. Then for any $\rho > 0$ there exist M and q depending only on A and ρ such that if $F(t, x) \in (M, q, \rho)$, a sufficiently small neighborhood of the zero element of $\mathfrak{B}$ will contain manifolds $\mathfrak{M}^-$ and $\mathfrak{M}^+$ with the following properties.*

a) *The manifold $\mathfrak{M}^\pm$ is homeomorphic to a neighborhood of zero of the subspace $\mathfrak{B}_\pm$.*

b) *The manifolds $\mathfrak{M}^+$ and $\mathfrak{M}^-$ have exactly one point z in common.*

c) *The solution $x_0(t)$ of equation* (4.4) *satisfying the condition $x_0(t_0) = z$ is bounded on the real line.*

d) *The solutions $x(t)$ of equation* (4.4) *satisfying the condition $x(t_0) \in \mathfrak{M}^+$ ($\mathfrak{M}^-$) exponentially approach $x_0(t)$ as $t \to +\infty$ ($t \to -\infty$) and exponentially recede from it as t tends indefinitely far in the opposite direction.*

REMARK 4.6. If one of the manifolds $\mathfrak{M}^\pm$ reduces to a single point belonging to $x_0(t)$, this solution is uniformly and asymptotically stable over the corresponding halfline. But if the dimensions of both $\mathfrak{M}^+$ and $\mathfrak{M}^-$ are not equal to zero, we say that this solution is conditionally stable.

REMARK 4.7. The above results extend without difficulty to equation (4.1) in the case when its nonstationary principal part admits an e-dichotomy of the solutions.

4. *Orbital stability of a periodic solutions of the autonomous equation.* We consider the autonomous equation

$$dx/dt = f(x) \tag{4.14}$$

with a continuously differentiable function $f(x)$ and having a periodic solution $\phi(t) = \phi(t + T)$ $(-\infty < t < \infty)$.

We have already indicated that in this case the variational equation

$$d\tilde{x}/dt = A(t)\tilde{x} \qquad (A(t) = f'(\phi(t))) \tag{4.15}$$

has the periodic solution

$$\tilde{x}(t) = \phi'(t) \tag{4.16}$$

and hence (we assume that $\phi(t) \not\equiv$ const) the spectrum of the monodromy operator $U(T)$ of this equation contains the point $\lambda = 1$.

We assume that this point is an isolated simple eigenvalue while the rest of the spectrum of $U(T)$ lies in the interior of the unit disk and does not surround the origin.

In this case there exists a periodic operator function $Q(t)$ that reduces (4.15) to the form $d\tilde{x}/dt = \Lambda\tilde{x}$, where Λ is a constant operator with zero as an isolated simple eigenvalue and the rest of its spectrum $\sigma_-(\Lambda)$ lying in the interior of the left halfplane. We take t_0 as the initial point and assume that $Q(t_0) = I$. The change of variable

$$x(t) = \phi(t) + Q(t)z(t) \tag{4.17}$$

takes (4.14) into the quasilinear equation

$$dz/dt = \Lambda z + F(t, z) \qquad (F(t, 0) = 0), \tag{4.18}$$

the function $F(t, z)$ satisfying in a sufficiently small neighborhood of zero the Lipschitz condition

$$\|F(t, z_2) - F(t, z_1)\| \leqq q\,\|z_2 - z_1\|$$

with a sufficiently small constant $q > 0$.

We denote by P_0 and P_- the spectral projections of Λ corresponding to the parts $\sigma_0 = \{0\}$ and σ_- of its spectrum. Let $-\nu = \sup_{\lambda\in\sigma_-} \operatorname{Re}\lambda$. The substitution

$$z = e^{-(\nu/2)(t-t_0)}\tilde{z}$$

converts equation (4.18) into the equation

$$\frac{d\tilde{z}}{dt} = \left(\Lambda + \frac{\nu}{2}I\right)\tilde{z} + \tilde{F}(t, \tilde{z}) \tag{4.19}$$

where the function $\tilde{F}(t, \tilde{z}) = e^{(\nu/2)(t-t_0)}F(t, e^{-(\nu/2)(t-t_0)}\tilde{z})$ has the same properties as $F(t, z)$. But the principal part of this equation is e-dichotomic and we can therefore apply the results of the preceding subsections to it.

Since the only solution of equation (4.19) that is bounded on the real line is the zero solution, the solutions issuing from the manifold $\mathfrak{M}^+$ (described in Remark 4.5) tend to zero as $t \to \infty$. Returning to equation (4.18), we obtain the following estimate for any solution of it satisfying the condition $z(t_0) \in \mathfrak{M}^+$:

$$\|z(t)\| \leqq c_\mu e^{-\mu(t-t_0)}\|P_-z(t_0)\|, \tag{4.20}$$

where $\mu < \nu$ but can be made arbitrarily close to ν by requiring that $z(t_0)$ lie in a sufficiently small neighborhood of zero.

These solutions $z(t)$ are determined from an equation of type (4.12), which, after one returns from (4.19) to (4.18), takes the form

$$z(t) = e^{\Lambda(t-t_0)}P_-z(t_0) + \int_{t_0}^{\infty} G_\Lambda(t - s)F(s, z(s))ds, \tag{4.21}$$

where

$$G_A(t)\begin{cases} e^{At}P_- & \text{for } t > 0, \\ -e^{-At}P_0 = -P_0 & \text{for } t < 0. \end{cases}$$

We note that although $\|G_A(t-s)\| = \|P_0\|$ $(s > t)$ does not decrease as $s \to \infty$, expression (4.21) is applicable to the solutions under consideration, since by virtue of estimate (4.20)

$$\|F(s, z(s))\| \leqq q\|z(s)\| \leqq c_\mu q e^{-\mu(s-t_0)} \|P_- z(t_0)\|.$$

After this, from (4.21) we obtain the estimate

$$\frac{\|z(t_0) - P_- z(t_0)\|}{\|P_- z(t_0)\|} \leqq c_\mu \frac{\|P_0\|}{\mu} q,$$

which shows that in a sufficiently small neighborhood of zero the manifold $\mathfrak{M}^+$ described by the points $z(t_0)$ lies outside a cone whose axis of rotation is the one-dimensional subspace $\mathfrak{B}_0 = P_0\mathfrak{B}$. We note that $\mathfrak{M}^+$ is projected in a one-to-one manner onto a neighborhood of zero in the subspace $\mathfrak{B}_- = P_-\mathfrak{B}$ and therefore divides a sufficiently small ball with center at zero in the space $\mathfrak{B}$.

We proceed, finally, to a consideration of the solutions of equation (4.14). The periodic solution $\bar{x}(t)$ of equation (4.15) has the form

$$\bar{x}(t) = Q(t)e^{A(t-t_0)}P_0\bar{x}(t_0) = Q(t)P_0\bar{x}(t_0),$$

and by virtue of (4.16)

$$\phi'(t_0) = \bar{x}(t_0) = P_0\bar{x}(t_0) \in \mathfrak{B}_0.$$

Thus the periodic solution $\phi(t)$ is tangent to the one-dimensional space $\mathfrak{B}_0$ at $t = t_0$.
On the other hand, the points

$$x(t_0) = \phi(t_0) + Q(t_0)z(t_0) = \phi(t_0) + z(t_0) \tag{4.22}$$

with $z(t_0) \in \mathfrak{M}^+$ constitute a surface $\mathfrak{M}_1^+$ that is obtained from $\mathfrak{M}^+$ by means of a parallel translation into the point $\phi(t_0)$, lies outside a cone with axis $\mathfrak{B}_0$ and divides a sufficiently small ball $B_\delta(\phi(t_0))$ with center at this point. Since the trajectory of $\phi(t)$ in the vicinity of t_0 obviously passes from one nappe of the cone to the other, for sufficiently small δ any trajectory passing through the ball under consideration also penetrates, by virtue of the continuity of $f(x)$, each of the nappes of the cone and consequently necessarily intersects the manifold $\mathfrak{M}_1^+$ at some instant t_1.

Thus, $\|x(t_0') - \phi(t_0)\| < \delta$ if at some time $t = t_0'$ there exists a t_1 such that $x(t_1) \in \mathfrak{M}_1^+$.

Let $x_1(t) = x(t - t_0 + t_1)$. Since equation (4.14) is autonomous, $x_1(t)$ is also a solution of it, this solution now intersecting $\mathfrak{M}_1^+$ when $t = t_0$. The corresponding solution $z_1(t)$ of equation (4.18) has the property

$$z_1(t_0) = x_1(t_0) - \phi(t_0) \in \mathfrak{M}^+$$

and is therefore subject to the estimate

$$\|z_1(t)\| \leqq c_\mu e^{-\mu(t-t_0)} \|P_- z_1(t_0)\|.$$

Hence by virtue of (4.17) we conclude that

$$\|x_1(t) - \phi(t)\| \leqq \bar{c}_\mu e^{-\mu(t-t_0)},$$

where $\bar{c}_\mu$ is a certain constant, or, finally, that

$$\|x(t - (t_1 - t_0)) - \phi(t)\| \leqq \bar{c}_\mu e^{-\mu(t-t_0)}. \tag{4.23}$$

Inequality (4.23) does not mean that $x(t)$ and $\phi(t)$ approach each other as $t \to \infty$, and thus it does not imply the stability of $\phi(t)$.

But this inequality does imply that if a solution $x(t)$ finds itself at some moment of time in a sufficiently small neighborhood of the set $\mathscr{L} = \{\phi(t) | 0 \leqq t \leqq T\}$, viz. the trajectory of the periodic solution $\phi(t)$, it will subsequently approach this set as $t \to \infty$.

The property described above is called *orbital asymptotic stability*. The shift in time $\tau_0 = t_1 - t_0$ after which the solutions asymptotically approach each other is called the *asymptotic phase*.

The arguments presented above lead to the following result.

THEOREM 4.5. *If the autonomous equation has a periodic solution (not identically constant) to which there corresponds a simple isolated eigenvalue of the monodromy operator of the variational equation while the rest of the spectrum of this operator is in the interior of the unit disk and does not surround the origin, the indicated periodic solution is orbitally asymptotically stable and, moreover, each solution that is sufficiently close to its trajectory has an asymptotic phase.*

§ 5. Stable integral manifolds

1. *Derivation of an integral equation for manifolds.* In this section the equation

$$dx/dt = Ax + F(t, x) \tag{5.1}$$

will be considered under less rigid restrictions concerning the operator A. We will no longer assume that the spectrum $\sigma(A)$ does not intersect the imaginary axis. This creates a more complicated situation.

We assume that the spectrum of A can be divided into two spectral sets $\sigma(A) = \sigma_1(A) \cup \sigma_2(A)$ such that for some $\alpha > 0$

$$|\operatorname{Re}\lambda| > \alpha \quad \text{for } \lambda \in \sigma_2(A),$$
$$|\operatorname{Re}\lambda| < \alpha \quad \text{for } \lambda \in \sigma_1(A).$$

Thus in our case the equation $dx/dt = Ax$ admits an exponential splitting of its solutions of generally third order.

We will say in this connection that *the spectrum $\sigma(A)$ admits an α-separation.*

As usual, P_1, P_2, $\mathfrak{B}_1$ and $\mathfrak{B}_2$ will denote the spectral projections and invariant subspaces of A corresponding to the spectral sets $\sigma_1(A)$ and $\sigma_2(A)$.

We will assume that $F(t, x)$ is of class (M, q), i.e. it is continuous and subject to the estimates

$$\|F(t, x)\| \leqq M, \tag{5.2}$$

$$\|F(t, x^{(2)}) - F(t, x^{(1)})\| \leqq q\|x^{(2)} - x^{(1)}\| \tag{5.3}$$

on the product space $(-\infty, \infty) \times \mathfrak{B}$.

The purpose of the following discussion is the construction of certain special integral manifolds for equation (5.1).

When the spectrum of A does not intersect the imaginary axis, these manifolds convert into the individual trajectories studied in the preceding section.

In order to explain the essence of the matter we consider as an elementary example the system of differential equations

$$\begin{cases} dx_1'/dt = 0, \\ dx_1''/dt = x_1', \\ dx_2/dt = -x_2, \end{cases}$$

or in vector form, setting $x = (x_1', x_1'', x_2)$,

$$\frac{dx}{dt} = \begin{pmatrix} 0 & 0 & 0 \\ 1 & 0 & 0 \\ 0 & 0 & -1 \end{pmatrix} x \qquad (= Ax).$$

The spectrum $\sigma(A)$ in this case consists of the two points $\lambda_0 = 0$ and $\lambda_1 = -1$, and the three-dimensional space $\mathfrak{B}$ decomposes into the direct sum $\mathfrak{B} = \mathfrak{B}_1 + \mathfrak{B}_2$ of the two-dimensional space $\mathfrak{B}_1$ consisting of the vectors $(x_1', x_1'', 0)$ and the one-dimensional space $\mathfrak{B}_2$ consisting of the vectors $(0, 0, x_2)$.

The solution of the system has, as can easily be seen, the form

$$x(t) = (x_{10}', x_{10}'', x_{10} + x_{10}' t, e^{-t}x_{20}).$$

It readily follows from this formula that the zero solution of the system, as well as any other solution initially in $\mathfrak{B}_1$, is unstable. But the whole set of solutions that are initially in $\mathfrak{B}_1$ (and, incidentally, remain in $\mathfrak{B}_1$) is stable in the sense that the distance between any other solution and $\mathfrak{B}_1$ tends to zero as $t \to +\infty$.

We can say that this set of solutions forms a stable integral manifold $\mathfrak{B}_1$.

An analogous situation exists, as we will see below, for more general nonlinear equations.

The individual solutions of such equations will not as a rule be stable or even conditionally stable because of the presence of a "critical" part of the spectrum lying on the imaginary axis. At the same time certain sets of solutions, which in general are no longer linear, can turn out to be stable (or conditionally stable) in the sense that near solutions will be "attracted" to them. The "dimension" of such a stable manifold of solutions always coincides with the multiplicity of the critical part of the spectrum of the coefficient of the principal linear part of the equation (and if the critical part is absent, the manifold reduces to a single trajectory).

Another example of this type of stable integral manifold is the orbit of a periodic solution of the autonomous equation considered in the preceding section.

In this section we deduce a functional equation for integral manifolds of a specific form and prove their existence, while in the next section we establish the stability of such a manifold under certain assumptions on the equation.

We note that after finding such a manifold, in order to determine the solutions composing it, we must consider, as will be seen, a differential equation in a space of dimension equal to the multiplicity of the critical part of the spectrum, i.e. less than the dimension of the initial phase space.

In particular, this dimension can be finite.

We proceed to more precise definitions.

By an *integral manifold of equation* (5.1) we will mean a set $\mathfrak{M}$ composed of the trajectories of this equation in the product space $(-\infty, \infty) \times \mathfrak{B}$.

Strictly speaking, we depart here from the usual meaning of the term "trajectory" as the set of values of a solution $x(t)$ $(-\infty<t<\infty)$ in the phase space $\mathfrak{B}$ and think of it rather as the set of points $(t, x(t))$ of the space $(-\infty, \infty) \times \mathfrak{B}$.

The component $x \in \mathfrak{B}$ of an element (t, x) will be called its *spatial projection*. From this point of view a trajectory in the usual sense of the word is the spatial projection of the trajectory being considered by us.

Thus, by definition, through each point of an integral manifold $\mathfrak{M}$ there passes a trajectory $x(t)$, and each trajectory having at least one point in $\mathfrak{M}$ has all of its points in $\mathfrak{M}$.

We will consider the integral manifolds of equation (5.1) that can be described by equations of the form

$$x_2 = \phi(t, x_1) \qquad (t \in (-\infty, \infty); \quad x_1 \in \mathfrak{B}_1) \tag{5.4}$$

in the "coordinates" t, $x_1 = P_1 x$ and $x_2 = P_2 x$.

Here we will assume that the function ϕ is continuous in t and satisfies a Lipschitz condition in x_1 with a constant not depending on t. The set of all functions is a linear space, which we denote by L.

Let L_η denote the subset of L consisting of the functions satisfying the condition

$$\begin{gathered}\|\phi(t, x_1^{(2)}) - \phi(t, x_1^{(1)})\| \leqq \eta \|x_1^{(2)} - x_1^{(1)}\| \\ (x_1^{(1)}, x_1^{(2)} \in \mathfrak{B}_1; \quad \eta = \text{const} > 0)\end{gathered} \tag{5.5}$$

(the *functions of class* L_η).

The integral manifold $\mathfrak{M}$ described by equation (5.4) is called a (ρ,η)-*manifold* if $\phi \in L_\eta$ and satisfies the condition

$$\|\phi(t,x_1)\| \leqq \rho \qquad (t \in (-\infty, \infty); \quad x_1 \in \mathfrak{B}_1). \tag{5.6}$$

Thus the spatial projection of a (ρ, η)-manifold is completely contained in the cylinder Π_ρ of $\mathfrak{B}$ whose base is the ball $B_\rho(\mathfrak{B}_2)$ in $\mathfrak{B}_2$ and whose "axis" is the subspace $\mathfrak{B}_1$.

If a trajectory $(t, x(t))$ lies on the considered manifold and $\psi(t) = P_1x(t)$ is its projection onto the subspace $\mathfrak{B}_1$, the whole trajectory can be expressed in terms of the function $\psi(t)$ by means of the formula

$$x(t) = \psi(t) + \phi(t, \psi(t)). \tag{5.7}$$

For constructing the manifolds of equation (5.1) that are of interest to us it is convenient to write this equation in the form of a system of two equations in the subspaces $\mathfrak{B}_1$ and $\mathfrak{B}_2$.

Multiplying (5.1) from the left by the projections P_1 and P_2 respectively, we obtain the required system

$$\begin{cases} \dfrac{dx_1}{dt} = A_1x_1 + F_1(t, x_1 + x_2), & (5.8a) \\ \dfrac{dx_2}{dt} = A_2x_2 + F_2(t, x_1 + x_2), & (5.8b) \end{cases}$$

where $x_k = P_kx$, $F_k = P_kF$, $A_k = P_kA$ $(k = 1, 2)$.

Let us derive an equation which the function $\phi(t, x_1)$ defining a (ρ, η)-manifold $\mathfrak{M}$ must satisfy.

If $x = x(t)$ is a trajectory lying on this manifold, the functions $x_1 = \psi(t) = P_1x(t)$ and $x_2 = \phi(t, \psi(t))$ must satisfy system (5.8).

Here equation (5.8a) takes the form

$$d\psi/dt = A_1\psi + F_1(t, \psi + \phi(t, \psi)). \tag{5.9}$$

Conditions (5.3) and (5.5) imply the condition

$$\|F_1(t, \psi'' + \phi(t, \psi'')) - F_1(t, \psi' + \phi(t, \psi'))\| \leqq q(1 + \eta)\|\psi'' - \psi'\|. \tag{5.10}$$

Therefore equation (5.9) satisfies for each given $\phi \in L$ the conditions of Theorem 1.2 and has for each $x_{10} \in \mathfrak{B}_1$ a unique solution $\psi(t) = \Psi(t, \tau, x_{10}|\phi)$ that is equal to x_{10} at $t = \tau$:

$$\Psi(\tau, \tau, x_{10}|\phi) = x_{10}. \tag{5.11}$$

We note that when t, τ and x_{10} are fixed the function $\Psi(t, \tau, x_{10}|\phi)$ is with respect to ϕ a nonlinear operator defined in the space L and taking its values in $\mathfrak{B}_1$.

On the other hand the function $x_2(t) = \phi(t, \psi(t))$ is a bounded solution on the real line of the equation

$$dx_2/dt = A_2x_2 + F_2(t, \psi(t) + x_2) \tag{5.12}$$

and must therefore satisfy the integral equation

$$x_2(\tau) = \int_{-\infty}^{\infty} G_2(\tau - t)F_2(t, \psi(t) + x_2(t))dt, \tag{5.13}$$

where $G_2(t)$ is a principal Green function of the operator A_2 in the subspace $\mathfrak{B}_2$.

We now choose for a fixed τ the solution $\psi(t)$ of equation (5.9) that is equal to a fixed element x_1 at $t = \tau$: $\psi(t) = \Psi(t, \tau, x_1|\phi)$.

From (5.13) we then obtain the integrofunctional equation

$$\phi(\tau, x_1) = \int_{-\infty}^{\infty} G_2(\tau - t)F_2(t, \Psi(t, \tau, x_1|\phi) + \phi(t, \Psi(t, \tau, x_1|\phi)))\, dt \tag{5.14}$$
$$(\tau \in (-\infty, \infty);\ x_1 \in \mathfrak{B}_1)$$

for the function $\phi(t, x_1)$.

We emphasize that this equation is not simply an integral equation, since ϕ enters into it not only as a function but as an element of the space L.

We have thus shown that the function $\phi(t, x_1)$ defining a (ρ, η)-manifold of equation (5.1) must satisfy equation (5.14).

On the other hand, a solution $\phi(t, x_1)$ of equation (5.14) satisfying conditions (5.5) and (5.6) defines a (ρ, η)-manifold of equation (5.1).

In fact, for any point (x_{10}, x_{20}, t_0) lying on $\mathfrak{M}$, i.e. satisfying the relation $x_{20} = \phi(t_0, x_{10})$, equation (5.9) has a solution

$$x_1 = \psi(t) = \Psi(t, t_0, x_{10}|\phi)$$

with the property $x_1(t_0) = x_{10}$.

It accordingly follows from (5.14) and the obvious relation

$$\Psi(t, \tau, \Psi(\tau, t_0, x_{10}|\phi)|\phi) = \Psi(t, t_0, x_{10}|\phi)$$

that the function $x_2 = \phi(t, \psi(t))$ is a solution of equation (5.12).

This result can be formulated in the following manner.

LEMMA 5.1. *A (ρ, η)-manifold of the differential equation* (5.1) *is described by a solution ϕ of the integrofunctional equation* (5.14) *that is of class L_η and satisfies condition* (5.6).

The trajectories lying on the manifold are determined by the solutions of the differential equation (5.9).

It will be proved below that for given $\rho > 0$ and $\eta > 0$, when M and q are sufficiently small, equation (5.14) has one and only one solution possessing the necessary properties, and that the integral manifold corresponding to it consists of all of the trajectories of equation (5.1) that lie in the cylinder $\|x_2\| \leqq \rho$.

2. *Proof of the existence of (ρ, η)-manifolds.* Let us first prove an auxiliary proposition. We note that if the operator A admits an α-separation, there exists a $\delta > 0$ for which the spectral set $\sigma_1(A)$ lies in the strip $|\mathrm{Re}\,\lambda| < \alpha - \delta$ and hence

$$\|e^{A_1 t}\| \leqq N_\delta e^{(\alpha-\delta)|t|}, \tag{5.15}$$

where $A_1 = P_1 A$.

On the other hand, it can be assumed that the spectral set $\sigma_2(A)$ lies in the domain $|\mathrm{Re}\,\lambda| > \alpha + \delta$ and hence that the estimate

$$\|G_2(t)\| \leqq N_\delta e^{-(\alpha+\delta)|t|} \tag{5.16}$$

is valid for the Green function $G_2(t)$.

LEMMA 5.2. *The operator $\Psi(t, \tau, x_1|\phi)$ representing a solution of equation* (5.9) *satisfies the conditions*

1) $\|\Psi(t, \tau, x_1^{(2)}|\phi^{(2)}) - \Psi(t, \tau, x_1^{(1)}|\phi^{(1)})\|$

$$\leqq N_\delta\Big\{\|x_1^{(2)} - x_1^{(1)}\|e^{(\alpha-\delta+\beta_q)|t-\tau|} \tag{5.17}$$

$$+ q\left|\int_\tau^t e^{(\alpha-\delta+\beta_q)|t-s|}\sup_{x_1}\|\phi^{(2)}(s, x_1) - \phi^{(1)}(s, x_1)\|ds\right|\Big\}$$

$$(\phi^{(k)} \in L_\eta;\quad x_1^{(k)} \in \mathfrak{B}_1;\quad k = 1,2),$$

where $\beta_q \to 0$ as $q \to 0$, and

$$2)\ \|\Psi(t, \tau, x_1|\phi)\| \leqq N_\delta e^{(\alpha-\delta)|t-\tau|}\left[\|x_1\| + \frac{M}{\alpha-\delta}\right] \quad (\phi \in L_\eta;\ x_1 \in \mathfrak{B}_1). \tag{5.18}$$

PROOF. Let

$$x_1^{(1)}(t) = \Psi(t, \tau, x_1^{(1)}|\phi^{(1)}),\quad x_1^{(2)}(t) = \Psi(t, \tau, x_1^{(2)}|\phi^{(2)}),$$
$$x_1(t) = \Psi(t, \tau, x_1^{(1)}|\phi^{(2)}).$$

These functions satisfy the integral equations

$$x_1^{(1)}(t) = e^{A_1(t-\tau)}x_1^{(1)} + \int_\tau^t e^{A_1(t-s)}F_1(s, x_1^{(1)}(s) + \phi^{(1)}(s, x_1^{(1)}(s)))\,ds, \tag{5.19}$$

$$x_1^{(2)}(t) = e^{A_1(t-\tau)}x_1^{(2)} + \int_\tau^t e^{A_1(t-s)}F_1(s, x_1^{(2)}(s) + \phi^{(2)}(s, x_1^{(2)}(s)))\,ds, \tag{5.20}$$

$$x_1(t) = e^{A_1(t-\tau)}x_1^{(1)} + \int_\tau^t e^{A_1(t-s)}F_1(s, x_1(s) + \phi^{(2)}(s, x_1(s)))\,ds. \tag{5.21}$$

From (5.19) and (5.21), using estimate (5.15) and the Lipschitz conditions (5.3) and (5.10) for the functions F and $\phi^{(1)}$, $\phi^{(2)}$, we obtain the estimate

$$\|x_1^{(1)}(t) - x_1(t)\|$$
$$\leqq \int_\tau^t \|e^{A_1(t-s)}\|\{\|F_1(s, x_1^{(1)}(s) + \phi^{(1)}(s, x_1^{(1)}(s))) - F_1(s, x_1(s) + \phi^{(1)}(s, x_1(s)))\|$$
$$+ \|F_1(s, x_1(s) + \phi^{(1)}(s, x_1(s))) - F_1(s, x_1(s) + \phi^{(2)}(s, x_1(s)))\|\}\,ds$$
$$\leqq N_\delta q(1+\eta)\int_\tau^t e^{(\alpha-\delta)(t-s)}\|x_1^{(1)}(s) - x_1(s)\|ds$$
$$+ N_\delta q\int_\tau^t e^{(\alpha-\delta)(t-s)}\sup_{x_1}\|\phi^{(1)}(s, x_1) - \phi^{(2)}(s, x_1)\|ds.$$

It can now be deduced from Remark III.2.1 that when $t \geqq \tau$

$$\|x_1^{(1)}(t) - x_1(t)\| \leqq u(t),$$

where $u(t)$ is a solution of the integral equation

$$u(t) = N_\delta q(1 + \eta) \int_\tau^t e^{(\alpha-\delta)(t-s)} u(s) ds$$

$$+ N_\delta q \int_\tau^t e^{(\alpha-\delta)(t-s)} \sup_{x_1} \|\phi^{(1)}(s, x_1) - \phi^{(2)}(s, x_1)\| \, ds,$$

which, as can easily be verified, has the form

$$u(t) = N_\delta q \int_\tau^t e^{[N_\delta q(1+\eta)+\alpha-\delta](t-s)} \sup_{x_1} \|\phi^{(1)}(s, x_1) - \phi^{(2)}(s, x_1)\| \, ds.$$

Thus

$$\|\Psi(t, \tau, x_1^{(1)} | \phi^{(1)}) - \Psi(t, \tau, x_1^{(1)} | \phi^{(2)})\|$$

$$\leqq N_\delta q \int_\tau^t e^{(\alpha-\delta+\beta_q)(t-s)} \sup_{x_1} \|\phi^{(1)}(s, x_1) - \phi^{(2)}(s, x_1)\| ds, \tag{5.22}$$

where the constant $\beta_q = N_\delta q(1 + \eta)$ tends to zero as $q \to 0$.

Further, (5.20) and (5.21) imply

$$\|x_1(t) - x_1^{(2)}(t)\| \leqq \|e^{A_1(t-\tau)}\| \, \|x_1^{(1)} - x_1^{(2)}\|$$

$$+ \int_\tau^t \|e^{A_1(t-s)}\| \, \|F_1(s, x_1(s) + \phi^{(2)}(s, x_1(s)))$$

$$- F_1(s, x_1^{(2)}(s) + \phi^{(2)}(s, x_1^{(2)}(s)))\| \, ds$$

$$\leqq N_\delta e^{(\alpha-\delta)(t-\tau)} \|x_1^{(1)} - x_1^{(2)}\|$$

$$+ N_\delta q \, (1 + \eta) \int_\tau^t e^{(\alpha-\delta)(t-s)} \|x_1(s) - x_1^{(2)}(s)\| \, ds.$$

Using Lemma III.2.1 once again, we readily deduce from this inequality the estimate

$$\|\Psi(t, \tau, x_1^{(1)} | \phi^{(2)}) - \Psi(t, \tau, x_1^{(2)} | \phi^{(2)})\|$$

$$\leqq N_\delta \|x_1^{(1)} - x_1^{(2)}\| e^{(\alpha-\delta+\beta_q)(t-\tau)}. \tag{5.23}$$

Combining (5.22) and (5.23), we obtain the required estimate (5.17) for $t \geqq \tau$. An analogous derivation exists for $t \leqq \tau$.

Inequality (5.18) follows directly from (5.2) and (5.15).

This lemma can be used to establish the existence of the integral manifolds that are of interest to us.

THEOREM 5.1. *Suppose the spectrum $\sigma(A)$ admits an α-separation. For any $\eta > 0$ and $\rho > 0$ there exist constants $M > 0$ and $q > 0$ depending only on η, ρ and the operator A such that if $F(t, x) \in (M, q)$ equation* (5.1) *has a unique (ρ, η)-manifold $\mathfrak{M}$. This manifold consists of all of the solutions of equation* (5.1) *that lie in the cylinder Π_ρ: $\|P_2x(t)\| \leqq \rho$ $(-\infty < t < \infty)$.*

PROOF. We consider, in the Banach space of functions $\phi(t, x) \in \mathfrak{B}_2$ that are continuous and bounded on $(-\infty, \infty) \times \mathfrak{B}_1$ with norm $|||\phi||| = \sup_{t,x_1}\|\phi(t, x_1)\|$, the closed subset $\mathscr{L}(\rho, \eta)$ consisting of the functions of class L_η that satisfy condition (5.6).

We will show with the use of the contraction principle that for $F \in (M, q)$ with sufficiently small q and M the operator S: $\phi \to \tilde{\phi}$ defined by the formula

$$\tilde{\phi}(\tau, x_1) = \int_{-\infty}^{\infty} G_2(\tau - t)F_2(t, \Psi(t, \tau, x_1|\phi) + \phi(t, \Psi(t, \tau, x_1|\phi)))\, dt,$$

acts in the space $\mathscr{L}(\rho, \eta)$ and has precisely one fixed point in it.

We first note that (5.2) and (5.16) imply the inequality

$$\|\tilde{\phi}(\tau, x_1)\| \leqq MN_\delta \int_{-\infty}^{\infty} e^{-(\alpha+\delta)|t|}\, dt = \frac{2MN_\delta}{\alpha + \delta},$$

which shows that the function $\tilde{\phi}$ satisfies condition (5.6) for sufficiently small $M > 0$.

Then, using inequalities (5.16) and (5.10) and Lemma 5.2, we obtain the estimate

$$\begin{aligned}
&\|\tilde{\phi}(\tau, x_1^{(1)}) - \tilde{\phi}(\tau, x_1^{(2)})\| \\
&\quad \leqq N_\delta \int_{-\infty}^{\infty} e^{-(\alpha+\delta)|\tau-t|}\|F_1(t, \Psi(t, \tau, x_1^{(1)}|\phi) + \phi(t, \Psi(t, \tau, x_1^{(2)}|\phi))) \\
&\qquad\qquad - F_1(t, \Psi(t, \tau, x_1^{(2)}|\phi) + \phi(t, \Psi(t, \tau\, x_1^{(2)}|\phi)))\|\, dt \\
&\quad \leqq N_\delta q(1 + \eta)\int_{-\infty}^{\infty} e^{-(\alpha+\delta)|\tau-t|}\|\Psi(t, \tau, x_1^{(1)}|\phi) - \Psi(t, \tau, x_1^{(2)}|\phi)\|\, dt \\
&\quad \leqq N_\delta^2 q(1 + \eta)\|x_1^{(1)} - x_1^{(2)}\| \int_{-\infty}^{\infty} e^{-(\alpha+\delta)|\tau-t| + (\alpha-\delta+\beta_q)|t-\tau|}\, dt \\
&\quad = \frac{2q(1 + \eta)N_\delta^2}{2\delta - \beta_q}\|x_1^{(1)} - x_1^{(2)}\|.
\end{aligned}$$

By taking $q > 0$ sufficiently small, the coefficient in the right side of this inequality can be made less than η. This proves that the function $\tilde{\phi}$ satisfies condition (5.5) and hence that the operator S acts in the space $\mathscr{L}(\rho, \eta)$.

In regard to a proof of the first assertion of the theorem it remains to show that the operator S is a contraction in this space.

From Lemma 5.2 it follows that

$$\|\bar{\phi}_2(\tau, x_1) - \bar{\phi}_1(\tau, x_1)\|$$
$$\leqq N_\delta \int_{-\infty}^{\infty} e^{-(\alpha+\delta)|\tau-t|} \|F_1(t, \Psi(t, \tau, x_1|\phi_2) + \phi_2(t, \Psi(t, \tau, x_1|\phi_2)))$$
$$- F_1(t, \Psi(t, \tau, x_1|\phi_1) + \phi_1(t, \tau, x_1|\phi_1)))\| \, dt$$
$$\leqq N_\delta q \int_{-\infty}^{\infty} e^{-(\alpha+\delta)|\tau-t|}(R_1 + R_2)\, dt,$$

where

$$R_1 = \|\Psi(t, \tau, x_1|\phi_2) - \Psi(t, \tau, x_1|\phi_1)\| \leqq q \frac{e^{(\alpha-\delta+\beta_q)|t-\tau|}}{\alpha - \delta + \beta_q} |||\phi_2 - \phi_1|||$$

and

$$\begin{aligned} R_2 &= \|\phi_2(t, \Psi(t, \tau, x_1|\phi_2)) - \phi_1(t, \Psi(t, \tau, x_1|\phi_1))\| \\ &\leqq \|\phi_2(t, \Psi(t, \tau, x_2|\phi_2)) - \phi_1(t, \Psi(t, \tau, x_1|\phi_2))\| \\ &\quad + \|\phi_1(t, \Psi(t, \tau, x_1|\phi_2)) - \phi_1(t, \Psi(t, \tau, x_1|\phi_1))\| \\ &\leqq |||\phi_2 - \phi_1||| + \eta\|\Psi(t, \tau, x_1|\phi_2) - \Psi(t, \tau, x_1|\phi_1)\| \\ &\leqq \left\{ 1 + q\eta \frac{e^{(\alpha-\delta+\beta_q)|t-\tau|}}{\alpha - \delta + \beta_q} \right\} |||\phi_2 - \phi_1|||. \end{aligned}$$

Using these estimates, we obtain after an integration the inequality

$$|||\bar{\phi}_2 - \bar{\phi}_1||| \leqq \frac{2N_\delta q}{\alpha + \delta} \left[1 + \frac{q(1 + \eta)(\alpha + \delta)}{(2\delta - \beta_q)(\alpha - \delta + \beta_q)} \right] |||\phi_2 - \phi_1|||,$$

which shows that the operator S is indeed a contraction for sufficiently small $q > 0$.

It remains to prove that the constructed manifold contains every solution $x(t)$ of equation (5.1) that satisfies the condition $\|P_2 x(t)\| \leqq \rho$ $(-\infty < t < \infty)$.

Suppose $x(t)$ is a solution with this property and let $\psi(t) = P_1 x(t)$, $\chi(t) = P_2 x(t)$. These functions satisfy the system

$$\begin{cases} d\psi/dt = A_1\psi + F_1(t, \psi(t) + \chi(t)), \\ \chi(t) = \int_{-\infty}^{\infty} G(t - s) F_2(s, \psi(s) + \chi(s))\, ds, \\ \psi(t)|_{t=t_0} = \psi(t_0). \end{cases}$$

It is not difficult to show by the same method as was used above that this system has a unique solution satisfying the condition $\|\chi(t)\| \leqq \rho$ $(-\infty < t < \infty)$.

On the other hand, the same system is satisfied by the functions $\psi_1(t)$ and $\chi_1(t)$, where $\psi_1(t)$ is a solution of equation (5.9) under the condition $\psi_1(t_0) = \psi(t_0)$ while $\chi_1(t) = \phi(t, \psi_1(t))$.

Therefore $\psi(t) \equiv \psi_1(t)$ and $\chi(t) \equiv \chi_1(t) \equiv \phi(t, \psi(t))$; but this means that the solution under consideration lies on our manifold. The theorem is proved.

REMARK 5.1. The result obtained in Theorem 5.1 bears a global character. We were able to obtain this result because we assumed that the condition $F \in (M, q)$ is satisfied on the whole space. But usually the function $F(t, x)$ is defined for $x \in B_\rho$, where ρ (> 0) is sufficiently small, and satisfies such a condition on $(-\infty, \infty) \times B_\rho$. It is not difficult in this case to extend $F(t, x)$ onto the whole space $\mathfrak{B}$ with preservation of the constants M and q. This can be done, for example, by letting $F(t, x)$ be constant on the extensions of the radii of the ball B_ρ.

After this one can assert on the basis of Theorem 5.1 that there exists a manifold $\mathfrak{M}$ which will serve as a "local" manifold for the original problem.

By solving equation (5.9) on this manifold, we can observe the component $x_1 = P_1 x$ of the solution and hence the whole solution as long as it remains in the ball B_ρ.

Such a "local" integral manifold exists, in particular, if the function $F(t, x)$ satisfies the quasilinearity condition

$$\|F(t, x)\| \leqq c(x)\|x\| \qquad (c(x) \to 0 \text{ as } x \to 0)$$

and the condition

$$\|F(t, x_2) - F(t, x_1)\| \leqq c_1(x)\|x_2 - x_1\| \qquad (c_1(x) \to 0 \text{ as } x \to 0). \tag{5.24}$$

REMARK 5.2. Under certain conditions the results of this (and the following) section carry over to equations of the form

$$dx/dt = A(t)x + F(t, x). \tag{5.25}$$

Suppose, for example, that the equation

$$dx/dt = A(t)x \tag{5.26}$$

is reducible to a stationary equation $dx/dt = Cx$ with an operator C whose spectrum admits an α-separation. Then equation (5.25) can be transformed by means of a change of variable into the equation studied above, and this permits one, as can easily be seen, to transfer the results obtained above to equation (5.25).

This is true, in particular, for a periodic reducible equation (5.26) with a monodromy operator whose spectrum admits an α-separation.

A more general case is when equation (5.25) admits a third order exponential splitting of its solutions, the Bohl spectrum of this equation admitting an α-separation. In this case Theorem 5.1 remains valid at least when the phase space $\mathfrak{B}$ is a Hilbert space $\mathfrak{H}$.[3]

In fact, by virtue of the results of § IV.6 the equation can in this case be reduced by means of a kinematic similarity transformation to a form analogous to (5.8), with variable operators $A_1(t)$ and $A_2(t)$. The second of these equations turns out to be e-dichotomic, a Green function exists for it, and therefore all of the subsequent calculations will be valid.

[3] We do not know the extent to which this assumption is essential.

REMARK 5.3. In the applications one frequently encounters the equation

$$dx/dt = Ax + \varepsilon F(t, x)$$

with a small parameter ε and a bounded function $F(t, x)$ satisfying a Lipschitz condition. The results presented above are applicable to equations of this type since for any $M > 0$ and $q > 0$, no matter how small, the condition $F(t, x) \in (M, q)$ will be satisfied if ε is sufficiently small.

An analogous remark can be made in regard to equations of the form

$$dx/dt = Ax + \varepsilon F(t, x) + F_1(t, x),$$

where $F_1(t, x)$ is quasilinear and satisfies condition (5.24).

§ 6. Integral manifolds bounded for $t \to +\infty$ (or $t \to -\infty$)

In this section we consider the integral manifolds of equation (5.1) that are composed of trajectories bounded for t tending to infinity in one direction only.

The integral manifold of the differential equation (5.1) defined by the equation

$$x_2 = \phi(t, x_1) \qquad (t \in (t_0, \infty);\ x_1 \in \mathfrak{B}_1)$$

will be called a $(\rho, \eta, +)$-*manifold* if there exists a number $\bar{t}$ such that when $t > \bar{t}$ the function $\phi(t, x_1)$ is of class L_η and satisfies the inequality $\|\phi(t, x_1)\| \leqq \rho$.

The $(\rho, \eta, -)$-*manifolds* composed of trajectories with the component x_2 bounded on a negative halfline are defined analogously.

In order to construct these manifolds we consider in more detail the separation of the spectrum of the operator A.

Let $\sigma_{2,+}(A)$ and $\sigma_{2,-}(A)$ be the spectral sets of A consisting respectively of the points lying to the right of the straight line $\operatorname{Re} \lambda = \alpha$ and to the left of the straight line $\operatorname{Re} \lambda = -\alpha$. Thus

$$\sigma_2(A) = \sigma_{2,+}(A) \cup \sigma_{2,-}(A).$$

We again consider a principal Green function $G_2(t)$ of the operator $A_2 = P_2 A$ in the subspace $\mathfrak{B}_2$. By carrying out arguments analogous to those that led us to equation (5.14) and using Theorem 4.3, we can show that the $(\rho, \eta, +)$-manifolds are described by functions

$$x_2 = \phi(t, x_1, y) \qquad (t \in (t_0, \infty);\ x_1 \in \mathfrak{B}_1;\ y \in \mathfrak{B}_{2,-};\ x_2 \in \mathfrak{B}_2)$$

satisfying the integral equation

$$\begin{aligned} \phi(\tau, x_1, y) & \qquad\qquad\qquad\qquad\qquad\qquad\qquad\qquad\qquad\qquad (6.1) \\ &= e^{A(\tau - t_0)} y + \int_{t_0}^{\infty} G_2(\tau - t) F_2(t, \Psi(t, \tau, x_1 | \phi) + \phi(t, \Psi(t, \tau, x_1 | \phi), y)) dt \\ & \qquad\qquad\qquad\qquad\qquad\qquad\qquad\qquad (t_0 \leqq \tau < \infty). \end{aligned}$$

The function $x_1 = \Psi(t, \tau, x_{10} | \phi)$ here is the solution of the differential equation

$$dx_1/dt = A_1x_1 + F_1(t, x_1 + \phi(t, x_1, y)) \tag{6.2}$$

which satisfies the condition $x_1(\tau) = x_{10}$.

LEMMA 6.1. *Suppose*

$$\begin{aligned} &\|\phi(t, x_1^{(2)}, y^{(2)}) - \phi(t, x_1^{(1)}, y^{(1)})\| \\ &\qquad \leqq \eta\|x_1^{(2)} - x_1^{(1)}\| + Ke^{-\nu(t-t_0)}\|y^{(2)} - y^{(1)}\| \end{aligned} \tag{6.3}$$

$$(x_1^{(1)}, x_1^{(2)} \in \mathfrak{B}_1;\ y^{(1)}, y^{(2)} \in \mathfrak{B}_{2,-};\ t \geqq t_0;\ \alpha - \delta > \nu > 0).$$

Then for $\phi^{(1)} = \phi(t, x_1, y^{(1)})$ *and* $\phi^{(2)} = \phi(t, x_1, y^{(2)})$

$$\begin{aligned} &\|\Psi(t, \tau, x_1^{(2)}|\phi^{(2)}) - \Psi(t, \tau, x_1^{(1)}|\phi^{(1)})\| \\ &\quad \leqq N_\delta\Big\{\|x_1^{(2)} - x_1^{(1)}\|e^{(\alpha-\delta+\beta_q)|t-\tau|} \\ &\qquad + \frac{Kq\|y^{(2)} - y^{(1)}\|}{\alpha - \delta + \nu + \beta_q}[e^{-\nu(\tau-t_0)+(\alpha-\delta+\beta_q)|t-\tau|} - e^{-\nu(t-t_0)}]\Big\}. \end{aligned} \tag{6.4}$$

PROOF. We have

$$\begin{aligned} &\left|\int_\tau^t e^{(\alpha-\delta+\beta)|t-s|}\sup_{x_1}\|\phi(s, x_1, y^{(2)}) - \phi(s, x_1, y^{(1)})\|\right| ds \\ &\quad \leqq K\left|\int_\tau^t e^{(\alpha-\delta+\beta_q)|t-s|-\nu(s-t_0)}\,ds\right|\|y^{(2)} - y^{(1)}\| \\ &\quad \leqq \frac{K}{\alpha - \delta + \nu + \beta_q}\{e^{-\nu(\tau-t_0)+(\alpha-\delta+\beta_q)|t-\tau|} - e^{-\nu(t-t_0)}\}\|y^{(2)} - y^{(1)}\|. \end{aligned}$$

Substituting this result into inequality (5.17), we obtain the required estimate.

Let $\mathscr{L}(\rho, R, \eta, K, \nu, \bar{t})$ denote the space of functions $\phi(t, x_1, y)$ $(t \in [t_0, \infty);\ \|y\| \leqq R;\ x_1 \in \mathfrak{B}_1)$, satisfying condition (6.3) and the condition

$$\|\phi(t, x_1, y)\| \leqq \rho \qquad (t \geqq \bar{t} > t_0) \tag{6.5}$$

and with the norm

$$|||\phi||| = \sup_{t \geqq \bar{t};\, x_1;\, \|y\| \leqq R} \|\phi(t, x_1, y)\|.$$

We will investigate in this space the operator $\check{S}: \phi \to \check{\phi}$ defined by the equality

$$\begin{aligned} &\check{\phi}(\tau, x_1, y) \\ &\quad = e^{A(\tau-t_0)}y + \int_{t_0}^\infty G_2(\tau - t)F_2(t, \Psi(t, \tau, x_1|\phi) + \phi(t, \Psi(t, \tau, x_1|\phi), y))dt. \end{aligned} \tag{6.6}$$

LEMMA 6.2. *Suppose R, ρ and η are arbitrary positive numbers, $K > N_\delta$ and $\nu < \alpha - \delta$. There exist numbers $q_0 > 0$, $M_0 > 0$ and $t_1 > t_0$ having the property* *that the operator $\check{S}$ acts in the space $\mathscr{L}(\rho, R, \eta, K, \nu, \bar{t})$ when $q \leqq q_0$, $M \leqq M_0$ and $t > t_1$.*

PROOF. Inequalities (5.2) and (5.17) imply the inequality

$$\|\bar{\varphi}(\tau, x, y)\| \leqq N_\delta e^{-(\alpha-\delta)(\bar{t}-t_0)} R + \frac{M}{\alpha - \delta} \qquad (\tau \geqq \bar{t};\ \|y\| \leqq R;\ x_1 \in \mathfrak{B}_1),$$

which shows that when $M > 0$ is sufficiently small and $\bar{t}$ is sufficiently large

$$\|\|\bar{\phi}\|\| = \sup_{\tau \geqq \bar{t}; x_1; \|y\| \leqq R} \|\bar{\phi}(\tau, x_1, y)\| \leqq \rho.$$

Let us verify the preservation of condition (6.3). We have

$$\begin{aligned}
&\|\bar{\phi}(\tau, x_1^{(2)}, y^{(2)}) - \bar{\phi}(\tau, x_1^{(1)}, y^{(1)})\| \leqq \|e^{A(\tau-t_0)}(y^{(2)} - y^{(1)})\| \\
&\quad + \int_{t_0}^{\infty} \|G_2(\tau - t)\| \, \|F_2(t, \Psi(t, \tau, x_1^{(2)}|\phi^{(2)}) + \phi(t, \Psi(t, \tau, x_1^{(2)}|\phi^{2}), y^{(2)})) \\
&\qquad\qquad - F_2(t, \Psi(t, \tau, x_1^{(1)}|\phi^{(1)}) + \phi(t, \Psi(t, \tau, x_1^{(1)}|\phi^{(1)}), y^{(1)}))\| \, dt \\
&\leqq N_\delta e^{-(\alpha-\delta)(\tau-t_0)} \|y^{(2)} - y^{(1)}\| \\
&\quad + qN_\delta \int_{t_0}^{\infty} e^{-(\alpha+\delta)|\tau-t|} \|\Psi(t, \tau, x_1^{(2)}|\phi^{(2)}) - \Psi(t, \tau, x_1^{(1)}|\phi^{(1)})\| \\
&\quad + \|\phi(t, \Psi(t, \tau, x_1^{(2)}|\phi^{(2)}), y^{(2)}) - \phi(t, \Psi(t, \tau, x_1^{(1)}|\phi^{(1)}), y^{(1)})\| \, dt \\
&\leqq N_\delta e^{-(\alpha-\delta)(\tau-t_0)} \|y^{(2)} - y^{(1)}\| \\
&\quad + q(1 + \eta)N_\delta \int_{t_0}^{\infty} e^{-(\alpha+\delta)|\tau-t|} \|\Psi(t, \tau, x_1^{(2)}|\phi^{(2)}) - \Psi(t, \tau, x_1^{(1)}|\phi^{(1)})\| \, dt \\
&\quad + qN_\delta K \int_{t_0}^{\infty} e^{-(\alpha+\delta)|\tau-t|} e^{-\nu(t-t_0)} \, dt \cdot \|y^{(2)} - y^{(1)}\|.
\end{aligned}$$

Estimating the second summand with the use of Lemma 6.1 and calculating the integrals in this inequality, we obtain the estimate

$$\begin{aligned}
&\|\bar{\phi}(t, x^{(2)}, y^{(2)}) - \bar{\phi}(t, x_1^{(1)}, y^{(1)})\| \\
&\leqq 2q(1 + \eta)N_\delta^2 \frac{\|x_1^{(2)} - x_1^{(1)}\|}{2\delta - \beta_q} + \|y^{(2)} - y^{(1)}\| \Big\{ N_\delta + qN_\delta K \frac{2(\alpha + \delta)}{(\alpha + \delta)^2 - \nu^2} \\
&\qquad + \frac{2Kq^2(1 + \eta)N_\delta^2}{(\alpha - \beta + \nu + \beta_q)(2\delta - \beta_q)} \Big\} e^{-\nu(t-t_0)}.
\end{aligned}$$

It follows from this estimate that the function $\bar{\phi}$ satisfies condition (6.3) when $q > 0$ is sufficiently small. The lemma is proved.

THEOREM 6.1. *Suppose the spectrum $\sigma(A)$ admits an α-separation. For any positive constants η, ρ, R, $\nu < \alpha$ and sufficiently large K there exist positive numbers l, M and q depending only on these constants and the operator A such that if $F(t, x) \in$*

$(M, q, +)$, there corresponds to each $y \in \mathfrak{B}_{2,-}$ ($\|y\| \leqq R$) one and only one (ρ, η)-manifold $\mathfrak{M}^{+}_{\rho,\eta}(y)$ of the differential equation (6.1), *which is defined by the equation*

$$x_2 = \phi(t, x_1, y) \qquad (x_1 \in \mathfrak{B}_1;\ y \in \mathfrak{B}_{2,-};\ x_2 \in \mathfrak{B}_2;\ t_0 \leqq t < \infty),$$

where the function ϕ satisfies the conditions

$$\begin{aligned} &1)\ P_{2,-}\phi(t_0, x_1, y) = y; \\ &2)\ \|\phi(t, x_1, y)\| \leqq \rho \text{ for } |t - t_0| \geqq l,\ \|y\| \leqq R; \\ &3)\ \|\phi(t, x_1^{(2)}, y^{(2)}) - \phi(t, x_1^{(1)}, y^{(1)})\| \leqq \eta \|x_1^{(2)} - x_1^{(1)}\| + Ke^{-\nu(t-t_0)}\|y^{(2)} - y^{(1)}\|. \end{aligned} \tag{6.7}$$

In particular, all of the manifolds $\mathfrak{M}^{+}_{\rho,\eta}(y)$ exponentially approach each other as $t \to +\infty$.

The manifold $\mathfrak{M}^{+}_{\rho,\eta}(y)$ consists of all of the trajectories of the differential equation (6.1) *that satisfy the conditions*

$$P_{2,-}x(t_0) = y, \quad \|P_2 x(t)\| \leqq \rho \qquad (t - t_0 \geqq l).$$

PROOF. It follows from Lemma 6.2 that the operator $\check{S}$ acts in the space $\mathscr{L}(\rho, R, \eta, K, \nu, t_0 + l)$ when $M > 0$ is sufficiently small and $l > 0$ is sufficiently large. Let us show that it is a contraction in this space. Using Lemma 5.2, we get that

$$\begin{aligned} &\|\check{\phi}^{(2)}(\tau, x, y) - \check{\phi}^{(1)}(\tau, x, y)\| \\ &\quad \leqq N_\delta \int_{t_0}^{\infty} e^{-(\alpha+\delta)|\tau-t|} \|F_1(t, \Psi(t, \tau, x_1|\phi^{(2)}) + \phi^{(2)}(t, \Psi(t, \tau, x_1|\phi^{(2)}), y)) \\ &\qquad\qquad - F_1(t, \Psi(t, \tau, x_1|\phi^{(1)}) + \phi^{(1)}(t, \Psi(t, \tau, x_1|\phi^{(1)}), y))\| dt \\ &\quad \leqq N_\delta q \int_{t_0}^{\infty} e^{-(\alpha+\delta)|\tau-t|}(R_1 + R_2)dt, \end{aligned} \tag{6.8}$$

where

$$\begin{aligned} R_1 &= \|\Psi(t, \tau, x_1|\phi^{(2)}) - \Psi(t, \tau, x_1|\phi^{(1)})\| \\ &\leqq q\frac{e^{(\alpha-\delta+\beta_q)|t-\tau|}}{\alpha - \delta + \beta_q} \sup_{s,x_1}\|\phi^{(2)}(s, x_1, y) - \phi^{(1)}(s, x_1, y)\|, \\ R_2 &= \|\phi^{(2)}(t, \Psi(t, \tau, x_1|\phi^{(2)}), y) - \phi^{(1)}(t, \Psi(t, \tau, x_1|\phi^{(1)}), y)\| \\ &\leqq \|\phi^{(2)}(t, \Psi(t, \tau, x_1|\phi^{(2)}), y) - \phi^{(1)}(t, \Psi(t, \tau, x_1|\phi^{(2)}), y)\| \\ &\quad + \|\phi^{(1)}(t, \Psi(t, \tau, x_1|\phi^{(2)}), y) - \phi^{(1)}(t, \Psi(t, \tau, x_1|\phi^{(1)}), y)\| \\ &\leqq \sup_{s,x_1}\|\phi^{(2)}(s, x_1, y) - \phi^{(1)}(s, x_1, y)\| \left\{1 + q\eta\frac{e^{(\alpha-\delta+\beta_q)|t-\tau|}}{\alpha - \delta + \beta_q}\right\}. \end{aligned}$$

Substituting these estimates into (6.8) and integrating, we obtain the required result for sufficiently small $q > 0$ in exactly the same way as in the proof of Theorem 5.1.

The proof of the last assertion of the theorem is carried out in exactly the same way as in Theorem 5.1.

§ 7. Averaging principle

1. *Averaging principle for a finite interval.* We consider in a Banach space $\mathfrak{B}$ the differential equation

$$dx/dt = \varepsilon f(t, x) \tag{7.1}$$

containing a small parameter ε.

We assume that on the set $[0, \infty) \times B$, where B is a ball in $\mathfrak{B}$, the function $f(t, x)$ is bounded, continuous in t and satisfies the Lipschitz condition

$$\|f(t, x_2) - f(t, x_1)\| \leqq c\|x_2 - x_1\| \qquad (x_1, x_2 \in B) \tag{7.2}$$

with a constant not depending on t.

We assume further that the mean value of $f(t, x)$ with respect to t:

$$f_0(x) = \lim_{T\to\infty} \frac{1}{T} \int_0^T f(t, x)dt \tag{7.3}$$

exists for each $x \in B$. Under this assumption it is expedient to consider along with equation (7.1) the equation

$$dx/dt = \varepsilon f_0(x). \tag{7.4}$$

It follows from (7.3) that the function $f_0(x)$ also satisfies condition (7.2) in B.

Let $y(t)$ $(0 \leqq t \leqq T_0)$ be a vector function taking values in B that satisfies the equation

$$\frac{dy}{dt} = f_0(y) \qquad (0 \leqq t \leqq T_0). \tag{7.5}$$

Then, as can easily be seen, the function $x_0(t; \varepsilon) = y(\varepsilon t)$ will be a solution of equation (7.4) on the interval $0 \leqq t \leqq T_0/\varepsilon$.

We will establish below a connection between $x_0(t; \varepsilon)$ and the solution $x(t; \varepsilon)$ of equation (7.1) that satisfies the condition

$$x(0; \varepsilon) = y(0) \quad (= x_0(0; \varepsilon)). \tag{7.6}$$

More precisely, we will prove the following assertion.

THEOREM 7.1. *Suppose that on the set* $[0, \infty) \times B$ *the function* $f(t, x)$ *is bounded, continuous in* t *and satisfies the Lipschitz condition* (7.2). *Suppose further that its mean value* (7.3) *exists for each* $x \in B$.

If $y(t)$ *is a solution of equation* (7.5) *that takes the interval* $[0, T_0]$ *into* B, *there exists for any* $\eta > 0$ *an* $\varepsilon_0 > 0$ *such that when* $0 < \varepsilon < \varepsilon_0$ *the solution* $x(t; \varepsilon)$ *of equation* (7.1) *satisfying condition* (7.6) *is subject to the estimate*

$$\|x(t;\varepsilon) - y(\varepsilon t)\| \leqq \eta \qquad (0 \leqq t \leqq T_0/\varepsilon).$$

Theorem 7.1 is obtained as a corollary of a general theorem on the continuous dependence of solutions of differential equations on a parameter.

To this end we consider the more general (than (7.1)) equation

$$\frac{dx}{d\tau} = f(\tau, x; \varepsilon) \qquad (0 \leqq \tau \leqq T_0; 0 \leqq \varepsilon \leqq \varepsilon_0) \tag{7.7}$$

with a function $f(\tau, x; \varepsilon)$ satisfying the Lipschitz condition

$$\|f(\tau,x_2;\varepsilon) - f(\tau, x_1;\varepsilon)\| \leqq c\|x_2 - x_1\| \tag{7.8}$$

and the condition

$$\lim_{\varepsilon\to 0} \int_0^{\tau_0} f(\tau, x; \varepsilon)d\tau = \int_0^{\tau_0} f(\tau, x; 0)d\tau \qquad (0 \leqq \tau_0 \leqq T_0), \tag{7.9}$$

which is called the condition of *integral continuity* at $\varepsilon = 0$.

Equation (7.7) can be converted into (7.1) by putting

$$f(\tau, x; \varepsilon) = \begin{cases} f(\tau/\varepsilon, x) & \varepsilon > 0, \\ f_0(x), & \varepsilon = 0, \end{cases}$$

and making the change of variable $\tau = \varepsilon t$.

Condition (7.9) is satisfied in this case since it takes the form

$$\lim_{\varepsilon\to 0} \int_0^{\tau_0} f\left(\frac{\tau}{\varepsilon}, x\right)d\tau = \tau_0 f_0(x)$$

or, equivalently,

$$\lim_{\varepsilon\to 0} \frac{\varepsilon}{\tau_0} \int_0^{\tau_0/\varepsilon} f(t, x)dt = f_0(x),$$

which is equivalent to (7.3).

Let us prove an auxiliary assertion.

LEMMA 7.1. *Suppose $x(\tau)$ is a continuous function on* $[0, T_0]$. *Then* (7.8) *and* (7.9) *imply the relation*

$$\lim_{\varepsilon\to 0} \int_0^{T_0} f(\tau, x(\tau); \varepsilon)d\tau = \int_0^{T_0} f(\tau, x(\tau); 0)d\tau. \tag{7.10}$$

PROOF. From (7.9) it follows that for any $\tau_1, \tau_2 \in [0, T_0]$

$$\lim_{\varepsilon\to 0} \int_{\tau_1}^{\tau_2} f(\tau, x; \varepsilon)d\tau = \int_{\tau_1}^{\tau_2} f(\tau, x; 0)d\tau,$$

and hence that for any $0 \leqq \tau_1 < \cdots < \tau_{n-1} < \tau_n = T_0$, $x_k \in B$ $(k = 1,2,\cdots, n)$

$$\lim_{\varepsilon\to 0}\sum_{k=1}^{n}\int_{\tau_{k-1}}^{\tau_k} f(\tau, x_k; \varepsilon)d\tau = \sum_{k=1}^{n}\int_{\tau_{k-1}}^{\tau_k} f(\tau, x_k; 0)d\tau.$$

If we introduce the step function

$$\bar{x}(\tau) = x_k \qquad (\tau_{k-1} \leqq \tau < \tau_k) \quad (k = 1,2,\cdots,n),$$

we can write the latter equality in the form

$$\lim_{\varepsilon\to 0}\int_0^{T_0} f(\tau, \bar{x}(\tau); \varepsilon)d\tau = \int_0^{T_0} f(\tau, \bar{x}(\tau); 0)d\tau.$$

This proves relation (7.10) for step functions.

Consider a sequence $\{\bar{x}_n(\tau)\}$ of step functions that converge to $x(\tau)$ uniformly on $[0, T_0]$. Using the Lipschitz condition for the function $f(\tau, x; \varepsilon)$, we obtain the estimate

$$\begin{aligned}
&\left\|\int_0^{T_0} f(\tau, x(\tau); \varepsilon)d\tau - \int_0^{T_0} f(\tau, x(\tau); 0)d\tau\right\| \\
&\leqq \int_0^{T_0}\|f(\tau, x(\tau); \varepsilon) - f(\tau,\bar{x}_n(\tau); \varepsilon)\|d\tau \\
&\quad + \left\|\int_0^{T_0}[f(\tau, \bar{x}_n(\tau); \varepsilon) - f(\tau, \bar{x}_n(\tau); 0)]d\tau\right\| \\
&\quad + \int_0^{T_0}\|f(\tau, \bar{x}(\tau); 0) - f(\tau, x_n(\tau); 0)\|d\tau \\
&\leqq 2T_0 \sup\|x(\tau) - \bar{x}_n(\tau)\| \\
&\quad + \left\|\int_0^{T_0}[f(\tau, \bar{x}_n(\tau); \varepsilon) - f(\tau, \bar{x}_n(\tau); 0)]\, d\tau\right\|,
\end{aligned}$$

which immediately implies the required assertion.

We will now prove the theorem on continuous dependence on a parameter in the form needed for our purposes.

THEOREM 7.2. *Suppose that for* $\tau \in [0, T_0]$, $x \in B$ *and* $0 \leqq \varepsilon < \varepsilon_1$ *the function* $f(\tau, x; \varepsilon)$ *is defined, bounded, piecewise continuous in* τ, *integrally continuous at* $\varepsilon = 0$ *and satisfies the Lipschitz condition* (7.8).

Suppose further that the equation

$$dx/d\tau = f(\tau, x; \varepsilon) \tag{7.11}$$

has a solution $x(\tau; 0)$ *for* $\varepsilon = 0$ *that satisfies the condition* $x(\tau; 0) \in B$ $(0 \leqq \tau \leqq T_0)$.

Then for any $\eta > 0$ *there exists an* ε_0 $(0 < \varepsilon_0 < \varepsilon_1)$ *such that every solution* $x(\tau; \varepsilon)$ *of equation* (7.11) *for* $0 < \varepsilon < \varepsilon_0$ *that is defined on the interval* $[0, T_0]$ *and satisfies the condition* $x(0; \varepsilon) = x(0; 0)$ $(= x_0)$ *is subject to the estimate*

$$\|x(\tau;\varepsilon) - x(\tau;0)\| < \eta \qquad (0 \leqq \tau \leqq T_0). \tag{7.12}$$

PROOF. The functions $x(\tau;\varepsilon)$ and $x(\tau;0)$ satisfy the integral equations

$$x(\tau;\varepsilon) = x_0 + \int_0^\tau f(s, x(s;\varepsilon);\varepsilon)ds$$

and

$$x(\tau;0) = x_0 + \int_0^\tau f(s, x(s;0);0)ds.$$

Using the Lipschitz condition (7.8), we obtain the estimate

$$\begin{aligned}
&\|x(\tau;\varepsilon) - x(\tau;0)\| \\
&= \left\| \int_0^\tau [f(s, x(s;\varepsilon);\varepsilon) - f(s, x(s;0);0)]ds \right\| \\
&\leqq \int_0^\tau \| f(s, x(s;\varepsilon);\varepsilon) - f(s, x(s;0);\varepsilon)\| ds \\
&\quad + \left\| \int_0^\tau [f(s, x(s;0);\varepsilon) - f(s, x(s;0);0)]ds \right\| \\
&\leqq c \int_0^\tau \|x(s;\varepsilon) - x(s;0)\| ds + \phi(\varepsilon),
\end{aligned}$$

where

$$\phi(\varepsilon) = \left\| \int_0^{T_0} [f(s, x(s;0);\varepsilon) - f(s, x(s;0);0)]ds \right\|.$$

On the basis of Corollary III.2.1 we can now conclude that

$$\|x(\tau;\varepsilon) - x(\tau;0)\| \leqq e^{c\tau}\phi(\varepsilon),$$

and it remains only to note that $\lim_{\varepsilon\to 0} \phi(\varepsilon) = 0$ by virtue of the integral continuity condition (7.9) and Lemma 7.1. The theorem is proved.

Theorem 7.1 immediately follows from what has been proved, as follows from the arguments given at the beginning of the present subsection.

2. *Averaging principle for the real line.* We can obtain more complete results if we make additional assumptions on the right side of the averaged equation.

Suppose again

$$f_0(x) = \lim_{T\to\infty} \frac{1}{T} \int_0^T f(t, x)\, dt.$$

Let us assume that there exists a point $x_0 \in \mathfrak{B}$ at which

$$f_0(x_0) = 0. \tag{7.13}$$

In this case the differential equation

$$dx/dt = \varepsilon f_0(x) \tag{7.14}$$

has the stationary solution $x_0(t) \equiv x_0$.

We will see that under certain assumptions the equation

$$dx/dt = \varepsilon f(t, x) \tag{7.15}$$

has for sufficiently small $\varepsilon > 0$ a bounded solution on the real line that does not leave a small neighborhood of the point x_0.

We require that the following conditions be satisfied.

(C_1) The function $f(t, x)$ is defined and bounded for $t \in (-\infty, \infty)$ and $x \in B_\rho(x_0)$, where $B_\rho(x_0)$ is the ball of radius $\rho > 0$ with center at x_0.

The limit

$$f_0(x) = \lim_{T\to\infty} \frac{1}{T} \int_t^{t+T} f(s, x)ds \tag{7.16}$$

exists uniformly in both $t \in (-\infty, \infty)$ and $x \in B_\rho(x_0)$, the functions $f_x'(t, x)$ and $f_0'(x)$ being bounded for $t \in (-\infty, \infty)$ and $x \in B_\rho(x_0)$.

(C_2) The vector functions $f(t, x)$ and $f_0(x)$ have in $B_\rho(x_0)$ continuous bounded derivatives with respect to x of first and second order.

(C_3) Relation (7.16) can be termwise differentiated twice, i.e. the relation

$$f_0^{(k)}(x) = \lim_{T\to\infty} \frac{1}{T} \int_t^{t+T} f_x^{(k)}(s, x)ds \qquad (k = 1,2) \tag{7.17}$$

holds uniformly in t and x.

We recall that, by definition, for each $x \in B_\rho(x_0)$ the expression $f_0'(x)$ represents an operator from $[\mathfrak{B}]$ satisfying the condition

$$f_0(x + h) - f_0(x) = f_0'(x)h + R(x, h) \quad (x, x + h \in B_\rho), \tag{7.18}$$

where

$$\|R(x, h)\| = o(\|h\|). \tag{7.19}$$

Let $A = f_0'(x_0)$ and $S(h) = R(x_0, h)$. Then by virtue of (7.13) and (7.18)

$$f_0(x_0 + h) = Ah + S(h). \tag{7.20}$$

We note that in the ball $\|h\| \leqq \sigma$ the vector function $S(h)$ satisfies the Lipschitz condition

$$\|S(h_2) - S(h_1)\| \leqq C_\sigma\|h_2 - h_1\|, \qquad ((7.21)$$

where

$$\lim_{\sigma\to 0} C_\sigma = 0. \tag{7.22}$$

For (7.20) implies

$$S'(h) = f_0'(x_0 + h) - f_0'(x_0), \tag{7.23}$$

and by virtue of the continuity of f_0' the quantity

$$C_\sigma = \sup_{\|h\| \leqq \sigma} \|f_0'(x_0 + h) - f_0'(x_0)\| \tag{7.24}$$

satisfies condition (7.22). Inasmuch as

$$S(h_2) - S(h_1) = \int_0^1 S'(h_1 + s(h_2 - h_1))(h_2 - h_1)ds,$$

we obtain (7.21).

In order to investigate equation (7.15) we will transform it into a form permitting an application of the results of § 4. We first rewrite it with the use of (7.20) in the form

$$dh/dt = \varepsilon Ah + \varepsilon g(t, h), \tag{7.25}$$

where $h = x - x_0$ and

$$g(t, h) = f(t, x_0 + h) - f_0(x_0 + h) + S(h).$$

In equation (7.25) we will make the change of variable

$$h = z - \varepsilon v(t, z; \varepsilon), \tag{7.26}$$

where

$$v(t, z; \varepsilon) = \int_0^\infty V(s + t, z)e^{-\varepsilon s}ds \qquad (\varepsilon > 0) \tag{7.27}$$

is the Laplace transform of the function

$$V(t, z) = f(t, x_0 + z) - f_0(x_0 + z). \tag{7.28}$$

We will use the following well-known proposition to show that the transformation (7.26) has an inverse.

LEMMA 7.2. *Suppose a function $\phi(s; \lambda)$ $(0 \leqq s < \infty)$ taking values in a Banach space $\mathfrak{B}$ and depending on an abstract parameter $\lambda \in \Lambda$ satisfies the condition*

$$(M =) \sup\left\{\left\| \frac{1}{t} \int_0^t \phi(s; \lambda)ds \right\| t \geqq 0; \lambda \in \Lambda\right\} < \infty$$

and, uniformly in $\lambda \in \Lambda$, the condition

$$\lim_{T\to\infty} \frac{1}{T} \int_0^T \phi(s; \lambda)ds = 0.$$

Then the relation

$$\lim_{p \to 0} p \int_0^\infty \phi(s; \lambda) e^{-ps} ds = 0$$

holds uniformly in $\lambda \in \Lambda$.

PROOF. Integrating by parts, we obtain the relation

$$\begin{aligned} p \int_0^\infty e^{-ps} \phi(s; \lambda) ds &= p \int_0^\infty e^{-ps} d\left[\int_0^s \phi(t; \lambda) dt \right] \\ &= p \left\{ (se^{-ps}) \frac{1}{s} \int_0^s \phi(t; \lambda) dt \bigg|_0^\infty + p \int_0^\infty \left[\int_0^s \phi(t; \lambda) dt \right] e^{-ps} ds \right\} \\ &= p^2 \int_0^\infty \left[\int_0^s \phi(t; \lambda) dt \right] e^{-ps} ds. \end{aligned} \tag{7.29}$$

For any $\delta > 0$ there exists a $T > 0$ such that when $s \geqq T$

$$\left\| \frac{1}{s} \int_0^s \phi(s; \lambda) ds \right\| \leqq \delta.$$

Therefore (7.29) implies the estimate

$$\begin{aligned} &\left\| p \int_0^\infty \phi(s; \lambda) e^{-ps} ds \right\| \\ &= \left\| p^2 \left\{ \int_0^T s \left[\frac{1}{s} \int_0^s \phi(t; \lambda) dt \right] e^{-ps} ds + \int_T^\infty se^{-ps} \left[\frac{1}{s} \int_0^s \phi(t; \lambda) dt \right] ds \right\} \right\| \\ &\leqq M [1 - e^{-pT}(1 + pT)] + \delta e^{-pT}(1 + pT), \end{aligned}$$

which by virtue of the arbitrariness of $\delta > 0$ proves the required assertion.

LEMMA 7.3. *The following relations hold uniformly in* $t \in (-\infty, \infty)$ *and* $z \in B_\rho$:

$$\lim_{\varepsilon \downarrow 0} \varepsilon v(t, z; \varepsilon) = 0, \tag{7.30}$$

$$\lim_{\varepsilon \downarrow 0} \varepsilon v_z'(t, z; \varepsilon) = 0. \tag{7.31}$$

For by virtue of (7.16) (7.17) and (7.28) the bounded functions $V(s + t, z)$ and $V_z'(s + t, z)$ satisfy the conditions of Lemma 7.2, the role of the parameter λ being played by the pair (t, z).

From (7.31) it follows that when $\varepsilon > 0$ is sufficiently small the operator $1 - \varepsilon v_z'(t, z; \varepsilon)$ $(t \in (-\infty, \infty),\ z \in B_\rho)$ has a bounded inverse, and hence the mapping (7.26) is invertible. In addition, from (7.30) it follows that for any preassigned $\rho_0 \in (0, \rho]$ there exist an $\varepsilon_0 > 0$ and an $\eta > 0$ such that $\|h\| < \rho_0$ when $(0 <)$ $\varepsilon < \varepsilon_0$ and $\|z\| = \eta$. Thus the change of variable (7.26) can be carried out in a sufficiently small neighborhood of the origin and for sufficiently small $\varepsilon > 0$.

Representing the function υ in the form

$$\upsilon(t, z; \varepsilon) = e^{\varepsilon t} \int_t^\infty V(s, z)e^{-\varepsilon s} ds,$$

we find that

$$\upsilon_t'(t, z; \varepsilon) = \varepsilon\upsilon(t, z; \varepsilon) - V(t, z), \tag{7.32}$$

and consequently

$$\frac{dh}{dt} = \frac{dz}{dt} - \varepsilon\upsilon_z' \frac{dz}{dt} - \varepsilon^2\upsilon + \varepsilon V. \tag{7.33}$$

Using relations (7.26), (7.28) and (7.33), we reduce equation (7.25) to the form

$$\begin{aligned}(I - \varepsilon\upsilon_z')dz/dt &= \varepsilon[f(t, x_0 + h; \varepsilon) - f(t, x_0 + z; \varepsilon)] \\ &\quad + \varepsilon f_0(x_0 + z) + \varepsilon_2\upsilon(t, z; \varepsilon) \\ &= \varepsilon(Az + S(z)) + \varepsilon^2\upsilon(t, z; \varepsilon) \\ &\quad + \varepsilon[f(t, x_0 + z - \varepsilon\upsilon, \varepsilon) - f(t, x_0 + z; \varepsilon)]\end{aligned}$$

or, after multiplying by the bounded operator $(I - \varepsilon\upsilon_z')^{-1}$ and introducing the "slow" time $\tau = \varepsilon t$ as the independent variable, to the form

$$dz/dt = Az + F(z, \tau; \varepsilon), \tag{7.34}$$

where, as is easily seen, $F(z, \tau; \varepsilon)$ has a bounded first derivative $F_z'(z, \tau; \varepsilon)$ and satisfies the relation

$$F(z, \tau; \varepsilon) - S(z) = O(\varepsilon).$$

uniformly in $\tau \in (-\infty, \infty)$ and $z \in B_\rho$.

Hence for any $M > 0$ and $\mu > 0$ there exist an ε_0 and a β_0 such that when $(0 <)$ $\varepsilon < \varepsilon_0$ and $\|z\| < \beta_0$ we have

$$\|F(z, \tau; \varepsilon)\| \leqq M \tag{7.35}$$

and

$$\|F(z_2, \tau; \varepsilon) - F(z_1, \tau; \varepsilon)\| \leqq \mu\|z_2 - z_1\|. \tag{7.36}$$

Suppose now that the spectrum of the operator $A = f_0'(x_0)$ does not intersect the imaginary axis.

By virtue of the results of § 4 we can conclude that there exists for sufficiently small $\rho_0 > 0$ an $\varepsilon_0 > 0$ such that when $(0 <)$ $\varepsilon < \varepsilon_0$ equation (7.34) has a unique solution $z_0(\tau)$ $(-\infty < \tau < \infty)$ whose trajectory lies entirely in the ball $\|z(\tau)\| \leqq \rho_0$. This solution is almost periodic if the function $F(z, \tau; \varepsilon)$ is almost periodic.

Moreover, there exists an initial manifold $\mathfrak{M}_-^0$ with a one-to-one projection onto a neighborhood of the origin in the subspace $\mathfrak{B}_- = P_-\mathfrak{B}$ (P_- is a spectral projection of the operator A) such that when $z(\tau_0) \in \mathfrak{M}_-^0$ the corresponding solution $z(\tau)$ of equation (7.34) satisfies the condition

$$\|z(\tau) - z_0(\tau)\| \leqq c_0 e^{-\nu(\tau-\tau_0)} \|z(\tau_0) - z_0(\tau_0)\| \quad (\tau \geqq \tau_0). \tag{7.37}$$

All other solutions leave any ball of radius less than ρ_0 as $\tau \to \infty$.

If we return to the solutions of equation (7.15) by means of the inverse of the transformation (7.26), we obtain the following result.

THEOREM 7.3. *Suppose conditions* (C_1), (C_2), (C_3) *and* (7.13) *are satisfied and the spectrum of the operator* $A = f_0'(x_0)$ *does not intersect the imaginary axis. Then for sufficiently small* $\rho_0 > 0$ *there exists an* $\varepsilon_0 > 0$ *such that when* $(0 <)\ \varepsilon < \varepsilon_0$ *equation* (7.15) *has a unique solution* $\tilde{x}_0(t)$ *satisfying the condition*

$$\sup_t \|\tilde{x}_0(t) - x_0\| < \rho_0.$$

There exists an initial manifold $\mathfrak{M}_-$ *(with a one-to-one projection onto a neighborhood of the point* $P_- x_0$ *in the subspace* $\mathfrak{B}_- = P_- \mathfrak{B}$*)*[4] *such that when* $x(t_0) \in \mathfrak{M}_-$

$$\|x(t) - \tilde{x}_0(t)\| \leqq c e^{-\nu(t-t_0)} \|x(t_0) - \tilde{x}_0(t_0)\| \quad (t \geqq t_0). \tag{7.38}$$

If $x(t_0) \in \mathfrak{M}_-$, *the solution* $x(t)$ *leaves any ball* $B_\rho(x_0)$ *with* $\rho < \rho_0$ *as* $t \to \infty$.

The solution $\tilde{x}_0(t)$ *is almost periodic (periodic) if the functions* $f(t, z)$ *and* $f_z'(t, z)$ *are almost periodic (periodic).*

REMARK 7.1. If the whole spectrum of the operator A lies in the interior of the left halfplane, the manifold $\mathfrak{M}_-$ coincides with a neighborhood of the point x_0 in the space $\mathfrak{B}$. Formula (7.38) shows that in this case the solution $\tilde{x}_0(t)$ of equation (7.15) is uniformly and asymptotically stable.

If part of the spectrum $\sigma(A)$ (but not all of it) is located in the right halfplane, this solution is only conditionally stable.

EXERCISES

We cite certain more precise existence and uniqueness theorems for equation (1.1). The following propositions 1–3 were proved by M. A. Krasnosel'skiĭ and S. G. Krein [2].

1. *Suppose the function* $f(t, x)$ *is representable in the form*

$$f(t, x) = f_1(t, x) + f_2(t, x) \qquad (\|x - x_0\| \leqq r;\ |t - t_0| \leqq a),$$

where $f_1(t, x)$ *is completely continuous and* $f_2(t, x)$ *satisfies the condition*

$$\|f_2(t, x) - f_2(t, y)\| \leqq K(t)\|x - y\|.$$

If $0 < h \leqq a$ *is such that*

$$\int_{t_0-h}^{t_0+h} K(t)dt < 1$$

and

$$h\Big\{\sup_{\substack{\|x-x_0\|\leqq r;\\ |t-t_0|\leqq a}} \|f_1(t, x)\| + \sup_{\substack{\|x-x_0\|\leqq r;\\ |t-t_0|\leqq a}} \|f_2(t, x)\|\Big\} \leqq r,$$

[4] The property indicated in parentheses of the manifold $\mathfrak{M}_-$ generally requires further explanation.

there exists a solution of equation (1.1) *that is defined on the segment* $[t_0 - h, t_0 + h]$ *and satisfies the initial condition* $x(t_0) = x_0$.

2. *Suppose* $\|f(t, x)\| \leqq L(\Phi(x))\psi(t)$, *where* $\Phi(x)$ *is a nonnegative functional on* $\mathfrak{B}$ *satisfying a Lipschitz condition and such that* $\lim_{\|x\|\to\infty}\Phi(x) = \infty$, $L(u)$ *is continuous, nonnegative and such that* $\int^{\infty}[L(u)]^{-1}du = \infty$, *and* $\psi(t)$ *is nonnegative and integrable on each finite interval.*

Then every integral curve of equation (1.1) *can be extended indefinitely far to the right.*

3. *Suppose* $x = x_1(t)$ *and* $x = x_2(t)$ *are two solutions of equation* (1.1) *satisfying one and the same initial condition* $x(t_0) = x_0$ *and lying in the ball* $\|x - x_0\| \leqq r$. *Suppose*

$$\|f(t, x) - f(t, y)\| \leqq L(\Phi(y - x))\psi(t)$$

with $\int_0[L(u)^{-1}]du = \infty$. *Then these solutions coincide:* $x_1(t) = x_2(t)$.

4. Consider in the finite-dimensional Euclidean phase space R^n the autonomous equation

$$dx/dt = f(x) \tag{0.1}$$

with a continuously differentiable function $f(x)$.

This equation induces on the differentiable functions in R^n with compact support a linear (unbounded) operator

$$A\phi(x) = \frac{d}{dt}\,\phi(x(t))\Big|_{t=0},$$

where $x(t)$ is the solution of equation (0.1) that is initially (when $t = 0$) at the point x.

a) Show that the operator iA is symmetric[5] in the Hilbert space $\mathscr{L}_2(R^n)$ precisely when

$$\operatorname{tr} f'(x) = 0. \tag{0.2}$$

(We recall that when x is fixed the derivative $f'(x)$ is a linear operator in R^n.)

Hint. Make use of the Gauss-Ostrogradskiĭ formula.

b) The following nontrivial test for the indefinite extendability of solutions is due to A. Ja. Povzner [1].

In order for a system (0.1) *satisfying condition* (0.2) *to have a solution in the large for* $t > 0$ ($t < 0$) (*in the sense that each point* x^0, *except for the points of a set of zero measure, does not go to infinity within a finite period of time*) *it is necessary and sufficient that the deficiency index in the upper* (*lower*) *halfplane of the closure of the operator* iA *be equal to zero.*

The cited paper of A. Ja. Povzner also contains a number of generalizations of this result.

5. Extend Theorem 1.3 to the case of the differential equation (1.19):

$$\frac{dx}{dt} = f(t, x) + g(t, x) \qquad (0 \leqq t < \infty;\ \|x\| < r), \tag{0.3}$$

in which the functions $f(t, x)$ and $g(t, x)$ respectively satisfy the Lipschitz condition

$$\|f(t, x_2) - f(t, x_1)\| \leqq L(t)\|x_2 - x_1\|, \tag{0.4}$$

and the condition

$$\|g(t, x)\| \leqq q(t)\|x\| \tag{0.5}$$

with integrally bounded $L(t)$ and $q(t)$:

$$\sup_t \frac{1}{\tau}\int_t^{t+\tau} L(s)ds < \infty; \qquad \sup_t \frac{1}{\tau}\int_t^{t+\tau} q(s)ds < \infty \tag{0.6}$$

(see V. E. Germaidze [1]).

6. Extend to nonlinear equations the test for negativeness of the Bohl exponent contained in Theorem III.6.1.

---

[5] I.e. $(iA\phi, \psi) = (\phi, iA\psi)$ for any pair of differentiable functions ϕ and ψ with compact support.

7. We will say that the differential equation

$$dx/dt = f(t, x) \qquad (0 \leqq t < \infty;\ \|x\| < r) \tag{0.7}$$

has *property* $\Omega_\varepsilon(N, \phi)$, where $\phi(t)$ $(0 \leqq t < \infty)$ is a given continuous function (B. A. Ščerbakov [1]), if every solution $x(t)$ of this equation that is in the ball of radius $\varepsilon > 0$ at some initial moment $t_0 \geqq 0$ $(\|x(t_0)\| < \varepsilon)$ is subject for all $t \geqq t_0$ for which it exists to the estimate

$$\|x(t)\| \leqq N\, \|x(t_0)\| \phi(t - t_0). \tag{0.8}$$

a) Show that equation (0.7) has a negative Bohl exponent at zero if it has property $\Omega_\varepsilon(N, \phi)$ for a continuous $\phi(t)$ tending to zero at infinity: $\lim_{t\to\infty} \phi(t) = 0$.

Hint. Make use of the result of the preceding exercise.

b) Suppose equation (0.7) satisfies condition (0.4) with a bounded $L(t)$ and has property $\Omega_\varepsilon(N,\phi)$ for some $\varepsilon > 0$, some $N > 0$ and $\phi(t) \to 0$ as $t \to \infty$.

Suppose, further, $\psi(t)$ $(0 \leqq t < \infty)$ is a continuous bounded function satisfying the conditions

1) $\phi(t)/\psi(t) \to 0$ as $t \to \infty$,
2) $\phi(t) \leqq \psi(t)$ $(0 \leqq t < \infty)$,
3) $\psi(t_1)\, \psi(t_2) \leqq \psi(t_1 + t_2)$ $(0 \leqq t_1;\ t_2 < \infty)$.

Show (B. A. Ščerbakov [1]) that for any $M > N$ there exist numbers $\delta > 0$ and $q > 0$ such that for any function $g(t, x)$ satisfying the condition

$$\|g(t, x)\| \leqq q\|x\| \qquad (\|x\| < r;\ 0 \leqq t < \infty),$$

equation (0.3) has property $\Omega_\delta(M, \psi)$.

c) Extend the latter result to the case of integrally bounded $L(t)$ and $q(t)$ (see (0.6)).

8. Show that the origin of the two-dimensional phase space of the nonlinear system

$$\begin{cases} d\zeta_1/dt = -a\zeta_1 \quad (1/2 < a < 1/\sqrt{2}), \\ d\zeta_2/dt = (\sin \ln t + \cos \ln t - 2a)\zeta_2 + \zeta_1^2 \end{cases} \tag{0.9}$$

is unstable (O. Perron [2], page 706), whereas a system of first approximation to it is stable (see § III.4.5).

Compare this result with Theorem 3.2.

Hint. Verify that system (0.9) has the general solution

$$\begin{cases} \zeta_1 = C_1 e^{-at}, \\ \zeta_2 = e^{(\sin \ln t - 2a)t} \left(C_2 + C_1^2 \int_0^t e^{\tau \sin \ln \tau} d\tau \right), \end{cases}$$

and investigate the values of $\zeta = \{\zeta_k\}_1^2$ at $t = \exp(2n + \frac{1}{2})\pi$ $(n = 1, 2, \cdots)$.

9. Consider the quasilinear equation

$$dx/dt = A(t)x + F(t, x) \qquad (F(t, x) = o(\|x\|)) \tag{0.10}$$

with a bounded operator function $A(t)$, and along with it the equation

$$dy/dt = A(t)y + F(t, y) + \Phi(t, y), \tag{0.11}$$

where $\Phi(t, y)$ is a bounded continuous function: $\|\Phi(t, y)\| \leqq M$ $(0 \leqq t < \infty;\ \|x\| < r)$.

a) Show (P. Bohl [2]) that the equation $\dot{x} = A(t)x$ has a negative upper Bohl exponent precisely when there exist numbers $\mu > 0$, $\delta > 0$ and $k > 0$ such that for $M < \mu$ any solutions $x(t)$ and $y(t)$ of equations (0.10) and (0.11) that are initially at one and the same point of the ball $B_\rho = \{x \mid \|x\| < \delta\}$: $x(0) = y(0) \in B_\rho$, satisfy on their whole interval of existence the inequality $\|x(t) - y(t)\| < k\mu$.

b) Extend this result to the case of integrally bounded $A(t)$ and $\Phi(t, y)$.

Bohl's result implies that equation (0.10) with a negative upper Bohl exponent at zero has the following property.

For any $\varepsilon > 0$ *there exist numbers* $\mu > 0$ *and* $\delta > 0$ *such that when* $\|x_0\| < \delta$ *and* $M < \mu$ *the solutions of equation* (0.11) *satisfy the estimate* $\|y(t)\| < \varepsilon$.

This property was subsequently rediscovered under the name, *stability of the zero solution of equation* (0.10) *under constantly acting perturbations* (see I. G. Malkin [5]). For a generalization of the property of stability under constantly acting perturbations see V. E. Germaidze and N. N. Krasovskiĭ [1].

10. Prove the following theorem (D. L. Kučer [2]). *Suppose the equation* $\dot{x} = A(t)x$ *has property* $\mathscr{B}(\nu_0, N_0)$ *and*

$$\|F(t, x)\| \leqq f(t)\|x\| \qquad (\|x\| \leqq r;\ t \geqq 0),$$

with $\int_0^\infty f^p(t)dt < \infty$ *for some* $p > 1$. *Then for any* $\varepsilon > 0$ *there exists an* $N > 0$ *for which the equation*

$$dx/dt = A(t)x + F(t, x) \tag{0.12}$$

has property $\mathscr{B}(\nu_0 - \varepsilon, N)$.

If $p = 1$, *the assertion is true for* $\varepsilon = 0$.

11. Prove the following theorem (D. L. Kučer [2]). *Suppose the equation* $\dot{x} = A(t)x$ *satisfies the condition*

$$\sup_{0 \leq \tau \leq t < \infty} \|U(t, \tau)\| > \infty$$

and

$$\|F(t, x)\| \leqq f(t)\|x\| \ (\|x\| \leqq r;\ t \geqq 0), \quad \textit{where} \quad \int_0^\infty f^p(t)dt < \infty.$$

Then the solutions of equation (0.12) *are subject to the estimate* $\|x(t)\| \leqq N\|x(t_0)\|$ *for some* $N > 0$.

12. Prove the following theorem (D. L. Kučer [2]). *Let* $m > 1$ *and* $p > 1$. *In order for the zero solution of equation* (0.12) *to be uniformly stable for any* $F(t, x)$ *satisfying the condition*

$$\|F(t, x)\| \leqq f(t)\|x\|^m \qquad (\|x\| \leqq r;\ t \geqq 0),$$

where $\int_0^\infty f^p(t)dt < \infty$, *it is necessary and sufficient that the Bohl exponent of the equation* $\dot{x} = A(t)x$ *be negative.*

13. Prove a theorem analogous to the preceding one, assuming $m > 1$, $p = 1$ and replacing the requirement that the Bohl exponent be negative by the requirement $\sup_{0 \leq \tau \leq t < \infty} \|U(t, \tau)\| < \infty$.

14. Consider in a Hilbert phase space H the differential equation

$$dx/dt = f(t, x) \qquad (-\infty < t < \infty;\ x \in \mathfrak{H}), \tag{0.13}$$

where $f(t, x)$ is a continuous function of t and a continuously differentiable function of x for $-\infty < t < \infty$ and $x \in \mathfrak{H}$ that is bounded for $x = 0$ $(\sup_t \|f(t, 0)\| < \infty)$. Let $A_x(t)$ denote the linear operator $A_x(t) = f'_x(t, x) \in [\mathfrak{H}]$.

Show (for the finite-dimensional case see B. P. Demidovič [1, 3]) that under the condition

$$\operatorname{Re} A_x(t) \leqslant -\alpha I \qquad (\alpha = \text{const} > 0;\ t \in [-\infty, \infty];\ x \in \mathfrak{H})$$

equation (0.13) has the *convergence property*, i.e. it has a unique solution $\tilde{x}(t)$ that is bounded on the real line and any other solution of this equation is indefinitely extendable to the right and exponentially approaches $\tilde{x}(t)$.

We note that the indicated conditions are satisfied if $f(t, x) = f(t) + g(x)$, where $f(t)$ is a bounded function and $\operatorname{Re} g'(x) \ll 0$ uniformly in x.

Hint. Show that there exists an $R > 0$ such that every solution of equation (0.13) ends up in the ball $\|x\| < R$ as $t \to \infty$. Show that any two solutions $x_1(t)$ and $x_2(t)$ that are in this ball at some moment t_0 are subject to the estimate

$$\|x_1(t) - x_2(t)\| \leqq e^{-\alpha(t-\tau)} \|x_1(\tau) - x_2(\tau)\| \qquad (t \geqq \tau). \tag{0.14}$$

To prove the existence of a solution bounded on the real line consider the sequence $\{x_k(t)\}$ of solutions satisfying the conditions $x_k(-k) = 0$ $(k = 1, 2, \cdots)$ and show, using (0.14), that $\{x_k(0)\}$ is a Cauchy sequence in $\mathfrak{H}$. The desired solution $\tilde{x}(t)$ is then determined by the condition $\tilde{x}(0) = \lim_{k\to\infty} x_k(0)$.

15. Let A be an operator in a finite-dimensional space whose spectrum does not intersect the imaginary axis, let $f(x)$ be a function satisfying the Lipschitz condition with a sufficiently small constant in a neighborhood of the origin and suppose $f(0) = 0$. Then there exists in a neighborhood of the origin a homeomorphism $y = \Phi(x)$ that transforms the solutions of the equation

$$dx/dt = Ax + f(x) \tag{0.15}$$

into the solutions of the equation $dx/dt = Ax$. (D. M. Grobman; see the book [1] by B. F. Bylov et al.)

Hint. Extend the function $f(x)$ to the whole space with preservation of the norm supremum and Lipschitz constant in such a way that it vanishes outside some sphere. If $x_0 \in \mathfrak{H}$, consider the solution $x(x_0, t)$ of equation (0.15) satisfying the condition $x(x_0, 0) = x_0$ and put (see (4.8))

$$\Phi(x_0) = x_0 - \int_{-\infty}^{\infty} G_A(-\tau) f(x(x_0, \tau)) d\tau.$$

16. Suppose the equation $dx/dt = A(t)x$ is e-dichotomic on the real line and the operator function $A(t)$ is almost periodic.

Consider the equation

$$dx/dt = A(t)x + F(t, x),$$

where $F(t, x)$ satisfies the same conditions as in Theorem 4.2. Generalize this theorem by proving the existence of an almost periodic solution of this equation. Show that the frequency module of this solution is contained in the minimal module containing the spectra of the almost periodic functions $A(t)$ and $F(t, x)$ (assuming here that the module of $F(t, x)$ does not depend on x).

Hint. Make use of the results of § 4.2 and Exercise II.11.

17. a) Generalize Theorems 4.1, 4.3, 5.1 and 6.1, replacing the condition $F(t, x) \in (M, q)$ by the conditions

$$\|F(t, x)\| \leqq M(t), \quad \|F(t, x^{(2)}) - F(t, x^{(1)})\| \leqq q(t)\|x^{(2)} - x^{(1)}\| \tag{0.16}$$

with integrally bounded functions $M(t)$ and $q(t)$.

b) Show that if conditions (0.16) are satisfied with

$$\int_{-\infty}^{\infty} M(t)dt = \mu < \infty \quad \text{and} \quad \int_{-\infty}^{\infty} q(t)dt = k < \infty,$$

and, in addition, $\sup_t \|e^{At}P_1\| < \infty$, the integral manifold described in Theorem 5.1 lies in the ball B_ρ for sufficiently small $\mu > 0$ and $k > 0$. It can be assumed here that conditions (0.16) are satisfied only in some ball.

An analogous assertion holds with respect to Theorem 6.1.

c) Extend this result to the case when

$$\int_{-\infty}^{\infty} t^n M(t)dt < \infty, \qquad \int_{-\infty}^{\infty} t^n q(t)dt < \infty$$

and $\|e^{At}P_1\|$ behaves like a power of t as $t \to \infty$.

18. Consider in the two-dimensional phase space R^2 the autonomous equation $dx/dt = f(x)$. Assume that it has a periodic solution $x(t) = x(t + T)$. The following test is due to Poincaré. *The solution $x(t)$ is orbitally asymptotically stable whenever*

$$\int_0^T \operatorname{tr} f'(x(t))dt < \infty. \tag{0.17}$$

Hint. Derive, by making use of Jacobi's well-known formula for the determinant of a fundamental matrix of a system of differential equations, the formula

$$\operatorname{tr} \ln U(T) = \frac{1}{T} \int_0^T \operatorname{tr} f'(x(t))dt, \tag{0.18}$$

where $U(T)$ is the monodromy operator for the variational equation corresponding to the solution $x(t)$. From this formula it follows by virtue of (0.17) that the only eigenvalue of the monodromy operator that is different from $\lambda = 1$ lies in the interior of the unit disk.

We note that in the multidimensional (and infinite-dimensional) case: $n > 2$, when the spectrum of the monodromy operator for the variational equation contains points lying in both the interior and exterior of the unit disk, some trajectories will approach a periodic motion for $t \to \infty$ and some, for $t \to -\infty$. In this connection there can arise a complicated phenomenon discovered by H. Poincaré, the so-called *homoclinic motion*: a trajectory that is asymptotic to a periodic motion for both $t \to \infty$ and $t \to -\infty$. A deep study of this phenomenon in the multidimensional case has recently been made by Ju. I. Neĭmark in the paper [1] (which contains a bibliography of the literature on this question).

NOTES

The existence and uniqueness theorems of § 1 are contained in various textbooks and have been well known for a long time. The result of Theorem 1.3, viz. the stability of the Bohl exponent of a nonlinear equation under nonlinear perturbations, is essentially contained in the works of Bohl (see the Notes to Chapter III and Exercise 9). The proof given by us is based on a method applied by B. A. Ščerbakov [1] in a more general situation (see Exercise 5 and the comments to it). Theorems 2.1 and 3.1 (see M. G. Kreĭn [2] and the "Lectures") constitute more special (but somewhat more exact) results. The method employed in Theorem 3.1 makes use of integral inequalities (a similar method was essentially used by Bohl; see the Notes to Chapter III). The method of proof of Theorem 2.1 is a generalization of a method of Ljapunov, as is also the method of proof of Theorem 2.2 (see the "Lectures"), which generalizes another well-known theorem of Ljapunov, viz. his instability theorem. In the theorems of §§2 and 3 mentioned above the perturbing term has first order smallness. The stability of the solutions of nonlinear equations under such an assumption on the perturbations was apparently first studied by O. Perron. The other instability theorem cited by us (Theorem 2.3) was proved by M. A. Rutman [1] in a more complicated way and under more complicated assumptions. Our proof, which is based on the same idea, was obtained by Ju. L. Daleckiĭ after analyzing the original proof.

Nonlinear perturbations of linear e-dichotomic equations were first considered by P. Bohl (for the case when the original equation is stationary) and later by O. Perron (see the Notes to Chapter IV).

Almost periodic solutions have been studied in the finite-dimensional case by many authors (G. I. Birjuk [1], N. N. Bogoljubov [1], N. N. Bogoljubov and Ju. A. Mitropol'skiĭ [1] and B. P. Demidovič [2]). In this connection we have waived the usually imposed requirement that the almost periodicity of $F(t, x)$ in t be uniform in x.

Theorem 4.5 was obtained in the two-dimensional case by A. A. Andronov and A. A. Vitt [1]; a more exact finite-dimensional version due to E. A. Coddington and N. Levinson [1] (see also the book by B. P. Demidovič [3]) has been carried over in the present book to the infinite-dimensional case.

The method of investigating multidimensional integral manifolds and their application to the substantiation and development of the averaging principle were proposed by N. N. Bogoljubov in his deep investigation [1]. The ideas of this investigation were used in all of his subsequent works.

Various versions of results in the finite-dimensional case that are similar to those presented in §§ 5 and 6 have been given in the works of N. N. Bogoljubov and Ju. A. Mitropol'skiĭ (N. N. Bogoljubov [1] and N. N. Bogoljubov and Ju. A. Mitropol'skiĭ [1]). Ju. A. Mitropol'skiĭ and O. B. Lykova (Ju. A. Mitropol'skiĭ [2] and Ju. A. Mitropol'skiĭ and O. B. Lykova [1]) have carried over some of these results to the case of Hilbert space.

The present account is of a more general character and in a number of places contains simpler proofs. It was obtained as a result of analyzing the above mentioned works and is due to Ju. L. Daleckiĭ [6]. The material presented in Chapter IV has permitted us to carry over many of these results to the nonstationary case.

The results presented in § 7 are a development of the well-known method of van der Pol. A theorem of the type of Theorem 7.1 was proved for systems with periodic coefficients by L. I. Mandel'štamm and N. D. Papaleksi [1], and for arbitrary systems of finite order by N. M. Krylov and N. N. Bogoljubov [1].

I. I. Gihman [1] noted that this theorem is a corollary of a special theorem on the continuous dependence of the solutions of a differential equation on a parameter. M. A. Krasnosel'skiĭ and S. G. Kreĭn [1] generalized this result and carried it over to the case of an equation in Hilbert space.

There exists a number of other generalizations due to J. Kurzweil and Z. Vorel [1], H. A. Antosiewicz [1] and others. We have cited only the simplest version in order to avoid widening the circle of ideas considered in this book, and our presentation follows that of M. A. Krasnosel'skiĭ and S. G. Kreĭn.

The results presented in § 7.2 concerning the estimation of the behavior of a solution over an infinite interval of time were obtained in the finite-dimensional case by N. N. Bogoljubov [1]. They were carried over to the case of Hilbert space by Z. Sirčenko [1]. We have given a somewhat simpler proof.

An extensive survey of the results concerning integral manifolds up to 1961 inclusively is contained in the report of N. N. Bogoljubov and Ju. A. Mitropol'skiĭ [2], while a survey of more recent works can be found in the book of Ju. A. Mitropol'skiĭ and O. B. Lykova [2]. For a survey of the results connected with the method of averaging see the book of Ju. A. Mitropol'skiĭ [3].

CHAPTER VIII

ASYMPTOTIC REPRESENTATION OF THE SOLUTIONS OF A LINEAR DIFFERENTIAL EQUATION WITH A LARGE PARAMETER

In this chapter we find an asymptotic expansion of the solutions of a linear differential equation in the inverse powers of a *large* parameter appearing in the equation. In the simplest case, when the principal part of the coefficient of a finite-dimensional system has a spectrum consisting of eigenvalues whose multiplicity does not change over the whole interval of time, such expansions have been the object of extensive studies, beginning with the works of Birkhoff and Tamarkin. Usually these expansions are used to prove the completeness of the eigenfunctions of selfadjoint boundary problems. We do not cite such applications since they lie to the side of our main theme and, furthermore, have appeared in textbooks for some time now (see, for example, M. A. Naĭmark [1]).

Of greater relevance for us is the use of asymptotic methods for the approximate calculation of the monodromy operator of a periodic equation and for the determination of estimates for stability bands. Applications of this kind are partially touched upon in the examples of § 2 and in the Exercises at the end of the chapter. On the other hand, equations with a large parameter can be interpreted in an equivalent form as equations with slowly varying coefficients.

In § 1 we present a method of asymptotically decomposing an equation with a large parameter in accordance with a decomposition of the spectrum of the principal part of its coefficient into nonintersecting spectral sets. In § 2, under the assumption that the spectrum of the principal part of the coefficient lies in the left halfplane, we estimate the error sustained in replacing the solution of an equation by its nth approximation. In § 3 we consider the more complicated case when, after going over to slow time, the coefficients of the equation can rapidly oscillate against a background of slowly varying amplitudes. In this case there can arise distinctive resonance phenomena analogous to parametric resonance.

§ 1. Approximate decomposition of the equation

1. *Statement of the problem.* We consider in a phase space $\mathfrak{B}$ the differential equation

$$\frac{d\psi}{d\tau} = \lambda B(\tau;\lambda)\psi \qquad (0 \leqq \tau \leqq T), \tag{1.1}$$

in which λ is a sufficiently large parameter. More precisely, we will assume that λ varies in a domain Π_{λ_0} of the complex plane which lies outside the disk of radius λ_0 ($|\lambda| \geqq \lambda_0$) and contains the ray $\lambda \geqq \lambda_0$.

We assume that the following representation holds:

$$B(\tau;\lambda) = B_0(\tau) + (1/\lambda)B^{(1)}(\tau;\lambda) \qquad (\lambda \in \Pi_{\lambda_0}), \tag{1.2}$$

where the operators $B_0(\tau)$ and $B^{(1)}(\tau;\lambda)$ are continuous in τ on the interval $[0, T]$ and, in addition,

$$\|B^{(1)}(\tau;\lambda)\| \leqq c \qquad (\tau \in [0, T];\ \lambda \in \Pi_{\lambda_0}).$$

In the sequel some differentiability requirements will also be imposed on these operator functions.

It will be assumed everywhere in this chapter that the spectrum $\sigma(B_0(\tau))$ decomposes into several spectral sets:

$$\sigma(B_0(\tau)) = \bigcup_{k=1}^{n} \sigma_k(\tau), \tag{1.3}$$

which do not intersect for any $\tau \in [0, T]$ and continuously depend on τ. The latter means that a smooth contour Γ_k separating a spectral set $\sigma_k(\tau_0)$ from its complement in the spectrum $\sigma(B_0(\tau_0))$ for some $\tau_0 \in [0, T]$ has the same property for any other sufficiently close values of $\tau \in [0, T]$.

We recall that according to the formula

$$P_k(\tau) = -\frac{1}{2\pi i}\oint_{\Gamma} (B_0(\tau) - \mu I)^{-1} d\mu \qquad (k = 1,\cdots,n)$$

the spectral projections $P_k(\tau)$ corresponding to the decomposition (1.3) are continuously differentiable in τ as many times as the operator $B_0(\tau)$.

Equation (1.1) can be written in a different form if one introduces the new variable $t = \lambda\tau$, viz.

$$d\psi/dt = [B_0(\tau) + \varepsilon B^{(1)}(\tau|\varepsilon)]\psi \qquad (\varepsilon = 1/\lambda). \tag{1.4}$$

In the case when the variable t has the meaning of time, the variable $\tau = \varepsilon t$ (for real $\varepsilon > 0$) is called the "*slow time*". The equations of form (1.4) are called *equations with slowly varying coefficients.*

When $\varepsilon = 0$ equation (1.4) degenerates into the equation $d\psi/dt = B_0(0)\psi$, whose evolution operator is easily calculated:

$$U(t, s) = e^{B_0(0)(t-s)} = \sum_{k=1}^{n} P_k(0)e^{B_0(0)P_k(0)(t-s)}. \tag{1.5}$$

It is natural to expect that the basic structure of this expression is preserved for sufficiently small values of $\varepsilon \neq 0$ with the constant operators being replaced by slowly varying operators plus terms of order $O(\varepsilon)$.

We will therefore seek an approximate expression for the evolution operator of equation (1.4) in the form

$$\tilde{U}(t, t_0; \varepsilon) = \sum_{k=1}^{n} [P_k(\tau) + \varepsilon V_k(\tau;\varepsilon)] Y_k(t, t_0; \varepsilon),$$

where the operator $Y_k(t, t_0; \varepsilon)$ satisfies the equation

$$\frac{dY_k(t, t_0; \varepsilon)}{dt} = [P_k(\tau)B_0(\tau) + \varepsilon Q_k(\tau;\varepsilon)] Y_k(t, t_0; \varepsilon)$$

and the condition

$$P_k(\tau)Y_k(t, t_0; \varepsilon) = Y_k(t, t_0; \varepsilon). \tag{1.6}$$

Returning to the original notation and using (1.6), we rewrite the indicated relations in the form

$$\tilde{U}(\tau, \tau_0; \lambda) = \sum_{k=1}^{n} \left[I + \frac{1}{\lambda} V_k(\tau; \lambda) \right] Y_k(\tau, \tau_0; \lambda), \tag{1.7}$$

$$\frac{dY_k(\tau, \tau_0; \lambda)}{d\tau} = \lambda \left[P_k(\tau)B_0(\tau) + \frac{1}{\lambda} \Omega_k(\tau; \lambda) \right] Y_k(\tau, \tau_0; \lambda)$$
$$(\tau, \tau_0 \in [0, T]; \lambda \in \Pi_{\lambda_0}). \tag{1.8}$$

The operators V_k and Ω_k $(k = 1, 2, \cdots, n)$ are to be found. We will try to choose them so that condition (1.6) is satisfied, i.e. so that the range of Y_k lies in the subspace $\mathfrak{B}_k(\tau) = P_k(\tau)\mathfrak{B}$ $(k = 1, \cdots, n)$.

We assume that the operator $B^{(1)}(\tau; \lambda)$ has an asymptotic expansion in inverse powers of λ, i.e.

$$B(\tau; \lambda) = \sum_{s=0}^{p} \frac{B_s(\tau)}{\lambda^s} + \frac{1}{\lambda^{p+1}} B^{(p+1)}(\tau; \lambda), \tag{1.9}$$

where

$$\| B^{(p+1)}(\tau; \lambda) \| \leqq c. \tag{1.10}$$

We will seek the operator functions V_k and Ω_k in the form

$$\frac{1}{\lambda} V_k(\tau; \lambda) = \sum_{s=1}^{p} \lambda^{-s} V_{ks}(\tau), \tag{1.11}$$

$$\frac{1}{\lambda} \Omega_k(\tau; \lambda) = \sum_{s=1}^{p} \lambda^{-s} \Omega_{ks}(\tau) \tag{1.12}$$

and attempt to choose the coefficients of these expansions so that the operator function (1.7) satisfies the equation

$$\frac{d\tilde{U}(\tau, \tau_0; \lambda)}{d\tau} = \lambda B(\tau; \lambda)\tilde{U}(\tau, \tau_0; \lambda) + \frac{1}{\lambda^p} \Phi_p(\tau; \lambda), \tag{1.13}$$

which differs from the equation

$$dU/dt = \lambda B(\tau; \lambda)U$$

for the desired evolution operator by a term of order $O(\lambda^{-p})$.

Under certain conditions, which will be described below, a solution of equation (1.13) differs from the evolution operator of equation (1.1) by a quantity of order $O(\lambda^{-p})$.

Once we have indicated a method of determining the coefficients of the expansions (1.11) and (1.12), the approximate solution of the Cauchy problem for equation (1.1) will be reduced to the solution of equations (1.8) in the phase spaces $\mathfrak{B}_k(\tau_0)$.

Such a reduction is called *an approximate (asymptotic for $\lambda \to \infty$) decomposition of equation* (1.1).

2. *Some auxiliary propositions.*

LEMMA 1.1. *Let $Q(\tau, \tau_0)$ be a differentiable operator with respect to τ that "tracks" the subspaces $\mathfrak{B}_k(\tau)$:*

$$\left.\begin{aligned} P_k(\tau)Q(\tau, \tau_0) &= Q(\tau, \tau_0)P_k(\tau_0) \qquad (k = 1,\cdots,n),\\ Q(\tau_0, \tau_0) &= I. \end{aligned}\right\} \tag{1.14}$$

If the relation

$$\begin{aligned} \Omega_k(\tau; \lambda) = P_k(\tau)\Omega_k(\tau; \lambda)P_k(\tau) &+ Q'(\tau, \tau_0)Q^{-1}(\tau, \tau_0) \\ &- P_k(\tau)Q'(\tau, \tau_0)Q^{-1}(\tau, \tau_0)P_k(\tau) \end{aligned} \tag{1.15}$$

holds identically in τ, the relation

$$P_k(\tau_0)Y_k(\tau_0, \tau_0; \lambda) = Y_k(\tau_0, \tau_0; \lambda)P_k(\tau_0) = Y_k(\tau_0, \tau_0; \lambda) \tag{1.16}$$

implies the relation

$$P_k(\tau)Y_k(\tau, \tau_0; \lambda) = Y_k(\tau, \tau_0; \lambda)P_k(\tau_0) = Y_k(\tau, \tau_0; \lambda) \qquad (\tau \in [0,T]). \tag{1.17}$$

PROOF. The substitution $Y_k(\tau, \tau_0; \lambda) = Q(\tau, \tau_0)X_k(\tau, \tau_0; \lambda)$ takes equation (1.8) into the equation

$$\begin{aligned} dX_k/d\tau = Q^{-1}(\tau, \tau_0)\{&\lambda P_k(\tau)B_0(\tau) \\ &+ \Omega_k(\tau, \lambda) - Q'(\tau, \tau_0)Q^{-1}(\tau, \tau_0)\}Q(\tau, \tau_0)X_k, \end{aligned}$$

which by virtue of conditions (1.15) and (1.14) reduces to the form

$$\begin{aligned} dX_k/d\tau = P_k(\tau_0)Q^{-1}(\tau, \tau_0)\{&\lambda P_k(\tau)B_0(\tau) + \Omega_k(\tau; \lambda) \\ &- Q'(\tau, \tau_0)Q^{-1}(\tau, \tau_0)\}Q(\tau, \tau_0)P_k(\tau_0)X_k. \end{aligned}$$

It is not difficult to see that this equation is satisfied by not only $X_k(\tau, \tau_0; \lambda)$ but also the operators $X_k(\tau, \tau_0; \lambda)P_k(\tau_0)$ and $P_k(\tau_0)X_k(\tau, \tau_0; \lambda)$.

From (1.16) it follows that all three operators coincide for $\tau = \tau_0$ and hence for all $\tau \in [0, T]$.

Inasmuch as

$$\begin{aligned} P_k(\tau)Y_k(\tau, \tau_0; \lambda) &= P_k(\tau)Q(\tau, \tau_0)X_k(\tau, \tau_0; \lambda) \\ &= Q(\tau, \tau_0)P_k(\tau_0)X_k(\tau, \tau_0; \lambda) \end{aligned}$$

and

$$Y_k(\tau, \tau_0; \lambda)P_k(\tau_0) = Q(\tau, \tau_0)X_k(\tau, \tau_0; \lambda)P_k(\tau_0),$$

equality (1.17) is proved.

REMARK 1.1. Condition (1.15) is obviously satisfied for an operator $\Omega_k(\tau;\lambda)$ of form (1.12) if

$$\begin{aligned}\Omega_{k1}(\tau) = P_k(\tau)\Omega_{k1}(\tau)P_k(\tau) + Q'(\tau,\tau_0)Q^{-1}(\tau,\tau_0)\\ - P_k(\tau)Q'(\tau,\tau_0)Q^{-1}(\tau,\tau_0)P_k(\tau),\end{aligned} \tag{1.18}$$

$$\Omega_{ks}(\tau) = P_k(\tau)\Omega_{ks}(\tau)P_k(\tau) \qquad (s \geqq 2). \tag{1.19}$$

REMARK 1.2. In the sequel it will be convenient to take as $Q(\tau,\tau_0)$ the conjugation operator constructed in § III.1. It satisfies the differential equation

$$\frac{dQ(\tau,\tau_0)}{d\tau} = \sum_{j=1}^{n} P_j'(\tau)P_j(\tau)Q(\tau,\tau_0).$$

In this case equality (1.18) takes the form

$$\Omega_{k1}(\tau) = P_k(\tau)\Omega_{k1}(\tau)P_k(\tau) + \sum_{j=1}^{n} P_j'(\tau)P_j(\tau). \tag{1.20}$$

Condition (1.16) will be satisfied, for example, if $Y_k(\tau_0,\tau_0;\lambda) = P_k(\tau_0)$.

From Lemma 1.1 and Remarks 1.1 and 1.2 it follows that if the operators $\Omega_{ks}(\tau)$ are selected so that they satisfy conditions (1.19) and (1.20), the operators $Y_k(\tau,\tau_0;\lambda)$ determined from equations (1.8) will satisfy condition (1.6) for all τ if they satisfy it for $\tau = \tau_0$.

3. *Construction of the approximate solution.* We proceed to calculate the coefficients of expansions (1.11) and (1.12). Differentiating (1.7) and using these expansions and (1.8), we find

$$\begin{aligned}
&\frac{1}{\lambda}\bar{U}'(\tau,\tau_0;\lambda)\\
&= \frac{1}{\lambda}\sum_{k=1}^{n}\sum_{s=1}^{p}\lambda^{-s}V_{ks}'(\tau)Y_k(\tau,\tau_0;\lambda)\\
&\quad + \sum_{k=1}^{n}\Big[I + \sum_{s=1}^{p}\lambda^{-s}V_{ks}(\tau)\Big]\Big[P_k(\tau)B_0(\tau) + \sum_{s=1}^{p}\lambda^{-s}\Omega_{ks}(\tau)\Big]Y_k(\tau,\tau_0;\lambda)\\
&= \sum_{k=1}^{n}\Big\{B_0(\tau)P_k(\tau) + \sum_{s=1}^{p}\lambda^{-s}[V_{ks}(\tau)B_0(\tau)P_k(\tau) + V_{k,s-1}'(\tau) + \Omega_{ks}(\tau)]\\
&\qquad + \lambda^{-(p+1)}V_{kp}'(\tau) + \sum_{s=2}^{p}\lambda^{-s}\sum_{j=1}^{s-1}V_{k,s-j}(\tau)\Omega_{kj}(\tau)\\
&\qquad\qquad + \sum_{s=p+1}^{2p}\lambda^{-s}\sum_{j=s-p}^{p}V_{k,s-j}(\tau)\Omega_{kj}(\tau)\Big\}Y_k(\tau,\tau_0;\lambda).
\end{aligned} \tag{1.21}$$

On the other hand, the right side of equation (1.13), after dividing through by λ and using (1.9), (1.7) and (1.11), takes the form

$$
\begin{aligned}
&B(\tau;\lambda)\bar{U}(\tau,\tau_0;\lambda)+\lambda^{-(p+1)}\Phi_p(\tau;\lambda)\\
&\quad=\left[\sum_{s=0}^{p}\lambda^{-s}B_s(\tau)+\lambda^{-(p+1)}B^{(p+1)}(\tau;\lambda)\right]\\
&\quad\times\sum_{k=1}^{n}\left[I+\sum_{s=1}^{p}\lambda^{-s}V_{ks}(\tau)\right]Y_k(\tau,\tau_0;\lambda)+\lambda^{-(p+1)}\Phi_p(\tau;\lambda)\\
&\quad=\sum_{k=1}^{n}\Bigg\{B_0(\tau)+\sum_{s=1}^{p}\lambda^{-s}[B_s(\tau)+B_0(\tau)V_{ks}(\tau)] \qquad (1.22)\\
&\qquad+\sum_{s=2}^{p}\lambda^{-s}\sum_{j=1}^{s-1}B_{s-j}(\tau)V_{kj}(\tau)+\sum_{s=p+1}^{2p}\lambda^{-s}\sum_{j=s-p}^{p}B_{s-j}(\tau)V_{kj}(\tau)\\
&\qquad+\lambda^{-(p+1)}B^{(p+1)}(\tau;\lambda)\left[I+\sum_{s=1}^{p}\lambda^{-s}V_{ks}(\tau)\right]\Bigg\}Y_k(\tau,\tau_0;\lambda)\\
&\quad+\lambda^{-(p+1)}\Phi_p(\tau;\lambda).
\end{aligned}
$$

The right sides of equalities (1.21) and (1.22) must coincide. By equating the coefficients of like powers of λ, making use of relations (1.17) and discarding the common factor $Y_k(\tau,\tau_0;\lambda)$, we obtain the system of relations

$$
\begin{aligned}
&V_{k1}(\tau)B_0(\tau)P_k(\tau)+\Omega_{k1}(\tau)P_k(\tau)=B_1(\tau)P_k(\tau)+B_0(\tau)V_{k1}(\tau)P_k(\tau);\\
&V_{ks}(\tau)B_0(\tau)P_k(\tau)+V'_{k,s-1}(\tau)P_k(\tau)+\Omega_{ks}(\tau)P_k(\tau)+\sum_{j=1}^{s-1}V_{k,s-1}(\tau)\Omega_{kj}(\tau)P_k(\tau)\\
&\quad=B_s(\tau)P_k(\tau)+B_0(\tau)V_{ks}(\tau)P_k(\tau)+\sum_{j=1}^{s-1}B_{s-j}(\tau)V_{kj}(\tau)P_k(\tau) \qquad (2\leqq s\leqq p),
\end{aligned}
$$

which is more suitably written in the form

$$
\begin{aligned}
&B_0(\tau)V_{k1}(\tau)P_k(\tau)-V_{k1}(\tau)P_k(\tau)B_0(\tau)\\
&\qquad=\Omega_{k1}(\tau)P_k(\tau)-B_1(\tau)P_k(\tau);
\end{aligned}
\tag{1.23}
$$

$$
\begin{aligned}
&B_0(\tau)V_{ks}(\tau)P_k(\tau)-V_{ks}(\tau)P_k(\tau)B_0(\tau)\\
&\qquad=\Omega_{ks}(\tau)P_k(\tau)-T_{ks}(\tau)P_k(\tau) \qquad (2\leqq s\leqq p),
\end{aligned}
\tag{1.24}
$$

where

$$
T_{ks}(\tau)=\sum_{j=1}^{s-1}B_{s-j}(\tau)V_{kj}(\tau)+B_s(\tau)-V'_{k,s-1}(\tau)-\sum_{j=1}^{s-1}V_{k,s-j}(\tau)\Omega_{kj}(\tau). \tag{1.25}
$$

Moreover, a comparison of the terms of order $O(\lambda^{-(p+1)})$ in (1.21) and (1.22) leads to the equality

$$
\begin{aligned}
&\Phi_p(\tau,\lambda)\\
&\quad=\sum_{k=1}^{n}\Bigg\{V'_{kp}(\tau)+\sum_{s=p+1}^{2p}\lambda^{-(s-p-1)}\sum_{j=s-p}^{p}[V_{k,s-j}(\tau)\Omega_{kj}(\tau)-B_{s-j}(\tau)V_{kj}(\tau)] \qquad (1.26)\\
&\qquad-B^{(p+1)}(\tau,\lambda)\left[I+\sum_{s=1}^{p}\lambda^{-s}V_{ks}(\tau)\right]\Bigg\}Y_k(\tau,\tau_0,\lambda).
\end{aligned}
$$

The system of equations (1.23)–(1.24) is recursive in nature inasmuch as the expression (1.25) for $T_{ks}(\tau)$ contains only those operators $V_{kj}(\tau)$ and $\Omega_{kj}(\tau)$ with subscripts $j = 1,\cdots, s-1$ and the derivative $V'_{k,s-1}(\tau)$.

We will now show how to successively determine all of the operators Ω_{kj} and V_{kj} $(j = 1,\cdots,p; k = 1,\cdots,n)$ from these equations with the use of formula (I.3.14). It will be assumed in this connection that the operator function $B_j(\tau)$ $(j = 1,\cdots,p)$ has strongly continuous derivatives up to order $p+1-j$ inclusively.

We begin by considering the equation of first approximation (1.23). Multiplying it from the left by the projection $P_j(\tau)$, we obtain for the operator

$$X_{jk}^{(1)}(\tau) = P_j(\tau)V_{k1}(\tau)P_k(\tau)$$

the equation

$$\begin{aligned} B_0(\tau)X_{jk}^{(1)}(\tau) &- X_{jk}^{(1)}(\tau)B_0(\tau) \\ &= P_j(\tau)[\Omega_{k1}(\tau) - B_1(\tau)]P_k(\tau) \qquad (j = 1,2,\cdots,n). \end{aligned} \tag{1.27}$$

This equation is satisfied for $j = k$ if one puts

$$X_{kk}^{(1)}(\tau) = 0 \tag{1.28}$$

and

$$P_k(\tau)\Omega_{k1}(\tau)P_k(\tau) = P_k(\tau)B_1(\tau)P_k(\tau). \tag{1.29}$$

Equalities (1.20) and (1.29) permit us to determine the operator $\Omega_{k1}(\tau)$:

$$\Omega_{k1}(\tau) = P_k(\tau)B_1(\tau)P_k(\tau) + \sum_{r=1}^{n} P_r'(\tau)P_r(\tau). \tag{1.30}$$

When $j \neq k$ we have $Y(\tau) = P_j(\tau)Y(\tau)P_k(\tau)$, where $Y(\tau)$ denotes the right side of (1.27) and the projections $P_j(\tau)$ and $P_k(\tau)$ correspond to disjoint spectral sets of $B_0(\tau)$. We can therefore apply formula (I.3.14), which gives us

$$\begin{aligned} &X_{jk}^{(1)}(\tau) \\ &= -\frac{1}{4\pi^2}\oint_{\Gamma_k(\tau)}\oint_{\Gamma_j(\tau)} \frac{(B_0(\tau)-\nu I)^{-1}P_j(\tau)[\Omega_{k1}(\tau)-B_1(\tau)]P_k(\tau)(B_0(\tau)-\mu I)^{-1}}{\nu-\mu}\,d\nu\,d\mu \\ &= -\frac{1}{4\pi^2}\oint_{\Gamma_k(\tau)}\oint_{\Gamma_j(\tau)} \frac{(B_0(\tau)-\nu I)^{-1}P_j(\tau)[P_k'(\tau)-B_1(\tau)]P_k(\tau)(B_0(\tau)-\mu I)^{-1}}{\nu-\mu}\,d\nu\,d\mu. \end{aligned} \tag{1.31}$$

Finally, we find the operators $V_{k1}(\tau)$ by putting

$$V_{k1}(\tau) = V_{k1}(\tau)P_k(\tau) = \sum_{j\neq k} X_{jk}^{(1)}(\tau). \tag{1.32}$$

As has already been mentioned the projections $P_k(\tau)$ have as many continuous derivatives as the operator $B_0(\tau)$. Therefore, as can be seen from formulas (1.30) and (1.31), the operators $\Omega_{k1}(\tau)$ and $V_{k1}(\tau)$ have continuous derivatives of order p.

We consider the equations for second and higher approximations in a completely analogous manner. Multiplying (1.24) from the left by $P_j(\tau)$, we obtain for the operator

$$X_{jk}^{(s)}(\tau) = P_j(\tau)V_{ks}(\tau)P_k(\tau)$$

the equation

$$B_0(\tau)P_j(\tau)X_{jk}^{(s)}(\tau) - X_{jk}^{(s)}(\tau)B_0(\tau)P_k(\tau) = P_j(\tau)\,[\Omega_{ks}(\tau) - T_{ks}(\tau)]P_k(\tau), \quad (1.33)$$

which is investigated in exactly the same way as (1.27).

When $j = k$ we put

$$X_{kk}^{(s)}(\tau) = 0 \quad (1.34)$$

and

$$P_k(\tau)\Omega_{ks}(\tau)P_k(\tau) = P_k(\tau)T_{ks}(\tau)P_k(\tau)$$

and get, using (1.19),

$$\Omega_{ks}(\tau) = P_k(\tau)T_{ks}(\tau)P_k(\tau) \qquad (s = 2,\cdots, p). \quad (1.35)$$

After this we find

$$X_{jk}^{(s)}(\tau) = -\frac{1}{4\pi^2}\oint_{\Gamma_k(\tau)}\oint_{\Gamma_j(\tau)}\frac{(B_0(\tau) - \nu I)^{-1}P_j(\tau)T_{ks}(\tau)P_k(\tau)(B_0(\tau) - \mu I)^{-1}}{\nu - \mu}\,d\nu\, d\mu \quad (1.36)$$

and put, as before,

$$V_{ks}(\tau) = \sum_{j\neq k} X_{jk}^{(s)}(\tau). \quad (1.37)$$

We note that the operator function

$$T_{k2}(\tau) = B_1(\tau)V_{k1}(\tau) + B_2(\tau) - V_{k1}'(\tau) - V_{k1}(\tau)\Omega_{k1}(\tau)$$

is continuously differentiable $p - 1$ times, and hence so also are the operators $V_{k1}(\tau)$ and $\Omega_{k2}(\tau)$.

It is not difficult to show by induction that the operators $V_{kr}(\tau)$ and $\Omega_{kr}(\tau)$ are continuously differentiable $p + 1 - r$ times. For if this is true when $r = s - 1$, the operator $T_{ks}(\tau)$ is according to formula (1.25) continuously differentiable $p - s + 1$ times, and it follows from formulas (1.35) and (1.37) that this property is also possessed by the operators $V_{kr}(\tau)$ and $\Omega_{kr}(\tau)$.

In particular, the operator $V_{kp}(\tau)$ has a continuous derivative.

Thus all of the coefficients of expansions (1.11) and (1.12) have been determined so that the operator function (1.7) satisfies equation (1.13). In this connection, as follows from formula (1.26), the operator function $\Phi_p(\tau; \lambda)$ is continuous in τ for $\tau \in [0, T]$ and $\lambda \in \Pi_{\lambda_0}$.

§ 2. Estimate of the error

1. *Basic lemmas.* We proceed to estimate the error sustained in using the above method for constructing the evolution operator of equation (1.1).

This estimate is obtained under certain additional assumptions connected with the location of the spectrum of the operator $B_0(\tau)$.

One of these additional assumptions is formulated as follows.

($C_{\mathfrak{B}}$) *The spectrum $\sigma(B_0(\tau))$ of $B_0(\tau)$ lies in the interior of the left halfplane for all values of $\tau \in [0, T]$.*

In certain important cases the spectrum $\sigma(B_0(\tau))$ can fall on the imaginary axis. Such a situation can be analyzed under certain additional assumptions described by the following condition.

($C_{\mathfrak{H}}$) *The phase space $\mathfrak{B}$ is a Hilbert space $\mathfrak{H}$ and there exists a uniformly positive continuously differentiable operator $W(\tau)$ $(0 \leqq \tau \leqq T)$ such that the operator $\mathrm{Re}(W(\tau)B_0(\tau))$ is nonpositive.*

We note that it automatically follows from condition ($C_{\mathfrak{H}}$) that the spectrum $\sigma(B_0(\tau))$ lies in the closure of the left halfplane. In fact, for any $\delta > 0$ the operator $B_0(\tau) - \delta I$ is obviously uniformly $W(\tau)$-dissipative and hence its spectrum lies in the interior of the left halfplane (see Theorem I.5.1).

An example of an operator satisfying condition ($C_{\mathfrak{H}}$) is an operator of the form $B_0(\tau) = S(\tau)H(\tau)$, where $H(\tau)$ is a strongly continuously differentiable uniformly positive operator and the operator $S(\tau)$ is skew Hermitian: $S^*(\tau) = -S(\tau)$. In fact, in this case it is possible to put $W(\tau) = H(\tau)$. Then

$$\mathrm{Re}(W(\tau)B_0(\tau)) = H(\tau)S(\tau)H(\tau) - H(\tau)S(\tau)H(\tau) = 0$$

and thus condition ($C_{\mathfrak{H}}$) is satisfied.

This case includes, in particular, the canonical equations ($S = i\mathcal{J}$).

Our subsequent analysis will be based on two auxiliary propositions.

LEMMA 2.1. *Suppose the continuous operator function $B_0(\tau)$ with values in $[\mathfrak{B}]$ satisfies condition ($C_{\mathfrak{B}}$) while the operator function $B_1(\tau; \lambda)$ is uniformly bounded in the domain $\Pi_{\lambda_0} = \Pi'_{\lambda_0,\theta_0} = \{\lambda = \rho e^{i\theta} | \rho \geqq \lambda_0; |\theta| < \theta_0\}$:*

$$\|B_1(\tau; \lambda)\| \leqq c \qquad (\lambda \in \Pi'_{\lambda_0,\theta_0};\ \tau \in [0, T)].$$

Then the evolution operator $U(\tau, \tau_0; \lambda)$ of the equation

$$d\psi/d\tau = [\lambda B_0(\tau) + B_1(\tau; \lambda)]\psi \qquad (\lambda \geqq \lambda_0;\ \tau \in [0, T]) \tag{2.1}$$

is uniformly bounded in $\Pi'_{\lambda_0,\theta_0}$ for sufficiently small $\theta_0 > 0$:

$$\|U(\tau, \tau_0; \lambda)\| \leqq c_1 \qquad (\lambda \in \Pi'_{\lambda_0,\theta_0};\ \tau \in [0, T]). \tag{2.2}$$

If the number λ_0 is sufficiently large, the stronger condition

$$\|U(\tau, \tau_0; \lambda)\| \leqq Ne^{-|\lambda|\nu_0(\tau-\tau_0)} \quad (\lambda \in \Pi'_{\lambda_0,\theta_0};\ \tau \in [0, T]), \tag{2.3}$$

where $N, \nu_0 > 0$, is satisfied.

PROOF. The totality of values of the continuous operator function $e^{i\theta}B_0(\tau)$

($\tau \in [0, T]$; $|\theta| < \theta_0$) is a precompact set of operators in $[\mathfrak{B}]$ the spectra of which for sufficiently small θ_0 lie in the interior of the left halfplane and hence to the left of a straight line $\operatorname{Re} \mu = -\nu$ ($\nu > 0$). It follows from Lemma III.6.2 in this case that there exists a constant $N > 0$ for which

$$\|e^{\lambda B_0(\tau)s}\| = \|e^{e^{i\theta}B_0(\tau)s|\lambda|}\| \leqq Ne^{-|\lambda|\nu s} \qquad (\lambda \in \Pi'_{\lambda_0,\theta_0};\ \tau \in [0, T];\ s \geqq 0). \tag{2.4}$$

We consider a number $\delta > 0$ having the property that when $|\tau - \tau_0| \leqq \delta$ ($\tau, \tau_0 \in [0, T]$)

$$\|B_0(\tau) - B_0(\tau_0)\| \leqq \nu/2N. \tag{2.5}$$

Writing the equation for the evolution operator in the form

$$\frac{dU(\tau, \tau_0; \lambda)}{d\tau} = \lambda B_0(\tau_0)U(\tau, \tau_0; \lambda) + \lambda[B_0(\tau) - B_0(\tau_0)]U(\tau, \tau_0; \lambda) + B_1(\tau; \lambda)U(\tau, \tau_0; \lambda),$$

we obtain the representation

$$U(\tau, \tau_0; \lambda) = e^{\lambda B_0(\tau_0)(\tau-\tau_0)} + \lambda \int_{\tau_0}^{\tau} e^{\lambda B_0(\tau_0)(\tau-s)}[B_0(s) - B_0(\tau_0)]U(s, \tau_0; \lambda)ds + \int_{\tau_0}^{\tau} e^{\lambda B_0(\tau_0)(\tau-s)}B_1(s; \lambda)U(s, \tau_0; \lambda)ds.$$

When $|\tau - \tau_0| < \delta$ it follows from this representation and (2.4) and (2.5) that

$$\|U(\tau, \tau_0; \lambda)\| \leqq Ne^{-|\lambda|\nu(\tau-\tau_0)} + N|\lambda|\frac{\nu}{2N}\int_{\tau_0}^{\tau} e^{-|\lambda|\nu(\tau-s)}\|U(s, \tau_0; \lambda)\|ds + cN\int_{\tau_0}^{\tau} e^{-|\lambda|\nu(\tau-s)}\|U(s, \tau_0; \lambda)\|ds.$$

Using Corollary III.2.2, we conclude that when $|\tau - \tau_0| < \delta$

$$\|U(\tau, \tau_0; \lambda)\| \leqq Ne^{-(|\lambda|\nu/2-cN)(\tau-\tau_0)}.$$

We obtain the desired estimate for arbitrary $\tau_0, \tau \in [0, T]$ by subdividing the interval $[\tau_0, \tau]$ by the points $\tau_0 < \tau_1 < \cdots < \tau_{m-1} < \tau_m = \tau$ so that $|\tau_k - \tau_{k-1}| < \delta$.

It then follows from the representation

$$U(\tau, \tau_0; \lambda) = \prod_{k=1}^{m} U(\tau_k, \tau_{k-1}; \lambda)$$

that

$$\|U(\tau, \tau_0; \lambda)\| \leqq \prod_{k=1}^{m}\|U(\tau_k, \tau_{k-1}; \lambda)\| \leqq N^{[(\tau-\tau_0)/\delta]+1}e^{-((1/2)|\lambda|\nu-cN)(\tau-\tau_0))} \qquad (\lambda \in \Pi'_{\lambda_0,\theta_0}). \tag{2.6}$$

We at once obtain estimate (2.2) from this inequality since for $\lambda \in \Pi'_{\lambda_0,\theta_0}$ and $0 \leqq \tau_0 \leqq \tau \leqq T$

$$\|U(\tau, \tau_0; \lambda)\| \leqq N^{[T/\delta]+1} e^{cNT}.$$

Moreover, (2.6) implies

$$\begin{aligned}\|U(\tau, \tau_0; \lambda)\| &\leqq Ne^{-(1/2)(|\lambda|\nu - cN - (1/\delta)\ln N)(\tau-\tau_0)} \\ &\leqq Ne^{-|\lambda| 2(\nu - cN/\lambda_0 - \ln N\delta\lambda_0)(\tau-\tau_0)}.\end{aligned}$$

It remains to note that the quantity $\nu_0 = \frac{1}{2}(\nu - cN/\lambda_0 - \ln N/\lambda_0\delta)$ is positive for sufficiently large λ_0. The lemma is proved.

COROLLARY 2.1. *If the coefficients of equation* (2.1) *are periodic and satisfy the conditions of Lemma* 2.1, *the spectrum of the monodromy operator lies in the interior of the unit disk for sufficiently large in modulus* $\lambda \in \Pi'_{\lambda_0,\theta_0}$ *and sufficiently small* $\theta_0 > 0$.

In fact, if the coefficients of equation (2.1) are periodic with period T, inequality (2.3) can be used to estimate the monodromy operator of this equation: $\|U(T, 0; \lambda)\| \leqq Ne^{-|\lambda|\nu_0 T}$.

It follows from this estimate that the inequality $\|U(T, 0; \lambda)\| < 1$ is valid for sufficiently large $|\lambda|$.

LEMMA 2.2. *If the operator* $B_0(\tau)$ *satisfies condition* $(C_{\mathfrak{H}})$ *while the operator* $B_1(\tau; \lambda)$ *is uniformly bounded in the strip* $\Pi_{\lambda_0} = \Pi''_{\lambda_0,\alpha} = \{\lambda | \operatorname{Re}\lambda \geqq \lambda_0; |\operatorname{Im}\lambda| \leqq \alpha\}$:

$$\|B_1(\tau; \lambda)\| \leqq c \qquad (\tau \in [0, T]; \lambda \in \Pi''_{\lambda_0,\alpha}),$$

the evolution operator of equation (2.1) *is also uniformly bounded in* $\Pi''_{\lambda_0,\alpha}$.

PROOF. We consider for fixed $x \in \mathfrak{H}$ the nonnegative function

$$\phi(\tau) = (W(\tau)U(\tau, \tau_0; \lambda)x, U(\tau, \tau_0; \lambda)x).$$

The following estimate holds for its derivative:

$$\begin{aligned}\phi'(\tau) &= (W'(\tau)U(\tau, \tau_0; \lambda)x, U(\tau, \tau_0; \lambda)x) \\ &\quad + (W(\tau)[\lambda B_0(\tau) + B_1(\tau; \lambda)]U(\tau, \tau_0; \lambda)x, U(\tau, \tau_0; \lambda)x) \\ &\quad + (W(\tau)U(\tau, \tau_0; \lambda)x, [\lambda B_0(\tau) + B_1(\tau; \lambda)]U(\tau, \tau_0; \lambda)x) \\ &= ([W'(\tau) + 2\operatorname{Re}\lambda \operatorname{Re} W(\tau)B_0(\tau) - 2\operatorname{Im}\lambda \operatorname{Im} W(\tau)B_0(\tau) \\ &\qquad + 2\operatorname{Re} W(\tau)B_1(\tau; \lambda)]U(\tau, \tau_0; \lambda)x, U(\tau, \tau_0; \lambda)x) \\ &\leqq |([W'(\tau)U(\tau, \tau_0; \lambda)x, U(\tau, \tau_0; \lambda)x)]| \\ &\quad + |([W(\tau)B_1(\tau; \lambda) + 2\alpha \operatorname{Im} W(\tau)B_0(\tau)]U(\tau, \tau_0; \lambda)x, U(\tau, \tau_0; \lambda)x)|\end{aligned}$$

$$\leqq \{ \| W^{-1/2}(\tau)W'(\tau)W^{-1/2}(\tau)\| + 2\alpha \| W^{-1/2}(\tau)(\mathrm{Im}\, WB_0)W^{-1/2}(\tau)\| + \| W^{1/2}(\tau)B_1(\tau;\lambda)W^{-1/2}(\tau)\| \}\phi(\tau) \leqq c_2\phi(\tau)$$
$$(0 \leqq \tau \leqq T;\ \lambda \in \Pi''_{\lambda_0\alpha}).$$

Integrating this inequality, we get

$$\phi(\tau) \leqq \phi(\tau_0)e^{c_2(\tau-\tau_0)} \leqq \phi(\tau_0)e^{cT}$$

and, finally,

$$\| U(\tau, \tau_0; \lambda)x\|^2 \leqq \| W^{-1}\| \phi(\tau) \leqq \| W^{-1}\| e^{cT}(Wx, x) \leqq \| W^{-1}\| \| W\| e^{cT}\| x\|^2.$$

The lemma is proved.

2. *Estimate of the solutions of the decomposition equations.* The result obtained above will first be used to estimate the operator $Y_k(\tau, \tau_0; \lambda)$ $(k = 1, \cdots, n)$. This operator satisfies the differential equation (1.8), the coefficients of which have been selected so that condition (1.15) is satisfied.

It was shown in the proof of Lemma 1.1 that the substitution

$$Y_k(\tau, \tau_0; \lambda) = Q(\tau, \tau_0)X_k(\tau, \tau_0; \lambda)$$

reduces equation (1.8) to the equation

$$dX_k/d\tau = [\lambda B_0^{(k)}(\tau) + B_1^{(k)}(\tau; \lambda)]X_k, \tag{2.7}$$

where

$$B_0^{(k)}(\tau) = P_k(\tau_0)Q^{-1}(\tau, \tau_0)P_k(\tau)B_0(\tau)P_k(\tau)Q(\tau, \tau_0)P_k(\tau_0) \tag{2.8}$$

and

$$B_1^{(k)}(\tau; \lambda) = P_k(\tau_0)Q^{-1}(\tau, \tau_0)\{\Omega_k(\tau; \lambda) - Q'(\tau, \tau_0)Q^{-1}(\tau, \tau_0)\}\, Q(\tau, \tau_0)P_k(\tau_0). \tag{2.9}$$

The operators $B_0^{(k)}(\tau)$ and $B_1^{(k)}(\tau; \lambda)$, as can be seen from expressions (2.8) and (2.9), act in the subspace $\mathfrak{B}_k(\tau_0)$. These operators are continuous in τ and the operator $B_1^{(k)}(\tau; \lambda)$ is bounded for $\tau \in [0, T]$ and $\lambda \in \Pi_{\lambda_0}$.

The spectrum of the operator $B_0^{(k)}(\tau)$ in the subspace $\mathfrak{B}_k(\tau_0)$ coincides with the spectral set $\sigma_k(\tau)$. Therefore condition $(C_{\mathfrak{B}})$ is satisfied by $B_0^{(k)}(\tau)$ if it is satisfied by $B_0(\tau)$.

Suppose now that the operator $B_0(\tau)$ satisfies condition $(C_{\mathfrak{G}})$. Then this condition is also satisfied by the operator $B_0(\tau)P_k(\tau)$. In fact, along with the operator $W(\tau)$ the operator

$$\tilde{W}_k(\tau) = P_k^*(\tau)W(\tau)P_k(\tau) + (I - P_k^*(\tau))\, W(\tau)(I - P_k(\tau))$$

is also uniformly positive.

It is not difficult to verify that the operator

$$\operatorname{Re} \bar{W}_k(\tau)P_k(\tau)B_0(\tau) = P_k^*(\tau)\operatorname{Re}[W(\tau)B_0(\tau)]P_k(\tau)$$

is nonpositive.

Thus in this case Lemma 2.2 is applicable to equation (1.8). We have arrived at the following proposition.

LEMMA 2.3. *Suppose that the operator* $B_0(\tau)$ *satisfies condition* ($C_{\mathfrak{B}}$) *or condition* ($C_{\mathfrak{H}}$). *Suppose further, that the operator* $B_k(\tau)$ *has a continuous derivative of order* $p+1-k$ $(k=0,1,\cdots,p)$.

Then the solution $Y_k^{(p)}(\tau, \tau_0; \lambda)$ *of the equation*

$$\frac{dY_k}{d\tau} = \lambda\left[B_0(\tau)P_k(\tau) + \sum_{s=1}^{p} \lambda^{-s}Q_{ks}(\tau)\right]Y_k \qquad (\lambda \in \Pi_{\lambda_0};\ \tau, \tau_0 \in [0, T]), \tag{2.10}$$

where the $Q_{ks}(\tau)$ $(k = 1,\cdots, n;\ s = 1,\cdots, p)$ *are the operators* (1.30) *and* (1.35), *that satisfies the initial condition* $Y_k(\tau_0, \tau_0; \lambda) = P_k(\tau_0)$ *is bounded*:

$$\| Y_k^{(p)}(\tau, \tau_0; \lambda)\| \leqq c \qquad (\tau_0, \tau \in [0, T];\ \lambda \in \Pi_{\lambda_0};\ k = 1,\cdots,n).$$

COROLLARY 2.2. *If in addition to the conditions of the lemma the operator* $B^{(p+1)}(\tau; \lambda)$ *is bounded for* $\lambda \in \Pi_{\lambda_0}$ *and* $\lambda \in [0, T]$, *so is the operator* (1.26):

$$\|\Phi_p(\tau; \lambda)\| \leqq c \qquad (\tau \in [0, T];\ \lambda \in \Pi_{\lambda_0}). \tag{2.11}$$

3. *Asymptotic estimates of the approximations.* We consider in conjunction with the operator (1.7) the operator

$$\tilde{U}_p(\tau, \tau_0; \lambda) = \sum_{k=1}^{n} \left\{I + \sum_{k=1}^{p} \lambda^{-s}V_{ks}(\tau)\right\} Y_k^{(p)}(\tau, \tau_0; \lambda)C(\tau_0; \lambda), \tag{2.12}$$

which differs from it by a factor $C(\tau_0; \lambda)$ that is independent of τ. If this factor is bounded for $\lambda \in \Pi_{\lambda_0}$, the operator $\tilde{U}_p(\tau, \tau_0; \lambda)$ satisfies an equation of the same type as (1.13).

We choose $C(\tau_0; \lambda)$ so as to provide a suitable initial value for the operator $\tilde{U}_p(\tau, \tau_0; \lambda)$, thanks to which it will differ little from the evolution operator $U(\tau,\tau_0; \lambda)$ of equation (1.1). Inasmuch as the operator $\tilde{U}_p(\tau, \tau_0; \lambda)$ must by construction serve as a pth approximation to $U(\tau, \tau_0; \lambda)$ it is natural to require that

$$\tilde{U}_p(\tau_0, \tau_0; \lambda) = I + O(\lambda^{-(p+1)}). \tag{2.13}$$

This can be done as follows. From (2.12) we get

$$\tilde{U}_p(\tau_0, \tau_0; \lambda) = \left\{I + \sum_{k=1}^{n} \sum_{s=1}^{p} \lambda^{-s}V_{ks}(\tau_0)\right\}C_p(\tau_0; \lambda).$$

For sufficiently large λ the operator in braces has an inverse which can be calculated by summing an infinite geometric series. We take as $C_p(\tau_0; \lambda)$ the expression

$$C_p(\tau_0; \lambda) = I + \sum_{r=1}^{p} (-1)^r\left[\sum_{k=1}^{n} \sum_{s=1}^{p} \lambda^{-s}V_{ks}(\tau_0)\right]^r, \tag{2.14}$$

which is obtained by replacing this series by its pth partial sum. As a result, condition (2.13) will be satisfied.

LEMMA 2.4. *Suppose the operator function $\tilde{U}(\tau, \tau_0; \lambda)$ satisfies the differential equation*

$$\frac{d\tilde{U}(\tau, \tau_0; \lambda)}{d\tau} = \lambda B(\tau; \lambda)\tilde{U}(\tau, \tau_0; \lambda) + \lambda^{-p}\Psi_p(\tau; \lambda) \qquad (\tau, \tau_0 \in [0, T]; \lambda \in \Pi_{\lambda_0}), \tag{2.15}$$

where $\|\Psi_p(\tau; \lambda)\| \leqq c_1$ $(\tau \in [0, T]; \lambda \in \Pi_{\lambda_0})$.

If the evolution operator $U(\tau, \tau_0; \lambda)$ of equation (1.1) *is bounded*:

$$\|U(\tau, \tau_0; \lambda)\| \leqq c_2 \qquad (\tau, \tau_0 \in [0, T]; \lambda \in \Pi_{\lambda_0}), \tag{2.16}$$

the condition

$$\|\tilde{U}(\tau_0, \tau_0; \lambda) - I\| \leqq c_3\lambda^{-p} \qquad (\tau_0 \in [0, T]; \lambda \in \Pi_{\lambda_0}) \tag{2.17}$$

implies the estimate

$$\|\tilde{U}(\tau, \tau_0; \lambda) - U(\tau, \tau_0; \lambda)\| \leqq c_4\lambda^{-p} \qquad (\tau, \tau_0 \in [0, T]; \lambda \in \Pi_{\lambda_0}). \tag{2.18}$$

And if the operator $U(\tau, \tau_0; \lambda)$ is subject to the stronger condition

$$\|U(\tau, \tau_0; \lambda)\| \leqq Ne^{-\lambda\nu(\tau-\tau_0)} \qquad (\tau, \tau_0 \in [0, T]; \lambda \in \Pi_{\lambda_0}), \tag{2.19}$$

the inequality

$$\|\tilde{U}(\tau_0, \tau_0; \lambda) - I\| \leqq c_3'\lambda^{-(p+1)} \qquad (\tau_0 \in [0, T]; \lambda \in \Pi_{\lambda_0}) \tag{2.20}$$

implies the estimate

$$\|\tilde{U}(\tau, \tau_0; \lambda) - U(\tau, \tau_0; \lambda)\| \leqq c_4'\lambda^{-(p+1)} \qquad (\tau, \tau_0 \in [0, T]; \lambda \in \Pi_{\lambda_0}). \tag{2.21}$$

PROOF. A solution of equation (2.15) can be written in the form

$$\tilde{U}(\tau, \tau_0; \lambda) = U(\tau, \tau_0; \lambda)\tilde{U}(\tau_0, \tau_0; \lambda) + \lambda^{-p}\int_{\tau_0}^{\tau} U(\tau, s; \lambda)\Psi_p(s; \lambda)ds,$$

which implies the inequality

$$\|U(\tau, \tau_0; \lambda) - \tilde{U}(\tau, \tau_0; \lambda)\| \leqq \|U(\tau, \tau_0; \lambda)\|\|\tilde{U}(\tau_0, \tau_0; \lambda) - I\| + c_1\lambda^{-p}\int_{\tau_0}^{\tau}\|U(\tau, s; \lambda)\|ds. \tag{2.22}$$

If conditions (2.16) and (2.17) are satisfied, inequality (2.22) implies that

$$\|U(\tau, \tau_0; \lambda) - \tilde{U}(\tau, \tau_0; \lambda)\| \leqq \lambda^{-p}(c_2c_3 + c_1c_2T).$$

And if conditions (2.19) and (2.20) hold,

$$\|U(\tau, \tau_0; \lambda) - \tilde{U}(\tau, \tau_0; \lambda)\| \leqq Nc_3'\lambda^{-(p+1)} + c_1\lambda^{-p}N\int_{\tau_0}^{\tau} e^{-\lambda\nu(\tau-s)}ds \leqq \lambda^{-(p+1)}N\left(c_3' + \frac{c_1N}{\nu}\right).$$

The lemma is proved.

The final result now follows from Lemmas 2.1–2.4.

THEOREM 2.1. *Suppose the operator* $B_k(\tau)$ $(0 \leqq \tau \leqq T;\ k = 0, 1, 2,\cdots,p)$ *has a continuous derivative of order* $p + 1 - k$ *while the operator* $B^{(p+1)}(\tau;\lambda)$ *is continuous and bounded for* $\tau \in [0, T]$ *and* $\lambda \in \Pi_{\lambda_0}$.

Suppose the operators $V_{ks}(\tau), \Omega_{ks}(\tau)$ $(k = 1,\cdots,n;\ s = 1,\cdots,p)$ *have been calculated by formulas* (1.30), (1.35), (1.32) *and* (1.37) *while the operators* $Y_k(\tau, \tau_0;\lambda)$ $(k = 1,2,\cdots,n)$ *satisfy equations* (2.10) *and the initial condition*

$$Y_k(\tau_0, \tau_0;\lambda) = P_k(\tau_0) \qquad (k = 1,2,\cdots,n).$$

Then, if condition ($C_{\mathfrak{B}}$) *is satisfied, for sufficiently large* λ_0 *and sufficiently small* θ_0 *the operator*

$$\tilde U_p(\tau,\tau_0;\lambda) = \sum_{k=1}^{n}\Big[I + \sum_{s=1}^{p}\lambda^{-s}V_{ks}(\tau)\Big]Y_k^{(p)}(\tau,\tau_0;\lambda) \times\Big\{I + \sum_{r=1}^{p}(-1)^r\Big[\sum_{k=1}^{n}\sum_{s=1}^{p}\lambda^{-s}V_{ks}(\tau_0)\Big]^r\Big\} \tag{2.23}$$

is subject in the domain $\Pi_{\lambda_0} = \Pi'_{\lambda_0,\theta_0}$ *to the estimate*

$$\|\tilde U_p(\tau,\tau_0;\lambda) - U(\tau,\tau_0;\lambda)\| \leqq c_1\lambda^{-(p+1)} \qquad (\tau,\tau_0 \in [0, T];\ \lambda \in \Pi'_{\lambda_0,\theta_0}). \tag{2.24}$$

And if condition ($C_{\mathfrak{H}}$) *is satisfied, for sufficiently large* λ_0 *and* $\alpha > 0$ *the operator*

$$\tilde U_p'(\tau,\tau_0;\lambda) = \sum_{k=1}^{n}\Big[I + \sum_{s=1}^{p}\lambda^{-s}V_{ks}(\tau)\Big]Y_k^{(p)}(\tau,\tau_0;\lambda) \times\Big\{I + \sum_{r=1}^{p-1}(-1)^r\Big[\sum_{k=1}^{n}\sum_{s=1}^{p}\lambda^{-s}V_{ks}(\tau_0)\Big]^r\Big\} \tag{2.25}$$

is subject in the domain $\Pi_{\lambda_0} = \Pi''_{\lambda_0,\alpha}$ *to the estimate*

$$\|\tilde U_p'(\tau,\tau_0;\lambda) - U(\tau,\tau_0;\lambda)\| \leqq c_2\lambda^{-p} \qquad (\tau,\tau_0 \in [0, T];\ \lambda \in \Pi''_{\lambda_0,\alpha}). \tag{2.26}$$

REMARK 2.1. Inequality (2.26) obviously continues to hold if one simplifies expression (2.25) by discarding all terms in it of order not less than $O(\lambda^{-p})$.

REMARK 2.2. In the simplest case $p = 1$, which is the most important for the applications, formula (2.23) takes the form

$$\begin{aligned}\tilde U_1(\tau,\tau_0;\lambda) &= \sum_{k=1}^{n}\Big[I + \frac{1}{\lambda}V_{k1}(\tau)\Big]Y_k^{(1)}(\tau,\tau_0;\lambda)\Big[I - \frac{1}{\lambda}\sum_{k=1}^{n}V_{k1}(\tau_0)\Big] \\ &= \sum_{k=1}^{n}Y_k^{(1)}(\tau,\tau_0;\lambda) \\ &\quad + \frac{1}{\lambda}\sum_{j,k=1}^{n}[V_{j1}(\tau)Y_k^{(1)}(\tau,\tau_0;\lambda) - Y_k^{(1)}(\tau,\tau_0;\lambda)V_{j1}(\tau_0)] + O\Big(\frac{1}{\lambda_2}\Big),\end{aligned} \tag{2.27}$$

while formula (2.25) takes the form

$$\tilde{U}_1'(\tau, \tau_0; \lambda) = \sum_{k=1}^{n} \left(I + \frac{1}{\lambda} V_{k1}(\tau)\right) Y_k^{(1)}(\tau, \tau_0; \lambda).$$

After making the simplification indicated in the preceding remark, we obtain the formula

$$\tilde{U}_1'(\tau, \tau_0; \lambda) = \sum_{k=1}^{n} Y_k^{(1)}(\tau, \tau_0; \lambda) + O(1/\lambda). \tag{2.28}$$

Under condition ($C_{\mathfrak{B}}$) the operator (2.27) satisfies the estimate

$$U(\tau, \tau_0; \lambda) = \tilde{U}_1(\tau, \tau_0; \lambda) + O(1/\lambda^2),$$

while under condition ($C_{\mathfrak{H}}$) the operator (2.28) satisfies the estimate

$$U(\tau, \tau_0; \lambda) = \tilde{U}_1'(\tau, \tau_0; \lambda) + O(1/\lambda).$$

In the latter case to obtain a more exact estimate with error of order $O(\lambda^{-2})$ it is necessary to consider a second approximation, which after the appropriate simplifications takes the form

$$\tilde{U}_2'(\tau, \tau_0; \lambda) = \sum_{k=1}^{n} Y_k^{(2)}(\tau, \tau_0; \lambda)$$
$$+ \frac{1}{\lambda} \sum_{j,k=1}^{n} \{V_{j1}(\tau) Y_k^{(2)}(\tau, \tau_0; \lambda) - Y_k^{(2)}(\tau, \tau_0; \lambda) V_{j1}(\tau_0)\} + O\left(\frac{1}{\lambda^2}\right).$$

4. *Canonical case.* We assume that $B_0(\tau) = i\mathcal{J}\mathcal{H}(\tau)$, where $\mathcal{J}$ is an invertible Hermitian operator satisfying the condition $\mathcal{J}^2 = I$ and $\mathcal{H}(\tau)$ $(0 \leqq \tau \leqq T)$ is a uniformly positive operator in a Hilbert space $\mathfrak{H}$. We have already noted above that the operator $B_0(\tau)$ satisfies condition ($C_{\mathfrak{H}}$) in this case.

Here the evolution operator $U(\tau, \tau_0; \lambda)$ of the equation

$$d\psi/dt = \lambda B_0(\tau)\psi$$

is $\mathcal{J}$-unitary for real λ.

We will now show that under the above conditions the operator $\sum_{k=1}^{n} Y_k^{(1)}(\tau, \tau_0; \lambda)$, viz. the first approximation to the evolution operator, is also $\mathcal{J}$-unitary for real $\lambda \geqq \lambda_0$.

We first note that the operator $Q(\tau, \tau_0)$, which tracks the invariant subspaces $\mathfrak{B}_k(\tau)$ of the operator $iB_0(\tau)$, is $\mathcal{J}$-unitary (see § IV.1), i.e. satisfies the condition

$$Q^*(\tau, \tau_0)\mathcal{J}Q(\tau, \tau_0) = \mathcal{J}. \tag{2.29}$$

Let

$$Y_k^{(1)}(\tau, \tau_0; \lambda) = Q(\tau, \tau_0) X_k(\tau, \tau_0; \lambda).$$

The operator $X_k(\tau, \tau_0; \lambda)$ is a solving operator in the subspace $\mathfrak{B}_k(\tau_0)$ of equation (2.7):

$$\frac{dX_k(\tau, \tau_0; \lambda)}{d\tau} = Q^{-1}(\tau, \tau_0)P_k(\tau)B_0(\tau)P_k(\tau)Q(\tau, \tau_0)X_k(\tau, \tau_0; \lambda). \tag{2.30}$$

The operator $B_0(\tau)$ is by assumption $\mathscr{J}$-skew-Hermitian, i.e.

$$\mathscr{J}B_0(\tau) = -B_0^*(\tau)\mathscr{J}.$$

Multiplying this equality from the left and from the right by the projections $P_k^*(\tau)$ and $P_k(\tau)$ respectively, we reduce it to the form

$$P_k^*(\tau)\mathscr{J}P_k(\tau)B_0(\tau) = -B_0^*(\tau)P_k^*(\tau)\mathscr{J}P_k(\tau).$$

With the use of this relation and the relations

$$Q(\tau, \tau_0)P_k(\tau_0)Q^{-1}(\tau, \tau_0) = P_k(\tau); \qquad Q^{-1*}(\tau,\tau_0)P_k^*(\tau_0)Q^*(\tau, \tau_0) = P_k^*(\tau)$$

we deduce that

$$\begin{aligned}
&P_k^*(\tau_0)\mathscr{J}P_k(\tau_0)Q^{-1}(\tau, \tau_0)P_k(\tau)B_0(\tau)P_k(\tau)Q(\tau, \tau_0)\\
&= P_k^*(\tau_0)Q^*(\tau, \tau_0)\mathscr{J}Q(\tau, \tau_0)P_k(\tau_0)Q^{-1}(\tau, \tau_0)P_k(\tau)B_0(\tau)P_k(\tau)Q(\tau, \tau_0)\\
&= P_k^*(\tau_0)Q^*(\tau, \tau_0)\mathscr{J}P_k(\tau)B_0(\tau)P_k(\tau)Q(\tau, \tau_0)\\
&= Q^*(\tau, \tau_0)P_k^*(\tau)\mathscr{J}P_k(\tau)B_0(\tau)P_k(\tau)Q(\tau, \tau_0)\\
&= -\,Q^*(\tau, \tau_0)P_k^*(\tau)B_0^*(\tau)P_k^*(\tau)\mathscr{J}P_k(\tau)Q(\tau, \tau_0)\\
&= -\,Q^*(\tau, \tau_0)P_k^*(\tau)B_0^*(\tau)P_k^*(\tau)Q^{-1*}(\tau, \tau_0)P_k^*(\tau_0)Q^*(\tau, \tau_0)\mathscr{J}Q(\tau, \tau_0)P_k(\tau_0)\\
&= -\,[Q^{-1}(\tau, \tau_0)P_k(\tau)B_0(\tau)P_k(\tau)Q(\tau, \tau_0)]^*P_k^*(\tau_0)\mathscr{J}P_k(\tau_0).
\end{aligned}$$

This equality means that the coefficient of equation (2.30) is a $P_k^*(\tau_0)\mathscr{J}P_k(\tau_0)$-Hermitian operator. Therefore the evolution operator $X_k(\tau, \tau_0; \lambda)$ of this equation is $P_k^*(\tau_0)\,\mathscr{J}P_k(\tau_0)$-unitary for real λ, i.e.

$$X_k^*(\tau, \tau_0; \lambda)P_k^*(\tau_0)\mathscr{J}P_k(\tau_0)X_k(\tau, \tau_0; \lambda) = P_k^*(\tau_0)\mathscr{J}P_k(\tau_0) \qquad (\lambda \geqq \lambda_0). \tag{2.31}$$

We note that the $\mathscr{J}$-Hermitianness of the operator $iB_0(\tau_0)$ implies the $\mathscr{J}$-Hermitianness of its spectral projections.

Therefore

$$P_j^*(\tau_0)\mathscr{J}P_k(\tau_0) = \mathscr{J}P_j(\tau_0)P_k(\tau_0) = 0 \qquad (j \neq k).$$

Taking into account these relations, we deduce from (2.31) the equality

$$\sum_j X_j^*(\tau, \tau_0; \lambda)\mathscr{J} \sum_k X_k(\tau, \tau_0; \lambda) = \mathscr{J},$$

denoting the $\mathscr{J}$-unitariness of the operator $\sum_k X_k(\tau, \tau_0; \lambda)$. Since a product of $\mathscr{J}$-unitary operators is $\mathscr{J}$-unitary, the operator

$$\sum_k Y_k^{(1)}(\tau, \tau_0; \lambda) = Q(\tau, \tau_0) \sum_k X_k(\tau, \tau_0; \lambda)$$

is $\mathcal{J}$-unitary.

5. *Example*. We consider as an example the equation

$$d\psi/dt = i\lambda \mathcal{J}\mathcal{H}_0(t)\psi \tag{2.32}$$

in a Hilbert space $\mathfrak{H}$, assuming that $\mathcal{H}_0(t)$ is a twice continuously differentiable in t uniformly positive periodic operator function: $\mathcal{H}_0(t+T) = \mathcal{H}_0(t)$.

The spectrum of the $\mathcal{J}$-positive operator $\mathcal{J}\mathcal{H}_0(t)$ lies the real axis. We assume that this spectrum consists of a finite number of points:

$$\sigma(\mathcal{J}\mathcal{H}_0(\tau)) = \{\mu_1(\tau), \cdots, \mu_n(\tau)\},$$

satisfying for each $\tau \in [0, T]$ the condition

$$\mu_j(\tau) \neq \mu_k(\tau) \qquad (j \neq k). \tag{2.33}$$

In this elementary case we will find under the assumption of large λ an approximate expression for the monodromy operator $U(T; \lambda)$.

In the invariant subspace $\mathfrak{H}_k(\tau)$ corresponding to a point $\mu_k(\tau)$ the spectrum of the operator $\mathcal{J}\mathcal{H}_0(\tau)$ consists of just this point. Since this operator is $\mathcal{J}$-positive,

$$\mathcal{J}\mathcal{H}_0(\tau)P_k(\tau) = \mu_k(\tau)P_k(\tau).$$

Therefore

$$\mathcal{J}\mathcal{H}_0(\tau) = \sum_{k=1}^{n} \mu_k(\tau)P_k(\tau).$$

Equation (2.30) now takes the form

$$\frac{dX_k}{d\tau} = i\lambda\mu_k(\tau)Q^{-1}(\tau, \tau_0)P_k(\tau)Q(\tau, \tau_0)X_k = i\lambda\mu_k(\tau)P_k(\tau_0)X_k.$$

Consequently,

$$X_k(\tau, \tau_0; \lambda) = e^{i\lambda\int_{\tau_0}^{\tau} \mu_k(s)\,ds} P_k(\tau_0)$$

and, finally,

$$\tilde{U}_1'(\tau, \tau_0; \lambda) = Q(\tau, \tau_0)\sum_{k=1}^{n} e^{i\lambda\int_{\tau_0}^{\tau} \mu_k(s)\,ds} P_k(\tau_0) + O\left(\frac{1}{\lambda}\right).$$

From this result we obtain the following representation for the monodromy operator:

$$U(T; \lambda) = Q(T, 0)\sum_{k=1}^{n} e^{i\lambda M_k} P_k(0) + O\left(\frac{1}{\lambda}\right) \quad (\mathrm{Re}\lambda \geqq \lambda_0;\ |\mathrm{Im}\lambda| \leqq \alpha), \tag{2.34}$$

where $M_k = \int_0^T \mu_k(s)ds$.

When the space $\mathfrak{H}$ is finite dimensional, this representation permits one to investigate the spectrum of the monodromy operator for large λ and to establish the presence of stability bands of equation (2.32) for arbitrarily large real λ.

The fact that the representation (2.34) is also valid for complex λ in a neighborhood of the ray $\lambda \geqq \lambda_0$ plays an important role in such an investigation.

Another important problem that arises in connection with equation (2.32) is the boundary problem with a condition of the form $E\psi(0) + F\psi(T) = 0$. Such a problem is regarded (for reasons which we do not cite here; see I. C. Gohberg and M. G. Kreĭn [2]) as *selfadjoint* if

$$E^* \mathscr{J} E = F^* \mathscr{J} F. \tag{2.35}$$

When the operator F is invertible it is possible without loss of generality to take $F = I$, in which case (2.35) becomes the $\mathscr{J}$-unitariness condition $E^* \mathscr{J} E = \mathscr{J}$ for the operator E.

We thus arrive at a boundary problem for equation (2.32) with the condition

$$E\psi(0) + \psi(T) = 0, \tag{2.36}$$

where E is a $\mathscr{J}$-unitary operator.

Substituting the expression $\psi(T) = U(T;\,\lambda)\,\psi(0)$ into (2.36) and using (2.34), we obtain the relation

$$\left[E + Q(T,0)\sum_{k=1}^{n} e^{i\lambda M_k} P_k(0)\right]\psi(0) = O(1/\lambda). \tag{2.37}$$

This expression can be used to investigate the spectrum of the boundary problem in question.

For example, in the finite-dimensional case it is known that the leading term in an asymptotic expansion of the spectrum does not depend on the choice of the $\mathscr{J}$-unitary operator E. This permits one to put $E = Q(T,\,0)$ and reduce equation (2.37) to the particularly simple form

$$\psi(0) + \sum_{k=1}^{n} e^{i\lambda M_k} P_k(0)\,\psi(0) = O(1/\lambda)$$

or, after multiplying by $P_j(0)$,

$$(1 + e^{i\lambda M_j})\,P_j(0)\,\psi(0) = O(1/\lambda).$$

In order for this relation to be satisfied when $\psi(0) \neq 0$ it is necessary (though not sufficient) that the following equality hold for some j ($j = 1, \cdots, n$):

$$1 + e^{i\lambda M_j} = O(1/\lambda).$$

Thus our boundary problem has n possible series of eigenvalues, which are defined by the asymptotic formula

$$\lambda_{jk} = \frac{2k+1}{\int_0^T \mu_j(s)\,ds}\,\pi + O\left(\frac{1}{\lambda}\right). \tag{2.38}$$

An additional investigation would be required to prove the fact that formula (2.38) actually represents eigenvalues for large k. Such an investigation, which essentially makes use of the assumption (2.33) and the asymptotic expression (2.34), has been carried out in the book of I. C. Gohberg and M. G. Kreĭn [2].

§ 3. The equation with a rapidly oscillating coefficient

1. *Description of the method.* In this section we again consider the equation

$$d\psi/d\tau = \lambda B(\tau;\lambda)\psi, \tag{3.1}$$

but under somewhat different assumptions on the coefficient.

We assume that

$$B(\tau;\lambda) = B_0 + \sum_{k=1}^{p} \lambda^{-k} B_k(\tau;\lambda) + \lambda^{-(p+1)} B^{(p+1)}(\tau;\lambda). \tag{3.2}$$

It is assumed for the sake of simplicity that the operator B_0 is a constant. As before, we will suppose that it satisfies one of the conditions ($C_{\mathfrak{B}}$) or ($C_{\mathfrak{H}}$) and, in addition, has a spectrum with a decomposition $\sigma(B_0) = \bigcup_{k=1}^{n} \sigma_k$ into spectral sets.

We further assume that the following representation is valid for sufficiently large real $\lambda \geqq \lambda_0$:

$$B_k(\tau;\lambda) = \sum_m B_k^{(m)}(\tau) e^{im\lambda\omega\tau} \qquad (k = 1, \cdots, p), \tag{3.3}$$

the coefficients of which are continuously differentiable a sufficient number of times. In order to avoid convergence questions we will suppose that only a finite number ν of terms in the expansion (3.3) are different from zero.

Finally, we assume that the operator function $B^{(p+1)}(\tau;\lambda)$ is continuous in τ and bounded:

$$\|B^{(p+1)}(\tau;\lambda)\| \leqq c \qquad (\tau \in [0, T];\ \lambda \geqq \lambda_0). \tag{3.4}$$

As in §1, we will seek an approximate expression for the evolution operator $U(\tau, \tau_0; \lambda)$ of equation (3.1) in the form

$$\tilde{U}_p(\tau, \tau_0; \lambda) = \sum_{k=1}^{n} \left\{ \sum_{s=0}^{p} \lambda^{-s} V_{ks}(\tau;\lambda) \right\} Y_k^{(p)}(\tau, \tau_0; \lambda), \tag{3.5}$$

where $Y_k^{(p)}(\tau, \tau_0; \lambda)$ satisfies in the subspace $\mathfrak{B}_k = P_k\mathfrak{B}$ the equation

$$\frac{dY_k^{(p)}}{d\tau} = \lambda \left\{ P_k B_0 + \sum_{s=1}^{p} \lambda^{-s} \Omega_{ks}(\tau;\lambda) \right\} Y_k^{(p)} \tag{3.6}$$

and the initial condition

$$Y_k(\tau_0, \tau_0; \lambda) = P_k. \tag{3.7}$$

But in accordance with expansion (3.3) we assume that the following analogous expansions hold:

$$V_{ks}(\tau;\lambda) = \sum_m V_{ks}^{(m)}(\tau)e^{im\lambda\omega\tau} \qquad (\lambda \geqq \lambda_0). \tag{3.8}$$

We will see that these expansions consist of a finite number of terms.

The coefficients $V_{ks}^{(m)}$ and Ω_{ks} must be chosen so that the operator $\tilde{U}(\tau, \tau_0; \lambda)$ satisfies an equation of the form

$$\frac{d\tilde{U}(\tau, \tau_0; \lambda)}{d\tau} = \lambda B(\tau; \lambda)\, \tilde{U}(\tau, \tau_0; \lambda) + \lambda^{-p}\Phi_p(\tau; \lambda). \tag{3.9}$$

Proceeding in the same way as in § 1 and, in addition, isolating the coefficients of the functions $e^{im\lambda\omega\tau}$, we obtain the system of relations

$$(B_0 - im\omega I)\, V_{k0}^{(m)}(\tau)P_k - V_{k0}^{(m)}(\tau)P_kB_0 = 0 \tag{3.10}$$

$$(k = 1, 2, \dots, n;\ m = 0, \pm 1, \dots);$$

$$(B_0 - im\omega I)V_{k1}^{(m)}(\tau)P_k - V_{k1}^{(m)}(\tau)P_kB_0$$

$$= V_{k0}^{(m)}(\tau)\Omega_{k1}(\tau)P_k + V'^{(m)}_{k0}(\tau)P_k - \sum_r B_1^{(m-r)}(\tau)V_{k0}^{(r)}(\tau)P_k \tag{3.11}$$

$$(k = 1, 2, \cdots, n;\ m = 0, \pm 1, \cdots);$$

$$(B_0 - im\omega I)\, V_{ks}^{(m)}(\tau)P_k - V_{ks}^{(m)}(\tau)P_kB_0$$

$$= V_{k0}^{(m)}(\tau)\Omega_{ks}(\tau)P_k + V'^{(m)}_{k,s-1}(\tau)P_k + \sum_{j=1}^{s-1} V_{k,s-j}^{(m)}(\tau)\Omega_{kj}(\tau)P_k \tag{3.12}$$

$$- \sum_{j=0}^{s-1}\sum_r B_{s-j}^{(m-r)}(\tau)\, V_{kj}^{(r)}(\tau)P_k$$

$$(s = 2, \cdots, p;\ k = 1, 2, \cdots, n\ ;\ m = 0, \pm 1, \cdots).$$

The following expression is obtained for the function $\Phi_p(\tau; \lambda)$ in this connection:

$$\Phi_p(\tau; \lambda) = \sum_{k=1}^{n}\Big\{ \sum_m V_{kp}^{(m)}(\tau)e^{im\lambda\nu\tau}$$

$$+ \sum_{s=p+1}^{2p} \lambda^{-(s-p-1)} \sum_{j=1}^{p} V_{ks-j}(\tau; \lambda)\Omega_{kj}(\tau)$$

$$- \sum_{s=p+1}^{2p} \lambda^{-(s-p-1)} \sum_{j=1}^{p} B_{s-j}(\tau; \lambda)V_{kj}(\tau; \lambda) \tag{3.13}$$

$$- B^{(p+1)}(\tau; \lambda) \sum_{s=0}^{p} \lambda^{-s}V_{ks}(\tau; \lambda)\Big\}\ Y_k^{(p)}(\tau, \tau_0; \lambda).$$

If we manage to determine all of the coefficients $V_{ks}^{(m)}$ and Ω_{ks} from equations (3.10)–(3.12), a subsequent investigation and estimate of the deviation of the operator $\tilde{U}(\tau, \tau_0; \lambda)$ from the solving operator of equation (3.1) can be carried out in exactly the same way as in § 2. We will therefore confine our attention to equations (3.10)–(3.12).

Our conduct of this investigation depends on whether or not the spectrum $\sigma(B_0)$ contains points that are congruent modulo $i\omega$ (the resonance case).

2. *Nonresonance case.* Equation (3.10) is satisfied for $m = 0$ if we put $V_{k0}^{(0)}(\tau) \equiv P_k$ $(k = 1,\cdots,n)$. But if $m \neq 0$, the fact that the spectra of the operators B_0 and $B_0 - im\omega I$ do not intersect implies by virtue of the results of § I.3 that this equation has only the trivial solution, and we put

$$V_{k0}^{(m)}(\tau) \equiv V_{k0}^{(m)}(\tau)P_k \equiv 0 \qquad (m \neq 0; k = 1,\cdots,n).$$

From relation (3.11) for $m = 0$ we obtain, after multiplying it from the left by a projection $P_{k'}$, the equation

$$B_0P_{k'}V_{k1}^{(0)}(\tau)P_k - P_{k'}V_{k1}^{(0)}(\tau)P_kB_0 = P_{k'}\Omega_{k1}(\tau)P_k - P_{k'}B_1^{(0)}(\tau)P_k.$$

If $k = k'$, this equation is satisfied for

$$\Omega_{k1}(\tau) = P_k\Omega_{k1}(\tau)P_k = P_kB_1^{(0)}(\tau)P_k \quad \text{and} \quad P_kV_{k1}^{(0)}P_k = 0.$$

When $k \neq k'$ it has a unique solution, which can be found by using the formulas of § I.3 and then putting

$$V_{k1}^{(0)}P_k = \sum_{k'} P_{k'}V_{k1}^{(0)}P_k.$$

At present the right side of equation (3.11) has been determined. When $m \neq 0$ equation (3.11) has a unique solution (since there is no resonance), which can be found by using the same formulas of § I.3.

The treatment of equations (3.12) is completely analogous.

3. *Resonance case.* We consider this case under additional simplifying assumptions. We assume that if the sets $\sigma_k(B_0) + im\omega$ and $\sigma_{k'}(B_0)$ intersect for some k, m and $k' = \phi(k, m)$, each of the spectral sets $\sigma_k(B_0)$ and $\sigma_{k'}(B_0)$ consists of a single point λ_k and $\lambda_{k'}$ respectively, so that

$$\lambda_{k'} - \lambda_k = im\omega. \tag{3.14}$$

In addition, we assume that the invariant subspaces of B_0 corresponding to these spectral sets are eigenspaces:

$$B_0P_k = \lambda_kP_k, \quad B_0P_{k'} = \lambda_{k'}P_{k'}. \tag{3.15}$$

Let us multiply equation (3.10) from the left by the projection $P_{k'}$. In this way we obtain the equation

$$(B_0 - im\omega I)P_{k'}V_{k0}^{(m)}P_k - P_{k'}V_{k0}^{(m)}P_kB_0 = 0, \tag{3.16}$$

which is treated in exactly the same way as in the resonance case if $k' \neq \phi(k, m)$.

Thus we put

$$V_{k0}^{(0)} = P_k \qquad (k = 1,\cdots,n) \tag{3.17}$$

and

$$P_{k'}V_{k0}^{(m)}P_k = 0 \qquad (m \neq 0;\, k' \neq \phi(k, m)). \tag{3.18}$$

But when $k' = \phi(k, m)$ equation (3.16) becomes an identity, and the operators

$$Z_{k'k}^{(m)} = P_{k'}V_{k0}^{(m)}P_k$$

remain indeterminate for the time being.

Multiplying equation (3.11) from the left by $P_{k'}$, we now obtain the equation

$$\begin{aligned}(B_0 - im\omega I)&P_{k'}V_{k1}^{(m)}(\tau)P_k - P_{k'}V_{k1}^{(m)}(\tau)P_kB_0 \\ &= (Z_{k'k}^{(m)})' - \sum_{r\neq 0}\Big[\sum_{k''=1}^{n} P_{k'}B_1^{(m-r)}(\tau)P_{k''}Z_{k''k}^{(r)}\Big] \\ &\quad - P_{k'}B_1^{(m)}(\tau)P_k + Z_{k'k}^{(m)}\Omega_{k1}P_k.\end{aligned} \tag{3.19}$$

When $m = 0$ and $k' = k$ we put, as before,

$$P_kV_{k1}^{(0)}P_k = 0 \tag{3.20}$$

and obtain the relation

$$\Omega_{k1} = P_k\Omega_{k1}P_k = \sum_{r\neq 0}\Big(\sum_{k''} P_kB_1^{(-r)}P_{k''}Z_{k''k}^{(r)}\Big) + P_kB_1^{(0)}(\tau)P_k. \tag{3.21}$$

When $m = 0$ and $k' \neq k$ equation (3.19) reduces to the equation

$$B_0P_{k'}V_{k1}^{(0)}(\tau)P_k - P_{k'}V_{k1}^{(0)}(\tau)P_kB_0 = -\sum_r P_{k'}B_1^{(-r)}(\tau)V_{k0}^{(r)}(\tau)P_k, \tag{3.22}$$

which, if the right side is known, has a unique solution determined, as above, by the formulas of § I.3. An equation of the same type is also obtained when $m \neq 0$ if $k' \neq \phi(k, m)$:

$$(B_0 - im\nu I)P_{k'}V_{k1}^{(m)}P_k - P_{k'}V_{k1}^{(m)}P_kB_0 = -\sum_r P_{k'}B_1^{(m-r)}V_{k0}^{(r)}P_k. \tag{3.23}$$

It remains to consider the case when $k' = \phi(k, m)$. In this case, as in the case of equation (3.16), the left side of equation (3.19) vanishes identically. We therefore obtain the relation

$$(Z_{k'k}^{(m)})' - \sum_{r\neq 0}\Big(\sum_{k''=1}^{n} P_{k'}B_1^{(m-r)}(\tau)P_{k''}Z_{k''k}^{(r)}\Big) - P_{k'}B_1^{(m)}(\tau)P_k + Z_{k'k}^{(m)}\Omega_{k1}P_k = 0,$$

which together with (3.21) leads to a system of differential equations for the unknown operators $Z_{k'k}^{(m)}$:

$$\begin{aligned}(Z_{k'k}^{(m)})' &- \sum_{r\neq 0}\Big[\sum_{k''=1}^{n} P_{k'}B_1^{(m-r)}(\tau)P_{k''}Z_{k''k}^{(r)}\Big] \\ &+ Z_{k'k}^{(m)}P_kB_1^{(0)}(\tau)P_k + Z_{k'k}^{(m)}\sum_{r\neq 0}\Big[\sum_{k''=1}^{n} P_kB_1^{(-r)}(\tau)P_{k''}Z_{k''k}^{(r)}\Big] \\ &- P_{k'}B_1^{(m)}(\tau)P_k = 0 \qquad (m = 0, \pm 1,\cdots;\, k' = \phi(k, m)).\end{aligned} \tag{3.24}$$

System (3.24) is a system of Riccati operator equations for fixed k with $m = 0, \pm 1, \cdots$ and $k' = \phi(k, m)$.

In order for the principal asymptotic part of $\bar{U}(\tau_0, \tau_0; \lambda)$ to be the identity operator:

$$\lim_{\lambda\to\infty} \bar{U}(\tau_0, \tau_0; \lambda) = I,$$

it suffices, as is easily verified, to take $Z_{k'k}^{(m)}(\tau_0) = 0$.

Thus system (3.24) should be considered under zero initial conditions. We remark that it then generally has a nonzero solution provided

$$P_{k'}B_1^{(m)}P_k \neq 0 \qquad (k' = \phi(k, m)).$$

We note that if for a given k there exists only one index k' that is representable in the form $k' = \phi(k, m_1)$, system (3.24) reduces to the single equation

$$\begin{aligned}(Z_{k'k}^{(m_1)})' + Z_{k'k}^{(m_1)}P_kB_1^{(-m_1)}P_{k'}Z_{k'k}^{(m_1)} + Z_{k'k}^{(m_1)}P_kB_1^{(0)}P_k \\ - P_{k'}B_1^{(0)}P_{k'}Z_{k'k}^{(m_1)} - P_{k'}B_1^{(m_1)}P_k = 0.\end{aligned} \tag{3.25}$$

In general, system (3.24) need not have solutions that are extendable onto the whole interval $[0, T]$; but if a solution exists, then, after finding all of the operators $Z_{k'k}^{(m)}$, we can determine the operators

$$V_{k0}^{(m)} = V_{k0}^{(m)}P_k = \sum_{k'} Z_{k'k}^{(m)}. \tag{3.26}$$

Furthermore, the right sides of equations (3.22) and (3.23) become known and it thus becomes possible to determine all of the operators $P_{k'}V_{k1}^{(m)}P_k$ except those which correspond to $k' = \phi(k, m)$. The calculations at subsequent steps ($s = 2, \cdots, p$) are carried out in exactly the same way. At each step the operators $P_{k'}V_{k,s-1}^{(m)}P_k$ left undefined at the preceding step are determined. We note, incidentally, that the differential equations for these operators turn out, as is easily verified, to be linear and not of second degree, as in the case $s = 1$. The last step ($s = p$) can be carried out by choosing the remaining indeterminate operators arbitrarily, for example, by putting

$$P_{k'}V_{kp}^{(m)}P_k = 0 \qquad (k' = \phi(k, m)).$$

For these operators are not subject to any other relations, as was the case for $s < p$.

The final construction of an operator approximating the evolution operator $U(\tau, \tau_0; \lambda)$ is carried out in the same way as in § 2.

We will therefore construct only the first approximation. Consider the operator

$$U_1(\tau, \tau_0; \lambda) = \bar{U}_1(\tau, \tau_0; \lambda)C(\tau_0). \tag{3.27}$$

From (3.5) we get

$$U_1(\tau_0, \tau_0; \lambda) = \sum_{k=1}^{n}\left\{V_{k0}(\tau_0, \tau_0; \lambda) + \frac{1}{\lambda} V_{k1}(\tau_0, \tau_0; \lambda)\right\} P_k C(\tau_0)$$

$$= \sum_{k=1}^{n}\left\{\sum_{m\neq 0} V_{k0}^{(m)}(\tau_0, \tau_0; \lambda)e^{im\lambda\nu\tau}\right.$$

$$\left. + V_{k0}^{(0)}(\tau_0, \tau_0; \lambda) + \frac{1}{\lambda} V_{k1}(\tau_0, \tau_0; \lambda)\right\} P_k C(\tau_0)$$

$$=\left(I + \frac{1}{\lambda}\sum_{k} V_{k1}(\tau_0, \tau_0; \lambda)P_k\right) C(\tau_0).$$

It therefore suffices to take

$$C(\tau_0) = I - \frac{1}{\lambda}\sum_{k} V_{k1}(\tau_0, \tau_0; \lambda)P_k.$$

Substituting this expression into (3.27) and neglecting terms of order $O(\lambda^{-1})$, we obtain the formula

$$U(\tau, \tau_0; \lambda) = \sum_{k=1}^{n}\left[P_k + \sum_{m\neq 0} V_{k0}^{(m)}(\tau)e^{im\omega\lambda\tau}\right] Y_k^{(1)}(\tau, \tau_0; \lambda) + O\left(\frac{1}{\lambda}\right). \tag{3.28}$$

We note that this expression contains (in contrast to the nonresonance case) rapidly oscillating terms with amplitudes that do not tend to zero as $\lambda \to \infty$.

4. *Example.* Let us illustrate the method presented above with a simple example. We consider the scalar equation of second order

$$y''(t) + [1 - \varepsilon\phi(\varepsilon t) \cos 2t]y(t) = 0, \tag{3.29}$$

which contains a small, slowly varying coefficient $\varepsilon\phi(\varepsilon t)$. When this coefficient is a constant, equation (3.29) is a special case of Mathieu's equation.

We reduce (3.29) to a system of form (3.1). To this end we introduce a new variable $\tau = \varepsilon t$ and put $\lambda = 1/\varepsilon$. Then

$$\frac{d^2y}{d\tau^2} + \lambda^2\left[1 - \frac{1}{\lambda}\phi(\tau)\cos 2\lambda\tau\right]y = 0. \tag{3.30}$$

Equation (3.30) is equivalent to the system

$$\begin{cases} \dfrac{dy}{d\tau} = \lambda y_1, \\ \dfrac{dy_1}{d\tau} = -\lambda\left[1 - \dfrac{1}{\lambda}\phi(\tau)\cos 2\lambda\tau\right]y, \end{cases}$$

which can be written in the form of a single equation with matrix coefficients for the vector $\omega = \begin{pmatrix} y \\ y_1 \end{pmatrix}$:

$$\frac{d\omega}{d\tau} = \left\{\lambda\begin{pmatrix} 0 & 1 \\ -1 & 0 \end{pmatrix} + \phi(\tau)\begin{pmatrix} 0 & 0 \\ 1 & 0 \end{pmatrix}\cos 2\lambda\tau\right\}\omega. \tag{3.31}$$

In this equation we make the substitution

$$\omega = \frac{1}{\sqrt{2}}\begin{pmatrix}1 & 1\\ i & -i\end{pmatrix}\psi$$

in order to reduce its principal part to diagonal form. We write the resultant equation in the form

$$\frac{d\psi}{d\tau} = \left\{\lambda i\begin{pmatrix}1 & 0\\ 1 & -0\end{pmatrix} + i\frac{\phi(\tau)}{4}\begin{pmatrix}-1 & -1\\ 1 & 1\end{pmatrix}(e^{2i\lambda\tau} + e^{-2i\lambda\tau})\right\}\psi. \tag{3.32}$$

This equation is of form (3.1) for $p = 1$, with

$$\left.\begin{gathered}\nu = 2,\ B_0 = i\begin{pmatrix}1 & 0\\ 0 & -1\end{pmatrix},\ B_1^{(1)} = B_1^{(-1)} = i\frac{\phi(\tau)}{4}\begin{pmatrix}-1 & -1\\ 1 & 1\end{pmatrix},\\ B_1^{(0)} = 0,\quad B^{(2)}(\tau;\lambda) = 0.\end{gathered}\right\} \tag{3.33}$$

The operator B_0 obviously satisfies condition (C_δ). Its spectrum consists of the two points $\lambda_1 = i$ and $\lambda_2 = -i$, to which correspond the spectral projections

$$P_1 = \begin{pmatrix}1 & 0\\ 0 & 0\end{pmatrix},\qquad P_2 = \begin{pmatrix}0 & 0\\ 0 & 1\end{pmatrix}. \tag{3.34}$$

Since $\lambda_1 - \lambda_2 = 2i = \omega i$, we have only the resonance case to deal with.

We will find the first approximation for the solving operator of equation (3.32). Using formulas (3.17) and (3.18), we get

$$\left.\begin{gathered}V_{10}^{(0)} = P_1 = \begin{pmatrix}1 & 0\\ 0 & 0\end{pmatrix},\quad V_{20}^{(0)} = P_2 = \begin{pmatrix}0 & 0\\ 0 & 1\end{pmatrix},\\ P_1V_{10}^{(1)}P_1 = 0,\ P_2V_{10}^{(1)}P_1 = 0,\ P_2V_{20}^{(-1)}P_2 = 0,\\ P_1V_{20}^{(-1)}P_2 = 0\end{gathered}\right\} \tag{3.35}$$

and thus

$$V_{10}^{(1)} = 0,\quad V_{20}^{(-1)} = 0. \tag{3.36}$$

In addition, from (3.18) it follows that

$$P_1V_{12}^{(-1)}P_1 = 0,\quad P_2V_{20}^{(1)}P_2 = 0.$$

We put $P_2V_{10}^{(-1)}P_1 = Z_1$ and $P_1V_{20}^{(1)}P_2 = Z_2$. Then

$$V_{10}^{(-1)} = Z_1,\quad V_{20}^{(1)} = Z_2. \tag{3.37}$$

Equation (3.25) for the operator Z_1 has the form

$$Z_1' + Z_1P_1B_1^{(1)}P_2Z_1 - P_2B_1^{(-1)}P_1 = 0. \tag{3.38}$$

The operator Z_1 maps the subspace $\mathfrak{B}_1$ into $\mathfrak{B}_2$. Its matrix must therefore be proportional to the matrix $\left(\begin{smallmatrix}0 & 0\\ 1 & 0\end{smallmatrix}\right)$. Setting

$$Z_1(\tau,\tau_0) = iz(\tau,\tau_0)\begin{pmatrix}0 & 0\\ 1 & 0\end{pmatrix}, \tag{3.39}$$

we obtain from (3.38) and (3.39), after some simple calculations, the equation

$$z'(\tau, \tau_0) + \phi(\tau)[z^2(\tau, \tau_0) - 1] = 0.$$

Its solution satisfying the condition $z(\tau_0, \tau_0) = 0$ has the form

$$z(\tau, \tau_0) = -\tanh \int_{\tau_0}^{\tau} \phi(s)ds. \tag{3.40}$$

It can be shown in an analogous way that

$$Z_2(\tau, \tau_0) = iz(\tau, \tau_0)\begin{pmatrix} 0 & -1 \\ 0 & 0 \end{pmatrix}, \tag{3.41}$$

where $z(\tau, \tau_0)$ is the same function as (3.40).

Using formula (3.21), we get

$$\Omega_{11} = \phi(\tau)z(\tau)\begin{pmatrix} 1 & 0 \\ 0 & 0 \end{pmatrix}, \quad \Omega_{21} = \phi(\tau)z(\tau)\begin{pmatrix} 0 & 0 \\ 0 & 1 \end{pmatrix}. \tag{3.42}$$

Thus equation (3.6) takes the form

$$dY_1^{(1)}/d\tau = (i\lambda + \phi(\tau)z(\tau, \tau_0))P_1Y_1^{(1)},$$

which implies that

$$Y_1^{(1)}(\tau, \tau_0) = e^{i\lambda(\tau-\tau_0)}e^{\int_{\tau_0}^{\tau}\phi(s)z(s,\tau_0)ds}P_1 = \frac{e^{i\lambda(\tau-\tau_0)}}{\cosh \int_{\tau_0}^{\tau}\phi(s)ds}P_1. \tag{3.43}$$

Analogously, we get

$$Y_2^{(1)}(\tau, \tau_0) = \frac{e^{-i\lambda(\tau-\tau_0)}}{\cosh \int_{\tau_0}^{\tau}\phi(s)ds}P_2. \tag{3.44}$$

Substituting these expressions into (3.28), we obtain the formula

$$U(\tau, \tau_0; \lambda) = (P_1 + Z_1e^{-2i\lambda\tau})\frac{e^{i\lambda(\tau-\tau_0)}}{\cosh \int_{\tau_0}^{\tau}\phi(s)ds}$$
$$+ (P_2 + Z_2e^{2i\lambda\tau})\frac{e^{-i\lambda(\tau-\tau_0)}}{\cosh \int_{\tau_0}^{\tau}\phi(s)ds} + O\left(\frac{1}{\lambda}\right),$$

which, after returning to the original notation, permits us to approximately calculate the fundamental matrix $W(t, t_0; \varepsilon)$ for the solutions of the original equation (3.29):

$$W(t, t_0; \varepsilon)$$
$$= \left\{\begin{pmatrix} 1 & 0 \\ 0 & 1 \end{pmatrix}\cos(t - t_0) + \begin{pmatrix} 0 & 1 \\ -1 & 0 \end{pmatrix}\sin(t - t_0)\right.$$
$$\left. - \tanh \varepsilon \int_{t_0}^{t} \phi(\varepsilon s)ds \left[\begin{pmatrix} 0 & 1 \\ 1 & 0 \end{pmatrix}\cos(t + t_0) + \begin{pmatrix} 1 & 0 \\ 0 & -1 \end{pmatrix}\sin(t + t_0)\right]\right\}$$
$$\times \frac{1}{\cosh \varepsilon \int_{t_0}^{t}\phi(\varepsilon s)ds} + O(\varepsilon) \qquad (0 \leqq t_0; t \leqq T/\varepsilon).$$

We note that if we had considered in place of (3.29) the equation

$$y'' + [1 - \varepsilon\phi(\varepsilon t)\cos 2\omega t]y(t) = 0$$

with $\omega \neq 1$, we would have encountered the nonresonance case. Some simple calculations show that in this case

$$W(t, t_0; \varepsilon) = \begin{pmatrix} 1 & 0 \\ 0 & 1 \end{pmatrix}\cos(t - t_0) + \begin{pmatrix} 0 & 1 \\ -1 & 0 \end{pmatrix}\sin(t - t_0) + O(\varepsilon).$$

EXERCISES

1. Consider the inhomogeneous equation

$$d\psi/d\tau = \lambda B(\tau;\lambda)\psi + f(\tau;\lambda)e^{i\lambda\theta(\tau)}, \tag{0.1}$$

where $B(\tau;\lambda)$ is the same operator as (1.9) and

$$f(\tau;\lambda) = \sum_{k=0}^{\infty} f_k(\tau)\lambda^{-k} + O(\lambda^{-(p+1)}).$$

Find an asymptotic representation for the solutions of this equation.

The construction of asymptotic approximations for such an equation essentially depends on whether the values of the function $k(\tau) = \theta'(\tau)$ for certain τ can appear in the spectrum of the operator $B_0(\tau)$ (resonance) or not (absence of resonance) (see S. F. Feščenko, N. I. Škil' and L. D. Nikolenko [1], Ju. L. Daleckiĭ and S. G. Kreĭn [1] and Ju. L. Daleckiĭ [1, 2]).

Hint. If the points of the curve $\lambda = k(\tau)$ $(0 \leqq \tau \leqq T)$ appear in the spectral sets $\sigma_j(\tau)$ $(j = 1,\cdots, s)$ but not in any other spectral sets of $B_0(\tau)$, a solution should be sought in the form

$$q(\tau;\lambda) = e^{i\lambda\theta(\tau)}\Big[\sum_{j=1}^{s} V_j(\tau;\lambda)\eta_j(\tau;\lambda) + g(\tau;\lambda)\Big],$$

where the $\eta_j(\tau;\lambda)$ are vector functions satisfying the equations

$$\frac{d\eta_j}{d\tau} = \lambda\Big[P_j(\tau)B_0(\tau) + \frac{1}{\lambda}\Omega_j(\tau;\lambda)\Big]\eta_j + b_j(\tau;\lambda) \qquad (j = 1,\cdots,s),$$

the operator functions V_j, Ω_j and vector functions b_j, g being sought in the form of expansions of type (1.11), (1.12).

2. Obtain under condition (C_a) or (C_b) the following estimate for the pth approximation to a solution of the equation of the preceding exercise:

$$\|x(\tau) - x^{(p)}(\tau)\| = O(\lambda^{-p}) \qquad (0 \leqq \tau \leqq T).$$

Obtain the following improved estimate for the solution with zero initial value in the nonresonance case:

$$\|x(\tau) - x^{(p)}(\tau)\| = O(\lambda^{-(p+1)}).$$

3. Show that the expression

$$\int_0^t Y_k(\tau, s;\lambda)f(\tau;\lambda)e^{i\lambda\theta(\tau)}d\tau$$

which arises when equation (0.1) is solved by making use of the general formula (III.1.2) and the expansion (1.7) for the evolution operator is of order $O(\lambda^{-})$ if the values of $\theta'(\tau)$ do not appear in the spectral set $\sigma_k(\tau)$ and the function $f(\tau;\lambda)$ is differentiable (Ju. L. Daleckiĭ and S. G. Kreĭn [1]).

4. Construct asymptotic approximations for the equation of second order

$$A(\tau;\varepsilon)\,d^2\psi/d\tau^2 + \varepsilon B(\tau;\varepsilon)\,d\psi/d\tau + C(\tau;\varepsilon)\,\psi = 0,$$

assuming that the coefficients have expansions of type (1.9) for $\varepsilon = \lambda^{-1}$, the operator $A_0(\tau)$ is invertible and the spectrum of the operator $A_0^{-1}(\tau)B_0(\tau)$ decomposes into nonintersecting spectral sets.

Obtain an estimate of the approximations, assuming that the operators $A_0(\tau)$ and $B_0(\tau)$ are positive (Ju. L. Daleckiĭ and S. G. Kreĭn [1]).

5. Consider the boundary problem

$$\begin{cases} d\psi/dt = i\lambda \mathscr{J}\mathscr{H}_0(t)\psi, \\ E\psi(0) + \psi(T) = 0, \end{cases} \tag{0.2}$$

where $\mathscr{H}_0(\tau)$ is a twice continuously differentiable uniformly positive periodic operator function.

A point λ for which this problem has only the trivial solution is called a *regular point of the problem*. The complement of the set of regular points is called, as usual, the *spectrum of the problem*.

a) Show that if the spectrum of the operator $\mathscr{J}\mathscr{H}_0(\tau)$ is of form (2.33), the spectrum of problem (0.2) under the special choice $E = Q(T, 0)$ is contained in the system of intervals with endpoints

$$\lambda_{jk} \pm O\left(\frac{1}{k}\right) = \frac{2k+1}{\int_0^T \mu_j(s)ds}\,\pi \pm O\left(\frac{1}{k}\right) \qquad (j = 1,2,\cdots,n;\ k = 1,2,\cdots).$$

b) Extend the preceding result to the case when the operator $\mathscr{J}\mathscr{H}_0(\tau)$ has a countable number of noncoincident eigenvalues $\mu_k(\tau)$ $(k = 1,2,\cdots)$ for $\tau \in [0, T]$ and the corresponding system of eigenspaces is complete.

Hint. Make use of the arguments of § 2.5. In the case of a countable collection of eigenvalues consider the decomposition of the spectrum

$$\sigma(\mathscr{J}\mathscr{H}_0(\tau)) = \{\mu_1(\tau)\} \cup \cdots \cup \{\mu_{n-1}(\tau)\} \cup \sigma_n(\tau),$$

where $\sigma_n(\tau) = \{\mu_n(\tau), \cdots, \mu_{n+p}(\tau), \cdots\}$.

6. Consider in a Hilbert space H the boundary problem

$$\begin{cases} \dfrac{d}{d\tau}\left(A(\tau)\dfrac{du}{dt}\right) + B(\tau)u + \lambda^2 C(\tau)u = 0, \\ u(0) = u(T) = 0. \end{cases}$$

Suppose that the operators $A(\tau)$ and $C(\tau)$ are uniformly positive and twice continuously differentiable, and that the operator $B(\tau)$ is bounded and continuous. Suppose, further, that the operator $A^{-1}(\tau)C(\tau)$ has a spectrum consisting of a sequence of positive numbers $\mu_j^2(\tau)$, with $\mu_j^2(\tau) \neq \mu_k^2(\tau)$ $(j \neq k, 0 \leqq \tau \leqq T)$, and that the sequence of corresponding eigenspaces $\mathfrak{B}_j(\tau)$ is complete in $\mathfrak{H}$.

Obtain the asymptotic formula

$$\lambda_{jk} \pm O\left(\frac{1}{k}\right) = \frac{k\pi}{\int_0^T \mu_j(\tau)d\tau} \pm O\left(\frac{1}{k}\right)$$

for the endpoints of a system of intervals containing the spectrum of the problem.

Hint. Reduce the equation to a canonical equation of first order in the space $\mathfrak{H}^{(2)} = \mathfrak{H} \dotplus \mathfrak{H}$.

7. By developing the arguments in § 2.5, prove the existence of stability bands of the finite-dimensional equation (2.32) for arbitrarily large λ under the assumption that the spectrum of the operator $\mathscr{J}\mathscr{H}_0(\tau)$ is of form (2.33).

Hint. Use formula (2.34) to obtain an asymptotic formula for the spectrum of the monodromy operator $U(T;\lambda)$. To prove the existence of eigenvalues near the values obtained from the asymptotic formula, use Rouché's theorem.

To find the stability bands, determine those values of λ for which the characteristic multipliers are all of the same kind (see M. G. Kreĭn [6]; for the definition of characteristic multipliers see Chapter V).

NOTES

Asymptotic methods for solving differential equations containing a parameter were first used by Liouville to investigate expansions in the eigenfunctions of a selfadjoint boundary problem.

These methods were developed for arbitrary equations of nth order and systems of equations of first order in the fundamental papers of G. Birkhoff [1] and J. Tamarkin [1] and later in the works of W. Trjitzinsky [1], V. S. Pugačev [1] and others. In all of these works it was either assumed that the eigenvalues of the principal part of the equation are simple or assumed that their multiplicities do not vary over the entire interval $[0, T]$.

A number of results on the asymptotic representation of solutions of ordinary differential equations were obtained in connection with problems in the spectral theory of differential operators by I. M. Rapoport [1].

The case when the spectrum of the matrix of the principal part of an equation decomposes into nonintersecting (for all t) spectral sets in the interiors of which the multiplicities of the eigenvalues can vary arbitrarily was first considered in the finite-dimensional case by S. F. Feščenko [1].

S. F. Feščenko indicated a procedure for asymptotically decomposing a finite-dimensional system of form (1.4) into subsystems corresponding to nonintersecting (for all t) spectral sets lying in the left halfplane.

These and other results obtained by combining the methods applied in the above mentioned papers with the methods developed by N. M. Krylov and N. N. Bogoljubov [1] (see also N. N. Bogoljubov [1]) can be found in the subsequently published book of S. F. Feščenko, N. I. Škil' and L. D. Nikolenko [1] (in particular, this book contains a detailed account of the history of the question and a bibliography).

We note that in the original papers of S. F. Feščenko the validity of the method employed was not completely established, since he did not prove the differentiability (needed for carrying out the next step) of certain expressions arising at each step of the process.

S. G. Kreĭn and Ju. L. Daleckiĭ extended these results to the case of equations in Hilbert space giving a complete substantiation of the method (Ju. L. Daleckiĭ and S. G. Kreĭn [1] and Ju. L. Daleckiĭ [1, 2]).

The results of these papers are presented in §§1 and 2 for the case of bounded operators. We note that the assumptions underlying the error estimate in §2 have been generalized.

The results presented in § 3 are due to Ju. L. Daleckiĭ [5].

The example given in § 2.5 was taken from the book of I. C. Gohberg and M. G. Kreĭn [2] while the example in § 3.4 is from the article of M. K. Belkin and Ju. L. Daleckiĭ [1].

BIBLIOGRAPHY

R. ABRAHAM AND J. ROBBIN

1. *Transversal mappings and flows*, Benjamin, New York, 1967. MR **39** #2181.

V. M. ADAMJAN

1. *Canonical differential operators in Hilbert space*, Dokl Akad. Nauk SSSR **178** (1968), 9–12 = Soviet Math. Dokl. 9 (1968), 1—5. MR **37** #550.

S. AGMON AND L. NIRENBERG

1. *Properties of solutions of ordinary differential equations in Banach space*, Comm. Pure Appl. Math. **16** (1963), 121—239. MR **27** #5142.

A. A. ANDRONOV AND A. A. VITT

1. *On Ljapunov stability*, Ž. Èksper. Teoret. Fiz. **4** (1933), 606–608. (Russian)

D. V. ANOSOV

1. *Geodesic flows on closed Riemannian manifolds of negative curvature*, Trudy Mat. Inst. Steklov. **90** (1967) = Proc. Steklov Inst. Math. **90** (1967). MR **36** #7157.

D. V. ANOSOV AND JA. G. SINAĬ

1. *Certain smooth ergodic systems*, Uspehi Mat. Nauk **22** (1967), no. 5 (137), 107–172 =Russian Math. Surveys **22** (1967), no 5, 103–167. MR **37** #370.

H. A. ANTOSIEWICZ

1. *Continuous parameter dependence and the method of averaging*, Proc. Internat. Sympos. Analytic Methods in the Theory of Non-Linear Vibrations (1961), Izdat. Akad. Nauk Ukrain. SSR, Kiev, 1963, pp. 51–58. (Russian) MR **28** #2274.

E. A. BARBAŠIN

1. *Introduction to the theory of stability*, "Nauka", Moscow, 1967; English translation, Wolters-Noordhoff, Groningen, 1970. MR **37** #515; **40**#8737.

M. K. BELKIN AND JU. I. DALECKIĬ

1. *On the theory of parametric amplification*, Radiotehnika (Har′kov) **21** (1966), no. 8, 22–28. (Russian)

R. E. BELLMAN

1. *On an application of a Banach-Steinhaus theorem to the study of the boundedness of solutions of non-linear differential and difference equations*, Ann. of Math. (2) **49** (1948), 515–522. MR **10**, 121.
2. *Stability theory of differential equations*, McGraw-Hill, New York, 1953; Russian transl., IL, Moscow, 1954. MR **15**, 794; **17**, 734.
3. *On the existence and boundedness of solutions of non-linear differential-difference equations*, Ann. of Math. (2) **50** (1949), 347–355. MR **10**, 715.

G. I. BIRJUK

1. *On an existence theorem for almost periodic solutions of certain systems of nonlinear differential equations with a small parameter*, Dokl. Akad. Nauk SSSR **96** (1954), 5–7. (Russian) MR **16**, 130.

G. D. BIRKHOFF

1. *On the asymptotic character of the solutions of certain linear differential equations containing a parameter*, Trans. Amer. Math. Soc. **9** (1908), 219–231.

N. N. BOGOLJUBOV

1. *On some statistical methods in mathematical physics*, Izdat. Akad. Nauk Ukrain. SSR, L'vov, 1945. (Russian) MR **8**, 37.

N. N. BOGOLJUBOV AND JU. A. MITROPOL'SKIĬ

1. *Asymptotic methods in the theory of nonlinear oscillations*, 3rd rev. ed., Fizmatgiz, Moscow, 1963; English transl. of 2nd ed., Internat. Monographs on Advanced Math. and Physics, Hindustan, Delhi; Gordon and Breach, New York, 1962. MR **25** #5242; **28** #1356.
2. *The method of integral manifolds in non-linear mechanics*, Proc. Internat. Sympos. Analytic Methods in the Theory of Non-Linear Vibrations (1961), Izdat. Akad. Nauk Ukrain. SSR, Kiev, 1963, pp. 93–154; English transl., Contributions to Differential Equations **2** (1963), 123—196. MR **28** #5235; **26** #6532.

P. BOHL (P. BOL')

1. *Sur certaines équations différentielles d'un type général utilisables en méchanique*, Dissertation, Tartu, 1900; French transl., Bull. Soc. Math. France **38** (1910), 5—138.
2. *Über Differentialungleichungen*, J. Reine Angew. Math. **144** (1913), 284–318.
3. *Selected works*, Izdat. Akad. Nauk Latv. SSR, Riga, 1961. (Russian)

M. L. BRODSKIĬ

1. *On properties of an operator mapping the non-negative part of a space with indefinite metric into itself*, Uspehi Mat. Nauk **14** (1959), no. 1 (85), 147–152. (Russian) MR **21** #5145.

B. F. BYLOV et al.

1. *Theory of Ljapunov exponents and its application to problems of stability*, "Nauka", Moscow, 1966. (Russian) MR **34** #6234.

L. CESARI

1. *Asymptotic behavior and stability problems in ordinary differential equations*, Ergebnisse der Mathematik und ihrer Grenzgebiete, Heft 16, Springer-Verlag, Berlin, 1959; Russian transl., "Mir", Moscow, 1964. MR **22** #9673.

E. A. CODDINGTON AND N. LEVINSON

1. *Theory of ordinary differential equations*, McGraw-Hill, New York, 1955; Russian transl., IL, Moscow, 1958. MR **16**, 1022.

W. A. COPPEL

1. *Stability and asymptotic behavior of differential equations*, D.C. Heath, Boston, Mass., 1965. MR **32** #7875.
2. *Almost periodic properties of ordinary differential equations*, Ann. Mat. Pura Appl. (4) **76** (1967), 27–49. MR **36** #4076.
3. *Dichotomies and reducibility*, J. Differential Equations **3** (1967), 500–521. MR **36** #6699.

W. A. COPPEL AND A. HOWE

1. *On the stability of linear canonical systems with periodic coefficients*, J. Austral. Math. Soc. **5** (1965), 169–195. MR **31** #4957.

R. COURANT AND D. HILBERT

1. *Methoden der Mathematischen Physik*. Vol. 1, Springer, Berlin, 1931; English transl., Interscience, New York, 1953; Russian transl., GITTL, Moscow, 1951. MR **5**, 97; **13**, 800; **16**, 426.

G. G. DAHLQUIST

1. *Stability and error bounds in the numerical integration of ordinary differential equations*, Kungl. Tekn. Högsk. Handl. Sockholm No. 130 (1959). MR **21** #1706.

JU. L. DALECKIĬ

1. *On the asymptotic solution of a vector differential equation*, Dokl. Akad. Nauk SSSR **92** (1953), 881–884. (Russian) MR **15**, 624.

2. *On some equations with closed operators*, Izv. Kiev. Politehn. Inst. **19** (1956), 157–177. (Russian)

3. *On continuous rotation of subspaces in a Banach space*, Uspehi Mat. Nauk **12** (1957), no. 3 (75), 147–154. (Russian) MR **20** #1199.

4. *A linear equation relative to elements of a normed ring*, Uspehi Mat. Nauk **14** (1959), no. 1 (85), 165–168. (Russian) MR **21** #2926.

5. *An asymptotic method for certain differential equations with oscillating coefficients*, Dokl. Akad. Nauk SSSR **143** (1962), 1026–1029 = Soviet Math. Dokl. **3** (1962), 520–523. MR **24** #A2726.

6. *Stable integral varieties of a nonlinear differential equation in a Banach space*, Ukrain. Mat. Ž. **20** (1968), 376–381. (Russian) MR **37** #6574.

7. *Functional integrals connected with operator evolution equations*, Uspehi Mat. Nauk **17** (1962), no. 5 (107), 3–115 = Russian Math. Surveys **17** (1962), no. 5, 1–107. MR **28** #4389.

JU. L. DALECKIĬ AND I. K. KOROBKOVA

1. *A certain operator differential equation with regular singular point*. Dopovidi Akad. Nauk Ukrain. RSR Ser. A **1968**, 972–977. (Ukrainian) MR **39** #1778.

JU. L. DALECKIĬ AND M. G. KREĬN

1. *Some results and problems of the theory of stability and asymptotic behavior of solutions of differential equations in Banach space*, Proc. Fifth Internat. Congress on Nonlinear Vibrations (Kiev, 1969), vol. 1, Akad. Nauk Ukrain. SSR Inst. Mat., Kiev, 1970, pp. 332–347. (Russian)

JU. L. DALECKIĬ AND S. G. KREĬN

1. *On differential equations in Hilbert space*, Ukrain. Mat. Ž. **2** (1950), no. 4, 71–91. (Russian) MR **13**, 954.

2. *Properties of operators depending on a parameter*, Dopovidi. Akad. Nauk Ukrain. RSR **6** (1950), 433–436. (Ukrainian)

B. P. DEMIDOVIČ

1. *On bounded solutions of a certain nonlinear system of ordinary differential equations*, Mat. Sb. **40** (**82**) (1956), 73–94. (Russian) MR **18**, 738.

2. *On a case of almost periodicity of a solution of an ordinary differential equation of 1st order*, Uspehi Mat. Nauk **8** (1953), no. 6 (58), 103–106. (Russian) MR **15**, 872.

3. *Lectures on the mathematical theory of stability*, "Nauka", Moscow, 1967. (Russian) MR **37** #1716.

V. I. DERGUZOV

1. *On the stability of the solutions of the Hamilton equations with unbounded periodic operator coefficients*, Mat. Sb. **63** (**105**) (1964), 591–619. (Russian) MR **29** #325.

2. *Sufficient conditions for the stability of Hamiltonian equations with unbounded periodic operator coefficients*, Mat. Sb. **64** (**106**) (1964), 419–435. (Russian) MR **29** #4965.

J. DIEUDONNÉ

1. *Foundations of modern analysis*, Pure and Appl. Math., vol. 10, Academic Press, New York, 1960. MR **22** #11074.

S. P. DILIBERTO

1. *On stability of linear mechanical systems*, Proc. Internat. Sympos. Analytic Methods in the Theory of Nonlinear Vibrations (1961), Izdat. Akad. Nauk Ukrain. SSR, Kiev, 1963, pp. 189–203. (Russian) MR **28** #4218.

JU. I. DOMŠLAK

1. *Behavior at infinity of solutions of the evolution equation with unbounded operator*, Izv.

Akad. Nauk Azerbaĭdžan. SSR Ser. Fiz.-Mat. Tehn. Nauk **1961**, no. 3, 9–22. (Russian) MR **25** #3253a.

2. *On the theory of differential equations in a Banach space with constant unbounded operators*, Akad. Nauk Azerbaĭdžan. SSR Dokl. **18** (1962), no. 5, 3–6. (Russian) MR **26** #4010.

R. J. Duffin

1. *A minimax theory for overdamped networks*, J. Rational Mech. Anal. **4** (1955), 221–233. MR **16**, 979.
2. *The Rayleigh-Ritz method for dissipative or gyroscopic systems*, Quart. Appl. Math. **18** (1960/61), 215–221. MR **22** #12775.

N. P. Erugin

1. *Reducible systems*, Trudy Mat. Inst. Steklov. **13** (1946). (Russian) MR **9**, 509.
2. *Linear systems of ordinary differential equations with periodic and quasi-periodic coefficients*, Izdat. Akad. Nauk BSSR, Minsk, 1963; English transl., Math. in Science and Engineering, vol. 28, Academic Press, New York, 1966. MR **27** #2685; **34** #6175.

J. Favard

1. *Leçons sur les fonctions presque-périodiques*, Gauthier-Villars, Paris, 1933.

S. F. Feščenko

1. *On the asymptotic reduction of integrals of linear ordinary differential equations containing a parameter*, Dopovidi Akad. Nauk Ukrain. RSR **1949**, 11–16. (Ukrainian)

S. F. Feščenko, N. I. Škil' and L. D. Nikolenko

1. *Asymptotic methods in the theory of linear differential equations*, Akad. Nauk Ukrain. SSR Inst. Mat., "Naukova Dumka", Kiev, 1966. (Russian) MR **33** #6057.

G. Floquet

1. *Sur les équations différentielles linéaires à coefficients périodiques*, Ann. École Norm. Sup. (2) **12** (1883), 47–89.

V. N. Fomin

1. *The perturbation method in the theory of dynamic stability of systems with distributed parameters*, Dokl. Akad. Nauk SSSR **163** (1965), 830–833 = Soviet Math. Dokl. **6** (1965), 1036–1039. MR **36** #490.
2. *The parametric resonance of elastic systems with distributed parameters*, Dokl. Akad. Nauk SSSR **164** (1965), 58–61 = Soviet Math. Dokl. **6** (1965), 1179–1182. MR **33** #909.
3. *Parametric resonance of elastic systems with an infinite number of degrees of freedom*. I, II, Vestnik Leningrad. Univ. **20** (1965), no. 13, 73–87; ibid. **20** (1965), no. 19, 74–86. (Russian) MR **33** #911; 912.
4. *Regions of dynamical instability of parametrically excited systems with an infinite number of degrees of freedom*, Problems Math. Anal. Boundary Value Problems Integral Equations, Izdat. Leningrad. Univ., Leningrad, 1966, pp. 135–165. (Russian) MR **35** #5167.
5. *Resonant oscillations of linear systems under real almost-periodic perturbations*, Problems of Math. Anal., no. 2, Izdat. Leningrad. Univ., Leningrad, 1969, pp. 28–79. (Russian)

G. Frobenius

1. *Über die Integration der linearen Differentialgleichungen durch Reihen*, J. Reine Angew. Math. **76** (1873), 214–235.

L. Fuchs

1. *Zur Theorie der linearen Differentialgleichungen mit veränderlichen Koeffizienten*, J. Reine Angew. Math. **66** (1866), 121–160; ibid. **68** (1868), 354–385.

F. R. Gantmaher

1. *The theory of matrices*, 2nd ed., "Nauka", Moscow, 1966; English transl. of 1st ed., Chelsea, New York, 1959. MR **16**, 438.

N. I. Gavrilov

1. *Methods of the theory of ordinary differential equations*, "Vysš. Škola", Moscow, 1962. (Russian) MR **30** #4008.

I. M. Gel'fand and V. B. Lidskiĭ

1. *On the structure of the regions of stability of linear canonical systems of differential equations with periodic coefficients*, Uspehi Mat. Nauk **10** (1955), no. 1 (63), 3–40; English transl., Amer. Math. Soc. Transl. (2) **8** (1958), 143–181. MR **17**, 482; **19**, 960.

I. M. Gel'fand and G. E. Šilov

1. *Generalized functions*. Vol. 3: *Some questions in the theory of differential equations*, Fizmatigiz, Moscow, 1958; English transl., Academic Press, New York, 1967. MR **21** #5142b; **36** # 506.

V. E. Germaidze

1. *On asymptotic stability of the first approximation*, Prikl. Mat. Meh. **21** (1957), 133–135. (Russian)

V. E. Germaidze and N. N. Krasovskiĭ

1. *On stability under constant perturbations*, Prikl. Mat. Meh. **21** (1957), 769–774. (Russian) MR **20** #3353.

I. I. Gihman

1. *Concerning a theorem of N. N. Bogoljubov*, Ukrain. Mat. Ž. **4** (1952), 215–219. (Russian) MR **17**, 738.

Ju. P. Ginzburg

1. *On $\mathcal{J}$-nonexpansive operators in Hilbert space*, Naučn. Zap. Odessk. Ped. Inst. **22** (1958), 1–13. (Russian)
2. *Projection in Hilbert space with a bilinear metric*, Dokl. Akad. Nauk SSSR **139** (1961), 775–778 = Soviet Math. Dokl. **2** (1961), 980–983. MR **25** #4336.
3. *Multiplicative representations of J-contractive operator functions*. I, Mat. Issled. **2** (1967), no. 2, 52–83; English transl., Amer. Math. Soc. Transl. (2) **96** (1970), 189–221. MR **38** #1551a.
4. *On subspaces of a Hilbert space with an indefinite metric*, Naučn. Zap. Odessk. Ped. Inst. **25** (1961), no. 2, 3–9. (Russian)

Ju. P. Ginzburg and I. S. Iohvidov

1. *A study of the geometry of infinite dimensional spaces with bilinear metric*, Uspehi Mat. Nauk **17** (1962), no. 4 (106), 3–56 = Russian Math. Surverys **17** (1962), no. 4, 1–51. MR **26** #2850.

I. C. Gohberg and M. G. Kreĭn

1. *Introduction to the theory of linear nonselfadjoint operators in Hilbert space*, "Nauka", Moscow, 1965; English transl., Transl. Math. Monographs, vol. 18, Amer. Math. Soc. Providence, R. I., 1969. MR **36** #3137.
2. *Theory and applications of Volterra operators in Hilbert space*, "Nauka", Moscow, 1967; English transl., Transl. Math. Monographs, vol. 24, Amer. Math. Soc., Providence, R. I., 1970. MR **36** #2007.
3. *The basic propositions on defect numbers, root numbers and indices of linear operators*, Uspehi Mat. Nauk **12** (1957), no. 2 (74), 43–118; English transl., Amer. Math. Soc. Transl. (2) **13** (1960), 185–264. MR **20** #3459; **22** #3984.
4. *Transformations of kernels of integral equations and their influence on the spectra of these equations*, Ukrain. Mat. Ž. **13** (1961), no. 3, 12–38; English transl., Amer. Math. Soc. Transl. (2) **35** (1964), 263–295. MR **27** #1788.

J. Hadamard

1. *Sur l'itération et les solutions asymptotiques des équations différentielles*, Bull. Soc. Math. France **29** (1901), 224–228.

Z. I. Halilov

1. *Stability of solutions of an equation in a Banach space*, Dokl. Akad. Nauk SSSR **137** (1961), 797–799 = Soviet Math. Dokl. **2** (1961), 362–364. MR **26** #430.

2. *On the stability of solutions of evolutionary equations with unbounded operator*, Akad. Nauk Azerbaĭdžan. SSR Dokl. **17** (1961), 91–95. (Russian)

P. R. Halmos and G. Lumer

1. *Square roots of operators*. II, Proc. Amer. Math. Soc. **5** (1954), 589–595. MR **16**, 48.

P. Hartman

1. *Ordinary differential equations*, Wiley, New York, 1964. MR **30** #1270.

E. Hille

1. *Linear differential equations in Banach algebras*, Proc. Internat. Sympos. Linear Spaces (Jerusalem, 1960), Jerusalem Academic Press, Jerusalem; Pergamon Press, Oxford, 1961, pp. 263–273. MR **24** #A2725.

I. S. Iohvidov and M. G. Kreĭn

1. *Spectral theory of operators in spaces with an indefinite metric*. I. Trudy Moskov. Mat. Obšč. **5** (1956), 367–432; English transl., Amer. Math. Soc. Transl. (2) **13** (1960), 105–175. MR **18**, 320; **22** #3983.

L. A. Ivanov

1. *Operators which realize evolutionary subspaces*, Voronež. Gos. Univ. Trudy Sem. Funkcional. Anal. **10** (1968), 44–50. (Russian)

V. A. Jakubovič

1. *Structure of the group of symplectic matrices and of the set of unstable canonical systems with periodic coefficients*, Mat. Sb. **44** (**86**) (1958), 313–352. (Russian) MR **20** #3325.
2. *The small parameter method for canonical systems with periodic coefficients*, Prikl. Mat. Meh. **23** (1959), 15–34 = J. Appl. Math. Mech. **23** (1959), 17–43. MR **21** #5774.
3. *A certain criterion for reducibility of a system of differential equations*, Dokl. Akad. Nauk SSSR **66** (1949), 577–580. (Russian) MR **11**, 110.
4. *Conditions of unlimited stability for a certain linear second-order differential equation*, Vestnik Leningrad. Univ. **16** (1961), no. 19, 83–92. (Russian) MR **25** #3244.
5. *On the asymptotic behavior of the solutions of a system of differential equations*, Mat. Sb. **28** (**70**) (1951), 217–240. (Russian) MR **12**, 611.
6. *Critical frequencies of quasicanonical systems*, Vestnik Leningrad. Univ. **13** (1958), no. 13, 35–63. (Russian) MR **21** #749.
7. *Criteria of stability for systems of two canonical equations with periodic coefficients*, Dokl. Akad. Nauk SSSR **78** (1951), 221–224. (Russian) MR **13**, 237.
8. *Arguments on the group of symplectic matrices*, Mat. Sb. **55** (**97**) (1961), 255–280. (Russian) MR **26** #5072.
9. *The structure of the function space of complex canonical equations with periodic coefficients*, Dokl. Akad. Nauk SSSR **139** (1961), 54–57 = Soviet Math. Dokl. **2** (1961), 887–890. MR **25** #5232.

P. Jonas

1. *Einige Betrachtungen zur Stabilität kanonischer Differentialgleichungen im Hilbertraum*, Diplomarbiet, Techn. Universität, Dresden, 1964.

L. V. Kantorovič

1. *Functional analysis and applied mathematics*, Uspehi Mat. Nauk **3** (1948), no. 6 (28), 89–185; English transl., Nat. Bur. Standards Report #1509, U. S. Dept. of Commerce, Washington, D.C., 1952. MR **10**, 380; **14**, 766.

P. L. Kapica

1. *A pendulum with vibrating suspension*, Uspehi Fiz. Nauk **44** (1951), 7–20. (Russian) AMR **5** #600.
2. *Dynamical stability of a pendulum whose point of suspension oscillates*, Ž. Èksper. Teoret. Fiz. **21** (1951), 588–597. (Russian) PA **55** #793.

T. M. Karaseva

1. *On the rate of growth and boundedness of solutions of second-order differential equations with periodic coefficients*, Izv. Akad. Nauk SSSR Ser. Mat. **29** (1965), 41–70; English Transl., Amer. Math. Soc. Transl. (2) **79** (1969), 181–214. MR **31** #3664.

T. Kato

1. *On the perturbation theory of closed linear operators*, J. Math. Soc. Japan **4** (1952), 323–337. MR **14**, 990.

I. I. Kovtun

1. *Some problems in the asymptotic solution of an operator differential equation*, Approximate Methods of Solving Differential Equations, Izdat. Akad. Nauk Ukrain. SSR, Kiev, 1963, pp. 33–39. (Russian) MR **29** #332.

M. A. Krasnosel'skiĭ and S. G. Kreĭn

1. *On the principle of averaging in nonlinear mechanics*, Uspehi Mat. Nauk **10** (1955), no. 3 (65), 147–152. (Russian) MR **17**, 152.
2. *On the theory of ordinary differential equations in Banach spaces*, Voronež. Gos. Univ. Trudy Sem. Funkcional. Anal. No. 2 (1956), 3–23. (Russian) MR **19**, 140.

N. N. Krasovskiĭ

1. *Certain problems in the theory of stability of motion*, Fizmatgiz, Moscow, 1959; English transl., Stanford Univ. Press, Stanford, Calif., 1963. MR **21** #5047; **26** #5258.

M. G. Kreĭn

1. *Sur la meilleure approximation des fonctions continues dérivables sur tout l'axe réel*, C. R. (Dokl.) Acad. Sci. URSS **18** (1938), 619–623.
2. *On some questions related to the ideas of Ljapunov in the theory of stability*, Uspehi Mat. Nauk **3** (1948), no. 3 (25), 166–169. (Russian) MR **10**, 128.
3. *A generalization of some investigations of A.M. Ljapunov on linear differential equations with periodic coefficients*, Dokl. Akad. Nauk SSSR **73** (1950), 445–448. (Russian) MR **12**, 100.
4. *The fundamental propositions of the theory of representations of Hermitian operators with deficiency index (m,m)*, Ukrain. Mat. Ž. **1** (1949), no. 2, 3–66; English transl., Amer. Math. Soc. Transl. (2) **97** (1970), 75–143. MR **14**, 56.
5. *On the application of an algebraic proposition in the theory of matrices of monodromy*, Uspehi Mat. Nauk **6** (1951), no. 1 (41), 171–177. (Russian) MR **14**, 277.
6. *The basic propositions of the theory of λ-zones of stability of a canonical system of linear differential equations with periodic coefficients*, In Memory of Aleksandr Aleksandrovič Andronov, Izdat. Akad. Nauk SSSR, Moscow, 1955, pp. 413–498. (Russian) MR **17**, 738.
7. *On criteria of stable boundedness of solutions of periodic canonical systems*, Prikl. Mat. Meh. **19** (1955), 641–680. (Russian) MR **17**, 1088.
8. *A new application of the fixed-point principle in the theory of operators on a space with indefinite metric*, Dokl. Akad. Nauk SSSR **154** (1964), 1023–1026 = Soviet Math. Dokl. **5** (1964), 224–227. MR **29** #6314.
9. *Lectures on the theory of stability of solutions of differential equations in a Banach space*, Izdat. Akad. Nauk Ukrain. SSR, Kiev, 1964. (Russian) MR **32** #1428.
10. *Introduction to the geometry of indefinite J-spaces and to the theory of operators in those spaces*, Second Math. Summer School, part 1, "Naukova Dumka", Kiev, 1965, pp. 15–92; English transl., Amer. Math. Soc. Transl. (2) **93** (1970), 103–176. MR **33** #574.
11. *The angular localization of the spectrum of a multiplicative integral in Hilbert space*, Funkcional. Anal. i Priložen. 3 (1969), no. 1, 89–90 = Functional Anal. Appl. 3 (1969), 73–74. MR **39** #3342.
12. *On some aspects of the theory of uniform W-contraction and its applications*, Mat. Issled. (to appear). (Russian)

13. *On a conjecture of A. M. Ljapunov*, Funkcional. Anal. i Priložen. **6** (1973), 45–54 = Functional Anal. Appl. **6** (1973) (to appear).

M. G. Kreĭn and V. A. Jakubovič

1. *Hamiltonian systems of linear differential equations with periodic coefficients*, Proc. Internat. Sympos. Analytic Methods in the Theory of Non-Linear Vibrations (1961), Izdat. Akad. Nauk Ukrain. SSR, Kiev, 1963, pp. 277–305. (Russian) MR **28** #312.

M. G. Kreĭn and G. K. Langer [H. Langer]

1. *A contribution to the theory of quadratic pencils of selfadjoint operators*, Dokl. Akad. Nauk SSSR **154** (1964), 1258–1261 = Soviet Math. Dokl. **5** (1964), 266–269. MR **29** #6315.
2. *Certain mathematical principles of the linear theory of damped vibrations of continua*, Proc. Internat. Sympos. Appl. Theory of Functions in Continuum Mechanics (Tbilisi, 1963), vol. II, Fluid and Gas Mechanics, Math. Methods, "Nauka", Moscow, 1965, pp. 283–322. (Russian) MR **33** #3493.

M. G. Kreĭn and G. Ja. Ljubarskiĭ

1. *Analytic properties of multipliers of periodic canonical differential systems of positive type*, Izv. Akad. Nauk SSSR Ser. Mat. **26** (1962), 549–572; English transl., Amer. Math. Soc. Transl. (2) **89** (1969), 1–28. MR **26** #400.

M. G. Kreĭn and F. E. Melik-Adamjan

1. *On the theory of S-matrices of canonical differential equations with summable potential*, Dokl. Akad. Nauk Armjan. SSR **26** (1968), no. 4, 150–155. (Russian)
2. *Some applications of the theory of factorizing unitary matrices*, Funkcional Anal. i Priložen. **4** (1970), no. 4, 73–75 = Functional Anal. Appl. **4** (1970), 327–329.

M. G. Kreĭn and M. A. Rutman

1. *Linear operators leaving invariant a cone in a Banach space*, Uspehi Mat. Nauk **3** (1948), no. 1 (23), 3–95; English transl., Amer. Math. Soc. Transl. (1) **10** (1962), 199–325. MR **10**, 256; **12**, 341.

M. G. Kreĭn and Ju. L. Šmul'jan

1. *Linear-fractional transformations with operator coefficients*, Mat. Issled. **2** (1967), no. 3, 64–96; English transl., Amer. Math. Soc. Transl. (2) **103** (1973), 125–152. MR 39 #813.

S. G. Kreĭn

1. **Linear differential equations in Banach space*, "Nauka", Moscow, 1967; English transl., Transl. Math. Monographs, vol. 29, Amer. Math. Soc., Providence, R. I., 1972. MR **40** #508.

N. M. Krylov and N. N. Bogoljubov

1. *Introduction to non-linear mechanics*, Izdat. Akad. Nauk SSSR, Moscow, 1937; English transl., Ann. of Math. Studies, no. 11, Princeton Univ. Press, Princeton, N. J., 1943. MR **4**, 142.

D. L. Kučer

1. *On some criteria for the boundedness of the solution of a system of differential equations*, Dokl. Akad. Nauk SSSR **69** (1949), 603–606. (Russian) MR **11**, 360.
2. *On stability and boundedness of solutions of some types of differential equations in normed linear spaces*, Candidate's Dissertation, Odessa, 1950. (Russian)
3. *Some questions on the integration of normal differential equations in Banach space*, Naučn. Zap. Odessk. Ped. Inst. **25** (1961), no. 2, 25–37. (Russian)

H. Kühne

1. *Eine Klasse J-selbstadjungierten Operatoren und stark gedämpfte Scharen*, Dissertation, Techn. Universität, Dresden, 1964.
2. *Minimaxprinzipe für stark gedämpfte Scharen*, Acta Sci. Math. Szeged **29** (1968), 39–68. MR 39 #6119.

*All page references are to the English translation.

J. KURZWEIL

1. *Exponentially stable integral manifolds, averaging principle and continuous dependence on a parameter*, Czechoslovak Math. J. **16 (91)** (1966), 380–423; ibid. **16 (91)** (1966), 463–492. MR **34** #6259.

J. KURZWEIL AND Z. VOREL

1. *Continuous dependence of solutions of differential equations on a parameter*, Czechoslovak Math. J. **7 (82)** (1957), 568–583. (Russian) MR **22** #2734.

L. D. LANDAU AND E. M. LIFŠIC

1. *Mechanics*, Fizmatgiz, Moscow, 1958; 2nd rev. ed., "Nauka", Moscow, 1965; English transl., *Course of theoretical physics*. Vol. I, Pergamon Press, Oxford; Addison-Wesley, Reading, Mass., 1960. MR **21** #985; **22** #11531; **32** #620.

H. LANGER

1. *Spektraltheorie linearer Operatoren in J-Räumen und einige Anwendungen auf die Schar* $\mathscr{L}(\lambda) = 2I + \lambda B + C$, Habilitationsschrift, Techn. Universität, Dresden, 1965.
2. *Über stark gedämpfte Scharen im Hilbertraum*, J. Math. Mech. **17** (1967/68), 685–705. MR **37** #4650.

B. JA. LEVIN

1. *Distribution of zeros of entire functions*, GITTL, Moscow, 1956; English transl., Transl. Math. Monographs, vol. 5, Amer. Math. Soc., Providence, R. I., 1964. MR **19**, 402; **28** #217.

B. M. LEVITAN

1. *Almost-periodic functions*, GITTL, Moscow, 1953. (Russian) MR **15**, 700.

V. B. LIDSKIĬ AND P. A. FROLOV

1. *The structure of the stability regions of a selfadjoint system of differential equations with periodic coefficients*, Mat. Sb. **71 (113)** (1966), 48–64. (Russian) MR **33** #7641.

A. M. LJAPUNOV

1. *Collected works*. Vol. 2, Izdat. Akad. Nauk SSSR, Moscow, 1956. (Russian) MR **27** #5980; errata, **30**, 1204.

N. JA. LJAŠČENKO

1. *On asymptotic stability of solutions of a system of differential equations*, Dokl. Akad. Nauk SSSR **96** (1954), 237–239. (Russian) MR **16**, 133.
2. *On asymptotic stability of solutions of nonlinear systems of differential equations*, Dokl. Akad. Nauk SSSR **104** (1955), 177–179. (Russian) MR **17**, 616.

JU. I. LJUBIČ

1. *Almost periodic functions in the spectral analysis of operators*, Dokl. Akad. Nauk SSSR **132** (1960), 518–520 = Soviet Math. Dokl. **1** (1960), 593–595. MR **22** #9863.

S. M. LOZINSKIĬ

1. *Error estimate for numerical integration of ordinary differential equations*. I, Izv. Vysš. Učebn. Zaved. Matematika **1958**, no. 5 (6), 52–90; errata, **1959**, no. 5 (12), 222 (Russian) MR **26** #3191.

A. D. MAĬZEL'

1. *On stability of solutions of systems of differential equations*, Ural. Politehn. Inst. Trudy **51** (1954), 20–50. (Russian) MR **17**, 738.

I. G. MALKIN

1. *Die Stabilitätsfrage bei Differentialgleichungen*, Sb. Naučn. Trudov. Kazan. Aviac. Inst. No. 2 (1934), 21–28. (Russian)
2. *On stability with respect to the first approximation*, Sb. Naučn. Trudov. Kazan. Aviac. Inst. No. 3 (1935), 7–17. (Russian)
3. *Certain questions on the theory of the stability of motion in the sense of Ljapunov*, Sb. Naučn.

Trudov. Kazan. Aviac. Inst. No. 7 (1937); English transl., Amer. Math. Soc. Transl. (1) **5** (1962), 1–174. MR **12**, 181.

4. *Über die Bewegungsstabilität nach der ersten Näherung*, C. R. (Dokl.) Acad. Sci. URSS **18** (1938), 159–164.
5. *Theory of stability of movement*, GITTL, Moscow, 1952; German transl., Oldenbourg, Munich, 1959. MR **15**, 873; **21** #2791.
6. *On almost periodic oscillations of nonlinear autonomous systems*, Prikl. Mat. Meh. **18** (1954), 681–704. (Russian) MR **16** 590.

L. I. MANDEL'ŠTAMM AND N. D. PAPALEKSI

1. *On the justification of a method for approximate solution of differential equations*, Ž. Èksper. Teoret. Fiz. **4** (1934), 117–122. (Russian)

M. MARCUS AND H. MINC

1. *A survey of matrix theory and matrix inequalities*, Allyn and Bacon, Boston, Mass., 1964. MR **29** #112.

A. A. MARKOV

1. *Quelques théorèmes sur les ensembles abélienes*, C. R. (Dokl.) Acad. Sci. URSS **10** (1938), 311–313.

L. MARKUS

1. *Continuous matrices and the stability of differential systems*, Math. Z. **62** (1955), 310–319. MR **17**, 37.

J. L. MASSERA AND J. J. SCHÄFFER

1. **Linear differential equations and function spaces*, Pure and Appl. Math., vol. 21, Academic Press, New York, 1966; Russian transl., "Mir", Moscow, 1969. MR **35** #3197.

V. M. MILLIONŠČIKOV

1. *On the spectral theory of nonautonomous linear systems of differential equations*, Trudy Moskov. Mat. Obšč. **18** (1968), 147–186 = Trans. Moscow Math. Soc. **1968**, 161–206. MR **39** #3109.

L. JA. MIROČNIK

1. *An estimate for multiplicators of the Sturm-Liouville differential equation*, Mat. Issled. **4** (1969), no. 1 (11), 90–97. (Russian) MR **41** #2106.

JU. A. MITROPOL'SKIĬ

1. *Problems of asymptotic theory of nonstationary oscillations*, "Nauka", Moscow, 1964; English transl., Israel Program for Scientific Translations, Jerusalem; Davey, New York, 1965. MR **32** #5984.
2. *A study of the integral manifold for a system of non-linear equations in Hilbert space which are close to equations with variable coefficients*, Ukrain. Mat. Ž. **16** (1964), 334–338. (Russian) MR **29** #6346.
3. *Lectures on the method of averaging in nonlinear mechanics*, "Naukova Dumka", Kiev, 1966; rev. ed., 1971; English transl., Transl. Math. Monographs, Amer. Math. Soc., Providence, R. I. (to appear). MR **36** #1740.

JU. A. MITROPOL'SKIĬ AND O. B. LYKOVA

1. *On the integral manifold of a non-linear system in Hilbert space*, Ukrain. Mat. Ž. **17** (1965), no. 5, 43–53; English transl., Amer. Math. Soc. Transl. (2) **76** (1968), 117–130. MR **32** #6012.
2. *Lectures on the method of integral manifolds*, Akad. Nauk Ukrain. SSR Inst. Mat., Kiev, 1968. (Russian) MR **40** #1655.

A. D. MYŠKIS AND I. M. RABINOVIČ

1. *Mathematician Piers Bohl from Riga*, "Zinatne", Riga, 1965. (Russian) MR **34** #4095.

*All page references are to the English original

M. A. NAĬMARK

1. *Linear differential operators*, 2nd ed., "Nauka", Moscow, 1969; English transl. of 1st ed., Ungar, New York, 1967. MR **35** #6885.

M. G. NEĬGAUZ AND V. B. LIDSKIĬ

1. *On the boundedness of the solutions of linear systems of differential equations with periodic coefficients*, Dokl. Akad. Nauk SSSR **77** (1951), 189–192. (Russian) MR **12**, 706.

JU. I. NEĬMARK

1. *Motions close to a doubly asymptotic motion*, Dokl. Akad. Nauk SSSR **172** (1967), 1021–1024 = Soviet Math. Dokl. **8** (1967), 228–231. MR **35** #498.

A. M. OSTROWSKI AND H. SCHNEIDER

1. *Some theorems on the inertia of general matrices*, J. Math. Anal. Appl. **4** (1962), 72–84. MR **26** #124.

O. PERRON

1. *Über eine Matrix Transformation*, Math. Z. **32** (1930), 465–473.
2. *Die Stabilitätsfrage bei Differentialgleichungen*, Math. Z. **32** (1930), 703–728.
3. *Die Ordnungszahlen linear Differentialgleichungssysteme*, Math. Z. **31** (1930), 748–766.

K. P. PERSIDSKIĬ

1. *On stability of motion with respect to the first approximation*, Mat. Sb. **40** (1933), 284–292. (Russian)
2. *On the theory of stability of integrals of systems of differential equations*, I, Izv. Fiz.-Mat. Obšč. Kazan. Univ. (3) **8** (1936/37), 47–85. (Russian)
3. *On the theory of stability of integrals of systems of differential equations*. II, Izv. Fiz.-Mat. Obšč. Kazan. Univ. (3) **11** (1938), 29–45. (Russian)
4. *On the stability of the solution of an infinite system of equations*, Prikl. Mat. Meh. **12** (1948), 597–612. (Russian) MR **10**, 299.

R. S. PHILLIPS

1. *The extension of dual subspaces invariant under an algebra*, Proc. Internat. Sympos. Linear Spaces (Jerusalem, 1960), Jerusalem Academic Press, Jerusalem; Pergamon Press, Oxford, 1961, pp. 366–398. MR **24** #A3512.
2. *Dissipative operators and hyperbolic systems of partial differential equations*, Trans. Amer. Math. Soc. **90** (1959), 193–254; Russian transl., Matematika **6** (1962), no. 4, 11–70. MR **21** #3669.

V. A. PLISS

1. *A reduction principle in the theory of stability of motion*, Izv. Akad. Nauk SSSR Ser. Mat. **28** (1964), 1297–1324. (Russian) MR **32** #7861.

V. P. POTAPOV

1. *The multiplicative structure of J-contractive matrix functions*, Trudy Moskov. Mat. Obšč. **4** (1955), 125–236; English transl., Amer. Math. Soc. Transl. (2) **15** (1960), 131–243. MR **17**, 958; **22** #5733.

A. JA. POVZNER

1. *A global existence theorem for a non-linear system and the defect index of a linear operator*, Sibirsk. Mat. Ž. **5** (1964), 377–386; English transl., Amer. Math. Soc. Transl. (2) **51** (1966), 189–200. MR **29** #336.

V. S. PUGAČEV

1. *On asymptotic representation of integrals of systems of linear differential equations containing a parameter*, Mat. Sb. **15** (57) (1944), 13–54. (Russian) MR **6**, 226.

I. M. RAPOPORT

1. *On asymptotic methods in the theory of differential equations*, Izdat. Akad. Nauk Ukrain. SSR, Kiev, 1954. (Russian) MR **17**, 734.

M. ROSENBLUM

1. *On the operator equation* $BX - XA = Q$, Duke Math. J. **23** (1956), 263–269. MR **18**, 54.

M. A. Rutman

1. *Spectral criteria of Ljapunov stability for some systems of linear partial differential equations*, Dokl. Akad. Nauk SSSR **101** (1955), 993–996. (Russian) MR **16**, 1113.

B. A. Ščerbakov

1. *Stability of the zero-solution of a differential equation in a Banach space*, Bul. Akad. Štiince RSS Moldoven. **1966**, no. 4, 111–115. (Russian) MR **35** #1903.

J. J. Schäffer

1. *More about invertible operators without roots*, Proc. Amer. Math. Soc. **16** (1965), 213–219. MR **30** #3374.

Z. S. Sirčenko

1. *Extension of a theorem of N. N. Bogoljubov to the case of a Hilbert space*, Ukrain. Mat. Ž. **16** (1964), 339–350. (Russian) MR **29** #6347.

V. M. Staržinskiĭ

1. *A survey of works on conditions of stability of the trivial solution of a system of linear differential equations with periodic coefficients*, Prikl. Mat. Meh. **18** (1954), 469–510; English transl., Amer. Math. Soc. Transl. (2) **1** (1955), 189–237. MR **16**, 249; **17**, 484.

I. Z. Štokalo

1. *Stability and instability criteria for solutions of linear differential equations with quasi-periodic coefficients*, Mat. Sb. **19** (**61**) (1946), 263–286. (Russian) MR **8**, 329.
2. *On the form of solutions of certain classes of linear differential equations with variable coefficients*, Ukrain. Mat. Ž. **4** (1952), 36–48. (Russian) MR **15**, 32.
3. *Operational methods and their development in the theory of linear differential equations with variable coefficients*, Izdat. Akad. Nauk Ukrain. SSR, Kiev, 1961. (Russian) MR **24** #A1434.

P. A. Švarcman

1. *On the analytic theory of linear differential equations in a Banach space*, Mat. Issled. **4** (1969), no. 3, 90–102. (Russian) MR **41** #2186.

B. von Sz.-Nagy

1. *On uniformly bounded linear transformations in Hilbert space*, Acta Univ. Szeged. Sect. Sci, Math. **11** (1947), 152–157. MR **9**, 191.

J. D. Tamarkin

1. *General problems in the theory of ordinary linear differential equations and series expansion of arbitrary functions*, Dissertation, Petrograd, 1917; abridged English transl., Math. Z. **27** (1927), 1–54.

O. Taussky

1. *A generalization of a theorem of Lyapunov*, J. Soc. Indust. Appl. Math. **9** (1961), 640–643. MR **24** #A3170.

W. J. Trjitzinsky

1. *Theory of linear differential equations containing a parameter*, Acta Math. **44** (1936), 1–50.

W. Wasow

1. *Asymptotic expansions for ordinary differential equations*, Pure and Appl. Math., vol. 14, Interscience, New York, 1965; Russian transl., "Mir", Moscow, 1969. MR **34** #3041.

H. Weyl

1. *Comment on the preceding paper*, Amer. J. Math. **68** (1946), 7–12. MR **7**, 382.

A. Wintner

1. *Asymptotic integration constants*, Amer. J. Math. **68** (1946), 553–559. MR **8**, 272.
2. *Asymptotic equilibria*, Amer. J. Math. **68** (1946), 125–132. MR **7**, 297.

NOTATION INDEX

$A_{\Re} = \operatorname{Re} A = \dfrac{A + A^*}{2}$

$A_{\Im} = \operatorname{Im} A = \dfrac{A - A^*}{2i}$

$\mathfrak{B}$ Banach space (real or complex)

$[\mathfrak{B}_1, \mathfrak{B}_2]$ Banach space of bounded linear operators acting from $\mathfrak{B}_1$ into $\mathfrak{B}_2$

$[\mathfrak{B}] = [\mathfrak{B}, \mathfrak{B}]$

C^n—n dimensional complex space with a positive definite Hermitian scalar product

$\mathfrak{H}$—Hilbert space (real or complex)

$H, \mathscr{P}, G$—Hermitian operators belonging to $[\mathfrak{H}]$

$H \gg 0$—uniformly positive operator: $\lambda_m(H) > 0$

$H \geqslant 0$—nonnegative operator: $\lambda_m(H) \geqq 0$

$I = I_{\mathfrak{B}}$—identity operator in $\mathfrak{B}$

$\mathscr{J}$—Hermitian operator in $\mathfrak{H}$ having the property $\mathscr{J}^2 = I$

J_Γ—skew-Hermitian operator ($J_\Gamma = -J_\Gamma^*$) in a real space $\mathfrak{H}$ having the property $J_\Gamma^2 = -I$

$\lambda_M(H) = \sup\limits_{x \in \mathfrak{H}} \dfrac{(Hx, x)}{\| x \|^2}$

$\lambda_m(H) = \inf\limits_{x \in \mathfrak{H}} \dfrac{(Hx, x)}{\| x \|^2}$

P—projection belonging to $[\mathfrak{B}]$: $P^2 = P$

R^n—n dimensional Euclidean space

$\sigma(A)$—spectrum of an operator A

$\sigma(A) = \sigma_+(A) \cup \sigma_-(A)$—decomposition of a spectrum into spectral sets lying respectively in the interiors of the right and left halfplanes

$\sigma(A) = \sigma_i(A) \cup \sigma_e(A)$—decomposition of a spectrum into spectral sets lying respectively in the interior and exterior of the unit disk

$\int\limits_a^b$—sign for a multiplicative integral with the indicated order of elementary factors

$\dot{x} = \dfrac{dx}{dt} = x'$

AUTHOR INDEX

Subject Index

Other Titles in This Series

148 **Vladimir I. Piterbarg,** Asymptotic methods in the theory of Gaussian processes and fields, 1996
147 **S. G. Gindikin and L. R. Volevich,** Mixed problem for partial differential equations with quasihomogeneous principal part, 1996
146 **L. Ya. Adrianova,** Introduction to linear systems of differential equations, 1995
145 **A. N. Andrianov and V. G. Zhuravlev,** Modular forms and Hecke operators, 1995
144 **O. V. Troshkin,** Nontraditional methods in mathematical hydrodynamics, 1995
143 **V. A. Malyshev and R. A. Minlos,** Linear infinite-particle operators, 1995
142 **N. V. Krylov,** Introduction to the theory of diffusion processes, 1995
141 **A. A. Davydov,** Qualitative theory of control systems, 1994
140 **Aizik I. Volpert, Vitaly A. Volpert, and Vladimir A. Volpert,** Traveling wave solutions of parabolic systems, 1994
139 **I. V. Skrypnik,** Methods for analysis of nonlinear elliptic boundary value problems, 1994
138 **Yu. P. Razmyslov,** Identities of algebras and their representations, 1994
137 **F. I. Karpelevich and A. Ya. Kreinin,** Heavy traffic limits for multiphase queues, 1994
136 **Masayoshi Miyanishi,** Algebraic geometry, 1994
135 **Masaru Takeuchi,** Modern spherical functions, 1994
134 **V. V. Prasolov,** Problems and theorems in linear algebra, 1994
133 **P. I. Naumkin and I. A. Shishmarev,** Nonlinear nonlocal equations in the theory of waves, 1994
132 **Hajime Urakawa,** Calculus of variations and harmonic maps, 1993
131 **V. V. Sharko,** Functions on manifolds: Algebraic and topological aspects, 1993
130 **V. V. Vershinin,** Cobordisms and spectral sequences, 1993
129 **Mitsuo Morimoto,** An introduction to Sato's hyperfunctions, 1993
128 **V. P. Orevkov,** Complexity of proofs and their transformations in axiomatic theories, 1993
127 **F. L. Zak,** Tangents and secants of algebraic varieties, 1993
126 **M. L. Agranovskiĭ,** Invariant function spaces on homogeneous manifolds of Lie groups and applications, 1993
125 **Masayoshi Nagata,** Theory of commutative fields, 1993
124 **Masahisa Adachi,** Embeddings and immersions, 1993
123 **M. A. Akivis and B. A. Rosenfeld,** Élie Cartan (1869–1951), 1993
122 **Zhang Guan-Hou,** Theory of entire and meromorphic functions: Deficient and asymptotic values and singular directions, 1993
121 **I. B. Fesenko and S. V. Vostokov,** Local fields and their extensions: A constructive approach, 1993
120 **Takeyuki Hida and Masuyuki Hitsuda,** Gaussian processes, 1993
119 **M. V. Karasev and V. P. Maslov,** Nonlinear Poisson brackets. Geometry and quantization, 1993
118 **Kenkichi Iwasawa,** Algebraic functions, 1993
117 **Boris Zilber,** Uncountably categorical theories, 1993
116 **G. M. Fel′dman,** Arithmetic of probability distributions, and characterization problems on abelian groups, 1993
115 **Nikolai V. Ivanov,** Subgroups of Teichmüller modular groups, 1992
114 **Seizô Itô,** Diffusion equations, 1992
113 **Michail Zhitomirskiĭ,** Typical singularities of differential 1-forms and Pfaffian equations, 1992
112 **S. A. Lomov,** Introduction to the general theory of singular perturbations, 1992
111 **Simon Gindikin,** Tube domains and the Cauchy problem, 1992
110 **B. V. Shabat,** Introduction to complex analysis Part II. Functions of several variables, 1992
109 **Isao Miyadera,** Nonlinear semigroups, 1992
108 **Takeo Yokonuma,** Tensor spaces and exterior algebra, 1992
107 **B. M. Makarov, M. G. Goluzina, A. A. Lodkin, and A. N. Podkorytov,** Selected problems in real analysis, 1992

Other Titles in This Series

106 **G.-C. Wen,** Conformal mappings and boundary value problems, 1992

105 **D. R. Yafaev,** Mathematical scattering theory: General theory, 1992

104 **R. L. Dobrushin, R. Kotecký, and S. Shlosman,** Wulff construction: A global shape from local interaction, 1992

103 **A. K. Tsikh,** Multidimensional residues and their applications, 1992

102 **A. M. Il'in,** Matching of asymptotic expansions of solutions of boundary value problems, 1992

101 **Zhang Zhi-fen, Ding Tong-ren, Huang Wen-zao, and Dong Zhen-xi,** Qualitative theory of differential equations, 1992

100 **V. L. Popov,** Groups, generators, syzygies, and orbits in invariant theory, 1992

99 **Norio Shimakura,** Partial differential operators of elliptic type, 1992

98 **V. A. Vassiliev,** Complements of discriminants of smooth maps: Topology and applications, 1992 (revised edition, 1994)

97 **Itiro Tamura,** Topology of foliations: An introduction, 1992

96 **A. I. Markushevich,** Introduction to the classical theory of Abelian functions, 1992

95 **Guangchang Dong,** Nonlinear partial differential equations of second order, 1991

94 **Yu. S. Il'yashenko,** Finiteness theorems for limit cycles, 1991

93 **A. T. Fomenko and A. A. Tuzhilin,** Elements of the geometry and topology of minimal surfaces in three-dimensional space, 1991

92 **E. M. Nikishin and V. N. Sorokin,** Rational approximations and orthogonality, 1991

91 **Mamoru Mimura and Hirosi Toda,** Topology of Lie groups, I and II, 1991

90 **S. L. Sobolev,** Some applications of functional analysis in mathematical physics, third edition, 1991

89 **Valeriĭ V. Kozlov and Dmitriĭ V. Treshchëv,** Billiards: A genetic introduction to the dynamics of systems with impacts, 1991

88 **A. G. Khovanskiĭ,** Fewnomials, 1991

87 **Aleksandr Robertovich Kemer,** Ideals of identities of associative algebras, 1991

86 **V. M. Kadets and M. I. Kadets,** Rearrangements of series in Banach spaces, 1991

85 **Mikio Ise and Masaru Takeuchi,** Lie groups I, II, 1991

84 **Dáo Trọng Thi and A. T. Fomenko,** Minimal surfaces, stratified multivarifolds, and the Plateau problem, 1991

83 **N. I. Portenko,** Generalized diffusion processes, 1990

82 **Yasutaka Sibuya,** Linear differential equations in the complex domain: Problems of analytic continuation, 1990

81 **I. M. Gelfand and S. G. Gindikin, Editors,** Mathematical problems of tomography, 1990

80 **Junjiro Noguchi and Takushiro Ochiai,** Geometric function theory in several complex variables, 1990

79 **N. I. Akhiezer,** Elements of the theory of elliptic functions, 1990

78 **A. V. Skorokhod,** Asymptotic methods of the theory of stochastic differential equations, 1989

77 **V. M. Filippov,** Variational principles for nonpotential operators, 1989

76 **Phillip A. Griffiths,** Introduction to algebraic curves, 1989

75 **B. S. Kashin and A. A. Saakyan,** Orthogonal series, 1989

74 **V. I. Yudovich,** The linearization method in hydrodynamical stability theory, 1989

73 **Yu. G. Reshetnyak,** Space mappings with bounded distortion, 1989

72 **A. V. Pogorelev,** Bendings of surfaces and stability of shells, 1988

71 **A. S. Markus,** Introduction to the spectral theory of polynomial operator pencils, 1988

70 **N. I. Akhiezer,** Lectures on integral transforms, 1988

69 **V. N. Saliĭ,** Lattices with unique complements, 1988

68 **A. G. Postnikov,** Introduction to analytic number theory, 1988

67 **A. G. Dragalin,** Mathematical intuitionism: Introduction to proof theory, 1988

(See the AMS catalog for earlier titles)